I0056036

Principles and Techniques of Chemical Engineering

Principles and Techniques of Chemical Engineering

Edited by **Serena Gibson**

CWILLFORD PRESS

New York

Published by Willford Press,
118-35 Queens Blvd., Suite 400,
Forest Hills, NY 11375, USA
www.willfordpress.com

Principles and Techniques of Chemical Engineering
Edited by Serena Gibson

© 2016 Willford Press

International Standard Book Number: 978-1-68285-031-2 (Hardback)

This book contains information obtained from authentic and highly regarded sources. Copyright for all individual chapters remain with the respective authors as indicated. All chapters are published with permission under the Creative Commons Attribution License or equivalent. A wide variety of references are listed. Permission and sources are indicated; for detailed attributions, please refer to the permissions page and list of contributors. Reasonable efforts have been made to publish reliable data and information, but the authors, editors and publisher cannot assume any responsibility for the validity of all materials or the consequences of their use.

The publisher's policy is to use permanent paper from mills that operate a sustainable forestry policy. Furthermore, the publisher ensures that the text paper and cover boards used have met acceptable environmental accreditation standards.

Trademark Notice: Registered trademark of products or corporate names are used only for explanation and identification without intent to infringe.

Printed in the United States of America.

Contents

Preface

Every book is a source of knowledge and this one is no exception. The idea that led to the conceptualization of this book was the fact that the world is advancing rapidly; which makes it crucial to document the progress in every field. I am aware that a lot of data is already available, yet, there is a lot more to learn. Hence, I accepted the responsibility of editing this book and contributing my knowledge to the community.

Chemical engineering is widely applied for the production of chemicals, raw material, petrochemicals, etc. This book covers in detail some existing theories and innovative concepts like researches on polymers, microfluids, separation processes, nanofabrication, etc. Also included in this book are some of the recent technological advances in the field of chemical engineering which have further expanded its scope. This book will be very useful for the students and professionals of chemical engineering, industrial chemistry and associated disciplines.

While editing this book, I had multiple visions for it. Then I finally narrowed down to make every chapter a sole standing text explaining a particular topic, so that they can be used independently. However, the umbrella subject sinews them into a common theme. This makes the book a unique platform of knowledge.

I would like to give the major credit of this book to the experts from every corner of the world, who took the time to share their expertise with us. Also, I owe the completion of this book to the never-ending support of my family, who supported me throughout the project.

Editor

Microwave Pyrolysis of Plastic

Elham Khaghanikavkani*, Mohammed M. Farid, John Holdem and Allan Williamson

Professor, Univeristy of Auckland, New Plymouth, New Zealand

Abstract

Microwave heating has been used in the chemical industry for many years for diverse applications due to advantages, such as volumetric heating, high power density, and fast and easier temperature control. The motivation of this work is to develop a methodology to use potential benefits of microwave as a heat source in plastic pyrolysis. A goal was to use the available knowledge to design and build a rotating microwave reactor, to attain homogeneous temperature distribution in anticipation of producing pyrolysis products with uniform molecular hydrocarbon distribution. Also, it was sought to investigate the effect of process parameters on products' yield and composition, and examine their suitability as fuel and useful chemicals. A rotating microwave reactor was designed and fabricated using a coaxial transmission structure without the limitations of the commonly-used enclosed glass quartz reactor, which made the design appealing for any future industrial-sized microwave reactor.

The optimum pyrolysis operating condition led to the production of a suitable product for a fuel application. However these products cannot be used directly as phase change materials due to their low latent heat and very broad melting range. Fractionation of the products may also be used as phase change materials, capable of storing/releasing heat at suitable application temperature. The microwave heating provided a more uniform heating distribution, although it did not alter product composition in comparison to the conventional pyrolysis of plastic. In the small microwave reactor used in this project, microwave benefit was not evident with regards to its effect in creating uniform temperature distribution. The benefit may be clearer in an industrial-sized reactor in which heat penetration in conventional thermal reactor can be a serious problem.

The pyrolysed products (oil/waxes) were quantitatively analysed using GC/FID. The results of microwave pyrolysis showed 73% oil/wax yield, with silicon carbide used as a microwave absorbent.

Keywords: Microwave reactor; Plastic; Pyrolysis; Gas Chromatography/Flame Ionization Detector (GC/FID)

Introduction

Pyrolysis can be described as a chemical process and thermal decomposition of organic components in an oxygen-free atmosphere to yield char, oil and gas. Microwave heating has been used for pyrolysis of biomass [1-4], scrap tires [5,6], wood [7-9], rubber [10], oil shales [11], coffee hulls [12,13] and production of other chemicals [14-16]; however, there have been few studies of plastic wastes pyrolysis [17-19]. The benefits of the pyrolysis are recycling some of the stored energy within the waste plastics, consequently diminishing imports of crude oil considering limited supply of natural resources (crude oil and gas), fluctuation of high price and availability of crude oil. Pyrolysis of plastic is an endothermic process, so that it requires a supply of heat. Two types of the heat sources can be provided here i.e. conventional heat source and microwave. The convectional plastic thermal decomposition has been established to operate in industrial scale around the world. Using microwave as a heat source would open a new horizon in this topic. Microwave energy can be delivered directly to the reacting or processing species by using their dielectric properties or by adding absorbers to materials which allows more volumetric heating of materials [20,21]. The high heating rate can be several orders of magnitude greater than with conventional heating [20-22]. Microwave generator can respond quickly to changes in process parameters with a feedback loop of an automated process [20]. Microwave heating results from induced currents so the heating tends to be volumetric however the penetration of microwaves is influenced by the properties of the material. The field penetrates it losses power and therefore the field intensity will decrease suggesting that heating may not be uniform. Due to the fact that materials are heated volumetrically, materials with more uniform microstructure can be produced, if energy losses are minimised using insulation [16] and if the material has the proper dielectric loss factor. Advantages of

microwave technology may facilitate moving forward to produce clean, fast and high quality product. Further study should be performed to get a clear picture of microwave pyrolysis of plastic process at high temperature. Challenges such as controlling electromagnetic field and uniformity, temperature measurements may require more sophisticated approaches to be tackled [23]. Albeit in the light of present technique and instrument of temperature measurement, it is not easy to acquire a precise result of the temperature distribution from the interior of the medium at high frequency and high temperature. The only question remains to be answered is whether it is possible to achieve microwave pyrolysis of plastic with uniform heating in much less time comparing to conventional heating on a reasonable scale.

A few studies have outlined the scale up challenges in microwave process [23,24]. To the knowledge of the author there is currently no industrial scale application of microwave pyrolysis of plastic. The reason may be explained by the difficulties of combining the chemical and electrical engineering technologies to meet the requirements for a high temperature microwave processing of plastic degradation. Due to the complex nature of pyrolysis of plastic a very particular and detailed design with the help of a robust electromagnetic simulation model is needed in order to achieve this goal.

*Corresponding author: Elham Khaghanikavkani, Professor, Univeristy of Auckland, New Plymouth, New Zealand

Microwave application is very broad. Hence, many studies have been carried out with different designs to investigate the process outcome in various chemical applications. Microwave system that are dedicated to chemical syntheses consist of a microwave power source (generator), a section of transmission line (different type of transmission lines are parallel wires, waveguide, coaxial and microstrip) that delivers microwaves from the generator into an applicator, and a microwave applicator (reactor). Industrial microwave heating systems typically use a variety of standardised waveguide components (different types of waveguide are circular, rectangular, elliptical or ridge cross section) with specific sizes and necessary functions. The physical size of waveguide determines frequency, bandwidth, power handling capacity and impedance of line. The most commonly used microwave applicators belong to three main types. There are travelling wave applicators, single mode cavity and multimode cavity, depending on physical size and design characteristic. The initial feature of a microwave cavity design is the ability to maximise the electric field intensity which increases the heating rate, operate the experiment at high temperatures over 800°C and allow the removal of pyrolysis products from the cavity [25]. Studies suggested that pyrolysis in a single mode reactor requires an order of magnitude lower input energy comparing to a multimode reactor [25]. This is simply because the electric field density in single mode reactor is high hence higher heating rate is achievable as the heating rate is proportional to the square of the electric field strength (heating rate αE^2). Despite the stronger field density in a single mode cavity, limitation on physical dimension of the sample due to highly non-uniform field configuration makes the industrial scale usage of a single mode cavity undesirable. The idea of passing the material through the high field intensity of a multimode cavity was a common approach in pilot scale microwave system [26,27].

In 2000 Esveld et al. [26] design a continuous pilot scale (10 to 100 kg/h) multimode microwave reactor and model the unit for dry and solvent free chemistry in esterification reaction (stearic acid with stearyl alcohol). In their study [26] the microwave absorption is improved by adding clay (50 wt%) to the sample. Temperature is measured using a fluoroptic probe within the sample. In the design 4.4 kW microwave power is fed to a multimode cavity by a diagonal slotted waveguide. The materials are closely packed in a Teflon coated glass fiber web conveyor moving at 17 cm/min. To avoid arcing and plasma formation the power is limited to lessen the field density which was 15 ± 4 kV/m measured by a Luxtron MEF probe.

Robinson et al. [27] develop a continuous microwave treatment system using a conveyor system for the remediation of contaminated drill cuttings at pilot scale (500 kg/h of material). In the oil industry drill cuttings are called the broken bits of solid material removed from oil or gas well borehole. The optimum geometry of the applicator is determined based on simulation outcome. Field uniformity is improved in the cavity with design of step positioning at an offset distance from the centre line of the incoming waveguide, a solution which yields both high and uniform power densities across the cavity geometry.

Generally, multimode applicators are commonly used for industrial applications due to greater size and operational flexibility. Acceptable results are achieved by using a multimode applicator for treatment of contaminated drill cutting under high microwave power between 10 to 15 kW [28]. Simulation of the field distribution and power density in microwave treatment of oil contaminated waste in Shang et al. study [29] also yield very useful information in respect of location of maximum electric field strength and dependency of dielectric loss factor and microwave absorption.

Simulation assessment is one of the efficient methods used to achieve the optimum design through a 3D visualisation of field distribution within the applicators and materials. This technique has been used in other studies using available electromagnetic design software [27,29] such as QuickWave-3D', COMSOL and Microwave Studio' in the market.

This paper was devoted to the microwave pyrolysis of High-Density Polyethylene (HDPE) for the production of fuel and other useful chemicals. In this research the design for a microwave unit with the use of a coaxial structure based on the simulation model was suggested. This was followed by the electromagnetic simulation conducted on the microwave unit, providing some insight into microwave behaviour with regards to energy dissipation and electromagnetic wave distribution. The microwave adjustments-including the building, tuning and calibrating of the microwave-induced pyrolysis unit-were then discussed, together with setting up the unit for operation. The experiments were then described and the factors influencing various operating conditions on pyrolysis products were explained. After that, condensable products were analysed quantitatively using GC/FID and Differential Scanning Calorimetry (DSC) in order to characterise their properties.

Materials and Methods

In this study HDPE in a pellet form with diameter of 2 mm to 3 mm was used with carbon blocks in a cubic form with dimensions of 30 mm×30 mm×30 mm as a microwave absorbent. Alternatively silicon carbide powder was also utilised.

In this study Microwave Studio' (MWS) was utilised to predict microwave field distribution, intensity and absorbed microwave power inside the waveguide and applicator. This tool can also help to optimise the unit dimensions in order to minimise the reflected power. MWS is powerful and easy to use software for high frequency range however it does not permit the evaluation of temperature dependent thermal properties or chemical phenomena such as latent heat or heat of reaction.

Experimental Apparatus

An experimental apparatus was designed, built and used (Figure 1)

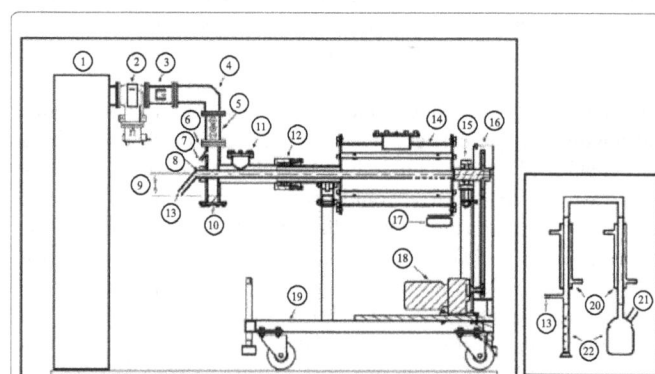

Figure 1: Schematic of the microwave pyrolysis of plastic apparatus in rotation mode

1) Generator, 2) Isolator (Circulator + Dummy load), 3) Directional coupler, 4) Bend, 5)3-stub tuner, 6) Quartz window, 7) Nitrogen inlet, 8) Central conductor, 9) Step height, 10) Step, 11) Rupture disk, 12) Gland, 13) Product outlet in microwave unit and product inlet in condensation sytem, 14) Reactor (Applicator), 15) Bearing, 16) Gear transition, 17) Wireless temperature data-logger, 18) Adjustable speed electrical machinery, 19) Stand , 20) Condensers, 21) Outlet of non-condensable gases, 22) Collection flasks.

based on the electromagnetic field simulation results which will be discussed below. The three main equipment components of the system were (a) the microwave generator (number 1), (b) the wave guide components (numbers 2 to 6), and (c) the applicator (number 14). The microwave generator was built by Keam Holdem Ltd. with a maximum of 6 kW output power operating at 2.45 GHz. The transmission line consists of the following parts:

1. An isolator consisting of a circulator and a dummy load, which was purchased from Richardson Electronics Pty Ltd.

2. A directional coupler with power readout devices, which was used to detect the reflected and forwarded power in the wave guide.

3. A three stub tuner, which was located close to the load in the waveguide section to tune the unit and minimise the reflected power in order to maximise the heating in the reactor.

4. The aluminium rectangular waveguide (WR340) with the dimension of 86.36 mm×43.18 mm which was constructed in the University of Auckland.

5. A quartz window (number 6 in Figure 1) was fixed between the flanges to keep the transmission line clean and protect the generator from the produced gases. Sealing around the glass quartz was very important and quite tricky, as it was necessary to prevent leakages of gases both into the generator and the atmosphere. To prevent any breakages, the quartz window was surrounded by a metal frame designed to protect it from cracking. Enough clearance was left for both the glass and metal frame expansion during the high temperature experiments. Two gaskets were placed above and beneath the metal frame to cease any leakages. And the gaskets edges were covered with copper shim to prevent arcing and provide good electrical cantact.

6. A step was attached to the end of the waveguide flange as shown in figure 1 with number 10. In order to ensure good contact between the step and waveguide's walls a row of beryllium copper teeth (0.5 mm thicknesses) were designed. The teeth (long fingers) touched the wall in such a way that they eliminated gaps and unexpected resistance between the metals' surfaces, so as to provide an effective electrical connection between the step and the waveguide.

7. A gland was designed to connect the rotated and fixed sections in the unit. Pure expanded braided graphite ring packing suitable for high temperatures (Uni-Pac1 compression packing with 98% purity from Novus sealing limited) was used to provide proper gas sealing and a good electrical contact.

In technical terms the cylindrical cavity (reactor) is a multimode resonator, used for batch processing. The reactor used was made of stainless steel (304) to accommodate high temperature process. The reactor was sized for the multimode type of the applicator in which the dimensions (length 39 cm and diameter 22 cm) were somewhat greater than the wavelength (12.2 cm). Increasing the dimension with respect to wavelength increases the number of excited modes in the cavity which has proved to be helpful in providing better heating uniformity.

The reactor was designed to rotate around its horizontal axis and had four longitudinal baffles bolted to the interior reactor wall in order to help the mixing process. Initially, the reactor was not rotated, therefore, for simplification; gaseous products were taken out via the top flange

of the reactor. In this study 10 experiments were conducted. All the experiments from numbers 1 to 8 were conducted in stationary mode of the reactor. However, both experiments 9 and 10 were conducted while the reactor was rotated. In this design the central conductor of coaxial had six holes in order to collecting the products from the microwave reactor during the reactor rotation. The coaxial structure provided flexibility and allowed for the collecting of the products and nitrogen passes through the unit. The schematic of the microwave system is shown in figure 1.

The reactor was filled and discharged via the top flange with HDPE pellets and microwave absorbents of either carbon blocks or silicon carbide (SiC). The reactor was operated under inert atmospheric pressure using nitrogen. In all experiments, the pyrolysis reactor was heated up to temperatures between 400°C and 550°C. In this thermo chemical process, plastic was pyrolysed into gases with no residue. The waxy liquid fraction of the gases products was collected using two glass coil condensers supplied by Technical Glass Products Limited. The temperature of the first condenser (coil length 250 mm, body length 320 mm, joints NS 14/23) was set at 60°C while cold tap water (17°C) was used as the circulating water in the second condenser (coil length 150 mm, body length 220 mm, joints NS 14/23). A small amount of waxy product was also collected from the interior parts of the unit when it cooled down. Non-condensable gases were vented without analysis, as their analysis was not part of this investigation. These gases are usually used as a source of the energy needed for the pyrolysis process. The liquid portion of the pyrolysed products can be used as a fuel source and to produce valuable chemicals with a high calorific value of 46.5 MJ/kg [30].

The central conductor was a hollow carbon steel tube (SEA 1045) with a diameter of 25.5 mm. A coaxial transmission line was utilised in this design to transmit energy from the associated electric and magnetic fields transverse to the axis of the line; this is illustrated in figure 2. Following this figure, the electric field vectors are normal to the inner and outer surfaces of the conductors. It also shows the magnetic field as a series of concentric circular vectors between the two conductors.

Simulation

The model structure for simulation of the microwave reactor discussed earlier was built using MWS as shown in figure 3. Microwave Studio' like other simulators follows three simple steps to perform calculation; firstly, model the geometry, secondly, run the solver to solve Maxwell's equations and lastly, plot the parameters. In this simulation the model's geometry was constructed by defining a cylindrical multimode applicator with the sample located inside the reactor, a coaxial transmission line including central conductor and outer conductor, waveguide structure, boundary conditions and excitation

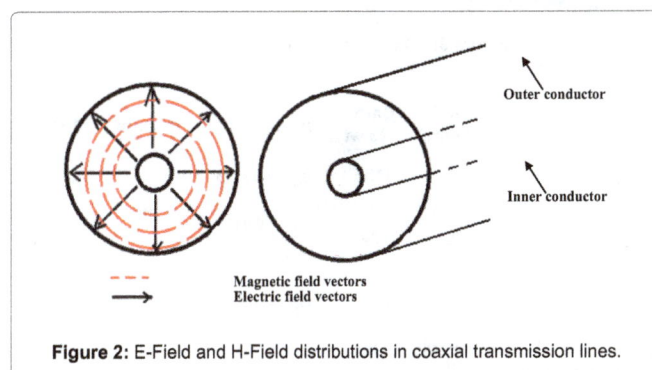

Figure 2: E-Field and H-Field distributions in coaxial transmission lines.

port. The material properties such as dielectric constant, loss tangent, density, thermal conductivity and specific heat capacity were assigned accordingly.

Physical properties

Electrical characteristic of the materials such as dielectric constant (real part) and dielectric loss factor (imaginary part) are terms used to specify the electromagnetic properties of material. Dielectric constant ($\acute{\varepsilon}$) represents the stored energy and loss factor ($\acute{\varepsilon}$) represents the lost energy in dielectric materials. Dielectric loss tangent (tan δ) is a term to quantify lossiness which defines the ability of materials to absorb and convert electromagnetic energy into heat at a specified temperature and frequency. The lossiness is the ratio of energy lost to energy stored in the dielectric material. The dielectric loss factor of plastic is too low to achieve effective heating using microwave energy, therefore the ratio of absorbent (carbon or silicon carbide) to plastic (HDPE) should be high enough to be able to pyrolyse the plastic in a multimode applicator. Two different ratios of 10 (5 kg of absorbent to 0.5 kg plastic) and 3.3 (1.66 kg of absorbent to 0.5 kg plastic) were assumed. The thermal physical properties of the mixture were calculated using the weight factor, as shown in table 1, for one of the simulation case studies.

The dielectric properties of the mixture were calculated using equation 1 which is known as the Landau & Lifshitz, Looyenga mixture Equation [31].

$$(\varepsilon_m)^{1/3} = v_1(\varepsilon_1)^{1/3} + v_2(\varepsilon_2)^{1/3} \qquad (1)$$

Where ε_m represents the complex permittivity of the mixture, ε_1 is the permittivity of the medium in which particles of permittivity ε_2 are dispersed, and v_1 and v_2 are the volume fractions of the respective components, where $v_1 + v_2 = 1$.

By substituting dielectric properties of plastic ($\varepsilon_1 = 1.95911 - 0.00027j$) and carbon blocks ($\varepsilon_1 = 2.35469 - 0.09584j$) in equation 1 with 5 kg of carbon and 0.5 kg of HDPE, the mixture dielectric properties will be $\varepsilon_m = 2.283 - 0.08j$ with a loss factor of 2.283, a dielectric constant of 0.08 and a tangent delta of 0.035.

Simulation results

The aim of conducting the simulations is to determine the correct

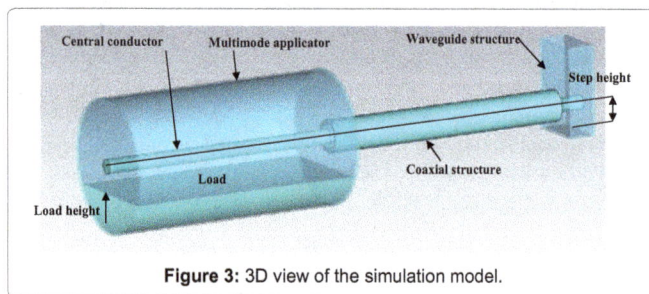

Figure 3: 3D view of the simulation model.

Property	HDPE (7 wt%)	Carbon (93 wt%)	property of the mixture	Mixture
Density (kg/m³)	960	2185	$\rho_m = 0.93\,\rho_c + 0.07\,\rho_{HDPE}$	2099
Thermal Conductivity (W/m.K)	0.25	0.14	$\dfrac{1}{k_m} = \dfrac{0.93}{k_c} + \dfrac{0.07}{k_{HDPE}}$	0.1444
Heat capacity (J/kg.K)	2200	1000	$C_{pm} = 0.93\,C_{pC} + 0.07\,C_{pHDPE}$	1084

Table 1: The physical properties of the samples used in Microwave Studio® (5 kg of carbon + 0.5 kg of HDPE).

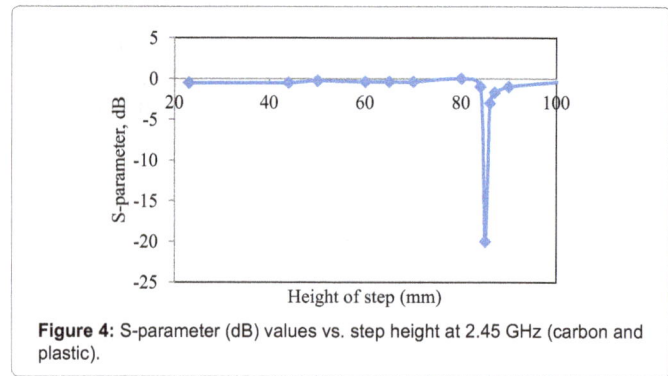

Figure 4: S-parameter (dB) values vs. step height at 2.45 GHz (carbon and plastic).

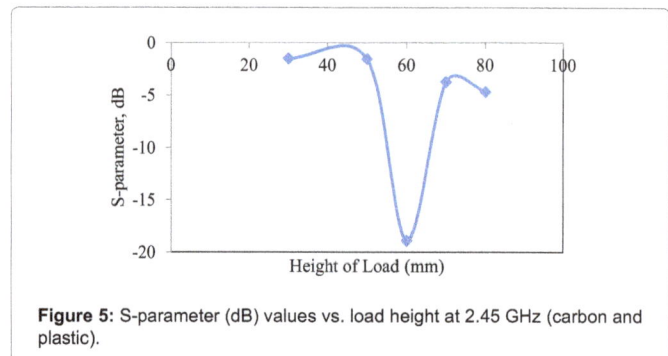

Figure 5: S-parameter (dB) values vs. load height at 2.45 GHz (carbon and plastic).

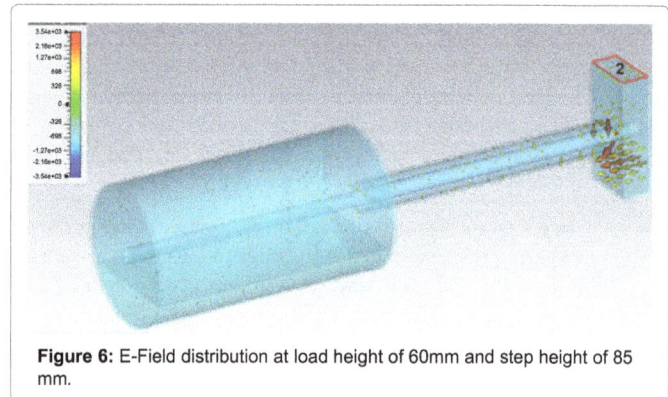

Figure 6: E-Field distribution at load height of 60mm and step height of 85 mm.

dimensions of the unit. Electromagnetic simulation is a useful theoretical way to design a microwave system and investigate the effect of material properties and system geometry [27,29]. This method was followed by optimising the step and load height. Step height indicates the distance between the wave guide flange to coaxial transition as shown in figure 3. The height of the step is one of key factors to minimise the reflected waves (maximise the S-parameter) and couple the impedance of sample with microwave source. S-parameter is a measure to identify how much energy is gained (or lost) in the system at given frequency and system impedance, and varies as a function of frequency. The simulation was run with different step heights. The sharp and high absorption in figure 4 designated the high sensitivity of the absorption in relation to the step height. It was evident that the lowest reflection (the highest absorption) was achievable at the step height of 85 mm. Afterwards the simulator was run to determine the optimised amount of sample. The best absorption took place at 60 mm load height as shown in figure 5.

Figure 6 and 7 show 3D views of electric and magnetic field patterns. It is evident that the intensity of the fields was maximised in the

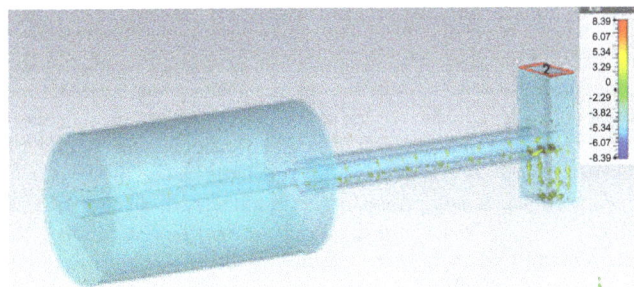

Figure 7: H-Field distribution at load height of 60 mm and step height of 85 mm.

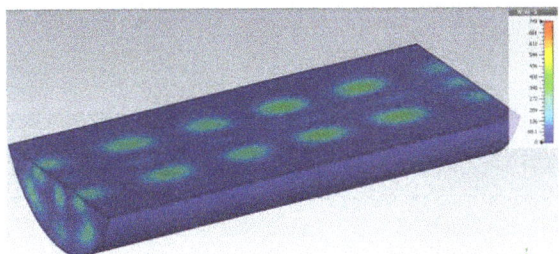

Figure 8: Power loss density within the sample in pyrolysis of HDPE using SiC as an absorbent.

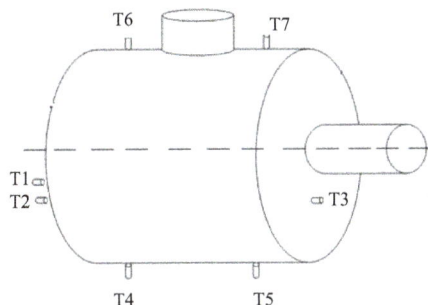

Figure 9: Location of the thermocouples on the reactor.

waveguide and decreased towards the kiln. Likewise the field intensity was maximised at the centre and decreased towards inner walls of the applicator. The field line patterns showed the existing intensity within the waveguide. Figure 8 illustrates the power loss density distribution in the sample in the stationary mode with scattered hot spots throughout the sample. This indicated a non uniformity in the heating process as the materials which were not in the hot spots will not be subjected to the same degree of microwave treatment [27]. It was expected that the scattered hot spots provide an efficient heating process due to a continuous mixing of material while the reactor rotated. Thereby, the located materials in the hot spots will be replaced by new cold materials repeatedly. This fact improves the heat transfer within the sample effectively and eliminates presence of the temperature gradient in the cavity. Furthermore, the changes of sample angle due to rotation have no sensible effect on power loss density variation within the sample based on the simulation outcome.

Operating procedure

The unit was assembled properly by placing the gaskets and quartz

window to isolate the waveguide and coaxial structure. Then it was calibrated and tuned after loading the reactor. The reactor was loaded with desirable amount of absorbent and then filled with a layer of plastic pellets on top. This arrangement of the sample loading allows the sample sink in the space between carbon blocks or absorbent pores in case of using SiC while it melts. Nitrogen was purged to provide the inert atmosphere in the unit before and during the experiment. Nitrogen was purged in the waveguide structure (from number 8 in Figure 7) for two reasons; firstly, to make sure of keeping the quartz window clean from any possible generated dust in the reactor and, secondly, to keep dust out of the waveguide and coaxial structure. The waveguide and coaxial both have much higher field intensity compared to the reactor due to their smaller dimensions. Afterwards access port holes were checked and sealed. Gas and microwave detectors were used to detect any gas and microwave leakage from the unit for safety matters. Heat accumulation can be identified by temperature rises with time very shortly after putting on the power. Plastic decomposition took place at above 300°C. Efficient coupling and controlled heating was achieved via tuning and power control. The three-stub tuner was used to maximise the amount of applied power absorbed by the sample with matching the impedance of the transmission line to the impedance of the load. Gaseous products were sent to the condenser system to condense the heavier components of the pyrolysed gases. The detail of tuning procedure and temperature measurement are explained below.

Calibration and tuning procedure

A network analyser (ENA_E5062A) was utilised to do the calibration and the initial tuning. The network analyser is an instrument that measures the electrical network parameters such as reflection (S–parameters) and transmission of electrical networks at high frequencies. Firstly, the unit was calibrated using three known standards, (a) zero-short, (b) offset- short, and (c) waveguide horn. Secondly, after loading the reactor with the desired amount of material, network analyser was connected to the waveguide just above the three-stub tuner. Then the calibration results based on standards were used to characterise the sample behaviour within the reactor using network analyser. The heights of the three-stub tuner's pins were adjusted in order to minimise the reflection at 2.45 GHz. Three markers were set at three frequencies, i.e. 2.44 GHz, 2.45 GHz and 2.46 GHz. These frequencies were set because 2.45 GHz is the nominal frequency and it is not expected that the operating frequency will drift by more than 10 MHz. This procedure allows us to provide efficient coupling at the starting point. Tuning the microwave unit is essential to achieve the best use of energy, protect the generator and also prevent high reflection and subsequent possible operating issues such as undesired overheating of the structure. However, during pyrolysis continuous tuning was required by adjusting the three-stub tuner pins height due to load phase changes. Changes of material volume within the cavity impact the electromagnetic field distribution and the power dissipation. It is worth noting that water has a high dielectric loss factor, therefore, relatively small variation in the sample's moisture contents would affect the final attained temperature during microwave heating [32]. For this reason calibration before each experiment was required to minimise the effect of daily fluctuation in moisture in the sample.

Temperature measuring system

Temperature measurement in microwave heating is highly controversial. Generally, three methods are suggested to do the measurements in microwave process. Choosing the right method depends on operating condition, material type, size and the limitation of measurement technique. These methods are optical fibre thermocouple, infrared thermography (IR) and conventional thermocouple. Each one has some advantages and drawbacks which reveal the suitability of the chosen method. The detailed discussion of different aspects and available measurements methods are thoroughly discussed in [33]. The thermocouple type K with or without protected sheaths in microwave heating causes some errors [34] due to provoking problems such as field distortion, enhanced energy absorption (leading to thermal runaway), heat loss through conduction particularly when it comes to measurements in low loss materials. However, in other studies by Janny et al. [35] and Haque [36], thermocouple type C [35] with molybdedeum sheath or type K thermocouple with an unground tip sheathed in Inconel 702 sheath [36] are suggested to perform the measurements within the sample, which is explained by accepting that microwaves cannot penetrate the metal shield. The majority of investigations agree to use fibre optic probes and pyrometer/IR sensors in microwave heating [20,34,37]. Optical pyrometers as a non-contact technique eliminate the interference in electrical field behaviour while they only record surface temperatures. The high temperature and strong agitation do not allow the use of fibre optics.

In the current study the temperature profile was logged continuously throughout the process using K type thermocouples inserted a certain short length (25 mm) inside the reactor. The reasons for choosing short length are, firstly, to eliminate the possibility of interference of the electromagnetic field by presence of K type thermocouple, and, secondly, over an extended specific length they could be burnt. The reason for overheating can be explained by induced currents at the thermocouple tip where currents are less but without the ability to get rid of heat on the stainless steel thermocouple sheet, because the stainless steel is a fairly poor conductor of both heat and current. The specific length of thermocouple in the reactor can be determined experimentally by trial and error. However, the temperature further inside the sample towards the central conductor is expected to be higher than the reactor wall as there is non-uniform electromagnetic field distribution in the reactor and field intensity increases towards central conductor. Figure 9 shows the mounted thermocouples in the reactor. A wireless data logger (WLS-TC) was utilised to monitor the temperature.

Analysis

All oil/wax products were analysed with the use of GC/FID and DSC for their compositions and thermal properties. In this study the reason for conducting DSC analysis is to determine the thermal characteristic (latent heat and melting range) of the products and investigate the suitability of the pyrolysed products for use as Phase Change Materials (PCMs). PCMs are capable of storing and releasing large amounts of energy during the phase changes. The scope of PCMs application is very wide and melting temperature of the material determines their suitability for the application. One of these applications is solar energy storage in building by encapsulating suitable PCMs. An ideal candidate for PCMs application should have proper melting ranges, large latent heat of melting, high thermal conductivity, high specific heat capacity, small volume change, and should be non-corrosive, non-toxic and exhibit little or no decomposition or super-cooling. Commercial paraffin waxes are cheap with moderate thermal energy storage density (200 kJ/kg) [38,39]. The paraffin waxes are straight chain hydrocarbons, similar to those products from pyrolysis plastic.

GC/FID

Oil/wax products were analysed using a Shimadzu GC-17A with flame ionisation detector, 0.53 mm tubular capillary column, 6 meters length (MXT-500 Sim Dist Cap) and split injection (split ratio of 0.5) with a volume of 1.0 μL. The injector and FID temperatures were 325°C and 395°C respectively and the oven was held at 50°C initially, then ramped to 380°C at 15°C/min. The column pressure was 3.6 kPa, which corresponded to a linear carrier gas (helium) velocity in the column of 25.5 cm/s. The total and column flows of helium were 5.7 and 3.17 ml/min respectively. Calibration was achieved using Restek Florida standard, which is comprised of even numbered n-alkanes from C_8 through to C_{40}. Alkenes and dienes could then be determined from their proximity to their corresponding n-alkane. Carbon disulphide was utilised as the solvent, due to its superior ability to dissolve heavier compounds. This was found by a preliminary solubility test performed on the collected samples from the unit before starting GC calibration. The GC would first be flushed once with pure solvent after each sample run to ensure all residues from the previous run was removed. Each sample was analysed 3 times to check for repeatability and consistency in the result.

DSC

The DSC (Shimadzu DSC-60) was calibrated using n-octadecane. Approximately 4.8-5 mg of sample was first melted at 80°C to achieve good contact with the aluminium pan surface, and then cooled beyond -135°C at low heating rate of 3°C/min. Measurements for all products (the first condenser, the second condenser and collected wax from the interior parts of the unit after cooling down) were then taken from -135°C to 130°C. The reason for starting from very low temperature was associated with the melting point of low carbon number of wax products i.e. -118.85°C for C_7 or -81.35°C for C_9. Therefore in order to get the correct base line in DSC curves, taking measurements from very low temperature is recommended.

Results and Discussion

The objective of the ten experiments conducted is to focus on the microwave pyrolysis reaction, through investigation of composition and yield under different operating procedures. Different nitrogen flow rates, absorbent type and mass ratios of plastic to absorbent, as well as the reactor in stationary and rotational mode with low and high rotation speed, were also investigated. The summary of the operating conditions is presented in table 2. The condensable gases were collected in waxy form via the condensation unit, and non-condensable gases were ventilated. The oil/wax products were then analysed using GC/FID and DSC to characterise their composition and heat flow.

Experiment 1 was designated as a trail run. In the first three experiments (Exp_2 to Exp_4) 0.5 kg HDPE was pyrolysed using 5

	Absorbent type &input mass	Mode	N₂ Flow rate (L/min)	Power (kW)	Duration (min)	Total Oil/wax Yield
Exp_1	Carbon (5Kg)	stationary	0-5.12	3-5	66	63.75
Exp_2	Carbon (5Kg)	stationary	0	5	88	54.92
Exp_3	Carbon (5Kg)	stationary	1.28	5	86	62.02
Exp_4	Carbon (5Kg)	stationary	9.60	5	86	58.24
Exp_5	SiC (5kg)	stationary	6.40	5	71	43.40
Exp_6	SiC (5kg)	stationary	1.28	3-5	73	34.67
Exp_7	SiC (1.66)	Stationary (failed)	3.81	3-5	31	37.32
Exp_8	SiC (1.66)	stationary	3.81	3-5	100	39.27
Exp_9	SiC (1.66)	Rotation (4rpm)	3.81	3-5	78	46.36
Exp_10	SiC (1.66)	Rotation (8-18rpm)	2.56	3-5	60	73.36

Table 2: Summary of experiment conditions.

kg carbon with different nitrogen flow rates and input microwave power of 5 kW. For the rest of the experiments (Exp_5 to Exp_10) silicon carbide was used as the absorbent according to the tabulated operating conditions. In the experiments with silicon carbide (6 to 10), the temperature was controlled by adjusting the power. The aim was to keep the temperature between 400°C and 500°C, as this temperature range is the optimum suggested range in the literature [40,41] to maximise the lighter liquid fraction. Hence, the input power was decreased when the average temperature in the reactor reached the proximity of 500°C. Experiment 7 was failed half way through the run due to leakage from the thermocouples; the run was stopped for safety concerns. The same operating conditions were repeated in experiment 8, however the products of the failed run (Exp_7) were still analysed and the results were reported. Also, in experiment 9 the rupture disk ruptured and it took about 25 minutes to remove the blockage and replace the aluminium foil. For this reason, during this period prior to commencing the product collection, the temperature profile showed a downward trend. However, the time wasted in experiment 9 has not been factored into the experiment duration. The reason for the rupture failure was the blockage of the condenser inlet by the travelled silicon carbide from the reactor during the reactor rotation. The effect of lower input power can be seen from the temperature profiles, with the recorded decline of the heating rates in experiments 7, 8 and 9. The reactor was rotated only in the last two experiments.

Figure 10 shows the temperature profile in experiment 8 (startionary mode) where the thermocouples 4 and 5 imersed in the sample. It is evident that there was a temperature gradient in the reactor. The temperature decreased from the microwave feeding location towards the far end of the reactor. In all stationary mode experiments (experiments 1 to 8) the temperature gradient was observed while very uniform temperature profile was monitored in experiment 10 when the reactor was rotated with high rpm (Figure 11).

Product yield

The yield of the products was calculated based on mass percentages

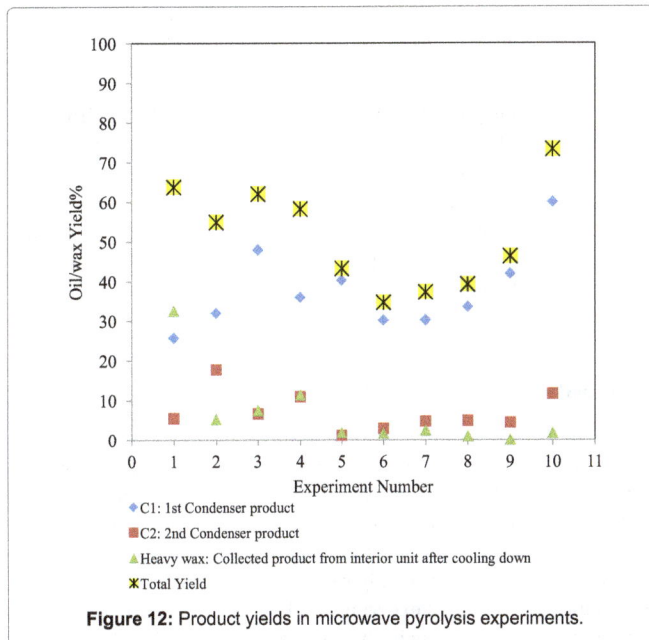

Figure 12: Product yields in microwave pyrolysis experiments.

of collected oil/wax products relative to the initial mass of plastic feed. The main portion of the product was collected from the first condenser in the form of oil/wax and the second abundant product was collected from the second condenser in a liquid form. A waxy residue was collected from the interior part of the unit after it cooled down. A smaller amount of waxy residue was obtained when using SiC as an absorbent and a larger amount when carbon was used. This suggests the possibility of higher cracking in plastic using SiC, which is a more efficient microwave absorbent.

From the reported data in figure 12 and table 2, it is evident that the total yield of products in the experiments using carbon (experiments 1 to 4) was around 60%, while in the experiments with the SiC the yield was around 40% (more cracking to non-condensable gases). The sole exception to this was the last experiment, which had very different operating conditions, as will be explained later in this section. Pyrolysis is a very temperature dependent process, so temperature plays the most crucial role in product yield and composition. Other factors such as the heating rate and nitrogen flow rate are important because these parameters influence the residence time. The residence time effect on the pyrolysed product yield and composition were discussed in the context of conventional thermal pyrolysis in reference [42].

The high gas yield results (Exp_5 to Exp_9) can be attributed to the use of silicon carbide as a microwave absorbent and its electrical physical properties (high loss factor) and thermal physical properties (high thermal conductivity). As mentioned earlier, in the experiments with silicon carbide the temperature was kept between 400°C and 500°C by controlling the input microwave power. This approach was chosen based on the decomposition mechanism of polyethylene, which initially requires a high temperature for breaking the polymer chain to overcome the bound-dissociation energy of carbon double bonds. Later the heating rate was kept lower than a specific value to extend secondary reactions. Furthermore, the volumetric nature of the microwave heating, coupled with dielectric materials such as SiC, led to the pyrolysis temperature being reached in a very short time (mostly less than 7-10 minutes), consequently exposing the pyrolysed products to heat for longer periods than in

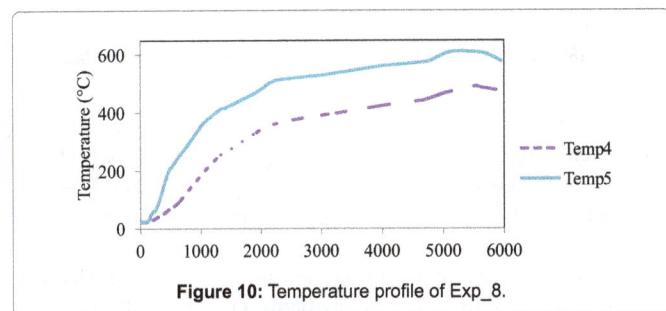

Figure 10: Temperature profile of Exp_8.

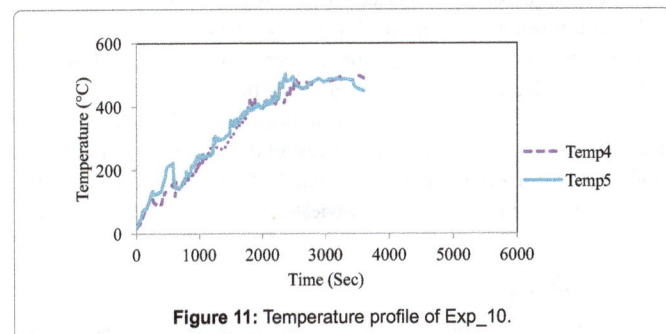

Figure 11: Temperature profile of Exp_10.

the experiments using carbon (over 35-40 minutes). This fact caused the severe decomposition of the polymer chain, resulting in a high portion of gaseous products.

The operating procedure in Exp_10 was very different from the rest of the runs. In this experiment the reactor was rotated with high rpm causing severe mixing and creating very uniform temperature distribution within the whole sample and, subsequently reaching the pyrolysis temperature of the whole sample at once. Having heating rate of 10°C/min and also, a high nitrogen flow rate led to shorten the pyrolysis duration and caused less cracking without significant secondary decomposition. This fact may justify the obtained high oil/wax yield in the Exp_10.

Composition of the products

Generally, the pyrolysis products of HDPE were hydrocarbons ranging from C_1 to C_{60}. In this study only condensable products, namely oil/wax, were analysed using GC/FID. The pyrolysis operations led to the production of hydrocarbons in the range of C_{10}-C_{35} with a maximum concentration at C_{19}, which approximately represented the collected oil/wax from the first condenser. Figure 13 shows the total composition of the product from the first experiment. The second condenser collected a very light fraction mostly in the range of C_8 to C_{15} (close to kerosene fraction which is within C_{10}-C_{18} range) with a maximum concentration of C_9. A heavy fraction was detected in the collected waxy residue from the interior part of the unit with the bulk of hydrocarbon in the range of C_{20}-C_{40} peaking at C_{30}. This product is named wax in figure 14a. The products of both condensers and also the residue waxes were analysed separately using GC/FID and afterwards the total compositions were calculated based on yield of each fraction. Figures 14a and 14b show the composition of the individual collected products and the GC chromatograms of used standard sample with the pyrolysis products from the first experiment. Similar trend was obtained in the rest of the experiments.

The objective of microwave experiments was to investigate if microwave pyrolysis leads to different product distribution and recognise the suitability of the products for use as fuel. As it is known

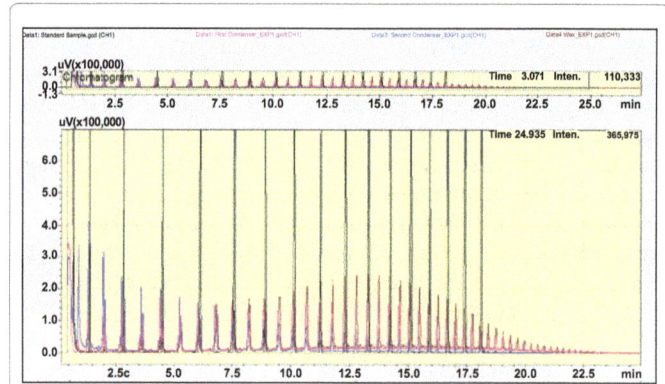

Figure 14b: GC chromatograms of used standard sample and pyrolysis products of Exp_1.

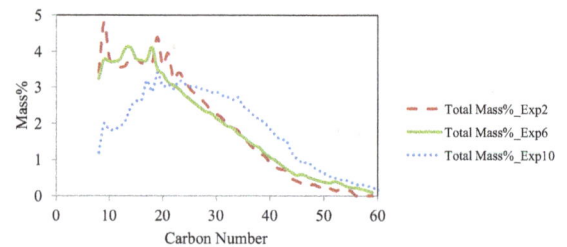

Figure 15: Composition of products in Exp_2, Exp_6 and Exp_10.

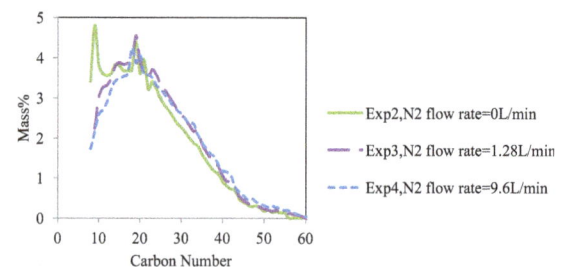

Figure 16: Composition of products in experiments 2 to 4 (Carbon and plastic).

the approximate carbon range of refinery fuels are [43]: Petrol C_5-C_{12} (~C_8), Kerosene C_{10}-C_{18} (~C_{15}), Diesel C_{15}-C_{25} (~C_{20}). This goal was achieved in the conducted experiments using both SiC and carbon. Based on analysed compositions the lightest fractions were detected in Exp_2 using carbon and in Exp_6 using SiC with the bulk production in the C_8-C_{20} range because of using a low nitrogen flow rate. Findings from this study show that the pyrolysis products with potential feedstock usage [44] were placed in the range of hydrocarbon series with the following application [43]; fuel for cars (~C_8), raw material for chemical and plastic (~C_{10}), fuel for aeroplane (~C_{15}), fuel for cars and lorries (~C_{20}), fuel for power station or lubricant or grease (~C_{35}), road surfing (~+C_{40}). The heaviest product composition was obtained in the last experiment (Exp_10). The rest of the runs (Exp_1 to Exp_9) produced hydrocarbon ranging between the two lightest fractions (red or green line in Figure 15) and heaviest fractions (blue line in Figure 15).

Effect of nitrogen flow rate

The change in residence time of the pyrolysed gases was due to

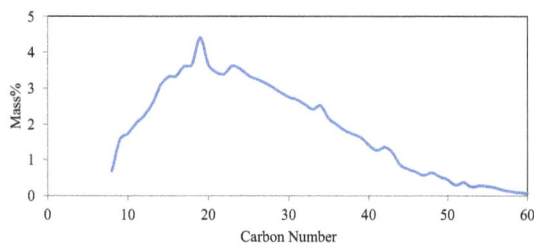

Figure 13: Total composition profile of Exp_1.

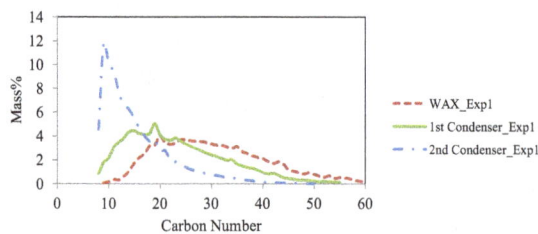

Figure 14a: Composition of each collected products of Exp_1.

alteration of nitrogen flow rate which made a significant shift in carbon number distribution. This fact was observed regardless of the type of the microwave absorbent (Figures 16 and 17). Clearly lower nitrogen flow rate led to higher residence time for the pyrolysed gases and caused more cracking of the polymer chain as a result of secondary reactions. Therefore the proportion of low carbon number products (C_8-C_{20}) was greater in the experiments with low nitrogen flow rate. A similar result was observed in conventional pyrolysis of plastic in reference [42]. The influence of nitrogen flow rate can be evidently seen in Exp_2 for carbon absorbent and in Exp_6 for SiC absorbent.

Effect of rotation

The general bulk composition in Exp_9 and Exp_10 lay within the C_{10}-C_{35} range similar to other experiments however variation in products distribution was observed with changes in rotation speed. The rotation of the reactor helped to achieve a uniform heating throughout the sample. Rotation enhanced the heat transfer and this fact made the temperature control easier. Furthermore, the rotation speed changed the residence time of the reaction; the faster it rotated the shorter time period for cracking was observed. The rotation speed was set constantly on 4 rpm in Exp_9 while higher speed was applied in Exp_10 starting at 8 rpm (in 12 minutes) and increased to 18 rpm after 40 minutes. According to the presented results in figure 18 increasing the rotation speed improved the heat transfer, shortened the decomposition time due to the whole plastic being heated at once and produced heavier

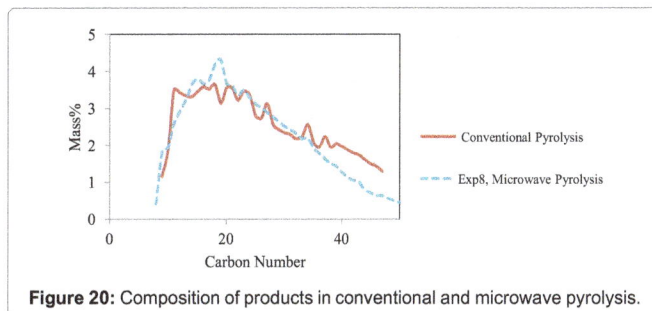

Figure 17: Composition of products in experiments 5 to 8 (SiC and plastic).

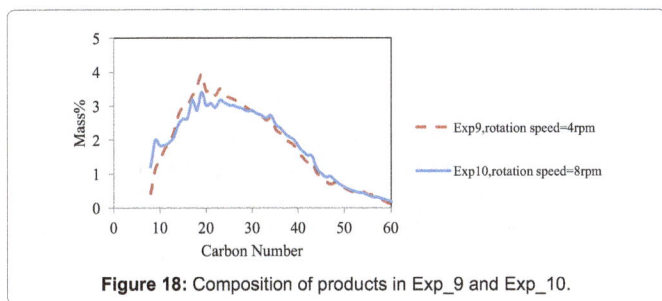

Figure 18: Composition of products in Exp_9 and Exp_10.

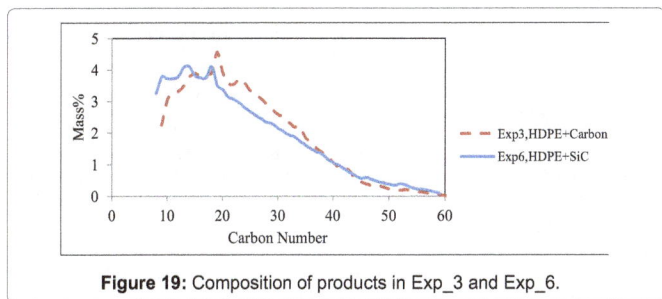

Figure 19: Composition of products in Exp_3 and Exp_6.

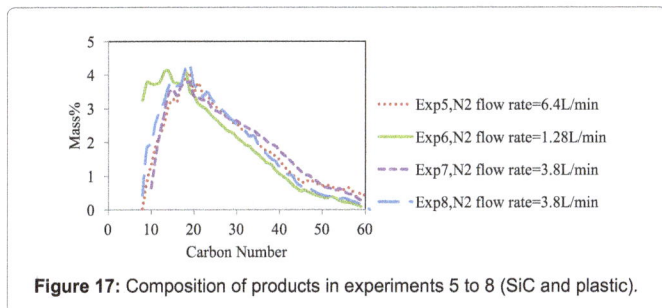

Figure 20: Composition of products in conventional and microwave pyrolysis.

fraction. And as expected this fact increased the oil/wax yield tremendously from 46% to 73%.

Effect of absorbent type

In comparison between the absorbent type it was found out that under the experiment conditions mentioned above for Exp_3 and Exp_6, carbon distribution was shifted to the lighter fraction when silicon carbide powder was used instead of carbon block (Figure 19). In both experiments the same nitrogen flow rate and mass of load were used but a lower input power was set in Exp_6. Occupied volume in the reactor due to density differences of the absorbents could be responsible for dissimilar microwave dissipations and field patterns. The geometry of the material, position and shape of the material set a new boundary condition for the field configuration. This fact is particularly true when the dielectric loss factor is high. Moreover, SiC is recommended as an absorbent over carbon due to its high density, making it less dusty and with a higher thermal conductivity and dielectric loss factor. Minimum arcing was seen when using SiC.

Comparison of microwave and conventional heating

One of the main objectives in this study was to investigate the variation of composition under microwave heating compared to conventional heating. Figure 20 shows the comparison between the compositions of products acquired in this study (using a reator with the volume of the 14825 cm³) with the compositions of conventional pyrolysis presented in literature (using a reator with the volume of the 352 cm³) [42]. A very similar product distribution was obtained in both pyrolysis methods. This result is in agreement with the reported results for microwave pyrolysis of HDPE using a semi batch microwave unit (with a maximum power output of 5 kW controlled by 4 magnetron) by Ludlow and Chase [17].

In this study microwave volumetric heating did not alter the product composition, however progressive improvement in heating uniformity was observed. It was found out that pyrolysis of plastic using microwaves provides a clean and efficient heating process with an easier and faster heating control on a reasonable scale (0.5 kg of plastic). Enhancement in heating uniformity as a result of volumetric microwave heating could result in a faster pyrolysis process, while in conventional thermal pyrolysis a large surface area is required for heat transfer.

DSC results

As mentioned earlier DSC analysis was conducted to measure the latent heat and melting range of pyrolysed products and investigate their suitability as useful chemicals such as PCM. Based on GC analyses, composition of the HDPE oil/waxes cover a very diverse hydrocarbon ranges from C_9 to C_{60} suggesting a wide heat flow patterns with endothermic behaviour as shown in figure 21. All measured latent heats and melting ranges of the product came out from the first condenser and collected waxes from the unit are summarised in figures 22 and 23. The latent heat of the second condenser product was reported only for experiments 5, 6, 8, 9, 10 and not for experiments 1 to 4 and 7. This is because for these products the DSC curve was very shallow in depth and no clear boarder for start and end points were recognised in curves, so accurate latent heats of melting could not be obtained.

According to GC/FID analyses, collected wax residue has the heaviest compounds. This may explain the higher latent heat measured with the average value of 130 J/g in figure 22 and the higher melting points with the average value of 60°C in figure 23, compared to the first condenser products with their average latent heat value of 112 J/g and 53°C melting point. Furthermore, the lowest latent heat was measured for the second condenser liquid product with an average latent heat value of 66 J/g. The measured values in this study were found below the theoretical values for the corresponding pure standard hydrocarbons. This may be associated with a high level of un-saturation in molecular structure and a very wide range of carbon numbers-as high as 160°C.

In this study it was observed that the latent heat of oil/wax products

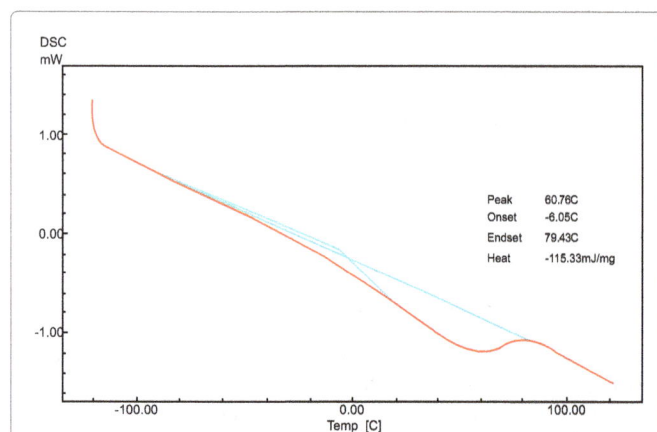

Figure 21: Heat flow (mW) vs. Temperature from DSC of the HDPE wax fraction (WG_Exp10).

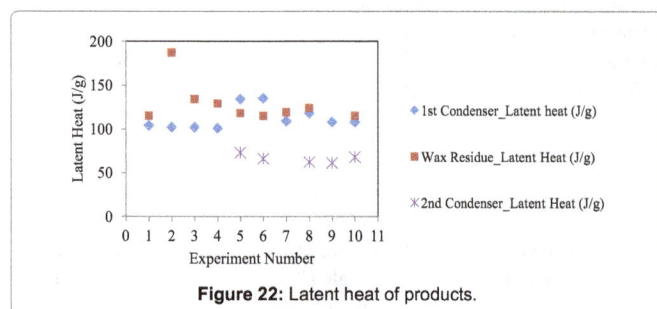

Figure 22: Latent heat of products.

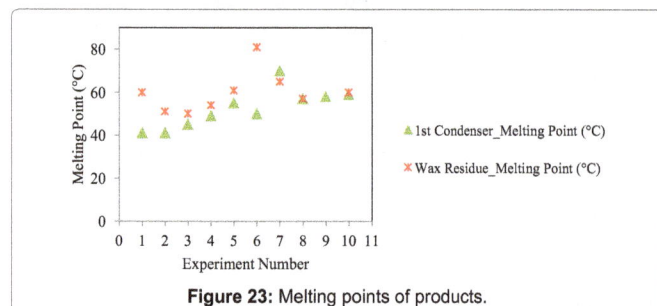

Figure 23: Melting points of products.

cannot be directly used as a PCMs application due to their low measured latent heat and very wide melting range. Hydrogenation is one of the effective methods that can be used to improve the latent heat of pyrolysed plastic products.

Conclusions

In this paper the design and performance of a microwave reactor were presented. Experiments were conducted and the results were analysed. The following items highlight the main achievements:

1. As a strong electromagnetic software Microwave Studio' software (MWS) provided very valuable information, furthering our understanding of electromagnetic field patterns and power loss density inside the microwave unit. The dimensions of the transmission line were optimised based on solving Maxwell's Equations using MWS.

2. The microwave pyrolysis of plastic on a reasonable scale is feasible using the proposed new design.

3. In a stationary microwave reactor, the total yield of the pyrolysis oil/wax products was around 60% and 40%, using carbon and silicon carbide as microwave absorbents, respectively. The oil/wax yield was increased to 73% with the use of silicon carbide in the rotating reactor. This is probably due to shorter residence time as a result of temperature distribution within the sample, which indicated that the whole plastic was being heated uniformly at the same time.

4. The microwave pyrolysis of plastic produced a very wide range of hydrocarbons-up to C_{60}. The large portion of the produced hydrocarbons is in the range of C_8 to C_{35}. Under the optimised pyrolysis operating condition the lower carbon numbers (C_8 to C_{20}) can be maximised and ultimately utilised as fuel.

5. Microwave heating (with reactor volume of 14825 cm³) did not alter the product composition in comparison to the conventional pyrolysis of plastic (with reactor volume of 352 cm³); however it provided very uniform heating distribution on a reasonable scale. This indicates that the microwave pyrolysis of plastic can be achieved much faster than thermal pyrolysis. Although both thermal and microwave reactors behaved the same here, this will not be the case in an industrial scale pyrolysis reactor, which would require a large heat transfer area, unlike in the microwave reactor.

6. Silicon carbide is recommended as a microwave absorbent over carbon due to its electrical and thermal physical properties, such as its higher loss factor making it able absorb microwaves more efficiently, its higher density meaning it creates less dust in the reactor, and higher thermal conductivity to aid in the heat transfer within the sample mixture.

References

1. Wang X, Chen H, Luo K, Shao J, Yang H (2008) The influence of microwave drying on biomass pyrolysis. Energy Fuels 22: 67-74.

2. Menendez JA, Inguanzo M, Pis JJ (2002) Microwave-induced pyrolysis of sewage sludge. Water Research 36: 3261-3264.

3. Dominguez A, Menendez JA, Inguanzo M, Pis JJ (2005) Investigations into the characteristics of oils produced from microwave pyrolysis of sewage sludge. Fuel Processing Technology 86: 1007-1020.

4. Heyerdahl P, Gilpin G (2012) Distributed biomass conversion.

5. Robinson S (1989) Wave goodbye to scrap tyres. European Rubber Journal 171: 32-33.

6. Lee A (2003) Waving aside tyre incineration. Engineer.

7. Miura M, Kaga H, Sakurai A, Kakuchi T, Takahashi K (2004) Rapid pyrolysis of wood block by microwave heating. J Anal Appl Pyrol 71: 187-199.

8. Robinson JP, Kingman SW, Barranco R, Snape CE, Al-Sayegh H (2009) Microwave pyrolysis of wood pellets. Ind Eng Chem Res 49: 459-463.

9. Rattanadecho P (2006) The simulation of microwave heating of wood using a rectangular wave guide: Influence of frequency and sample size. Chem Eng Sci 61: 4798-4811.

10. Soongprasit K, Aht-Ong D, Sricharoenchaik V (2007) Microwave induced thermal conversion of Ethylene-Vinyl Acetate copolymer/Natural rubber composite from shoe sole scrap. Fuel 79-733.

11. El harfi K, Mokhlisse A, Chanaa MB, Outzourhit A (2000) Pyrolysis of the Moroccan (Tarfaya) oil shales under microwave irradiation. Fuel 79: 733-742.

12. Dominguez A, Menendez JA, Fernandez Y, Pis JJ, Nabais JMV, et al. (2007) Conventional and microwave induced pyrolysis of coffee hulls for the production of a hydrogen rich fuel gas. J Anal Appl Pyrol 79: 128-135.

13. Menendez JA, Dominguez A, Fernandez Y, Pis JJ (2007) Evidence of self-gasification during the microwave-induced pyrolysis of coffee hulls. Energy and Fuels 21: 373-378.

14. Kim HY, Song YM, Yoon P (1992) Microwave-heated fluidized bed CVD process for continuous production of polysilicon. Toronto, Ont, Can: Publ by Electrochemical Soc Inc, Manchester, NH, USA.

15. Baysar A, Kuester JL, El-Ghazaly S (1992) Theoretical and experimental investigations of resonance frequencies in a microwave heated fluidized bed reactor. IEEE.

16. Gerdes T, Tap R, Bahke P, Willert-Porada M (2006) CVD-processes in microwave heated fluidized bed reactors. Advances in Microwave and radio frequency Processing 720-734.

17. Ludlow-Palafox C, Chase HA (2001) Microwave-induced pyrolysis of plastic wastes. Ind Eng Chem Res 40: 4749-4756.

18. Klepfer JS, Honeycutt TW, Sharivker V, Tairova G (2001) Process and reactor for microwave cracking of plastic material, US6184427 B1.

19. Van Dinh C, Roces S, Bacani F, Yimsiri P, Kubouchi M (2005) Design and fabrication of microwave pyrolysis system. Tokyo, Japan: Institute of Electrical and Electronics Engineers Computer Society, USA.

20. Cioni B, Lazzeri A (2008) Modeling and development of a microwave heated pilot plant for the production of SiC-based ceramic matrix composites. International Journal of Chemical Reactor Engineering 6: 1542-6580.

21. Jeni K, Yapa M, Rattanadecho P (2010) Design and analysis of the commercialized drier processing using a combined unsymmetrical double-feed microwave and vacuum system (case study: tea leaves). Chemical Engineering and Processing: Process Intensification 49: 389-395.

22. Menendez JA, Arenillas A, Fidalgo B, Fernandez Y, Zubizarreta L, et al. (2010) Microwave heating processes involving carbon materials. Fuel Processing Technology 91: 1-8.

23. Bykov YV, Rybakov KI, Semenov VE (2001) High-temperature microwave processing of materials. Journal of Physics D: Applied Physics 34: R55-R75.

24. Appleton TJ, Colder RI, Kingman SW, Lowndes IS, Read AG (2005) Microwave technology for energy-efficient processing of waste. Applied Energy 81: 85-113.

25. Robinson JP, Kingman SW, Snape CEH, Shang (2007) Pyrolysis of biodegradable wastes using microwaves. In Proceedings of the ICE - Waste and Resource Management 160: 97-103.

26. Esveld E, Chemat F, Van Haveren J (2000) Pilot Scale Continuous Microwave Dry-Media Reactor – Part 1: Design and Modeling. Chemical Engineering and Technology 23: 279-283.

27. Robinson JP, Kingman SW, Snape CE, Bradshaw SM, Bradley MSA, et al. (2010) Scale-up and design of a continuous microwave treatment system for the processing of oil-contaminated drill cuttings. Chemical Engineering Research and Design 88: 146-154.

28. Shang H, Snape CE, Kingman SW, Robinson JP (2006) Microwave treatment of oil-contaminated North Sea drill cuttings in a high power multimode cavity. Separation and Purification Technology 49: 84-90.

29. Shang H, Robinson JP, Kingman SW, Snape CE, Wu Q (2007) Theoretical study of microwave enhanced thermal decontamination of oil contaminated waste. Chemical Engineering & Technology 30: 121-130.

30. Scheirs J, Kaminsky W (2006) Feedstock recycling and pyrolysis of waste plastics: converting waste plastics into diesel and other fuels. John Wiley &Sons, Ltd.

31. Nelson SO (2004) Useful relationships between dielectric properties and bulk density of powdered and granular materials.

32. Diprose MF (2001) Some considerations when using a microwave oven as a laboratory research tool. Plant and Soil 229: 271-280.

33. Nuchter M, Ondruschka B, Bonrath W, Gum A (2004) Microwave assisted synthesis - A critical technology overview. Green Chemistry 6: 128-141.

34. Pert E, Carmel Y, Birnboim A, Tayo O, Gershon D, et al. (2001) Temperature measurements during microwave processing: The significance of thermocouple effects. Journal of the American Ceramic Society, 84: 1981-1986.

35. Janney MA, Kimrey HD, Allen WR, Kiggans JO (1997) Enhanced diffusion in sapphire during microwave heating. Journal of Materials Science 32: 1347-1355.

36. Haque KE (1999) Microwave energy for mineral treatment processes - A brief review. International Journal of Mineral Processing 57: 1-24.

37. Mullin J, Bows J (1993) Temperature measurement during microwave cooking. Food Additives and Contaminants 10: 663-672.

38. Albright G, Farid M, Al-Hallaj S (2010) Development of a model for compensating the influence of temperature gradients within the sample on DSC-results on phase change materials. Journal of Thermal Analysis and Calorimetry. 101: 1155-1160.

39. Khudhair AM, Farid MM (2004) A review on energy conservation in building applications with thermal storage by latent heat using phase change materials. Energy Conversion and Management 45: 263-275.

40. Walter Kaminsky JS (2006) Feedstock recyclig and pyrolysis of waste plastics,Converting waste plastics into diesel and other fuels. John Wiley & Sons,Ltd.

41. Castelino M (2008) Pyrolysis of waste plastics to produce waxes for use as a phase change material, University of Auckland, New Zealand.

42. khaghanikavkani E, Farid MM (2010) Pyrolysis of Plastics: Effects of Temperature and Residence Time on Product Yields and Compositions. Chemeca 2010 : Adelaide, Austarila.

43. http://www.moorlandschool.co.uk/earth/oilrefinery.htm

44. Williams PT, Williams EA (1999) Fluidised bed pyrolysis of low density polyethylene to produce petrochemical feedstock. J Anal Appl Pyrol 51: 107-126.

2

Adsorption of U(VI) from Aqueo0020us Solution by Chitosan Grafted with Citric Acid via Crosslinking with Glutraldehyde

Nguyen Van Suc Ho Thi Yeu Ly*

Ho Chi Minh City University of Technical Education, Vietnam

Abstract

This paper reports on the studied results of the sorption ability of the grafted chitosan with citric acid via crosslinking with glutraldehyde (C-Gch). The results obtained shown that after grafting citric acid on crosslinked chitosan, the adsorption capacity for U(VI) was significantly enhanced. Effects of the adsorption process including contact time, pH, initial concentration and some of metal ions were investigated. It was found that the maximum capacity obtained was 172 mg/g after 300 min of contact time and at pH 4. In the presence of cations such as Cu (II), Zn (II), Pb (II) and Cd (II), the adsorption capacity for U(VI) was reduced. The effect of these metals on the adsorption of U(VI) by C-Gch was arranged in the order: Cu (II) > Zn (II) > Cd (II) > Pb (II). The experimental data good fitted the pseudo second order model with correction coefficients $R^2 \geq 0.995$ for all U(VI) concentrations ranging from 10 to 80 mg/L. The Langmuir model was found to be an appropriate model for describing the equilibrium adsorption process. Based on the Langmuir model, the maximum adsorption capacity was found to be 172 mg U(VI)/g adsorbent.

Keywords: Crosslinked chitosan; Grafted chitosan; Citric acid; Uranium; Adsorption; Kinetic model; Isothermal model

Introduction

Recently, the development of nuclear industry along with activities in the mining, fertilizer production, combustion of fossil fuels, the water sources in many parts of the word has been contaminated by uranium [1-3]. Because of its toxic and radioactive, uranium exposure can cause adverse effects to human health [4,5]. Therefore, the need to remove uranium from the polluted water to the permitted level is essential.

Adsorption is a process that has been widely used to remove heavy metals including uranium from contaminated water environment [6,7]. Particularly, adsorption is a very effective process when dealing with low concentrations of pollutants. One of the critical components in the adsorption process is the adsorbent that is crucial to the efficiency of the process as well as treatment cost. In generally, active carbon, zeolite, etc. are adsorbents that have been commonly used in the adsorption. However, these adsorbents are relatively expensive. Therefore, finding alternative materials with reasonable cost to use in the adsorption is an urgent task for investigators [8,9].

Recent researches have been focused on studying alternative materials including agriculture by products for expensive materials in the adsorption. The results obtained have been proved that materials from agriculture by-products are capable of adsorbing heavy metals and shows great potential as adsorbent utilizing in the adsorption [7-13].

One of the byproducts of agriculture has been identified for the adsorption material is chitosan which is a natural product derived from chitin, a polysaccharide foundation in the exoskeletons of shellfish like shrimps and crabs. Because of owning different functional groups such as amino group (NH_2), hydroxyl group (OH), the major sites for adsorption, chitosan has very large adsorption capacity for metal ions [14,15].

To increase the applicability of chitosan which has been modified by cross linking with a number of chemical agents such as epichlorohydrin (EPI), glutaraldehyde [14-17]. In the most crosslinking case, chitosan becomes stability in an acidic environment. However, crosslinked chitosan with such agents was varied adsorption properties, its adsorption capacity can be reduced by blocking various functional groups in the polymer network.

To overcome this problem, crosslinked chitosan was grafted with new function groups which can be changed pH range for metal sorption and uptake mechanism in order to increase sorption selectivity for target metal. For example, sulfur compounds were grafted on chitosan via a cross-linking agent such as glutaraldehyde or epichlorhydrin to enhance the adsorption efficiency of mercury and noble metals [14,18].

Recently, low ionic carboxylic acid such as citric acid and its salts have been widely used to modify chitosan for pharmaceutical application [19,20]. In this study, we used citric acid, a chemical compound containing hydroxyl and carboxylic groups to graft onto chitosan by two steps consisting of pre-reaction of chitosan with glutaradehyde followed by reaction with citric acid. The resulting adsorbent was used to adsorb uranyl ions in contaminated water environment. The results reported here including effects of parameters such as pH, contact time, adsorbent dose and co-ions. Characteristics of U(VI) adsorption process described using the adsorption kinetic and isotherm models were also include.

Materials and Method

Chemicals

Commercial chitosan, 80% deacetylation from Center of Irradiation

---

***Corresponding author:** Nguyen Van Suc Ho Thi Yeu Ly, Ho Chi Minh City University of Technical Education, 01 Vo Van Ngan, Thu Duc disrt. Ho Chi Minh City, Vietnam

Technology, VINAGAMMA, Ho Chi Minh City, Vietnam was used to prepare the adsorbent. All chemicals used in this work were analytical grade. The standard uranium solution (1.0 mg.mL^{-1}) was prepared by dissolving 0.10 g of uranium metal (99.99%) in 15 mL of conc. HNO$_3$ and evaporated to dryness on a water bath. The residue was dissolved with 10 mL of 0.1M HNO$_3$ and transferred to a 100 mL measuring flask. The solution was diluted with water to the mark and shaken well. The arsenazo III solution (0.07%) was prepared by dissolving 0.07g arsenazo III (Merck Co.) in 100 ml of 0.05M HCl solution and kept in a polyethylene vessel.

Preparation of adsorbent

Chitosan (5g) was dissolved in 2 % (w/v) acetic acid and reprecipitated by adding dropwise of 2N NaOH solution into the mixture until pH of the solution reached to 6.5 -7. The pure mass precipitate of chitosan was washed with distilled water and air-dried. The obtained chitosan was further dried under vacuum at 60^{0}C and crushed in an agate mortar to fine powder. Crosslinking reactions were performed by adding 0.5 molar ratios of glutaraldehyde. Reactions were performed for a period of 24h under pH value of 3-5. The crosslinked chitosan was washed several time with distilled water and dried at room temperature and labeled as Gch. Citric grafted chitosan was obtained by immersing the Gch in aqueous 5.0% (w/v) citric acid solution at pH 5 and 4°C for 24 hrs. After grafting, modified chitosan was repeatedly washed with deionized water, thoroughly dried in air and under vacuum at 60°C and labeled as C-Gch. The FT-IR spectra of C-Gch and Gch samples were recorded by FT-IR -8400S-SHIMADU.

Uranium analysis

Determination of U(VI) in the liquid phase after equilibriums was based on measuring the color complex of U(VI) with arsenazo III in the medium of 10^{-3} M HClO$_4$ [21]. The procedure for determination of U(VI) can be briefly described as follows:

A test solution containing U(VI) was placed into a 50 mL measuring flask. 1 mL of 0.07 % arsenazo III and 10 mL of 3M HClO$_4$ were added. The mixture was shaken for two min and 3M HClO$_4$ solution was then added to the measuring volume. The mixture was shaken for three min. The absorbance of the absorbing solution was measured at 650 nm using a Libra S32 Perking Elmer spectrophotometer. Uranium content in the test solution was determined from a calibration curve, which was constructed using a uranium standard solution.

Adsorption experiment

For determination of contact time required for adsorption equilibrium, batch experimental mode was conducted. A desired amount of adsorbent were added to each 50 ml of solution containing a desired amount of U(VI). pH values of reaction solutions was adjusted to 4. The adsorption solutions were shaken at speed 200 rpm by mechanical shaker for 300 min at room temperature (30 ± 1°C). A series of 0.5mL were taken from the adsorption solutions at predefine time intervals for analysis of the U(VI) residual. For evaluation of pH effect, adsorption experiments were carried out as above mentioned but variation of pH of the adsorption solution and retaining the equilibrium time which was obtained from the first experiment. pH of the adsorption solutions was varied from 1 to 6 and adjusted by diluted solution of HNO$_3$ or NaOH. The obtained pH value at which the adsorption efficiency reached to maximum value was taken for the

next experiments. To investigate the effect of co-ions including Cu (II), Zn(II), Pb(II) and Cd(II), the experiments were conducted under optimum conditions such as pH, contact time. The amount of U(VI) was kept constants (50 mg/L) while variation of co-ion concentrations ranging from 100 to 900 mg/L. Similarly, the adsorption isothermal of U(VI) by C-Gch were studied by adding a desired amount of adsorbent into each 50 ml of solution containing U(VI) concentration in the range of 30, 50, 100, 150, 200, 250, 300 mg/l.

Concentration of U(VI) in the liquid phase of all experiment was determined by spectrophotometric method following the procedure which was mentioned above. The adsorption capacity for U(VI) in each experiment was calculated using the following equations:

$$q_e = \frac{C_0 - C_e}{m} \times V \tag{1}$$

Where q_e is adsorbent capacity, mg/g, %; C_0 is initial concentration of U(VI), mg/l; m is weigh of adsorbent, g; C_e is final concentration of U(VI) at equilibrium adsorption, mg/l; V is volume of U(VI) used for adsorption reaction, L.

Results and Discussion

FT-IR spectra of C-Gch

The FTIR spectra of chitosan (Ch), crosslinked chitosan with glutraldehyde (Gch) and crosslinked chitosan grafted with citric acid (C-Gch) are shown in Figure 1a, 1b and 1c, respectively. From the FTIR spectra of chitosan (Figure 1a), it can be observed the band amide 1 around 1634 cm^{-1} stretching frequency, characteristic of acetylated units. The FTIR spectra of Gch (Figure 1b) shows the more adsorption band in the region from 1600-1700 cm^{-1} for the crosslinked chitosan which is a characteristic of C=O groups from aldehyde proving the presence of glutaradehyde in crosslinked chitosan [15]. The appearance of adsorption 1717 cm^{-1} peak and more intensity of adsorption peaks at 1388 cm^{-1} (Figure 1c) indicated that citric acid was grafted into chitosan network [22].

Adsorption rate and kinetic model

Figure 2 is a graph represents the relationship of the adsorption capacity and contact time of C-Gch for U(VI). As can be seen from results, it was found that the U(VI) adsorption rate at the beginning increased up 300 min of contact time. After that, the adsorption rate was slowly increased and reached plateau values. These adsorption behaviors can be explanted by adsorption mechanisms which are governed by the surface adsorption on the pore wall of adsorbent for the rapid adsorption rate or by the membrane transport of U(VI) ions onto adsorbent for the slow adsorption rate [23]. Such a variation of the adsorption rate is also observed in the most case of biosorption studies for metal ions. The obtained results suggested that 300 min of contact time were sufficient to achieve equilibrium condition and would be applied for subsequence experiments.

For chemical adsorption, the kinetic model used for fitting the time-dependent experimental adsorption data is pseudo–second-order model [24]:

$$\frac{dq_t}{dt} = k(q_e - q_t)^2 \tag{2}$$

The equation (2) can be integrated and the linear form of this model equation is given as:

$$\frac{t}{q_t} = \frac{1}{h} + \frac{1}{q_e}t \tag{3}$$

$$h = kq_e^2 \qquad (4)$$

Where q_e is the amount of U(VI) adsorbed at equilibrium (mg/g), k is the equilibrium rate constant (g/mg.min), q_t is the amount of U(VI) adsorbed at time t(mg/g). h is the initial adsorption rate as t → 0. The plot of t/q_t against t from Eq. (3) gives a linear relationship, which allows calculated of k, h and predicted q_e. The linear plots for pseudo-second-order model at different initial concentration of U(VI) are presented in Figure 3. The parameter values of pseudo-second-order model are shown in Table 1. The regression coefficients R^2 values were found to be near 1 for all concentration ranging from 10 to 80 mg/L indicating that the experimental data good fitted the pseudo-second-order model.

Effect of pH

pH is one of the important parameters for the adsorption of metal ions because pH controls the charge of the adsorbent surface as well

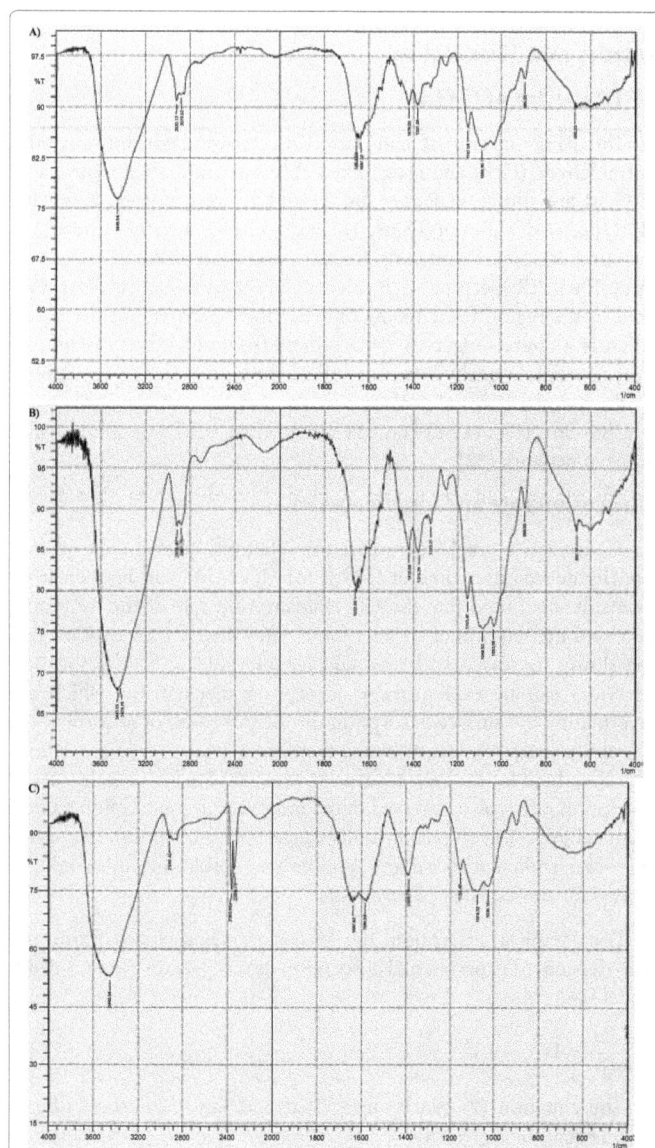

Figure 1: FTIR spectra of 1a. chitosan (ch) , 2b. crosslinked chitosan with glutraldehyde (Gch) and 1c. crosslinked chitosan grafted with citric acid (C-Gch).

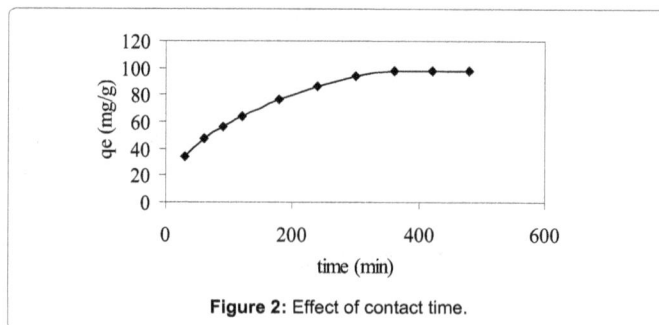

Figure 2: Effect of contact time.

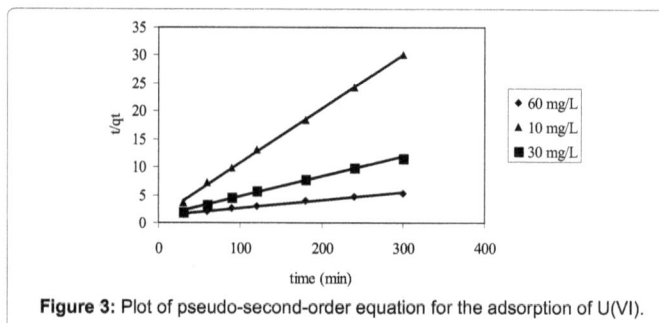

Figure 3: Plot of pseudo-second-order equation for the adsorption of U(VI).

as the existence in various forms of the sorbate in the solution. The effect of pH on the U(VI) adsorption by C-Gch is shown in Figure 4. For U(VI), the presence of different hydrolyzed species in the wide range of pH will be affected to the adsorption efficiency. In the range of acid to neutral of the solution, U(VI) ions are mainly exist in the hydrolyzed forms such as UO_2^{2+}, $(UO_2)(OH)_2^{2+}$, $UO_2(OH)^+$ and $(UO_2)_3(OH)_5^+$ [23, 25]. Thus, the negative charge sites on the adsorbent surface play an important role in the adsorption process of U(VI). As results obtained showing in Figure 3, the adsorption capacity for U(VI) by C-Gch increased with increasing pH and reached the maximum value at pH 3.5 - 4 and then gradually reduced when pH reached the value more than 4.5. The low capacity value at pH < 3 can be explained that hydrolyzed ions of U(VI) are strongly competed against with H^+ ions for protonation of negative charge sites on the adsorbent surface. At pH ≥ 3-4.5, the competition of H^+ ions is reduced leading the adsorption capacity increase. On the other hand, the presence of carboxylic groups in C-Gch leading to increasing adsorption sites on the adsorbent surface. According to Memon et al. [26], at pH from 3 to 4.5, the carboxylic group is deprotonated and became negatively charged, hence, increasing the availability of binding sites for positively charged of U(VI) ions. At pH > 5, U(VI) begins partly forming the anionic complexes with OH^- and CO_3^{2-} (formed by dissolution of CO_2 from air) such as $(UO_2)CO_3(OH)_3^-$, $UO_2(CO_3)_2^{2-}$, $UO_2(CO_3)_3^{4-}$ leading to slightly reduce adsorption capacity of U(VI).

Effect of initial U(VI) concentration

Results obtained in study of an effect of initial U(VI) concentration were illustrated in Figure 5. As can be seen from the Figure 5, the adsorption capacities were increased with increasing initial U(VI) concentration. At higher concentration of U(VI) (≥ 100 mg/L), the adsorption capacities slowly go to near a constant value. This phenomenon can be explained that at initial state, a large number adsorption vacant site were available on the adsorbent surface leading to increase the adsorption capacity. At near equilibrium, most vacant adsorption sites were filed resulting in slowdown of adsorption capacity.

Concentration (mg/L)	Equation	K (g/mg.min)	R^2	h (g/mg.min)	q_e (mg/g)	$q_{e\text{-}exp}$ (mg/g)
10	y = 0.096x + 1.085	$8.59 \cdot 10^{-3}$	0.999	0.921	10.35	9.94
30	y = 0.036x + 1.107	$1.18 \cdot 10^{-3}$	0.995	0.90	27.70	25.79
60	y = 0.014x + 1.303	$1.46 \cdot 10^{-4}$	0.995	0.767	72.46	68.87

Table 1: Parameters of pseudo-second-order model for the adsorption of U(VI).

Effect of adsorbent dose

Figure 6 shows results of the effect of the adsorbent dose on the adsorption capacity. The results showed that the adsorption capacity was reduced with increasing the adsorbent dose. It was found that at low dosage of C-Gch from 0.05 -0.1 g/ 50mL, the adsorption capacity reduced from 43.4 mg/g to 18.5 mg/g and rapidly reduced to 7.32 mg/g at dosage ranging from 0.1 to 0.4 g/5mL. This phenomenon is due to the fact that at higher dosage of adsorbent, the vacant adsorption sites on the adsorbent are remained constants resulting in reduced metal uptake.

Effect of other metal ions

Effects of cations including Cu (II), Pb(II), Zn(II) and Cd(II) on the adsorption of uranium (VI) byC-Gch are shown in Figure 7. It

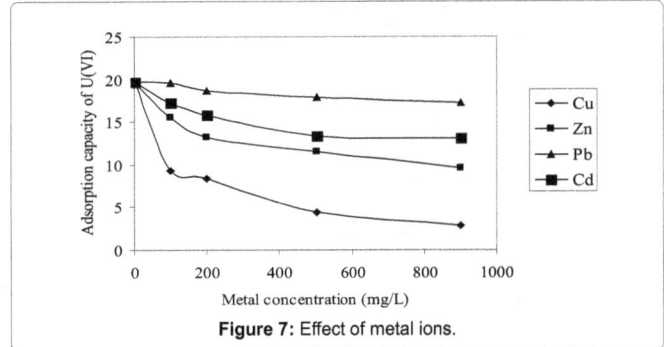

Figure 7: Effect of metal ions.

was found that the adsorption capacity of U(VI) was reduced with increasing concentrations of cations. As can be seen from results, in the presence of cations with concentration ranging from 100 to 900 mg/L, the adsorption capacity of U(VI) reduced from 20mg/g to 4.3, 11.6, 15.1 and 17.5 mg/g in the presence of Cu(II), Zn(II), Cd(II) and Pb(II), respectively. The effect of metal ions on the adsorption of uranium (VI) can be arranged in the order: Cu (II) > Zn (II) > Cd (II) > Pb (II).

Adsorption isotherm

The adsorption isotherm is the most important information, which indicates how the adsorbent molecules distribute between the liquid and the solid phases when the adsorption process reaches an equilibrium state [10,11]. In this study, two famous adsorption isotherm models were used to describe the adsorption process including Langmuir and Freundlich models [27].

Langmuir model:

$$q_{eq} = \frac{K_L q_{max} C_{eq}}{1 + q_{max} C_{eq}} \tag{5}$$

The linear equations of Langmuir model is presented as following:

$$\frac{C_{eq}}{q_{eq}} = \frac{C_{eq}}{q_{max}} + \frac{1}{K_L q_{max}} \tag{6}$$

Where:

q_{eq} – amount of U(VI) adsorbed per gram of dry adsorbent at equilibrium (mg/g)

C_{eq} – equilibrium concentration of U(VI) in solution (mg/L)

K_L – Langmuir constant (L/g)

q_{max} – maximum U(VI) to adsorb per gram of dry adsorbent (mg/g)

Freundlich model:

$$q_{eq} = K_F C_{eq}^{1/n}$$

The linearised form of the Freundlich model was used:

$$\log q_{eq} = \log K_L + 1/n \log C_{eq}$$

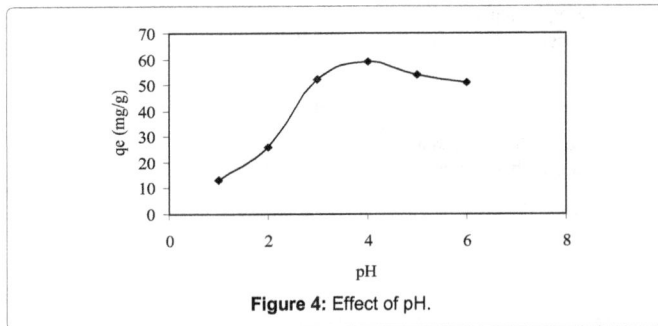

Figure 4: Effect of pH.

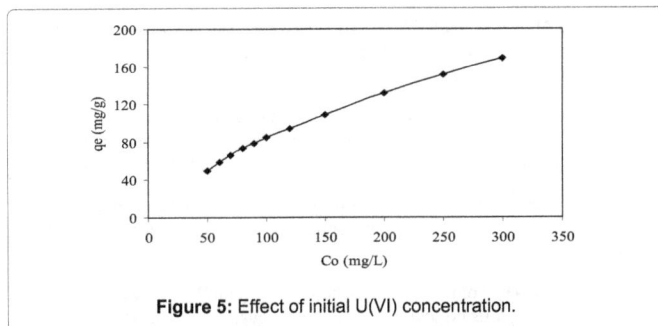

Figure 5: Effect of initial U(VI) concentration.

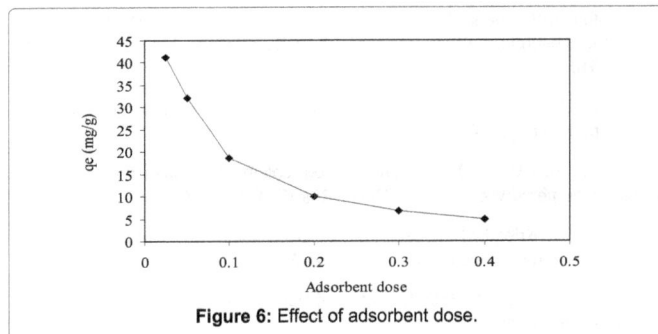

Figure 6: Effect of adsorbent dose.

Where:

q_{eq} – amount of U(VI) adsorbed per gram of dry adsorbent (mg/g)

C_{eq} – equilibrium concentration in solution (mg/L)

$1/n$ – Freundlich constant (mg/g)

K_F – Freundlich constant (g/L)

The plots of the linear Langmuir and Freundlich models were presented in Figure 8 and 9. The parameters in two models were calculated from slope and interpreted of the plots. The results obtained were given in Table 2. From the results in Table 2, the correction values R^2 of Langmuir equation and Freundlich equation were found to be 0.9693 and 0.9353, respectively. Based on the R^2 values, It suggested that the adsorption process of U(VI) by C-Gch was fixed to both models. However, the used of Langmuir model for description of U(VI) adsorption by C-Gch is better than that of Freundlich model due to higher value of R^2. Based on Langmuir model, the maximum adsorption capacity of U(VI) was found to be 172 mg/g for C-Gch. The high value of adsorption capacity of C-Gch shows that carboxyl groups of citric acid grafted on crosslinked chitosan could significantly contribute to the adsorption process of U(VI).

Conclusion

Research results on adsorption U(VI) by C-Gch showed that grafting citric acid on crosslinked chitosan with glutradehyde was improved the adsorption capacity. The equilibrium adsorption was

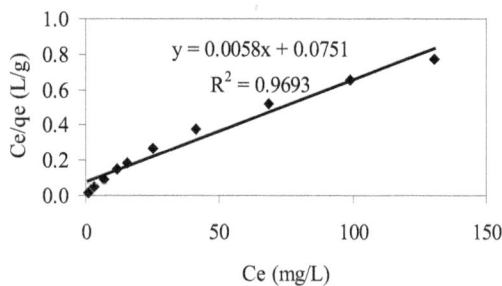

Figure 8: Plot of Langmuir isotherm model adsorption for U(VI).

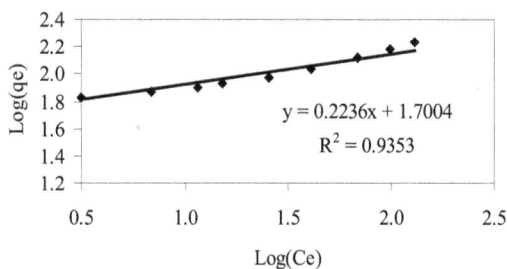

Figure 9: Plot of Freundlich isotherm model for U(VI).

Langmuir model			Freundlich model		
K_L (L/mg)	q_{max} (mg/g)	R^2	K_F (mg/g)(mg/l)^n	$1/n$	R^2
0.07	172	0.969	50.16	0.223	0.935

Table 2: The Langmuir and Freundlich adsorption parameters for uranium adsorption.

reached at 300 min of the contact time and maximum adsorption capacity was obtained at pH 4. Effect of co-ion such as Cu (II), Zn (II), Cd(II) and Pb(II) was studied. The results show that the presence of metal ions significantly affected on the adsorption of U(VI). Adsorption kinetic study of U(VI) on C-Gch showed that experimental data good fitted the pseudo second order. This means that the adsorption process of U(VI) is the chemical sorption. Two adsorption isothermal models were used to describe the adsorption equilibrium. It was found that the Langmuir model with correction value $R^2 = 0.9963$ was appropriate to use for description of the adsorption of U(VI) by C-Gch. Based on the Langmuir model, the maximum capacity of C-Gch for U(VI) was found to be 172mg/g. From the results obtained, it could be concluded that this material is potential adsorbent to use in the adsorption technology for remove U(VI) from contaminated water environment.

References

1. Alirezazadeh N, Garshasbi H (2003) A survey of natural uranium concentration in drinking water supplies in Iran. Iran J Radiat Res 1: 139-142.

2. Kurttio P, Auvinen A, Salonen L, Saha H, Pekkanen J, et al. (2002) Renal effects of uranium in drinking water. Environ Health Perspect 110: 337-342.

3. Miller AC, Xu J, Stewart M, Brooks K, Hodge S, et al. (2002) Observation of radiation-specific damage in human cells exposed to depleted uranium: dicentric frequency and neoplastic transformation as endpoints. Radiation prot Dosimetry 99: 275-278.

4. Igarashi Y, Yamakawa A, Ikeda N (1987) Plutonium and uranium in Japanese human tissues. Radioisotopes 36: 433-439.

5. Kalin M, Wheeler WN, Meinrath G (2005) The removal of uranium from mining waste water using algal/microbial biomass. J Environ Radioact 78: 151-177.

6. Grégorio C (2005) Recent developments in polysaccharide-based materials used as adsorbents in wastewater treatment. Prog Polym Sci 30: 38-70.

7. Wan Ngah WS, Endud CS, Mayanar R (2002) Removal of copper(II) ions from aqueous solution onto chitosan and cross-linked chitosan beads. React Funct Polym 50: 181-190.

8. Ibrahim SC, Hanafiah MAKM, Yahya MZA (2006) Removal of cadimium from aqueous solution by adsorption onto sugarcane bagasse. American-Eurasian J Agric & Environ Sci 1: 179-184.

9. Igwe JC, Abia AA, Ibeh CA (2008) Adsorption kinetics and intraparticulate diffusivities of Hg, As and Pb ions on unmodified and thiolated coconut fiber. Int J Environ Sci Tech 5: 83-92.

10. Augustine AA, Orike BD, Edidiong AD (2007) Adsorption kinetic and modeling of Cu(II) ion sorption from aqueous solution by mercatoacetic acid modified cassava (manihot sculenta cranz) wastes. EJEAF Che 6: 2221- 2234.

11. Ho YS, Porter JF, McKay G (2002) Equilibrium isotherm studies for the sorption of divalent metal ions onto peat: copper, nickel and lead single component systems. Water Air Soil Pollut 141: 1-33.

12. Abdel-Ghani NT, Hefny M, El-Chaghaby GAF (2007) Removal of lead from aqueous solution using low cost abundantly available adsorbents. J Environ Sci Tech 4: 67-73.

13. Preetha B, Viruthagiri T (2007) Application of response surface methodology for the biosorption of copper using Rhizopus arrhizus. J Hazard Mater 143: 506-510.

14. Eric G (2004) Interactions of metal ions with chitosan-based sorbent: a review. Sep purif Technol 38: 43-74.

15. Jeon C, Höll WH (2003) Chemical modification of chitosan and equilibrium study for mercury ion removal. Water Res 37: 4770-4780.

16. Schmuhl R, Krieg HM, Keizer K (2001) Adsorption of Cu(II) and Cr(VI) ions by chitosan: Kinetics and equilibrium studies 27: 1-8.

17. Deans JR, Dixon BG (1992) Uptake of Pb^{2+} and Cu^{2+} by novel biopolymers. Water Res 26: 469-472.

18. Hu XJ, Wang JS, Liu YG, Li X, Zeng GM, et al. (2011) Adsorption of chromium (VI) by ethylenediamine-modified cross-linked magnetic chitosan resin: isotherms, kinetics and thermodynamics. J Hazard Mater 185: 306-314.

19. Findon A, Mckay G, Blair HS (1993) Transport studies for the sorption of copper ions by chitosan. J Environ Sci and Health 28: 173-185.

20. Honary S, Hoseinzadeh B, Shalchian P (2010) The effect of polymer molecular weight on citrate crosslinked chitosan films for site-specific delivery of a non-polar drug. Tropical J Pharmaceu Res 9: 525-531.

21. Shjo K, Sakai K (1982) Rapid spectrophotometric determination of uranium (VI) in sea Water. Japan anal 31: 395-400.

22. Roy S, Panpalia SG, Nandy BC, Rai VK, Tyagi LK, et al. (2009) Effect of method of preparation on chitosan microspheres of mefenamic acid. IJPSDR 1: 36-42.

23. Khani MH, Keshtkar AR, Meysami B, Zarea MF, Jalali R (2006) Biosorption of uranium from aqueous solutions by nonliving biomass of marinealgae Cystoseira indica. Electronic J Biotechnol 9: 101-106.

24. Ho YS, McKay G (1999) Pseudo-second order model for sorption processes. Process Biochem 34: 451-465.

25. Baes CF, Mesmer RE (1976) The hydrolysis of cations. Wiley-Interscience, John Wiley and sons, New York, 512.

26. Memon JR, Memon SQ, Bhanger MI, Khuhawar MY (2008) Banana peel: a green and economical sorbent for Cr(III) removal. Pak J Anal Environ Chem 9: 20-25.

27. Sahmoune MN, Louhab K, Boukhiar A (2008) Kinetic and equilibrium models for the biosorption of Cr (III) on Streptomyces rimosus. J Appl Sci Res 3: 294-301.

3

Characterization of Polyvinyl acetate/Epoxy Blend Foam

Mohamed M El-Toony* and A.S. Al-Bayoumy

National center for radiation research and technology, Atomic energy authority, Nasr City, Cairo, Egypt

Abstract

Synthesis of polyvinyl acetate-epoxy blend by mixing of 40% PVA to Epoxy by weight and foaming by high rate of air were furthermore carried out to achieve light weight blend foam. It was found that; 5KGy irradiation dose by using Gamma irradiation was enough to attain finally compatiblization. Mechanical properties were studied using hardness tester, while surface morphology was studied as well by measuring scan electron microscope (SEM). Immersing the foam to serial dilution of different acids, alkalis and salts solution for different times was performed. Thermal behavior was discussed by measuring thermal gravimetric analysis (TGA). Investigation of electrical conductivity (Dc) for the soaked foam showed that; conductivity of acids up taken foam is more than alkali while salts is the least. Electrical conductivity of the acids soaked foam were arranged by the following: H_3PO_4 > H_2SO_4 > HCl > Citric acid, while alkalis soaked foam was as the following manner: KOH > NaOH > LiOH > Glycerin. Maximum electrical conductivity (9.63 x 10^{-2} Simon/cm) was achieved by 2N of H_3PO_4. Heating of glycerin soaked foam raise the electrical conductivity to 8.8 x10^{-3} Simon/cm. While electrical conductivity of salts solutions swelled foam were as the following: $CaCl_2$ > KCl > NaCl.

Keywords: Blend; Foam; Swelling; Irradiation; Hardness; SEM; Dc; TGA

Introduction

Materials that lack electron conduction are insulators if they lack other mobile charges as well. For example, if a liquid or gas contains ions, then the ions can be made to flow as an electric current, so the material is a conductor. Electrolytes and plasmas contain ions will act as conductors whether or not electron flow is involved. The history of the field has been recounted from several perspectives [1,2]. The first report on polyaniline goes back to the discovery of aniline. In the mid 1800's, Letheby reported the electrochemical and chemical oxidation products of aniline in acidic media, noting that reduced form was colorless but the oxidized forms were deep blue. Despite almost 30 years of history of polymeric electrolytes and their application in ion storage devices still some fundamental problems and questions from the initial times of development of these solid ionic conductors remain unsolved [3-5]. One of the most promising approaches was to prepare composite polymeric electrolytes [6,7]. This is due to their higher conductivity, improved cation transport numbers and enhanced electrolyte-lithium electrode stability compared to standard polyether based electrolytes. Composite electrolytes usually consist of three components: polymer matrix, dopant salt and filler. The role of the latter is to modify polymer-ion and ion-ion interactions leading to an improvement in the ion transport. A variety of models has been designed and used to describe ion transport phenomena in composite polymeric electrolytes with particular attention paid to the role of the filler [8-10]. These models are similar in many respects but some contradictions can also be found. Many polymer electrolytes showing high alkali metal ion conductivities have been developed. These electrolytes are polymers doped with alkali metal salts such as $LiClO_4$, $LiCF_3SO_3$, LiSCN, NaSCN, $CsHSO_4$ and KSCN, while the basic polymers include polyethylene oxide (PEO), polyvinyl acetate, polyvinyl alcohol, polymethyl acrylate, polypropylene oxide (PPO) and polymethylene oxide [7,11,12]. PEO doped with ammonium salts (NH_4HSO_4, NH_4I, $(NH_4)_2SO_4$, NH_4SCN) shows a conductivity ranging from 10^{-7} to 10^{-2} S cm^{-1} at room temperature [13,14]. Very few studies on PEO doped with acids (H_3PO_4, sulfonic acid, poly (thiphenylenesulfonic acid) have appeared in the literature, for the degradation of polymer by using the strong acid comes into a major problem [15-17]. It was reported

previously that: preliminary studies on PAAm based hydrogels doped with H_3PO_4 and H_2SO_4 have been performed. It has been shown that; the conductivity of hydrogels depends on the concentration of acid, water and cross-linking and gelation agents. Rooms' temperature conductivities up to 2×10^{-2} S cm^{-1} were measured for H_3PO_4 doped electrolytes; conductivities increase with an increase in temperature up to 10^{-1} S cm^{-1} at 100°C. Hydrogels doped with H_2SO_4 exhibit similar ambient temperature conductivities to those obtained for the same polymer doped with H_3PO_4. However, for gels containing a high concentration of H_2SO_4 a decrease in conductivities was observed at temperature exceeding 60°C. This decrease results from dehydration of the hydrogels or from degradation of the polymer matrix; both occur in the presence of a strong inorganic acid. Such tendency has not observed for the system doped with H_3PO_4 [18]. In our work; synthesizing hydrophilic foam (PVA/Epoxy) was carried out using Gamma irradiation. Characterization include hardness, morphological properties, thermal behavior and water uptake were performed. Insertion of the foam into serial dilution of acids, alkalis and salts for different times and measuring the electrical conductivity (Dc) at room temperature were done. The maximum conductivity was 9.63 x 10^{-2} simon/cm for 2N. H_3PO_4 uptaken by foam for 7 days soaking which exceed the previous work of hydrogel by 481.5%.

Experimental

Materials

The samples prepared using phenyl epoxy; polyvinyl acetate and an

*Corresponding author: Mohamed M El-Toony, National center for radiation research and technology, Atomic energy authority , 3 Ahmad El-Zomr street, P.O. Box 29- Nasr City, Cairo, Egypt

air compressor (used for foaming the blend) were commercial grade. The PVA paste was solved by (using magnetically stirred) distilled water for 3 hours at 80°C getting an emulsion with the concentration of 50%. Mixing of epoxy to PVA emulsion by 60% to 40% ratio respectively was performed. All the chemicals were used as received with no extra purification and it purchased from El-Gomhoria Co., Egypt.

Polyvinyl acetate (Structure 1)

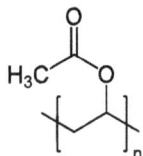

Phenyl epoxy (Structure 2)

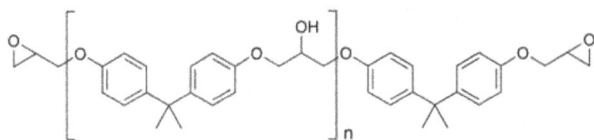

Hardener (Triethylenetetramine)(Structure 3)

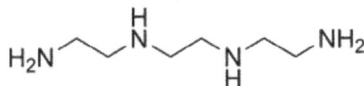

Methods of preparations

Epoxy/PVA blend was mixed to realize the optimum ratio which has good structure and optimal hydrophilic character. 40%, 60% and 70% by weigh of PVA were mixed with Epoxy and its solidifying agent (Triethylenetetramine, the amine (NH) groups react with the epoxide groups of the resin during polymerization). Heating of the mixture to 80°C for reaching the compatibility and then reduce the temperature of the mixture blend to 10°C till attain very viscous state. Foaming of the blend was for more than 5hours till solid form performed to achieve a highly porous solid blend. Irradiation of different doses; 25, 50, KGy by Gamma cell was carried out to realize the highly porous, hydrophilic and compatible material.

Water uptake

Respective water uptake behavior of different ratios of foam was studied in water as a function of pH. Swollen polymer was wiped

Figure 1: Scan Electron Microscope (SEM) of polyvinyl acetate-Epoxy blend foam with no treatment.

off with tissue paper and then weighed immediately to evaluate the swelling percent or water uptake percent, which was calculated by the following equation:

$$Water\ uptake\% = \frac{Weight\ of\ swollen\ hydrogel\ -\ Weight\ of\ dry\ hydrogel}{Weight\ of\ dry\ hydrogel} \times 100...Eq.(A.1)$$

Electrical conductivity (Box 1)

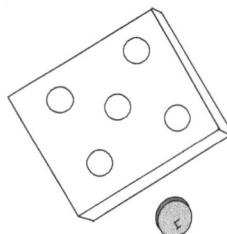

Diagrammatic cell used for measuring DC electrical conductivity of the foam

Teflon cell have 5 holes used as samples holder while the circular disk shown in figure made of red copper used for sandwiched the samples from below and upper sides. All disks were conducted to the electrometer for measuring the Dc electrical conductivity.

The samples were measured by 6514 Electrometer, Kiethly, USA

Digital Filter: Median and averaging (selectable from 2 to 100 readings).

Damping: User selectable on Amps function.

Environment:

Operating: 0°-50°C; relative humidity 70% non-condensing, up to 35°C.

Storage: −25° to +65°C.

Warm-Up: 1 hour to rated accuracy (see manual for recommended procedure).

Power: 90-125V or 210-250V, 50-60Hz, 60VA

Scientific equipments

Scanning electron microscope: Investigation and magnification of the polymer surface was carried out by SEM, JEOL-JSM-5400; Japan.

Thermal gravimetric analysis: Shimadzu TGA -50, Japan, was used to characterize the thermal stability of the porous blend.

Hardness tester: Samples were cut for 2.5 x 2.5 x 1 (l x w x h) cm for hardness test. The measurement was carried out according to (ASTM D2240, 2000) by manual analogue instrument with pin produced termed Baxio, UK. The unit of hardness is expressed in (Shore-D).

Gamma irradiation

Gamma irradiation was carried out by ^{60}Co gamma rays with a cylinder irradiation chamber. All irradiations were performed at ambient temperature (about 45°C at the chamber) and a dose rate of about 1.22 Gy/Sec.

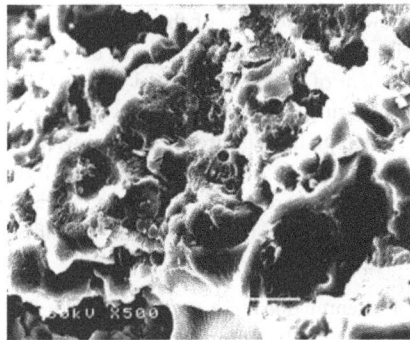

Results and Discussion

Morphological study

There are different modes of action by immersing the PVA/ Epoxy foam into different chemicals. Investigation by scan electron microscope (SEM) for foam (Figure 1) showed dispersion of PVA all over the epoxy. Vacant and pores are seen obviously in homogenous manner through the sample.

Figure 2 showed large pores comparing to the untreated sample. This appearance assures the role of phosphoric acid for reaction with the blend foam. Enlargement foam matrix around the pores is seen leading to pressing to widen the pore opening. Homogeneity is noticed apparently as no scattered spots, except for two white small dots cohered to the large pore. These dots may be approved that; phosphoric acid has two roles of action. First of which is simultaneous passage through the pores without any obstacles have performed by the foam. Second of which is diffusion into the net matrix of the foam leading to enlargement of the matrix bulk while press into the pore opening to be widen. All these notices interpret the high value of electrical conductivity occurred by dealing with phosphoric acid. Electrical conductivity may be performed as a result of sorption of hydrogen protons within phosphoric acid. Exchange of hydrogen proton of poly vinyl acetate is simply occurred, this exchange is enhanced by applying voltage. Simultaneous passage of PO_4^{3-} through foam pores raises the EC value as well.

Figure 3 showed the foam after dealing with 1N potassium hydroxide. Less pores number are scattered through the blend foam,

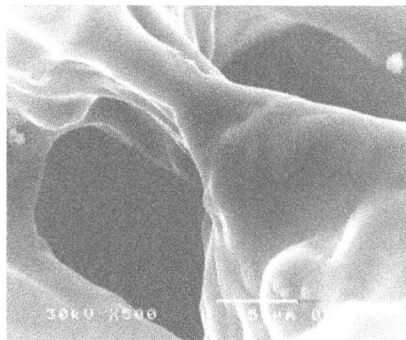

Figure 2: Scan Electron Microscope (SEM) of polyvinyl acetate-Epoxy blend foam soaked in 1 N phosphoric acid for 7 days.

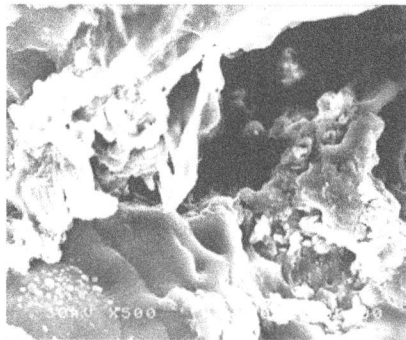

Figure 3: Scan Electron Microscope (SEM) of polyvinyl acetate-Epoxy blend foam soaked in 1N KOH for 7 days.

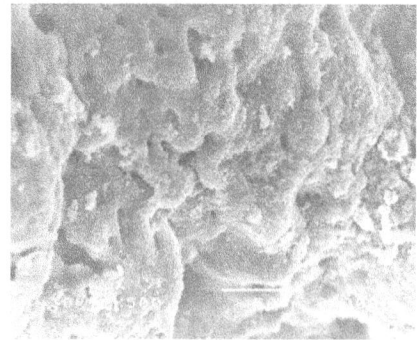

Figure 4: Scan Electron Microscope (SEM) of polyvinyl acetate-Epoxy blend foam soaked in 1N Ca Cl$_2$ for 7 days.

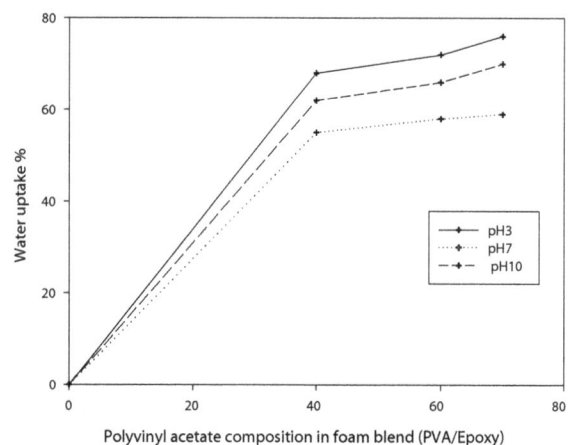

Figure 5: Effect of water uptake of PVA/ Epoxy blend foam at ambient temperature (25°C) and 50 KGy irradiation dose at different pH 3, 7 and 10.

may be resulted as action mode of potassium hydroxide. Moderate coherence of KOH into foam active sites is seen. PVA is showed to cohere and distribution over epoxy blended with them. Heterogeneity is seen to some extent which assured by irregular distribution of white spots through the sample. All these appearances assure the potassium hydroxide terms (K$^+$ and OH$^-$) passage through the foam pores while it preferred to pass through the net matrix. The electrical conductivity may be explained by sorption of potassium ion (K$^+$) and allow hydroxyl group (OH$^-$) to pass simultaneously through the pores leading to raising the electrical conductivity value.

Figure 4 showed the soaking of the foam sample into 1N CaCl$_2$ solution. Distribution of white and grey spots over the sample is seen. Regular arrangement of pores was noticed through the foam sample. Coherence of salt terms is obviously seen over the sample. Swellable sample shown attempt the affinity of foam towards the salt used. Sorption of calcium ion (Ca^{++}) is strongly suggested. It is suggested also the exchanging hydrogen proton of PVA by Ca^{++} leading to raising the EC value which is enhanced also by simultaneous diffusion of 2 Cl$^-$ of Ca Cl$_2$ through foam's pores.

Figures 2-4 attempt on; these appearances are preliminary step before foam degradation. It attempt also; the affinity to low pH while moderate to neutral and nearly small affinity towards high pH. Figure 4 has stable form than that of Figures 2 and 3.

Time	Acids											
	Hydrochloric acid				Sulfuric acid				Phosphoric acid			
	0.1N	0.5N	1N	2N	0.1N	0.5N	1N	2N	0.1N	0.5N	1N	2N
2 days	48	48	48	49	48	48	48	49	48	48	49	49
7 days	49	50	51	52	49	51	52	52	50	52	53	55
15 days	54	55	56	57	53	56	57	58	56	58	61	64
30 days	56	58	62	63	56	59	62	69	60	64	70	74
45 days	60	62	64	67	60	61	66	62	66	68	63	66
60 days	63	65	68	61	64	65	63	0	68	54	0	0

Table 1: Hardness (Shore-D) testing of polyvinyl acetate-Epoxy blend foam after soaking of different acids at different times.

Time	Alkali											
	Lithium hydroxide				Sodium hydroxide				Potassium hydroxide			
	0.1N	0.5N	1N	2N	0.1N	0.5N	1N	2N	0.1N	0.5N	1N	2N
2 days	49	49	49	49	48	48	48	48	48	48	48	49
7 days	50	52	52	54	49	50	52	53	50	51	53	54
15 days	53	55	56	57	53	55	56	56	54	55	56	57
30 days	57	58	59	58	55	56	57	58	55	56	57	59
45 days	59	59	60	60	58	59	60	60	57	59	61	61
60 days	58	59	60	61	59	60	61	61	58	60	61	60

Table 2: Hardness (shore-D) testing of polyvinyl acetate-Epoxy blend foam after soaking of different alkalis at different times.

Time	salts											
	Sodium chloride				Potassium chloride				Calcium chloride			
	0.1N	0.5N	1N	2N	0.1N	0.5N	1N	2N	0.1N	0.5N	1N	2N
2 days	48	48	48	48	48	48	48	48	48	48	48	48
7 days	51	52	52	53	51	51	51	52	51	51	52	53
15 days	54	54	54	55	53	54	54	54	54	55	55	55
30 days	59	59	59	58	58	58	59	59	59	58	58	59
45 days	60	60	59	60	58	59	59	59	59	59	59	59
60 days	60	60	60	60	59	59	59	60	59	59	60	60

Table 3: Hardness (Shore-D) testing of polyvinyl acetate-Epoxy blend foam after soaking of different salts at different times.

Hardness discussion

Hardness examination upon soaking PVA/Epoxy blend foam showed that; (Table 1-3) no significant difference within two days. Changes in hardness appear obviously after seven days soaking. Increase of hardness within phosphoric acid dealing with blend foam more than that of foam soaked through the least chemicals. Foam immersed in potassium hydroxide is less in hardness than phosphoric acid soaked and more than that of calcium chloride solution. Higher concentration of chemicals leads to higher hardness. Foam dealt with phosphoric acid has highest value than other acids such as hydrochloric acid and sulfuric acid blend soaked. Foam hardness immersed in acids is arranged as follow: $H_3PO_4 > H_2SO_4 > HCl$. Blend soaked in potassium hydroxide has higher value than LiOH and NaOH while NaOH > LiOH. Blend Soaking for two weeks have the same behavior while the value difference is pronounced. Foam hardness determined after two days, one, two, three four, six and 8 weeks. Hardness after 6 weeks resulted in maximum value while; 8 weeks soaking leads to complete hardness changes of foam dealt with 1N H_3PO_4 and H_2SO_4 (converted to powdered). While dealing with hydrochloric acid results in brittle form. Foam soaked in 1N alkali showed brittle form after 8 weeks. No significant difference of hardness upon dealing with salt solution with different concentrations. Glycerin soaked foam leading to reduction of hardness value which appears apparently with time. After 8 weeks hardness reduced to what resemble to gum form reached to 10 shore-D hardness value. Citric acid approximately does not affect the hardness of the blend foam.

Water uptake

Water uptake of the porous blend resin was studied as a function of composition, 40, 60 and 70% PVA composition of the blend showed water uptake begin to be stable in mechanical form at 40%. PVA is a hydrophilic polymer, blending them with epoxy which is considered as carrier of the functionalized polymer or filler. Increase PVA percentage increase water uptake while more percentage has no significant beside reduce their mechanical behavior. Blend foaming was remarkably improved water uptake may be due to an increase in pores number, pores size and/or increase in pore dimensions of the synthesized foam, i.e. increase in the allowed area surface of the resin net matrix.

Thermal behaviors

Thermograms of different chemicals (Phosphoric acid, Potassium hydroxide, Calcium chloride and glycerin) immersed through the understudied foam showed different scenarios.

Figure 6 could be characterized into 3 divisions, first of which showed no decrease of weight except for 4% by raising temperature to 177°C. This result confirmed the foam applicability through a wide range of temperature. By raising temperature, the second division of the thermogram has been appeared describing abrupt weight decrease into 60% by raising temperature to 388°C. The third division of the figure showed a regular weight decrease to 80% by increase the temperature to 574°C. There is no side peak via all the temperature range, which proved compatibility of the blend under study. It confirmed also no phosphoric acid residuals through the foam net matrix.

Figure 7 & 8 represents foam soaked into potassium hydroxide and calcium chloride in a regular manner. The 2 figures have the same behaviors; it could be characterized into 4 divisions. First division describes no weight decrease percent except for 5% which considered as operating temperature. Loss of weight of the first thermogram has occurred at 156°C, while the second curve showed the decrease of weight by temperature increase to 165°C, which proved their stability more than the previous. Second division of the thermograms represented gradual weight decrease by raising their temperature. The weight loss reached to 21% by increase the temperature to 305°C for the first case while the second showed 22% weight decrease as the temperature increased to 308°C. Third division of the figures represented dramatic weight loss by raising temperature to 382°C for the first curve and to 376°C through the second thermogram. The loss of weight reached to 59% for the first curve while it was 56% for the second one. The fourth division showed gradual weight decrease by

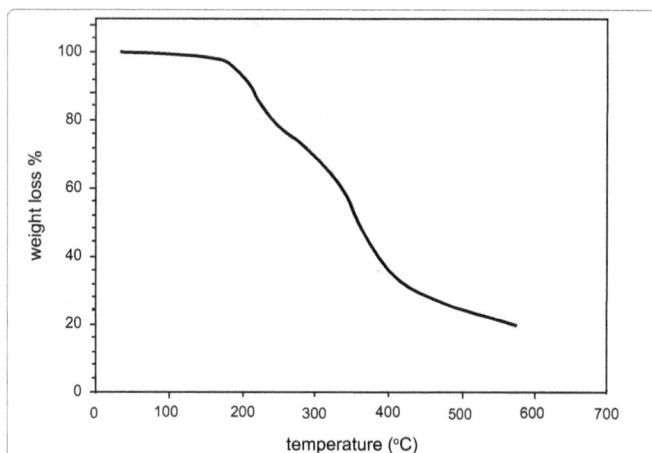

Figure 6: Thermogravimetric analysis of of polyvinyl acetate-Epoxy blend foam soaked in 1N phosphoric acid for 7 days.

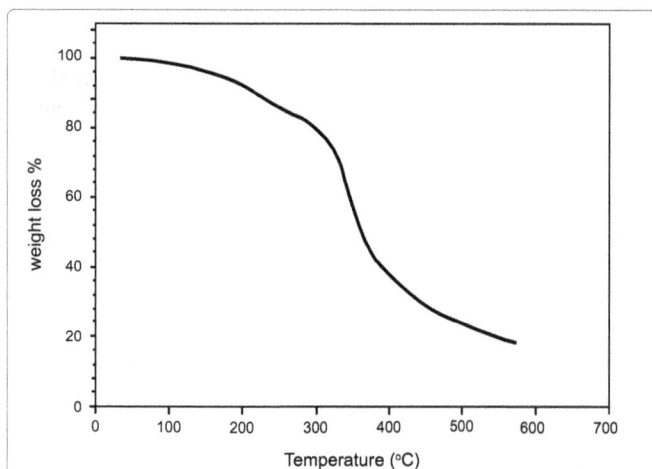

Figure 7: Thermogravimetric analysis of polyvinyl acetate-Epoxy blend foam soaked in 1N KOH for 7 days.

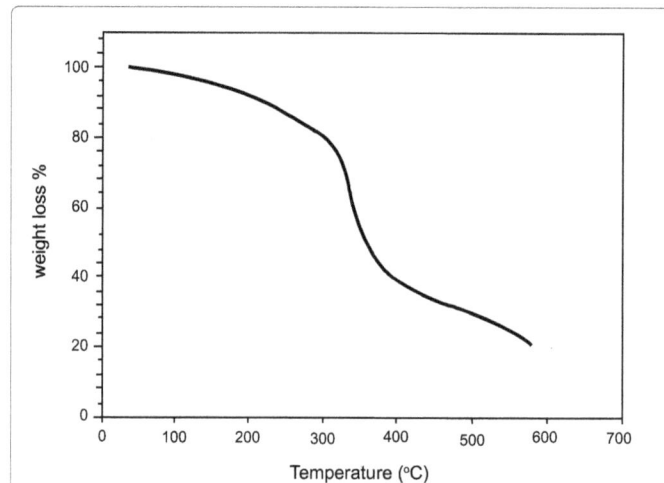

Figure 8: Thermogravimetric analysis of polyvinyl acetate-Epoxy blend foam soaked in 1N Ca Cl$_2$ for 7 days.

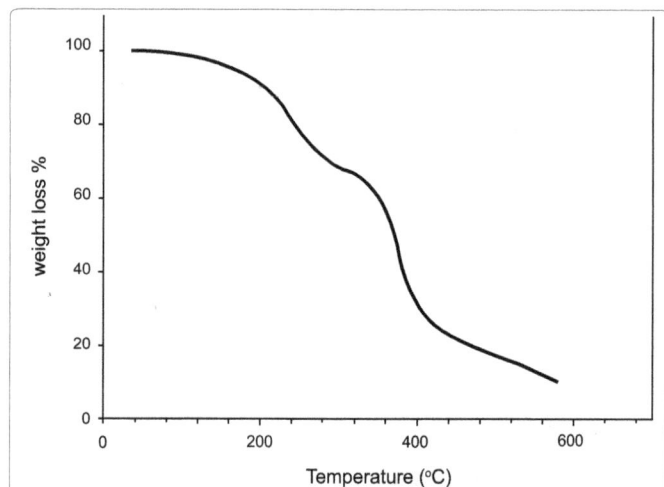

Figure 9: Thermogravimetric analysis of polyvinyl acetate-Epoxy blend foam soaked in glycerin for 7 days.

raising temperature to 575°C. The loss of weight reached to 82% for the first case while it was 89% for the second one. The foam soaked through calcium chloride reached to 11% from its original value which may due chemical reaction of hydroxyl group of PVA and that resulted in epoxy ring opening with calcium chloride. This suggested reaction may be summarized as the following:

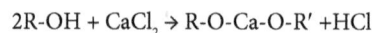

$$2R\text{-}OH + CaCl_2 \rightarrow R\text{-}O\text{-}Ca\text{-}O\text{-}R' + HCl$$

Where: R and R' are polyvinyl and/or epoxy opened ring.

This reaction make partial blocking of the active sites resulted in reduction of more salt terms (cations and anions) passage and so less electrical conduction.

Figure 9 (foam soaked in glycerin) have different scenario. The thermogram could be categorized into 4 divisions. First division described the low loss weight (operating temperature of the foam) of the foam which reached to 5% by increase the temperature to 169°C. The second division study the abrupt weight decrease reached to 30% at temperature 287°C. Furthermore temperature increase a side peak has been characterized the third division of the thermogram. This side peak confirmed more or less compatibility which may due to sorption of some glycerin particles through the foam net matrix. The side peak was through 287°C to 391°C temperature range while loss of weight was 56%. The last division of the thermogram represented gradual decrease of weight reached to 89% by raising the temperature to 575°C. The low weight value of the foam at the end of thermogram (11 % from the original weight) may be due to a chemical reaction of glycerin with alcohol's hydroxyl group of the blend. The reaction could be summarized as the follow;

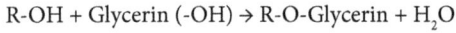

R-OH + Glycerin (-OH) → R-O-Glycerin + H$_2$O

Where: R is the net matrix of the blend foam.

This reaction confirmed partially blocking of the blend net matrix's active sites which reduce the glycerin passage and so less electrical conduction. These new matrix may be synthesized give good chance for further glycerin sorption and so further electrical conduction reduction.

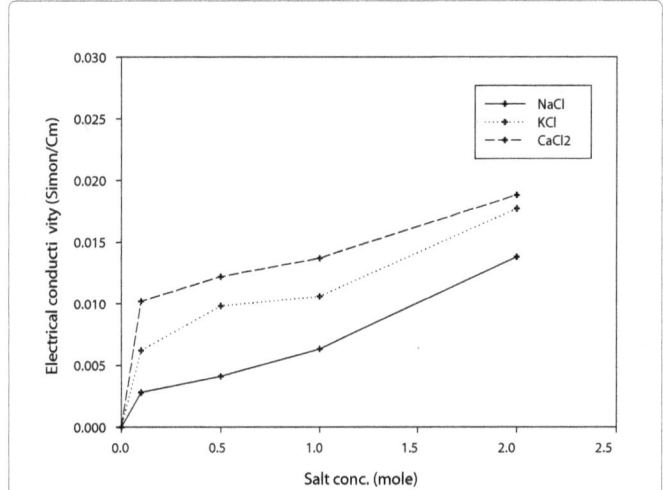

Figure 12: Effect of salts soaking of PVA/ epoxy foam on electrical conductivity (Dc)(S/Cm).

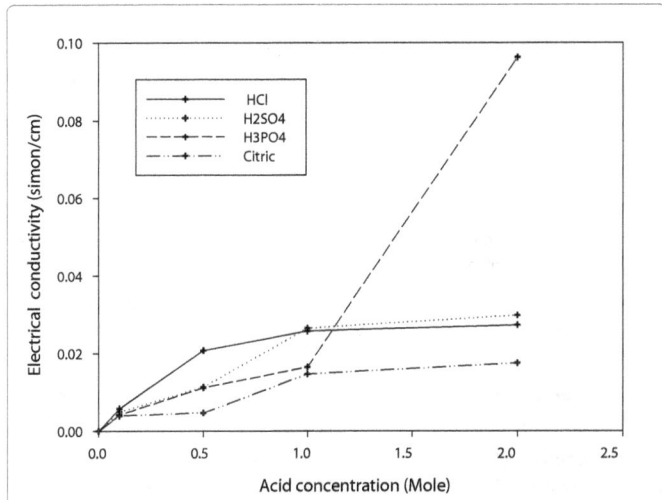

Figure 10: Effect of acids soaking of PVA/Epoxy for 7 days on electrical conductivity (DC) (Simon/Cm).

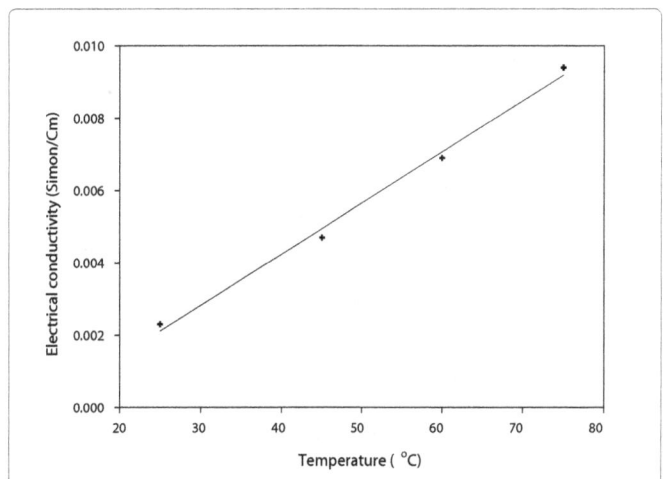

Figure 13: Effect of temperature on glycerin soaking of PVA/epoxy foam on electrical conductivity (Dc)(S/Cm).

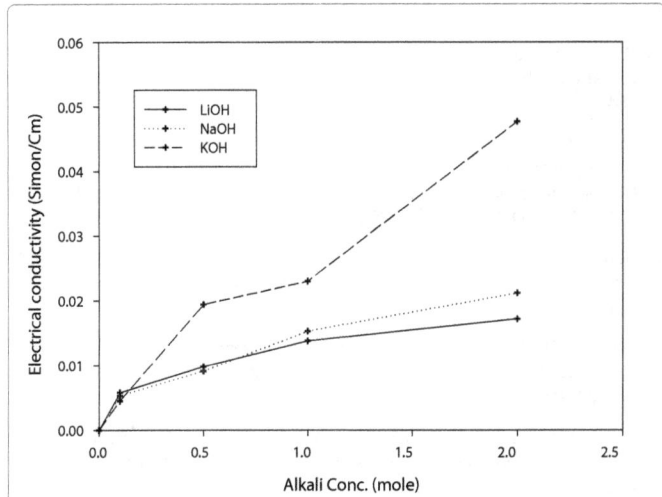

Electrical properties

As seen in Figures 10-12, the conductivity of foam increase with increase in water concentration. While acids showed higher conductivity as lightly mobile hydrogen protons and by increase hydrogen protons increase conductivity as it can be seen in Figure 3. Less conductivity was clearly seen at low concentration of phosphoric acid may be due to chain reaction with foam. Potassium hydroxide showed higher conductivity as in Figure 4 which may due to a special affinity to foam to be abstracted. More charge on cations more electrical conductivity as it seen in Figure 5. Therefore it can be assume that foam is a mixture of 2 phases. These are dry PVA/Epoxy phase with no water present and water / acids, bases or salts phase trapped in a polymer phase. The bulk conductivity of the foam is a result of interaction between these 2 phases. To make these calculations easier it assumed that; each has spherical symmetry. The general self-consistent equation for the multiphase materials is [19,20].

$$\Sigma \; wi \; \sigma i - \sigma^* / \; \sigma i (d-1) \; \sigma^* = 0$$

Where; d is a dimensionality parameter equal to 3 for spherical

Figure 11: Effect of alkali soaking of PVA/ epoxy foam on electrical conductivity (Dc)(S/Cm).

aggregates, σ_i (i =1, 2) are conductivities of the water phase, σ^* dry polymer phase and w_i is the volume fraction of the i phase.

The Lewis acid–base reactions model, widely employed in carboxyl-based composite solid polymer electrolytes is extended to the present composite polymer electrolytes. The carboxyl groups in PVA , hydroxyl group of basic (LiOH, NaOH and KOH) and anions of salts used (NaCl, KCl and $CaCl_2$) were served as the Lewis base centers and the cations while hydrogen protons donors such as acids (HCl, H_2SO_4, H_3PO_4 and citric acid) used as strong Lewis acid. Therefore, numerous acid-base complexes are expected in the composite system with two major groups of complexes, i.e., cations and Hydrogen proton with polymer foam. The conductivity in composite polymer electrolyte is not a linear function of the filler concentration [21,22]. At low content level, the dilution effect is efficiently contrasted by the specific interactions of the polymer, which promote fast ion transport, and the overall effect is progressive enhancement of the conductivity. On the other hand, at high filler content, the dilution effect predominates and the conductivity decays. Similarly, concave effect on conductivity with inorganic/polymer materials as fillers in solid polymer electrolyte [23-25] and ion pairing (insufficient salt dissociation) is identified as the origin of the later deterioration.

Temperature dependence

The different behavior of the temperature dependence in the conductivity suggests two different conducting pathways. The first conduction pathway via ion motion is facilitated by polymer segmental motions, and is quickly frozen upon lowering temperature. The second conducting pathway, similar to the ion transport on solid oxide electrolytes, is established by the ions on the surface (both interior and external) of porous blend which migrates by replacing the nearby vacancy. The presence of additional conducting channels in the composite polymer electrolyte delivers favorable low temperature conductivity over those without porous blend. The improvements are most pronounced in the conductivity at low temperatures. However, a flexible polymer is still essential for the second type of conduction, and becomes an important factor governing the conductivity in the high temperature region. Above the melting point of the composite the effect of the filler on conductivity was not so significant [26].

Electrical conductivity is strongly dependent on temperature. In metals, electrical conductivity decreases with increasing temperature, whereas in semiconductors, electrical conductivity increases with increasing temperature as it seen in Figure 13. Over a limited temperature range, the electrical conductivity can be approximated as being directly proportional to temperature. In order to compare electrical conductivity measurements at different temperatures, they need to be standardized to a common temperature.

At extremely low temperatures (not far from absolute zero), a few materials have been found to exhibit very high electrical conductivity in a phenomenon called superconductivity.

Conclusion

For the aim to prepare a new polymer electrolytes and /or modernization of fuel cells, synthesis of PVA/epoxy blend foam has been performed. Gamma irradiation proved to be an important factor influencing the blend synthesis. Characterization of the blend after soaking in different chemicals, confirmed their applicability from mechanical and thermal point of view. Measuring the EC of the blend upon soaking through different chemicals such as acids, alkalis and salts for seven days represent the EC of acids soaked blend is more than alkalis which exceed over salts. Phosphoric acid soaked porous blend have maximum value of Dc, while hydrochloric acid soaked blend have maximum mechanical behavior stability. Soaking the porous blend through phosphoric acid (2N) for 1week achieve 9.63 x 10^{-2} simon/cm which is more the previous work of hydrogel by 481.5%, while porous blend has more mechanical properties stability. Morphological properties of the blend focused on mode of action between them and chemicals under study, it put spots also on characters change of the blend through soaking time.

References

1. (2008) Conducting Polymers.

2. Hush NS (2003) An overview of the first half-century of molecular electronics. Ann N Y Acad Sci 1006: 1-20.

3. http://www.cambridge.org/gb/knowledge/isbn/item1156360/?site_locale=en_GB

4. http://onlinelibrary.wiley.com/doi/10.1002/pi.1994.210330323/abstract

5. http://onlinelibrary.wiley.com/doi/10.1002/%28SICI%291097-0126%28199805%2946:1%3C78::AID PI16%3E3.0.CO;2-I/abstract

6. Croce F, Curini R, Martinello A, Persi L, Ronci F, et al. (1999) Physical and Chemical Properties of Nanocomposite Polymer Electrolytes. J Phys Chem B 103: 10632-10638.

7. Croce F, Appetecchi GB, Persi L, Scrosati B (1998) Nanocomposite polymer electrolytes for lithium batteries. Nature 394: 456-458.

8. Wieczorek W, Such K, Wycislik H, Plocharski J (1989) Modifications of crystalline structure of peo polymer electrolytes with ceramic additives. Solid State Ionics 36: 255-257.

9. Wieczorek W, Such K, Florja´nczyk Z, Stevens JR (1995) Polyacrylamide based composite polymeric electrolytes. Electrochim Acta 40: 2417-2420.

10. Bhattacharyya AJ, Maier J (2004) Second Phase Effects on the Conductivity of Non-Aqueous Salt Solutions: SoggySandElectrolytes. Adv Mater 16: 811-814.

11. Imrie CT, Ingram MD, McHattie GS (1999) Ion Transport in Glassy Polymer Electrolytes. J Phys Chem B 103: 4132-4138.

12. Xiong HM, Zhao X, Chen JS (2001) New Polymer-Inorganic Nanocomposites: PEO-ZnO and PEO-ZnO-LiClO4 Films. J Phys Chem B 105: 10169-10174.

13. Chandra S, Hashmi SA, Prasad G (1990) Studies on ammonuim perchlorate doped polyethylene oxide polymer electrolyte. Solid States Ionics 40-41: 651-654.

14. Stainer M, Charles Hardy L, Whitmore DH, Shriver DF (1984) Stoichiometry of Formation and Conductivity Response of Amorphous and Crystalline Complexes Formed Between Poly(ethylene oxide) and Ammonium Salts: PEO_x · NH_4SCN and PEO_x · $NH_4SO_3CF_3$. J Electrochem Soc 131: 784-790.

15. Herranen J, Kinnunen J, Mattsson B, Rinne H, Sundholm F, et al. (1995) Characterisation of poly(ethylene oxide) sulfonic acids. Solid State Ionics 80: 201-212.

16. Mattsson B, Brodin A, Torell LM, Rinne H, Hamara J, et al. (1997) Raman scattering investigations of PEO and PPO sulphonic acids. Solid State Ionics 97: 309-314.

17. Miyatake K, Fukushima K, Takeoka S, Tsuchida E (1999) Nonaqueous Proton Conduction in Poly(thiophenylenesulfonic acid)/Poly(oxyethylene) Composite. Chem Mater 11: 1171-1173.

18. Wieczorek W, Steven JR (1997 Proton transport in polyacrylamide based hydrogels doped with H_3PO_4 or H_2SO_4. Polymer 38: 2057-2065.

19. Nan CW (1993) Physics of inhomogeneous inorganic materials. Prog Mater Sci 37: 1-116.

20. Vanheumen J, Wieczorek W, Siekierski M, Stevens JR (1995) Conductivity and Morphological-Studies of Tpu-Nh4Cf3So3 Polymeric Electrolytes. J Phys Chem 99: 15142-15152.

21. Croce F, Persi L, Scrosati B, Serriano-Fiory F, Plichta E, et al. (2001) Role of the ceramic fillers in enhancing the transport properties of composite polymer electrolytes. Electrochim Acta 46: 2457-2461.

22. Subba Reddy ChV, Wu GP, Zhao CX, Zhu QY, Chen W, et al. (2007) Characterization of SBA-15 doped (PEO + LiClO4) polymer electrolytes for electrochemical applications. J Non Cryst Solids 353: 440-445.

23. Wen Z, Itoh T, Ikeda M, Hirata N, Kubo M, et al. (2000) Characterization of composite electrolytes based on a hyperbranched polymer. J Power Sources 90: 20-26.

24. Hashmi SA, Thakur AK, Upadhyaya HM (1998) Experimental studies on polyethylene oxide–$NaClO_4$ based composite polymer electrolytes dispersed with Na_2SiO_3. Eur Polym J 34: 1277-1282.

25. Sekhon SS, Sandhar GS (1998) Effect of SiO_2 on conductivity of PEO-AgSCN polymer electrolytes. Eur Polym J 34: 435-438.

26. Hsien-Ming KT, Yi-Yuan T, Shih-Wei C (2005) Functionalized mesoporous silica MCM-41 in poly (ethylene oxide) -based polymer electrolytes : NMR and conductivity studies. Polymer 176: 1261-1270.

4

Study on the Assessment of Adsorption Potential of Dry Biomass of *Canna indica* with Reference to Heavy Metal Ions from Aqueous Solutions

Archana Dixit[1]*, Savita Dixit[1] and Goswami CS[2]

[1]*Department of Chemistry, Maulana Azad National Institute of Technology, Bhopal, India*
[2]*Department of Chemistry, Kamal Radha Girls College, Gwalior, India*

Abstract

Present paper is an attempt to evaluate the adsorption of heavy metals like Cadmium (Cd), chromium (Cr), zinc (Zn), and lead (Pb) by the dry biomass of terrestrial plants *Canna indica* commonly called Saka siri. Very less literature is available for the study on the absorption/adsorption of heavy metals by this plant. The present experimental study was conducted to assess the adsorption capacity of dry biomass of *Canna indica* to compare and identify their potential to improve the water quality by removing the impurities. The paper critically evaluates the water-purifying capacity of dry-biomass of this plant basically knows for its ornamental identity. Manuscript will be helpful in showing the water purifying capacity of dry biomass of *C. indica* and also will evaluate the best results of adsorption shown by varying quantity of the adsorbent.

Keywords: Dry biomass; *Canna indica;* Heavy metals; Adsorbent quantity; Adsorption; Isotherm

Introduction

The rapid technological advancement, industrialization, urbanization and population growth has resulted in the deterioration of water, air and land quality, making it unfit for human consumption, water is an essential component of life, but is getting polluted day by day and thus unsafe to consume. There are number of toxic elements known to exist in the environment, which directly or indirectly make their way into water bodies. Sewage, industrial chemicals, heavy metals from industrial processes, and household waste are examples of materials commonly discharged into water bodies. Natural sources of water are depleting fast and are polluted due to industrialization and urbanization in haphazard manner. The Potential toxic metal elements such as cadmium, chromium, lead, Copper, Zinc etc. are identified to cause health hazards in animals [1,2] these heavy metals are reported to be toxic and found associated with the occurrence of several health effects.

The lakes and reservoirs are under great environmental stresses, which are now gradually getting filled up by excessive sewage input, silting and growth of organic matter [3]. On the other hand heavy effluents discharge from the industries lead to intrusion of heavy metals in the water bodies which leads to further toxicity to man and environment. Waste water generated from residential and industrial day today activities must be treated before is released into surface water bodies or to environment. So that it does not cause further pollution of water sources. Due to the extreme consequences, environmental contamination with heavy metals is a topic of significant concern. Treatment processes for metals contaminated waste streams can be treated with so many methods but still a cost effective alternative technologies or sorbets for treatment of contaminated waste streams are needed, so in the present study an effort is made by utilizing dead or non-living biomass [4] of *Canna indica* plant available in large quantity as adsorbent for the removal of selected heavy metals from aqueous solution, the study is presently carried out at laboratory scale. The success of which can be further replicated for the field study i.e. for the treatment of the waste water.

Materials and Methods

The present study was mainly concentrated upon adsorption of four heavy metals viz. chromium, cadmium, lead and zinc by dry biomass of *C. indica* for which the experiments are conducted. The plants are collected from sahapura lake drainage basin. The plants are washed with Milli-Q water to eliminate the remains of lake sediments and particulate matter, and then the plants are cut into pieces and sun dried. After being completely dried/dehydrated they are grinded into powder. The powder was grounded to pass through 2 mm sieve. The heavy metal samples of 10, 50 and 100 mg/l concentration, for the analysis were prepared by standard method [5]. In 100 ml of each of the heavy metal samples i.e. chromium, cadmium, lead and zinc, varied quantity i.e. 0.5, 1.0, 1.5, 2.0, 2.5, 3.0, 3.5, 4.0 gm respectively of the *C. indica* powder was added and then put into shaker at 65 rpm and at temperature of 28°C, for time period of 30 min. The samples after attaining complete reaction period were filtered with whatmann no. 40 filter paper. All the experiments were set in duplicate, and for all the parameters a control set was also studied where there was no powder (dry biomass) added. All the parameters were analyzed by the method as mentioned in APHA.

Results

The results of the study indicate the effect of variation of the quantity of adsorbent (dry biomass powdered of *C. indica*) on the adsorption of selected heavy metals. Figure 1a-d, represents adsorption of Cd, Cr, Pb, and Zn ion on dry biomass of *C. indica* at various adsorbent concentration i.e. from 0.5 to 4.0 gm. Results after treating 10 ppm of standard solution of cadmium at different biomass conc. are very encouraging. Adsorption percent of 'Cd' solution was found in the range of 36 to 96.5% (Figure 1a-d), it shows that the dry biomass adsorbed

*Corresponding author: Archana Dixit, Research scholar, Department of Chemistry, Maulana Azad National Institute of Technology, Bhopal, India

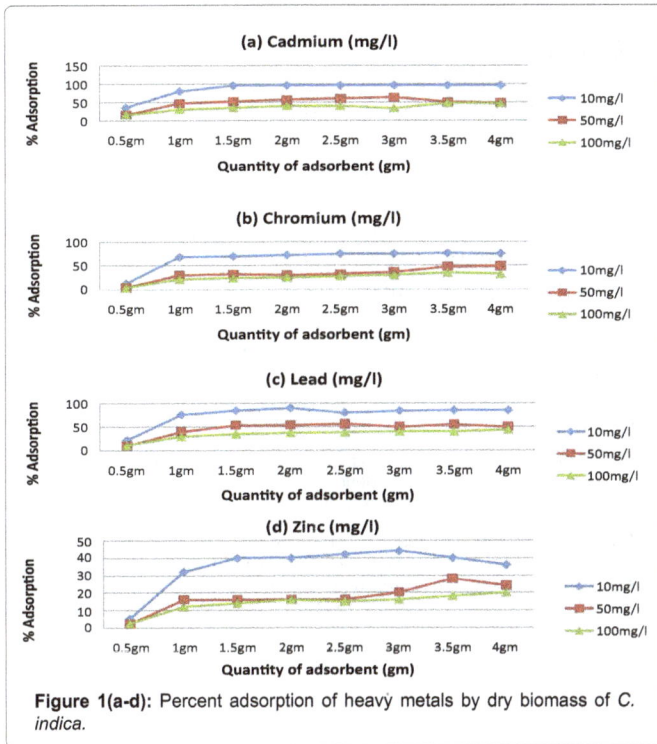

Figure 1(a-d): Percent adsorption of heavy metals by dry biomass of *C. indica.*

almost whole of the Cd metal ion. Here the adsorption concentration was found directly proportional to quantity of adsorbent. Cadmium is considered very much toxic [6,7] metal ion.

Dry biomass of *C. indica* efficiently adsorbed Chromium from 10, 50, 100 mg/l concentration of aqueous solution using about 2-2.5 gm of adsorbent. The adsorption percent increased with increasing quantity of adsorbent. The average percentage adsorption of chromium was found 76.0% for 10 mg/l solution. The result shows that at initial concentration the adsorption rate was high and by further increasing the quantity from 2.5 gm. The adsorption was almost stationary. Cr is very much toxic and carcinogenic for living beings.

The lead metal showed reduction in its concentration after treating with the dry biomass, the concentration of sample is decreased from 10 ppm to 1.0 ppm by treating the adsorbent of upto 4 gm. It shows that there is the adsorption of about 90% on an average for 10 ppm solution. With the increasing adsorbent quantity the original concentration in the solution decreased with slow rate and remained stationary after 4 gm. It showed adsorption of 56% for 50 ppm solution and even less for 100 ppm solution. Lead is very much toxic and carcinogenic for living beings.

Results obtained by treating 10 mg/l of zinc soln. with varied adsorbent quantity showed marked decrease in its concentration (Figure 1a-d) with only 2 gm of adsorbent and then there was reduction in adsorption percent after utilizing 3 gm of adsorbent. The % adsorption of Zn metal ion of 10 mg/l concentration calculated for adsorbent concentration of 1.0 gm, 2.0 gm, 3.0 gm and 4.0 gm were 32%, 40%, 44% and 36% respectively. Zinc is not very much toxic, the permissible limit is 30.0 ppb for aquatic life and 5.0 mg.L^{-1} for drinking water as per WHO [8].

Discussion

Increase in percent adsorption of Cd, Cr, Zn and Pb with increase in concentration of adsorbent is supported by considerable increase

in reaction rate. Probably at high sorbent dosage the available ions are inadequate to cover all the available sites on the sorbent. Thus, the time of contact required to reach saturation varied with the biomass quantity. The figure shows that within a short time a large fraction of the total amount of heavy metal ion was removed but the uptake capacity of heavy metal ion per unit amount of sorbent (mg/g) decreases with increase in biomass concentration [9,10]. It is observed that there is a sharp increase in percentage removal of heavy metals with adsorbent quantity for studied heavy metal ions upto 1-2 gm of adsorbent quantity but after that there is gradual percentage removal with increasing quantity, it is due to the greater availability of the exchangeable sites or surface area [11] and as the sites are occupied the adsorption slowly decreases.

Results for Cd. Cr, Zn and Pb ions removal indicate that when quantity of adsorbent is increased from 0.5 to 2.5 gm initially lesser time is required to attain equivalence concentration, while the same equivalence concentration is attained with more adsorbent at higher metal ion concentration. This can be interpreted as more removal of Cd. Cr, Zn and Pb contents with less amount of adsorbent, when longer contact period is permitted. In other words, increase in contact period and quantity of adsorbent affects economics of the adsorption process and enhances the removal Cd. Cr, Zn and Pb contributing components of the waste water.

The regression analysis and experimental observations shows that the experimental data fitted well in both the isotherms. However, the values of correlation coefficient 'R^2', were marginally better for Langmuir isotherm than the values of 'R^2' for Freundlich isotherm [12] (Table 1).

Two types of observations were collected after treating the aqueous solution with the dry- biomass of *C. indica*.

1. Reduction in concentration of studied heavy metals due to adsorption on the dry biomass of *C. indica* as adsorbent.

2. Increase in the rate of adsorption with increasing adsorbent quantity.

Conclusions

Observations obtained from the above mentioned batch study concluded that dry biomass of whole plant of *C. indica* can be used for reducing heavy metals impurities in the present context but it can be further considered even for waste water treatment of effluents generated from industrial activities with technological up gradation. The control samples merely showed any visible difference in the concentration to that of original concentration which signifies that the dry biomass of the plant has good potential for removal of selected heavy metals. It is well evaluated that the dry mass of the plant showed the required

Metal	Temperature °C	Langmuir Isotherm			Freundlich Isotherm		
		Q°(mg/g)	b×10³ (dm³/mg)	R²	1/n	K	R²
Cadmium	30	31.25	71.51	0.979	0.632	0.421	0.982
	60	16.39	7.37	0.987	0.511	0.013	0.953
Chromium	30	71.43	1930.50	0.963	0.742	1.073	0.931
	60	76.92	160.93	0.944	0.479	0.343	0.926
Lead	30	111.11	487.33	0.99	0.694	0.680	0.989
	60	166.67	400.64	0.996	0.843	0.176	0.999
Zinc	30	322.58	333.80	0.999	0.881	0.117	0.999
	60	1000.0	556.17	0.999	1.017	0.300	0.998

Table 1: Freundlich & Langmuir isotherm parameters for Cd, Cr, Pb, and Zn.

good results in case of heavy metals as it has shown great reduction from its original concentration for all the studied heavy metals. It was also noticed that many of the studied heavy metals viz. Cadmium, Chromium, Zinc and Lead showed reduction in their concentration with increasing adsorbent quantity but after 4 gm they showed stability in reduction percentage. This concludes that the dry biomass powder of C. indica can be effectively utilized for reducing heavy metals and it gives the results in short span of time with less amount of adsorbent (dry biomass of C. indica) and so upgrading the same can prove to be a powerful tool for waste water treatment.

References

1. Bryan GW (1976) Heavy metal contamination in the sea. In R. Johnston (Edn), Maarine pollution. London: Academic.

2. 2. Sivakumar K, Subbalah KV, Sai Gopal DVR (2001) Studies of certain trace elements in industrial efflutents sediments and their effect on plants physiology. Pollution Research 20: 99-102.

3. Bhatnagar GP (1982) Limnology of Lower lake Bhopal with reference to sewage pollution and eutrophication. Department of Limnology Bhopal University, MP-MAB UNESCO.

4. Schneider I, Rubio J (1999) Sorption of heavy metals by the nonliving biomass of freshwater macrophytes. Environ Sci Technol 33: 2213-2217.

5. APHA (1999) Standard methods for the examination of water and waste water (20th edn). American Public Health Association, American Water Works Association and Water Poll. Control Federation. American Public Health Association. Washington, DC.

6. Volesky B (1990) Biosorption by fungal biomass. In: Volesky B, editor. Biosorption of heavy metals. Florida: CRC press 140-171.

7. Wang JL (2002) Immobilization techniques for biocatalysts and water pollution control. Beijing: Science Press.

8. WHO (2002) The guideline for drinking water quality recommendations. World Health Organization, Geneva.

9. Bishnoi NR, Pant A, Grima (2004) Biosorption of copper from aqueous solution using algal biomass. J Scientific and Industrial Research 63: 813-816.

10. Babák L, Šupinová P, Zichová M, Burdychová R, Vítová E (2012) Biosorption of Cu, Zn and Pb by Thermophilic Bacteria-Effect of Biomass Concentration on Biosorption Capacity. Acta Universitatis Agriculturae Et Silviculturae Mendelianae Brunensis LX: 9-18.

11. Abdel-Ghani NT, Hefny M, El-Chaghaby GAF (2007) Removal of lead from aqueous solution using low cost abundantly available adsorbents. International Journal of Environmental Science & Technology 4: 67-73.

12. Okeola FO, Odebunmi EO (2010) Freundlich and Langmuir Isotherms Parameters for Adsorption of Methylene Blue by Activated Carbon Derived from Agrowastes. Advances in Natural and Applied Sciences 4: 281-288.

5

Utility of Benzimidazoles in Synthesis of New Bases of Nucleoside Moieties, and as Antioxidant in Lubricant Oils

Maher A El-Hashash[1], Sameh A Rizk[1]* and Maher I Nessim[2]

[1]Ain Shams University, Science Faculty, Chemistry Department, Cairo, Egypt
[2]Egyptian Petroleum Research Institute, Evaluation and Analysis Department, Cairo, Egypt

Abstract

The treatment of 4-(4-Chloro-3-methylphenyl)-4-oxobut-2-enoic acid with benzimidazole, and 2-mercapto benzimidazole afforded aza- and thia-Michael adduct as unnatural α-amino acid 1 and α-thiaacid 6 respectively. Micheal adducts 1 and 6 are used to synthesize some antioxidant heterocycles. TAN study can be confirmed, the fused heterocycles has a lower antioxidant expt that contain sulfur atom. Quantum chemical studies of 2-mercaptobezimidazole, and its amide long chain, confirmed that N-butyl-S-benzimidazol-2-ylthioglycolate at 400 ppm was more effective antioxidant heterocycles.

Keywords: 4-Aryl-4-oxobut-2-enoic acid; Unnatural amino acid; Antioxidant heterocycle; 2-marcapto/1-furo-4-yl/1,3-hiazolo/1,3-thiazinobenzimidazole; 4-benzimidazo-1-ylpyridazine/oxazine; Triazinobenzimidazole

Introduction

Amino acids have proven to play a significant role in the synthesis of novel drug candidate with the use of non-proteinogenic and unnatural amino acids [1-8]. Cytotoxic activity of eight thiazolobenzimidazole derivatives on sensitive HL60 and multidrug-resistant (MDR) (HL60R) leukemia cell lines can be reported [9]. Benzimidazoles have been identified as inhibitors of the microsomal NADPH-dependent lipid peroxidation (LP) levels [10] and anti-proliferative effect to human colorectal cancer cell line HT-29, breast cancer cells MDA-MB-231 [11]. From this point of view the authors try to investigate the reaction of 4-(4-chloro-3-methypheny) l-4-oxobut-2-enoic acid with benzimidazole, under aza-Michael reaction conditions to afford unnatural α-amino acid derivative. And so, to synthesize some heterocyclic compounds carrying benzimidazole moiety aiming at obtaining some interesting antioxidant materials as additives in lubricant oils.

Experimental Analysis

All melting points are uncorrected. Elemental analyses were carried out at the Micro analytical Center, National Research Center, Cairo, Egypt. By Elementar Viro El Microanalysis IR spectra (KBr) were recorded on infrared spectrometer FT-IR DOMEM Hartman Braun, Model: MBB 157, Canada and ^{1}H-NMR spectra recorded in DMSO-d6 on a Varian Gemini spectrophotometer at 200 MHz (Germany 1999) using TMS as internal standard. The mass spectra were recorded on Shimadzu GCMS-QP-1000 EX mass spectrometer at 70 eV using the electron ionization technique. Homogeneity of all compounds synthesized was checked by TLC.

3(4-chloro-3-methylphenyl)-2-(1H-benzimidazol-1-yl)-4-oxobutanoic acid (1)

A solution of acid 1 (3 g; 0.01 mol) and benzimidazole (1.38 g; 0.01 mol) in 30 ml ethanol and 4 drops piperidine was refluxed for 3 h. The reaction mixture was allowed to cool and the crude product was washed by petroleum ether (b.p 40- 60°C), and then, crystallized from dioxane. Yield 78%, m.p. 154-156°C, IR spectrum $\upsilon CO(1732-1685)$ cm^{-1} for carboxylic and ketone groups and 3437-3360, 3175 cm^{-1} attributable to υOH. The ^{1}HNMR spectrum DMSO-d6 for revealed

at δ 2.1(s, 1H,CH3), three protons for ABX spin system 2.7-2.9 (2dd, CH_2-C=O, dia stereo topic protons), 3.1-3.3 (dd,CH-CO, stereogenic methine proton), 7.2-7.5 (m, 8ArH aromatic protons), 9.2 (s,1H, acidic proton which exchanged in D_2O. Elemental anal. Found % C 56.55, % H 3.28, % N 7.62, % Cl 9.53, $C_{18}H_{15}N_2O_3Cl(342.5)$, Calc. % C 56.943, % H 3.28, % N 7.69, % Cl 9.80 (Figure 1).

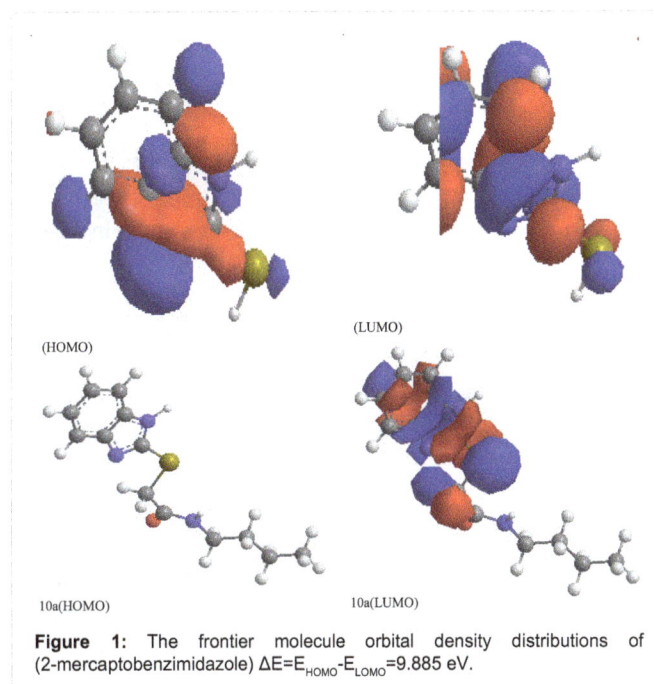

Figure 1: The frontier molecule orbital density distributions of (2-mercaptobenzimidazole) $\Delta E=E_{HOMO}-E_{LOMO}=9.885$ eV.

*Corresponding author: Sameh A. Rizk, Ain Shams University, Science Faculty, Chemistry Department, Cairo, Egypt

5-(4-chloro-3-methylphenyl)-3 (1H-benzimidazol-1-yl)-2- (3H) furanone (2)

A mixture of 1 (3.73 g; 0.01 mol) and acetic anhydride (9.4 mL, 0.1 mol) and then refluxed on water bath for 1h. The reaction mixture was allowed to cool and then pour into ice-H_2O and the separated product was filtered, dried and were re-crystallized from toluene. Yield 82%, m.p. 218-220°C, IR spectrum υCO (lactonic) 1780 cm^{-1}. EIMS m/z: 324.5, 15% (M+) and m/z: 140(2-phenyl furan molecular radical entity as a base peak. The ^{1}HNMR spectrum in DMSO-d6 exhibits Signals at δ 2.1(s, 1H, CH3), 2.7(dd, 1H, the methine proton), 7.0-7.9 (m, 8H, 7ArH and 1H furanone moiety). Elemental anal. Found %C 58.86, %H 3.15, %N 8.09, %Cl 10.53, $C_{18}H_{13}N_2O_2Cl$ (324.5), Calc. %C 58.95, %H 3.17, %N 8.09, %Cl 10.52 (Figure 2).

6-(4-chloro-3-methylphenyl)4-(benzimidazol-1-yl)2,3,4,5-tetrahydro-3(2H)pyridazinone (3)

A solution of 1 and/ or 2 (0.01 mol) in ethanol (40 mL) was treated with hydrazine hydrate 98% (1.5 mL; 0.04 mol) and then refluxed for 3 h. The solid that separated after concentration and cooling was re-crystallized. Yield 77%, m.p. 196-198°C. IR spectrum υCO) 1683 cm^{-1} ^{1}HNMR spectrum DMSO-d6 of revealed singlet at δ 2.1(s, 1H, CH$_3$), 2.8-3.0 (2 dd,1Ha and 1Hb methylene protons), 3,8-4.2(dd, CH-CO, methine proto), 7.2-7.8 (m, 8 ArH), 11.2 (s,1H, pyridazinone proton), EIMS m/z: 338.56% (M$^+$) and 3410.5%(M$^+$+2), m/z: 248 (M$^+$-benzimidazolyl radical) as a base peak. Elemental anal. Found %C 56.66, %H 3.57, %N 15.56, %Cl 9.67, $C_{18}H_{15}N_4OCl$ (338.5), Calc. %C 56.66, %H 3.61, %N 15.55, %Cl 9.72 (Figure 3).

4-(3,4-dichlorophenyl)-1-phenyl-4-(1H-benzo[d]imidazolo[1,2-c]1,2,4-triazin -3-one (4)

A mixture of 2 (3.7 g; 0.01 mol) and phenyl hydrazine hydrate (1

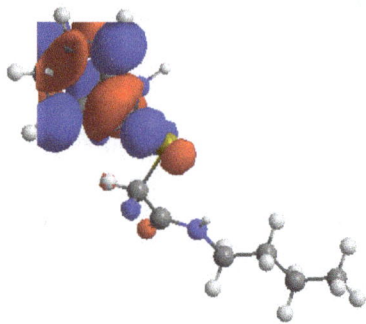

mL, 0.01 mol) in boiling ethanol and then refluxed on water bath for 2h. The reaction mixture was allowed to cool and then pour into ice-H2O and the separated product was filtered, dried and were re-crystallized from ethylacetate. Yield 64%, m.p. 152-154°C, IR spectrum υ CO 1672 cm^{-1} The ^{1}HNMR spectrum DMSO-d6 revealed δ 2.1(s, 1H, CH3), 3.0-3.3 (2 dd,1Ha and 1Hb methylene protons, CH_2-C=N, J=15.6, J=7.1 diastereotopic protons), 3,8-4.2(dd, CH-CO, methine proton), 6.2(bs,3H, 2NH and CH protons) , 7.2-7.8 (m,12H,ArH). Elemental anal. Found %C 560.86, %H 3.97, %N 12.56, %Cl 8.67, $C_{24}H_{21}N_4O_2Cl$ (433.5), Calc. %C 60.96, %H3.91, %N 12.55, %Cl 8.72 (Figure 4).

6-(4-chloro-3-methylphenyl)-4(benzimidazol-1-yl2,3,4,5 tetrahydro 3(2H) oxazinone (5)

A solution of 1 (0.01 mol) was treated with hydroxyl amine

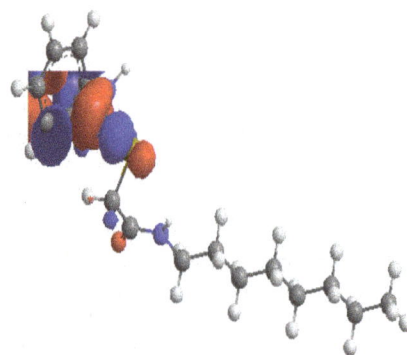

Figure 4: Optimized structure of (10bLUMO).

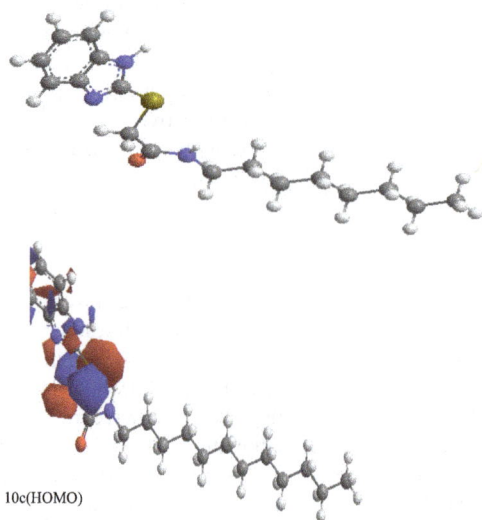

10c(HOMO)

10c(LUMO)

Figure 5: The frontier molecule orbital density distributions of (10c) ΔE = E$_{HOMO}$-E$_{LOMO}$=9.390 eV.

Figure 2: The frontier molecule orbital density distributions of (10a) ΔE=E$_{HOMO}$-E$_{LOMO}$=9.347 eV.

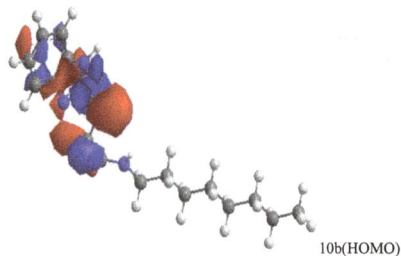

10b(HOMO)

Figure 3: The frontier molecule orbital density distributions of (10b ΔE = E$_{HOMO}$-E$_{LOMO}$= 9.377 eV) 10b(HOMO)

hydrochloride (1.5 g; 0.04 mol) in boiling pyridine (30 mL) and then refluxed for 3 h. The reaction mixture was pour into ice/HCl and the solid that separated was re-crystallized. Yield 74%, m.p. 174-176°C, IR spectrum υCO (1704 cm⁻¹). ¹HNMR spectrum DMSO-d6 of singlet at δ 2.1(s, 1H, CH₃), 2.8-3.0 (2 dd, 2H, diastereotopic protons), 3,8-4.2(dd, CH-CO, methine proton ,), 7.4-7.8 (m, 7ArH). Elemental anal. Found %C 56.50, %H 3.37, %N 11.56, %Cl l0.67, $C_{18}H_{15}N_3O_2Cl_2$ (339.5), Calc. %C 56.56, %H 3.31, %N 11.55, %Cl 10.72 (Figure 5).

4-(4-chloro-3-methylphenyl)-2-(1H-benzimidazol-2-ylmercapto)-4-oxobutanoic acids (6)

A solution of acid 1 (3 g; 0.01 mol) and 2-mercapto benzimidazole (1.70 g; 0.01 mol) in 50 ml benzene and 4 drops piperidine was allowed overnight at r.t. The crude product that formed was washed by petroleum ether (b.p 40-60°C), and then, crystallized from benzene. Yield 80 %, m.p. 122-124°C. IR spectrum υCO 1710, 1686 cm⁻¹, The ¹HNMR spectrum DMSO-d6 δ 2.1(s, 1H, CH3), 2.4-2.6 (2 dd, methylene protons, CH₂-C=O), 2,7-3.1(dd ,CH-CO, sterogenic methine proton), 6.9-7.3 (m, 7ArH and 1H of imidazole moiety), 8.2 (s,1H, acidic proton exchanged in D₂O). Elemental anal. Found %C 56.46, %H 3.27, %N 7.56, %Cl 10.67, $C_{18}H_{15}N_2O_3SCl$ (374.5), Calc. %C 56.46, %H 3.28, %N 7.55, % Cl 10.72 (Figure 6).

Formation of compounds 7 and 8

A mixture of 6 (3.73 g; 0.01 mol) and acetic anhydride (9.4 mL, 0.1mol) and then refluxed on water bath for 1h. The reaction mixture was allowed to cool and then pour into ice-H₂O and the separated product was filtered, dried and was fractional crystallized.

2-(4-chloro-3-methylbenzoyl) methyl-3-Oxo-1, 3-thiazolo [3,2-a]benzimidazole (7)

Yield 39%, m.p. 146-148 °C (toluene), IR spectrum υCO 1715, 1681 cm⁻¹. The ¹HNMR spectrum in DMSO-d6 exhibits Signals at δ 2.1(s, 1H, CH₃), 2.4-2.6(2dd, 1Ha and 1Hb CH₂-C=O, diastereotopic protons) 2.7(dd, 1H, proton of thiazole moiety), 7.0-7.9 (m, 7ArH). Elemental anal. Found %C 60.66, %H 3.64, %N 6.56, %S 7.32, %Cl 9.17, $C_{18}H_{13}N_2O_2SCl$ (356.5), Calc. %C60.66, %H 3.61, %N 6.55, %S 7.30, %Cl 9.12.

4-(4-chloro-3-methylphenyl)-1,3-thiazinoo[2,3-a]benzimidazole-2-carboxylic acid (8)

Yield 38% , m.p. 178-180°C (ethanol), IR spectrum υCO 1706 cm⁻¹

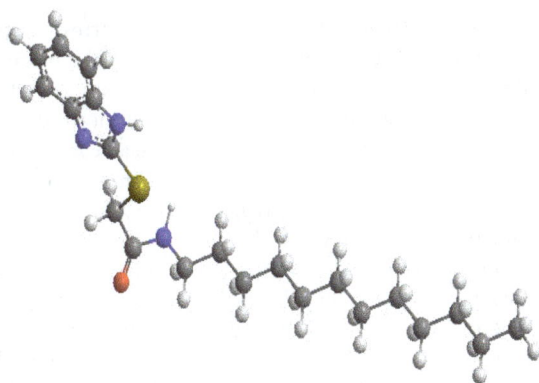

Figure 6: Optimized structure of (10c).

The ¹HNMR spectrum in DMSO-d6 exhibits Signals at δ 2.2 (s, 1H, CH₃), 2.7(dd,1H, methine proton), 7.0-7.9 (m, 8H of both 7ArH and 1H thiazine moieties). Elemental anal. Found %C 56.42, %H 3.59, %N 7.56, %S 7.84, %Cl 9.67, $C_{18}H_{13}N_2O_2SCl$ (356.5), Calc. %C 60.46, %H 3.64, %N 7.85, %S 7.80, %Cl 8.72.

Ethyl-2-benzoimidazol-2-ylthioglycolate (9)

A mixture of 16.7 g (0.1 mole) of 2-mercapto-benzoImidazole and 12.25 g of ethyl chloroacetate (0.1mole). This mixture is then refluxed for 3 h in ethanol. The ester is then collected and re-crystallized from n-pentane. 70% yield, m.p. 136-138°C. IR spectrum υCO (ester) 1746 cm⁻¹. The ¹HNMR spectrum in DMSO-d6 exhibits Signals at δ 1.23 (t, 3H, CH₃), 2.2 (s, 1H, CH₃), 3.7(q, 2H, CH₂), 4.1(s, 2H, CH₂), 7.4-7.7 (m, 4H, 4ArH). Elemental anal. Found %C 56.66, %H 3.57, %N 15.56, %Cl 9.67, $C_{11}H_{12}N_2O_2S$ (236), Calc. %C 56.66, %H 3.61, %N 15.55, %Cl 9.72.

Formation of additives, ethyl-2-(benzoimidazol-2-ylthio)-N-alkyl acetamide] (10a-c)

A mixture of 11.05 g (0.05 mole) of ester 9 and (0.05 mole) of N-alkyl amines, [(N-butylamine(a), N-octylamine(b) and N-dodecylamine (c)]. The mixture is cooled to zero °C in ethanolic KOH for one hour. The products were filtered and re-crystallized from ethanol. The ¹HNMR spectra in DMSO-d6 exhibit Signals at δ 1.33-1.48 (m, alkyl protons), 2.1 (s, 1H, CH₃), 4.0-4.1(s, 2H, CH₂), 7.3-7.8 (m, 4H, 4ArH). Elemental anal. For butyl deriv. Found %C 58.74, %H 7.17, %N 15.66, %S 12.17, $C_{13}H_{19}N_3OS$ (265), Calc. %C 58.86, %H 7.16, %N 15.84, %Cl 12.07.

Results and Discussion

With the aim of broading the synthetic potential of 4-Aryl-4-oxo-but-2-enoic acids [12-22], the authors can be reported the behavior of 4-(4-chloro-3-methylphenyl)-4-oxo-but-2-enoic acid was allowed to react with benzimidazole and 2 mercaptobenzimidazole afforded aza/thia-Michael adducts. The preference of nitrogen and sulfur nucleophiles at C₂ was due to stability of the primary zwitterionic adducts 1 and 6 (Chart 1).

2(3H) Furanone as a new antioxidant and anti-inflammatory agent, Cotelle et al. [23] and Weber et al. [24] synthesized new ascorbic acid analogues have resulted in obtaining antioxidant and anti-tumoral [25]. The most activating furanone when substituted in 2 and 5 positions by activating aryl group [26]. In continuation to our previous works to design and synthesize new furanones substituted in position 2 by 4-chlorophenyl moiety and benzimidazole in 5-position to increase its anti-oxidant activity. The synthesis can be achieved by the lactonization of the acid 1 on heating water bath for 1h with acetic anhydride, afforded 5-(4-chloro-3-methylphenyl)-3-(1H-benzimidazol-1-yl)-2-(3H) furan one (2) (Chart 1). Furthermore, reaction of the furanone 2 with hydrazine hydrate in boiling ethanol, afforded the pyridazinone derivative 3. So, the reaction is favor the route i versus the route ii that afford fused benzimidazolo [1,2-c]triazinone derivative 4a that reflect to us the pyridazinone isomer 3 is more thermodynamic stable. Otherwise, the treatment of furanone 2 with phenylhydrazine afforded benzimidazolo [1,2-c]triazinone derivative 4b. That can be confirmed, the presence of phenyl group increase the stability of fused heterocycle 4 (Chart 2).

A series of 2H-pyridazine-3-one and 1,2-oxazine derivatives have anti-inflammatory activity was tested in vitro on superoxide formation and effects on lipid peroxidation[27], as antioxidants in natural rubber [28,29], pyridazinone PDE inhibitors [30], 1,2-oxazine as PTP 1B inhibitors [31]. An authentic reaction was done by refluxing the acids

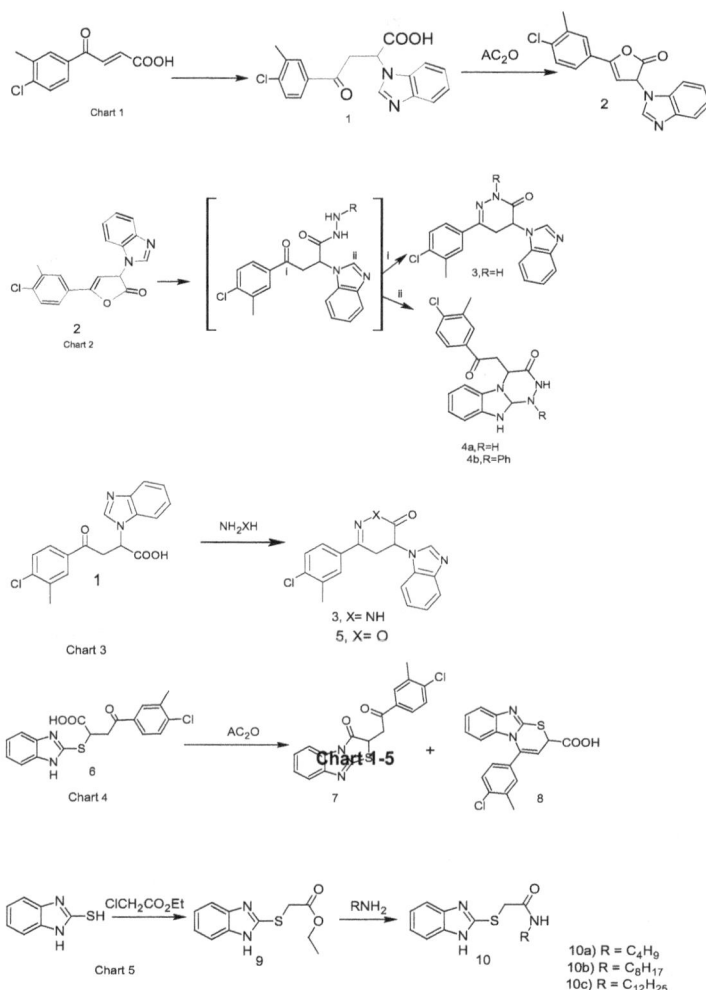

Chart 1

1

2

Chart 2

3, R=H

4a, R=H
4b, R=Ph

Chart 3

1

3, X= NH
5, X= O

Chart 4

6

Chart 1-5

7

8

Chart 5

9

10

10a) R = C_4H_9
10b) R = C_8H_{17}
10c) R = $C_{12}H_{25}$

Total Acid Numbers, mg KOH / g Sample x 10^2

	3	2	1	6	5	4	7	8	9	10b	10c	10a
0 ppm	197	197	197	197	197	197	197	197	197	197	197	197
200 ppm	182	101	117	92	51	134	77	102	79	74	86	83
400 ppm	167	99	81	68	70	157	84	99	86	65	82	78
500 ppm	184	121	173	167	133	171	101	150	99	73	74	74
1000 ppm	191	130	171	174	156	174	120	156	101	80	87	90

Table 1: TAN after 72 Hours x 10^2

1 with hydrazine hydrate and/or hydroxylamine hydrochloride in boiling pyridine afforded the pyridazinone, and oxazinone derivatives 3,5 respectively in good yield (Chart 3).

On the other hand, when the 4-(4-dichloro-3-methylphenyl)-4-oxo-but-2-enoic acid was allowed to react with 2-mercaptobenzimidazole in boiling benzene yielded the adduct 6. The charge density localized on the sulfur atom (0.266), was found to be greater than nitrogen atom (0.236). Consequently, the attack preferred via sulfur atom (Thiol tautomer). IR spectra of them reveal strong absorption bonds at (1710-1682 cm^{-1}) for υCO of adduct. Also, treatment of the adduct 6 with acetic anhydride gave the corresponding thiaazolobenzimidazole derivative 7, and 1, 3-thiazinoquinazoline derivative 8 (Chart 4).

An important class of compounds in the field of petroleum chemistry because of their broad spectrum antioxidant and anticorrosive activities [32-36]. The performance of engine oils and industrial lubricants are improved by the addition of specific types of additives. These additives are oil soluble chemicals and usually added to prevent the deposition of insoluble materials, lubricant oxidation and metal corrosion. Most antioxidant functions are either by thermal decomposition via C-C bond chain or reacting with free radical via hydroperoxide radical mechanism. In the present work, the authors can be reported the benzimidazole derivative 10, R=C4 at 400 ppm [37] has a higher stability due to the electron donating nature of the alkyl group butyl > octyl>dodecyl groups that facilitate to generate stable free radical. The results confirm the N-alkyl-S-benzimidazol-2-yl thioglycolamide 10 are better antioxidant than 2-mercaptobenzimidazole itself. Enhancement of the amide 10 as antioxidant prove that the process of thermal degradation of engine lubricants proceed via thermal decomposition of C-C bond chain (Chart 5).

At higher concentration of the hydrazide 10, gave higher % SO_2 concentration and formation of the diametric structures. The SO_2 moiety will be increased % sulfuric acid and therefore increase TAN that causes competition for antioxidant role and so optimum concentration of antioxidant hydrazine derivatives 10 were 400 ppm . The Correlation between the antioxidant character of the heterocyclic additives and their structure has been investigated, using Ab initio (HF/3-21G) and semi-empirical gas phase AM1(Austin model 1) calculations. Parameters as total energy, HOMO and LUMO energies, dipole moment and dipole-dipole interaction 2-mercaptobenzoimidazole derivatives 10 indicate the importance of the thiol structure as antioxidant and anticorrosive. To investigate the effect of substituent on the inhibition mechanism and efficiency, they computed the E_{HOMO}, E_{LUMO} energies and energy gap. According to the frontier molecular orbital theory, the formation of a transition state is due to an interaction between frontier orbital's (HOMO and LUMO) of reacting species [38,39]. Usually the total acid number of the oil increases by increasing the oxidation time. The increment of TAN value is due to oxidation processes which produce peroxides when subjected to heat and air. In presence of additives, the total acid numbers after thermal oxidation of the base oil for 24-72 h. The total acid numbers decrease by increasing the additive dose from 200 part per million to 1000 parts per million (Table 1).

From the Table 1 at 0 ppm, no additive in the oil lubricant, we can notice the total acid number(TAN) can be reflect the oxidation stability of the antioxidant organic materials 1-10, at different concentration that increased in compounds 1,3 and 6 because of the presence of acidic protons.

Conclusion

Cyclization of the acids 1, 6 afforded 2, 7 and 8 respectively that increase the oxidation stability and becomes act as good antioxidants. The presence of sulfur atom in the compounds 6, 7 and 8 can be afforded the higher oxidation stability than corresponding compounds 1 and 2 respectively. Finally, the fused heterocycles e.g. compounds 4, 7 and 8 (although the presence of phenyl group and sulfur atom) has a lower oxidation stability than separated heterocycles 2, and 5, as expected the stability of radicals appear in a large area and size of atom or compound.

References

1. Toshikazu K (2002) Tukuba Research lab Food. Food ingredients J Japan.

2. Barrett D, Tanaka A, Harada K, Ohki H, Watabe E, et al. (2001) Synthesis and biological activity of novel macrocyclic antifungals: acylated conjugates of the ornithine moiety of the lipopeptidolactone FR901469. Bioorg Med Chem Lett 11: 479-482.

3. Kovalainen JT, Christains JA, Kotisaati S, Laitinen JT, Mannisto PT, et al. (1999) Synthesis and in vitro pharmacology of a series of new chiral histamine H3-receptor ligands: 2-(R and S)-Amino-3-(1H-imidazol-4(5)-yl) propyl ether derivatives. J Med Chem 42: 1193-1202.

4. El-Faham A, El Massry AM, Amer A, Gohar YM (2002) A versatile synthetic route to chiral quinoxaline derivatives from amino acids precursors. Lett Pept Sci 9: 49-54.

5. Polyak F, Lubell WD (1998) Rigid Dipeptide Mimics: Synthesis of Enantiopure 5- and 7-Benzyl and 5,7-Dibenzyl Indolizidinone Amino Acids via Enolization and Alkylation of delta-Oxo alpha,omega-Di-[N-(9-(9-phenylfluorenyl))amino] azelate Esters. J Org Chem 63: 5937-5949.

6. Roy S, Lombart HG, Lubell WD, Hancock REW, Farmer SW (2002) Exploring relationships between mimic configuration, peptide conformation and biological activity in indolizidin-2-one amino acid analogs of gramicidin S. J Peptide Res 60: 198-214.

7. Marsham PR, Wardleworth JM, Boyle FT, Hennequin LF, Kimbell R, et al. (1999) Design and synthesis of potent non-polyglutamatable quinazoline antifolate thymidylate synthase inhibitors. J Med Chem 42: 3809-3820.

8. Xia Y, Yang ZY, Xia P, Bastow KF, Nakanishi Y, et al. (2000) Antitumor agents. Part 202: novel 2'-amino chalcones: design, synthesis and biological evaluation. Bioorg Med Chem Lett 10: 699-701.

9. Grimaudo S, Raimondi MV, Capone F, Chimirri A, Poretto F, et al. (2001) Apoptotic effects of thiazolobenzimidazole derivatives on sensitive and multidrug resistant leukaemic cells. Eur J Cancer 37: 122-130.

10. Canan K, Gulguim A, Benay C, Mumtaz S (2004) Arch Pharm Res 27: 156-163.

11. Goshev I, Mavrova A, Mihaylova B, Wesselinova DJ (2013) Cancer Res Therapy 1: 87-91.

12. Rizk SA (2011) Synthesis Some Fused Heterocycles and Spiro Compounds. American Journal of Chemistry 1: 65-71.

13. Rizk SA, El-Hashash MA, Mostafa KK (2008) Utility of β-Aroyl Acrylic acid in Heterocyclic Synthesis. Egypt J Chem 51: 116-121.

14. Suroor AK, Mullick P, Pandit S, Kaushik D (2009) Acta Poloniae Pharmaceutica-Drug Research (Pol. Pharmaceutical Soc). 66: 169-172.

15. El-Hashash MA, Rizk SA, Inter J Chem Petrochem Tech (IJCPT) 3: 2013-2012.

16. Yousef ASA, Marzouk MI, Madkour HMF, El-Soll AMA, El-Hashash MA (2003) Synthesis of Some Heterocyclic system of anticipated biological activities via 6-aryl-4-pyrazol-1-yl-pyridazin-3-one. Can J Chem 83: 251-259.

17. Dong F, Kai G, Zhenghao F, Xinili Z, Zuliang L (2009) A practical and efficient synthesis of quinoxaline derivatives catalyzed by task-specific ionic liquid. Catalysis Communication 9: 317.

18. Umpreti M, Pant S, Dandia A (1996) Phosphorous, Sufur, Silicon 113: 165.

19. Khachikyan RD, Karamyan NV, Panosyan GA, Indzhikyan MG (2005) IZV Akad Nauk Ser Khim 1923.

20. Elhashash MA, Rizk SA, Bakeer HM, Elbadwy A, Kowrany HM (2012) Utility of p-Bromo phenyl oxo but-2enoic Acid in the Synthesis of New α-Amino Acids and Using them as Building Blocks in Heterocyclic Synthesis; Formation of Benzodiazapine and Benzoxazapine moiety. J Pur Utility Reaction Environment 1: 24-43.

21. Rizk SA, El-Hashash MA (2011) Egypt J Chem 54: 3.

22. Rizk SA, EL-Hashash MA, Aburzeza MM (2011) Utility of p-Acetamidobenzoyl Prop-2enoic Acid in the Synthesis of New α-Amino Acids and Using them as Building Blocks in Heterocyclic Synthesis. Egypt J Chem 54: 1-10.

23. Cotelle P, Cotelle N, Teissier E, Vezin H (2003) Synthesis and antioxidant properties of a new lipophilic ascorbic acid analogue. Bioorg Med Chem 11: 1087-1093.

24. Weber V, Rubat C, Duroux E, Lartigue C, Madesclairea M, et al. (2005) New 3- and 4-hydroxyfuranones as anti-oxidants and anti-inflammatory agents. Bioorg Med Chem 13: 4552-4564.

25. Raic-Malic S, Svedruzic D, Gazivoda T, Marunovic A, Hergold-Brundic A, et al. (2000) Synthesis and antitumor activities of novel pyrimidine derivatives of 2,3-O,O-dibenzyl-6-deoxy-L-ascorbic acid and 4,5-didehydro-5,6- dideoxy-L-ascorbic acid. J Med Chem 4: 4806-4811.

26. Hashem AI, Youssef AS, Kandeel KA, Abou-Elmagd WS (2007) Conversion of some 2(3H)-furanones bearing a pyrazolyl group into other heterocyclic systems with a study of their antiviral activity. Eur J Med Chem 42: 934-939.

27. Caliskan-Ergün B, Süküroğlu M, Coban T, Banoğlu E, Suzen S (2008) Screening and evaluation of antioxidant activity of some pyridazine derivatives. J Enzyme Inhib Med Chem 23: 225-229.

28. Ladopoulou E, Matralis AN, Kourounakis AP (2013) New Multifunctional Di-tert-butylphenoloctahydro(pyrido/benz) oxazine Derivatives with Antioxidant, Antihyperlipidemic, and Antidiabetic Action. J Med Chem 56: 3330-3338.

29. Ismail MN, Younan AF, Yehia AA (1993) Effect of Chemical Structure of Some Pyridazine Derivatives as Antioxidants in Natural Rubber. J Elastomers Plastics 25: 266-274.

30. Allcock RW, Blakli H, Jiang Z, Johnston KA, Morgan KM (2011) Phosphodiesterase inhibitors. Part 1: Synthesis and structure–activity relationships of pyrazolopyridine–pyridazinone PDE inhibitors developed from ibudilast. Bioorg Med Chem Lett 21: 3307-3312.

31. Cho SY, Baek JY, Han SS, Kang SK, Ha JD, et al. (2006) PTP-1B inhibitors: Cyclopenta[d][1,2]-oxazine derivatives. Bioorg Med Chem Lett 16: 499-502.

32. Willermet PA (1998) Some engine oil additives and their effects on antiwear film formation. Tribology Letters 5: 41-47.

33. Wan Y (2008) Synergistic lubricating effects of ZDDP with a heterocyclic compound. Industrial Lubrication and Tribology 60: 317-320.

34. Saji VS (2010) A Review on Recent Patents in Corrosion Inhibitors. Recent Patents on Corrosion Science 2: 6-12.

35. Sangeetha M, Rajendran S, Muthumegala TS, Krishnaveni A (2011) Green corrosion inhibitors - An Overview. Zaštita Materijala 52: 3-19.

36. Fouda AS, Nazeer A, Ashour EA (2011) Amino acids as environmentally-friendly corrosion inhibitors for Cu10Ni alloy in sulfide-polluted salt water: Experimental and theoretical study. Zaštita Materijala 52: 21-34.

37. Nessim MI, Ahmed MHM, Ali AM, Bassoussi Salem AA, Attia SK (2012) The effect of Some benzothiazole derivatives as antioxidant for base stock. J Appl Sci 27: 243-258.

38. Feng Y, Chen S, Zhang H, Li P, Wu L, et al. (2006) Characterization of iron surface modified by 2-mercaptobenzothiazole self-assembled monolayers. Appl Surf Sci 253: 2812-2819.

39. Fukui K (1975) Theory of Orientation and Stereoselection. Springer-Verlag, NewYork 1-85.

Development and Characterization Salbutamol Sulphate Mouth Disintegrating Tablet

Basavaraj K Nanjwade*, Ritesh Udhani, Jatin Popat, Veerendra K Nanjwade and Sachin A Thakare

Department of Pharmaceutics, KLE University's College of Pharmacy, Belgaum-590010, Karnataka, INDIA

Abstract

An orodispersible dosage form has been developed as a user-friendly formulation that disintegrates in the mouth immediately. Thus an attempt was made to improve the onset of action of bronchodilator used commonly in the treatment of asthma. Formulation was optimized for type of disintegrant used and method of formulation. Disintegrants such as CCS (Croscarmellose Sodium), SSG (Sodium Starch Glycolate), L-HPC (Low-substituted Hydroxy Propyl Cellulose), and Crospovidone XL-10 were used and tablets were prepared by direct compression method and wet granulation. Wet granulation formulation were again sub-divided where disintegrant was added intragranularly in one type and was added both in intra and extra granulation in the other. Mint flavor was added to give good mouth feel. Out of all formulations prepared, the one prepared with Crospovidone XL-10 added both intra and extra granulation showed least disintegrating time (9 sec) with good flow property. Direct compression blends had poor flow. Tablets were also evaluated for various physicochemical parameters. All the tablets showed burst release of drug. Hence, it was concluded that out of all formulations, the one prepared with Crospovidone XL-10 added both intra and extra granulation was the best formulation as it showed the least disintegration time.

Keywords: Crospovidone XL-10; Intra and extra granular addition; Asthma; Mouth feel; Wetting time; Water absorption ratio

Abbreviations: L.O.D: Loss On Drying; MCC: Micro Crystalline Cellulose

Introduction

Convenience of administration and patient compliance are gaining significant importance in the design of dosage forms. Recently more stress is laid down on the development of organoleptically elegant and patient friendly drug delivery system for pediatric and geriatric patients [1,2]. One important innovation in this direction is the development of fast dissolving/disintegrating oral dosage forms that dissolve or disintegrate instantly upon contact with recipient's tongue or buccal mucosa [3,4]. A tablet which can rapidly disintegrate in saliva is an attractive dosage form and a patient-oriented pharmaceutical preparation [5].

Superdisintegrants are added in formulation to increase the dissolution characteristics thus increasing bioavailability of drug [6]. There are three methods of addition of disintegrant into the formulation, intragranular (Internal addition), extragranular (External addition), partly intragranular and extragranular addition [7]. The time for disintegration of orally disintegrating tablets is generally considered to be less than one minute [8,9,10,11] although patients can experience actual oral disintegration times that typically range from 5-30 sec. Many companies have developed various types of fast-disintegrating dosage forms. A freeze-dried porous wafer known as Zydis [12, 13], a molding tablet known as EMP [13], an effervescent tablet known as OraSolve [13], and a disintegrant addition [13] have all been developed.

Asthma is a chronic inflammatory disease, which affects over 5-10% of population in industrialized countries [14]. It affects approx. 53 million people across world mostly in United States, France, Germany, Italy, Spain, United Kingdom, and Japan [14,15]. Thus, an attempt was made for preparation of fast dissolving tablets of a model bronchodilator, salbutamol sulphate with an aim of reducing lag time and providing faster onset of action to relieve immediately acute asthmatic attack.

Material and Methods

Salbutamol sulphate was provided as a gift sample by Lincoln pharmaceuticals, Ahmedabad, Lactose was obtained from Ranbaxy Fine-Chem. Ltd, Delhi, MCC PH 101 and Magnesium Stearate were obtained from Loba chemicals Pvt. Ltd., Mumbai, CCS, SSG, Crospovidone XL-10 and L-HPC were obtained from Microlabs, Bangalore and Aerosil was obtained from Eonik Degussa, Mumbai. All other chemicals were of analytical grade.

Equipment used

UV Spectrophotometer: Systronic 2201 UV/Vis double beam Spectrophotometer.

Tablet Compression Machine: Rimek 10 Station Press, Cadmach Machinery Co. Pvt. Ltd., Ahmedabad, India.

Dissolution test apparatus: Dissolution test apparatus-TDT-06T, Electrolab, Mumbai, India.

Roche friabilator: Camp-bell Electronics, Mumbai, India

Hardness tester: Validated dial type, Model: 1101, Shivani Scientific Industries Pvt. Ltd., Mumbai.

Sartorious electronic balance: Model CP- 224 S, Labtronic.

The amount of disintegrant, lactose and MCC PH 101 was previously optimized and hence kept constant in all the formulations.

*Corresponding author: Dr. Basavaraj K. Nanjwade, Department of Pharmaceutics, KLE University College of Pharmacy, Belgaum-590010, Karnataka, India

Preparation of the mouth disintegrating tablet of salbutamol sulphate by direct compression method: Accurately weighed quantities of salbutamol sulphate, Lactose, MCC PH 101 and disintegrant, were mixed for 10 minutes and passed through sieve no. 40. The blend was mixed with mint flavor and Aerosil passed through sieve no. 40 for 5 minutes. Finally Magnesium Stearate passed from sieve no. 60 was mixed in above blend and was mixed for 3 minutes. The homogenized mixture was subjected for direct compression to produce 100 mg tablets by using Rimek 10 Station Press. The tablets obtained contained 4.8 mg of salbutamol sulphate. (equivalent to 4mg of salbutamol) The drug and composition of the ingredients are shown in Table 1.

Preparation of the mouth disintegrating tablet of salbutamol sulphate by wet granulation method where disintegrant is added intragranularly: Accurately weighed quantities of salbutamol sulphate, Lactose, MCC PH 101 and disintegrant, were mixed for 10 minutes and passed through sieve no. 40. The blend was subjected to granulation with the help of water as a granulating solvent. The granules were dried at 60°C till L.O.D reaches 2-2.5%. The dried granules were passed through sieve no. 16 and then mixed with mint flavor and Aerosil passed through sieve no. 40 for 5 minutes. Finally Magnesium Stearate passed from sieve no. 60 was mixed in above blend and was mixed for 3 minutes. The homogenized mixture was subjected for compression to produce 100 mg tablets by using Rimek 10 Station Press. The drug and composition of the ingredients are shown in Table 2.

Preparation of the mouth disintegrating tablet of salbutamol sulphate by wet granulation method where disintegrant is added both intragranular and extragranular: Accurately weighed quantities of salbutamol sulphate, Lactose, MCC PH 101 and disintegrant, were mixed for 10 minutes and passed through sieve no. 40. The blend was subjected to granulation with the help of water as a granulating solvent. The granules were dried at 60°C till L.O.D reaches 2-2.5%. The dried granules were passed through sieve no. 16 and then mixed with disintegrant, mint flavor and Aerosil passed through sieve no. 40 for 5 minutes. Finally Magnesium Stearate passed from sieve no. 60 was mixed in above blend and was mixed for 3 minutes. The homogenized mixture was subjected for compression to produce 100 mg tablets by using Rimek 10 Station Press. The drug and composition of the ingredients are shown in Table 3.

Ingredients	Batch Code			
	IG1	IG2	IG3	IG4
Salbutamol Sulphate	4.8	4.8	4.8	4.8
Lactose Monohydrate	52.2	52.2	52.2	52.2
MCC PH 101*	35	35	35	35
CCS*	5	-	-	-
SSG*	-	5	-	-
Crospovidone XL-10	-	-	5	-
L-HPC*	-	-	-	5
Water	Q.S.	Q.S.	Q.S.	Q.S.
Mint flavor	1	1	1	1
Aerosil	1	1	1	1
Magnesium Stearate	1	1	1	1

Salbutamol sulphate 4.8 mg is equivalent to 4 mg of salbutamol
*MCC – Microcrystalline Cellulose, CCS – Croscarmellose Sodium, SSG – Sodium Starch Glycolate, L-HPC – Low substituted Hydroxy Propyl Cellulose

Table 2: Composition of Tablet Prepared by Wet Granulation Method where Disintegrant was added Intragranularly.

Ingredients	Batch Code			
	IG1	IG2	IG3	IG4
Salbutamol Sulphate	4.8	4.8	4.8	4.8
Lactose	52.2	52.2	52.2	52.2
MCC PH 101*	35	35	35	35
CCS*	2.5	-	-	-
SSG*	-	2.5	-	-
Crospovidone XL-10	-	-	2.5	-
L-HPC*	-	-	-	2.5
Water	Q.S.	Q.S.	Q.S.	Q.S.
CCS*	2.5	-	-	-
SSG*	-	2.5	-	-
Crospovidone XL-10	-	-	2.5	-
L-HPC*	-	-	-	2.5
Mint flavor	1	1	1	1
Aerosil	1	1	1	1
Magnesium Stearate	1	1	1	1

Salbutamol sulphate 4.8 mg is equivalent to 4 mg of salbutamol
*MCC – Microcrystalline Cellulose, CCS – Croscarmellose Sodium, SSG – Sodium Starch Glycolate, L-HPC – Low substituted Hydroxy Propyl Cellulose

Table 3: Composition of Tablet Prepared by Wet Granulation Method where Disintegrant was added both Intra and Extragranularly.

Evaluation of tablets

Pre-compressional evaluation [16]:

a) Angle of repose (θ)

Angle of repose (θ) was determined using funnel method. The blend was poured through a funnel that can be raised vertically until a maximum cone height (h) was obtained. The radius of the heap (r) was measured and angle of repose was calculated.

$$\theta = \tan^{-1}(h/r)$$

b) Bulk density

Apparent bulk density (δb) was determined by placing presieved drug excipients blend into a graduated cylinder and measuring the volume (Vb) and weight (M) "as it is".

$$\delta b = M/Vb$$

c) Tapped density

The measuring cylinder containing a known mass of blend was tapped for a fixed time. The minimum volume (Vt) occupied in the

Ingredients	Batch Code			
	DC1	DC2	DC3	DC4
Salbutamol sulphate	4.8	4.8	4.8	4.8
Lactose Monohydrate	52.2	52.2	52.2	52.2
MCC PH 101*	35	35	35	35
CCS*	5	-	-	-
SSG*	-	5	-	-
Crospovidone XL-10	-	-	5	-
L-HPC*	-	-	-	5
Mint flavor	1	1	1	1
Aerosil	1	1	1	1
Magnesium Stearate	1	1	1	1

Salbutamol sulphate 4.8 mg is equivalent to 4 mg of salbutamol
*MCC – Microcrystalline Cellulose, CCS – Croscarmellose Sodium, SSG – Sodium Starch   Glycolate, L-HPC – Low substituted Hydroxy Propyl Cellulos

Table 1: Composition of Direct Compression Tablets.

cylinder and the weight (M) of the blend was measured. The taped density was calculated using following formula.

$$\delta t = M/Vt$$

d) Compressibility index

The simplest way of measurement of free flow property of powder is compressibility, an indication of the ease with which a material can be induced to flow is given by % compressibility which is calculated as follows:

$$C = (\delta t - \delta b) / \delta t * 100$$

e) Hausner's ratio

Hausner's ratio is an index of ease of powder flow, it is calculated by following formula. Hausner's ratio = $\delta t / \delta b$

Post compression parameters

a) Measurement of the tablet tensile strength and friability [16]

Six tablets of each formulation were picked randomly and dimensions were determined. Hardness of tablets was examined using a hardness tester to measure the crushing strength of the tablets (Validated dial type). The mean hardness was calculated and expressed as Kg/cm^2. The friability of tablets was determined using Roche Friabilator (USP) at 25rpm for 4 minutes. It is expressed in percentage (%).

b) Drug content uniformity [17]

Tablets containing 4.8 mg of drug is dissolved in 100 ml of simulated gastric fluid (SGF) pH 1.2. The drug is allowed to dissolve in the solvent, the solution was filtered, and 1ml of filtrate was suitably diluted with simulated gastric fluid pH 1.2 and analyzed spectrophotometrically at 276 nm. The amount of salbutamol sulphate was estimated by using standard calibration curve of the drug. Drug content studies were carried out in triplicate for each batch of formulation.

c) In vitro disintegration time [17]

Tablet disintegration was carried out by placing one tablet in each tube of the basket and top portion of the each tube was closed with a disc. The apparatus was run with pH 1.2 SGF (simulated gastric fluid) maintained at 37 ± 2^0C as the immersion liquid. The assembly was raised and lowered upto 30 cycles per minute. The time taken for complete disintegration of the tablet with no palpable mass remaining in the apparatus was measured and recorded. The experiment was carried out in triplicate.

d) Wetting time and Water absorption ratio [18]

The tablet was placed in a Petri dish having a 6.5 cm in diameter, containing 10 ml of water and the time for complete wetting was recorded. The experiment was carried out in triplicate at room temperature. Water absorption ratio was calculated by keeping the tablet on a piece of tissue paper folded twice in a small Petri dish containing 6ml of distilled water. Time for complete wetting of tablet was recorded. The wetted tablet was then weighed. Water absorption ratio R, was determined using equation, $R = 10 \times [(W_a - W_b) \div W_b]$, Where, W_b = weight of the tablet before water absorption and W_a = weight of the tablet after water absorption.

e) Mouth feel and in vivo disintegration time

To know the mouth feel, taste and disintegration of the tablets, formulations were given to six healthy human volunteers. The mouth feel, taste and in vivo disintegration was evaluated.

Results and Discussion

Precompression parameters

Granules ready for compression containing drug and various excipients was subjected for pre-compression parameters (Micromeritic properties) to study the flow properties of granules, to achieve uniformity of tablet weight. The data obtained for angle of repose for all the formulations were tabulated in Table 4 and the values were found to be in the range of 19° to 33°. The formulations of direct compression revealed poor flow property and formulations of wet granulation had good flow property. Loose bulk density (LBD) and tapped bulk density (TBD) for the blend is shown in Table 4. The loose bulk density and tapped bulk density for all the formulations blend varied from 0.56 gm/cm^3 to 0.76 gm/cm^3 and 0.73 gm/cm^3 to 0.89 gm/cm^3 respectively. The results of Carr's consolidation index or compressibility index (%) for all the formulations blend ranged from 9.52 to 24.32. The results for all the formulations were recorded in Table 4.

Post-compression parameters

The tablets prepared were subjected for evaluation according to various official specifications and other parameters. Hardness, friability, weight variation, wetting time, in vitro water absorption ratio, drug content, disintegration time, in vivo taste, mouth feel and disintegration were performed. Formulations prepared were randomly picked from

FMC*	Angle of Repose (θ)	Loose Bulk Density (gm/cm³)	Tapped Bulk Density (gm/cm³)	% Compressibility	Hausner's ratio
DC1	33	0.58	0.75	22.66	1.29
DC2	31	0.56	0.74	24.32	1.32
DC3	31	0.56	0.73	23.28	1.3
DC4	32	0.57	0.74	22.97	1.29
IG1	24	0.73	0.86	15.11	1.17
IG2	22	0.76	0.84	9.52	1.10
IG3	22	0.72	0.89	13.25	1.23
IG4	24	0.75	0.84	10.71	1.12
IEG1	21	0.69	0.80	13.75	1.15
IEG2	19	0.71	0.82	13.41	1.15
IEG3	20	0.72	0.84	14.28	1.16
IEG4	20	0.71	0.81	12.34	1.14

FMC*- Formulation code.

Table 4: Angle of repose, loose bulk density, tapped bulk density, Carr's Compressibility Index, Hausner's ratio.

FMC*	Weight Variation	Drug content (%)	Hardness (Kg/cm²)	Friability (%)
DC1	99±0.002	98.32±.860	4.5±0.246	0.84
DC2	99±0.001	97.39±0.124	4.8±0.131	0.83
DC3	100±0.002	98.36±0.679	4.6±0.346	0.75
DC4	98±0.002	99.00±0.374	4.5±0.456	0.73
IG1	100±0.002	102.13±0.659	4.3±0.283	0.63
IG2	101±0.002	99.36±0.980	4.2±0.546	0.59
IG3	101±0.001	98.33±0.618	4.2±0.244	0.56
IG4	99±0.001	98.66±0.231	4.2±0.224	0.59
IEG1	100±0.002	98.79±0.679	4.3±0.248	0.63
IEG2	100±0.002	99.03±0.776	4.3±0.218	0.69
IEG3	101±0.001	99.64±0.235	4.4±0.216	0.66
IEG4	101±0.002	101.33±0.577	4.5±0.212	0.67

*FMC- Formulation code.
All values are indicated as Mean ± S.D (n=3).

Table 5: Characterization of salbutamol sulphate Tablets.

each batch examined under lens for shape and in presence of light for color. Tablets showed flat, circular shape and were white in color. The hardness of the tablets was found in the range of 4.2±0.224 to 4.8±0.131 Kg/cm^2, respectively. The mean hardness test results are tabulated in Table 4.Friability of the all the formulation was in the range of 0.56% to 0.84%. The obtained results were found to be well within the approved range (<1%) in all designed formulations. The results are shown in Table 4.The content uniformity was performed for all the formulations and results are tabulated in Table 5. The drug content was found to be 97.39±0.124% to 102.13±0.659%. The results were within the range and that indicated uniformity of mixing of the drug with excipients in the developed formulations. The weight variation for all the formulations is shown in Table 5. All the tablets passed the weight variation test; average percentage weight variation was found within the pharmacopoeial limits of ±7.5%. The obtained results were found to be 98±0.002mg to 101±0.002mg.

In vitro disintegration time: The disintegration time recorded of all the formulation found in the range of 7.01±0.04 to 18.15±0.05 seconds. The results are shown in Table 6 and Figure 1.

FMC*	In Vitro DT (sec)*	Wetting Time (Sec)	Water Absorption Ratio	In Vivo DT (Sec)*	Mouth feel
DC1	8.02±0.06	11.23±2.13	71.6±2.14	14.56±3.19	+
DC2	8.52±0.02	12.34±3.15	73.45±2.54	15.65±2.14	+
DC3	7.01±0.04	10.24±2.16	76.25±2.14	13.52±1.65	+
DC4	7.21±0.03	10.98±3.15	73.25±1.65	13.65±2.58	+
IG1	18.15±0.05	22.42±3.12	74.35±2.30	25.46±2.65	+
IG2	17.53±0.04	23.12±1.32	71.26±2.89	26.41±2.24	+
IG3	14.89±0.01	18.35±3.32	78.52±3.45	22.41±3.31	+
IG4	16.14±0.04	19.25±2.35	76.54±3.68	25.12±2.16	+
IEG1	12.02±0.05	15.25±1.35	77.0±2.46	19.25±1.95	+
IEG2	11.54±0.03	15.62±3.45	76.4±2.09	18.8±2.89	+
IEG3	9.06±0.05	12.42±1.45	78.2±3.15	13.25±1.56	+
IEG4	13.02±0.08	13.10±3.15	77.2±3.54	15.06±2.78	+

*FMC-Formulation code, DT- Disintegration, '+' good palatable mouth feel, '-' poor palatable mouth feel
All values are indicated as Mean ± S.D (n=3).

Table 6: Characterization of salbutamol sulphate Tablets.

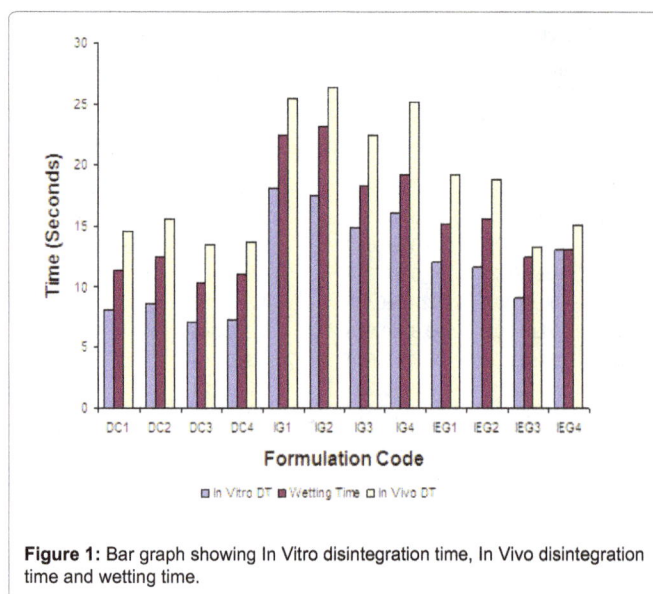

Figure 1: Bar graph showing In Vitro disintegration time, In Vivo disintegration time and wetting time.

Wetting time and Water absorption ratio: Wetting time of all the formulations recorded was found to be 10.24±2.16 to 23.12±1.32 seconds. The result of wetting time is shown in Table 5. Wetting time is closely related to the inner structure of the tablet. The obtained results mimic the action of saliva in contact with the tablet to illustrate the water uptake and subsequent wetting of the tablet. The wetting process was very rapid in all the formulations. The hydration studies demonstrated that all tablets were characterized by very similar hydration profiles. They hydrated quickly and showed high hydration percentage 71.26±2.89 to 78.52±3.45. Tablet hydration capacity is a very important parameter in the design of a new drug fast disintegrable dosage forms because of a strict relationship between water absorption and the drug release mechanism. The obtained results are shown in Table 5 and Figure 1.

Mouth feel and in-vivo disintegration: In the formulations, mint flavor and lactose were added to improve the mouth feel and taste of salbutamol sulphate tablets. . Volunteers felt that the tablets had a good taste and good palatable mouth feel. In-vivo disintegration times of tablet were found to be 13.25±1.56 to 26.41±2.24 seconds. The results are shown in Table 6.

Conclusion

From the present study, it can be concluded that orodispersible/ mouth disintegrating tablets can be prepared using different techniques such as direct compression method, wet granulation where disintegrant is added intragranularly and wet granulation method where disintegrant is added both intra and extragranularly. However, the direct compression method shows some problems of poor flow. Thus, wet granulation method where disintegrant is added both intra and extragranularly can be considered as the most promising technique as it overcomes the poor flow problem and shows short disintegration time. Out of all the disintegrants used Crospovidone XL-10 showed the least disintegration time in all formulation techniques, hence it can be concluded to be the best superdisintegrant.

Declaration of Interest

The authors report no declarations of interest.

References

1. Bhusan SY, Sambhaji SP, Anant RP, Kakasaheb RM (2000) New drug delivery system for elderly. Indian Drugs 37: 312-318.

2. Wadhwani AR, Prabhu NB, Nandkarni MA, Amin PD (2004) Consumer friendly mucolytic formulations. Indian J Pharm Sci 7: 506-507.

3. Shishu A Bhatti, T Singh (2007) Preparation of tablets rapidly disintegrating in saliva containing bitter taste-masked granules by compression method. Indian J Pharm Sci 69: 80-84.

4. Kuno Y, Kojima M, Ando S, Nakagami H (2003) Proceedings of the controlled Release Society 30th annual meeting, Glasgow, United Kingdom Pp: 19-23.

5. Ishikawa T, Watanabe Y, Utoguchi N, Matsumoto M (1999) Preparation and evaluation of tablets rapidly disintegrating in saliva containing bitter taste-masked granules by compression method. Chem Pharm Bull (Tokyo) 47: 1451-1454.

6. Kaushik D, Durega H, Saini TR (2004) Formulation and evaluation of olanzipine mouth dissolving tablet by effervescent formulation approach. Indian Drugs 41: 410-412.

7. Sekar V, Chellan VR (2008) Immediate release tablets of telmisartan using superdisintegrant- Formulation, Evaluation, and Stability studies. Chem Pharm Bull (Tokyo) 56: 575-577.

8. Liang AC, Chen LH (2001) Fast-dissolving intraoral drug delivery systems. Expert Opin Ther Patent 11: 981-986.

9. Morita Y, Tsushima Y, Yasui M, Termoz R, Ajioka J, et al. (2002) Evaluation of disintegration time of rapidly disintegrating tablets via a novel method utilizing a CCD Camera. Chem Pharm Bull (Tokyo) 50: 1181-1186.

10. Schiermeier S, Schmidt PC (2002) Fast dispersible ibuprofen tablets. Eur J Pharm Sci 15: 295-305.

11. Siewert M, Dressman J, Brown CK, Shah VP (2003) FIP/AAPS Gidelines for Dissolution/In vitro release testing of novel/special dosage forms, Dissolution Technologies. AAPS PharmSciTech 4: article 7.

12. Seager H (1995) Drug delivery products and the Zydis fast dissolving dosage form. J Pharm Pharmacol 50: 375-382.

13. Mizumoto T, Tamura T, Kawai H, Kajiyama A, Itai S (2008) Formulation design of an oral, fast-disintegrating dosage form containing taste-masked particles of Famotidine. Chem Pharm Bull (Tokyo) 56: 946-950.

14. Chang RK, Guo X, Burnside B, Couch R (2000) Fast-dissolving tablets. Pharm Technol 24: 52-58.

15. Biradar SS, Bhaavati ST, Kuppasad IJ (2006) Fast dissolving drug delivery systems: A brief overview Int J Pharmacol 4: 2.

16. Banker GS, Anderson NR (1987) Tablets. In: Lachman L, Lieberman HA, Kanig JL, ed. The theory and practice of industrial pharmacy. 3rd ed. Mumbai: Varghese Publishing House, 182-184, 296-303, 311-312.

17. Indian Pharmacopoeia (1996) New Delhi, Controller or Publication 2: 555-556.

18. Gohel MC, Bansal G, Bhatt N (2005) Formulation and evaluation of orodispersible taste masked tablets of Famotidine. Pharma Biol World 3: 75-80.

7

Degradation Study of Phenazin Neutral Red from Aqueous Suspension by Paper Sludge

Elaziouti Abdelkader[1]* Laouedj Nadjia[2] and Bekka Ahmed[1]

[1]LCPCE Laboratory, Faculty of sciences, Department of industrial Chemistry, University of the Science and Technology of Oran (USTO M.B). BP 1505 El M'naouar 31000 Oran, Algeria

[2]Dr. Moulay Tahar University, Saida, Algeria

Abstract

The potential ability of paper sludge to remove neutral red (NR) from aqueous solution was investigated using UV-visible and FTIR spectroscophotometers. The stability of the optical properties of NR was assessed in terms of evolution of the main absorption bands of NR in aqueous solution and in NR/incinerated sludge suspension as a function of pH and exposure time. The results showed that, the adsorption kinetic of NR onto incinerated paper sludge was fast as a result of 60.09% removal efficiency obtained within 80 minutes at pH 5. The adsorption reaction was perfectly described by pseudo- second-order kinetic model. The profile of isotherm adsorption of NR onto incinerated sludge was S-shape type. Incinerated sludge exhibited excellent performance for adsorption of NR with a maximum of 374. 98 mg/g. The equilibrium adsorption data were well fitted with Freundlich model. FTIR analysis revealed the strong interaction forces operating on heterogeneous surface of the incinerated sludge between the dissociation of the oxygenated groupings, which are in general acid functional, and dimethylamine goup $[-N^+(CH_3)_2]$ of the NR dye which could be used to explain the high adsorption capacity of cationic dye onto incinerated sludge. These findings can support the design of remediation processes and also assist in predict their fate in the environment.

Keywords: Incinerated sludge; Kinetic; Isotherm; Neutral red; UV visible; FTIR

Introduction

The industry of paper rejects significant amounts of the waste whose setting in discharge was the means the elimination of simplest and cheapest. One distinguishes several dies from assessment of sludge including: the setting as a cover for discharges, the sludge incineration at a prohibitory cost and present a risk related to the pollutant gas impact on the environment such as that of the dioxin (ADEME, 1999 /CE) [1], energy valorization (production of biogas like source of heat and of electricity), biological or agricultural valorization (production; manure and of compost) and valorization in the sector of building. The choice of a die must be dependent on the cost of installation, origin of sludge, added-value of the product which results from it and the impact which the die retained on the environment could have. The setting discharges some (also called storage) proves to be a technique little developing and is legally prohibited in many countries (directive 1999/31/CE) [2].

Dyes effluents released (approximately 7. 10^5 tons) into the environment by technological activities pose a serious threat to the environment. Their presence in water, even at very low concentrations, may significantly affect photosynthetic activity in aquatic life due to reduced light penetration.

Due to their synthetic origin and complex aromatic molecular structures, which make them more stable, non-biodegrade, conventional methods of treatment of the aqueous solutions containing dye (the oxido-reduction and the exchanging resins of ions [3], coagulation/ flocculation [4], membrane separation [5], the biological methods [6] and more recently the advanced processes of oxidation [7], do not allow obtaining threshold of pollution lower or equal to the maximum permissible concentration (MPC) imposed by the environmental recommendations. Adsorption is one of the highly efficient methods to remove colored textile contaminants from wastewaters. The adsorption appeared a very effective method for the reduction of the color in aqueous mediums. Many low-cost adsorbents for dye removal from mineral waste [8], agricultural wastes, microbial biomass [9], higher plant biomass [10], tree fern [11], orange peel [12], date pits [13], palm kernel fiber, sawdust [14], peanut hull [15], neem leaf [16], de-oiled soya [17], moss Rhytidiadelphus squarrosus [18], activated carbon [19] rice husk-based porous carbon [20] magnetic particles [21] and paper sludge [22]. The aim of the present study was to assess the potential ability of locally available and highly efficient paper sludge for removal of neutral red as dye model from the aqueous solutions. The stability of the absorption band maxima of NR in aqueous solution and in NR/ incinerated sludge suspension were assessed in terms of evolution the main absorption bands and adsorption efficiency as a function of pH solution using UV-visible spectroscopy monitoring. The equilibrium and kinetics of adsorption of Neutral red from aqueous solution to incinerated sludge were investigated. Adsorption kinetic reaction was determined quantitatively by the pseudo-first- and second-order models. The equilibrium adsorption isotherm data were analyzed with Langmuir and Freundlich. The incinerated sludge untreated and treated with NR were characterized by FTIR spectroscopy.

***Corresponding author:** Elaziouti Abdelkader, LCPCE Laboratory, Faculty of sciences, Department of industrial chemistry, University of the Science and the technology of Oran (USTO M.B). BP 1505 El M'naouar 31000 Oran, Algeria

Materials and Methods

The sludge of purification plant of industrial water "workshop 22" of production unit GIPEC (Saida, Algeria) was used in this study. The sludge sample of paper mill was washed, dried with the air during several days and then incinerated at 250°C during 2 hours in a muffle furnace. The material resulting from the incineration of sludge was filtered. A phenazin dye, Neutral red, NR (CI 50040, MW=319.50 g.mol^{-1}, λ max=520 nm, ε =25000 cm^{-1} mole^{-1}dm^{-3}) and a thiazin group cation, Methylene blue, MB (MW= 319,5 g.mol^{-1}, λ_{max}=665 nm, ε =95000cm^{-1} mole^{-1}dm^3) from Across product for microbiological analysis and used without any further purification. Molecular structures of basic dyes: (a) Methylene blue (MB) and of Neutral red (NR) are shown in Figure 1.

Aqueous dye solution stock was prepared by dissolving accurately weighed neat dye in distilled water to the concentration of 0.1g/L. Experimental solutions were obtained by successive dilutions.

The initial pH of the dye solutions was adjusted by buffer solutions of NaOH /HCl (0.1M) using (WTN: WISSENSCHAFLLICH TECHNISECHE WERKSTÄTTEN; weilehein Allemagne pH-330) digital pH-meter. The incineration of the paper sludge sample was carried out in a muffle furnace (Nabertherm, ZAH 2002). X-ray fluorescence was performed on a spectrometer of mark (Oxford). The sample was prepared (pearl borated) and was subjected to a source of X-radiation of fluorescence characteristic of its chemical composition. The separation of adsorbent /adsorbate suspension was performed by the centrifuge (EBA-Hetlich) at 4000 rpm for 10 min. Vis-absorption spectra of the dye in aqueous solution and adsorbed on incinerated sludge were obtained by (Model: UV – 2401 (PC) SHIMADZU – corporation spectrometer) in the range 350-800 nm, using 1cm optical pathway cells at maximum wavelength of dye. Fourier transformed infrared (FTIR) spectra were measured in dispersed incinerated paper sludge in KBr pellets (1/200 w/w) with Perkin-Elmer spectrometer in the range 4000- 400 cm^{-1}, with resolution of 4°.

The effect of initial pH was performed on incinerated paper sludge suspension in 50 mg/L of dye, (solid /liquid ratio of 1g/L) and over a range of pH values from 2 to 12. The initial pH values of solutions were adjusted with 0.1M HCl / NaOH solutions. The suspensions were stirred for 80 mn, and then separated by centrifugation. Dye concentrations in the supernatant solution were estimated by UV-visible spectrophotometer at maximum wavelength of NR dye.

Batch kinetic adsorption experiment was carried out in flasks containing 1L of NR solutions with defined concentrations. pH value of solutions was adjusted at optimum pH solution (pH=5). An amount of 1g/L of incinerated paper sludge was then added and flask was agitated at 298 K. Samples were taken from flasks at the predeterminated time intervals and adsorbent was separated from the NR/incinerated sludge

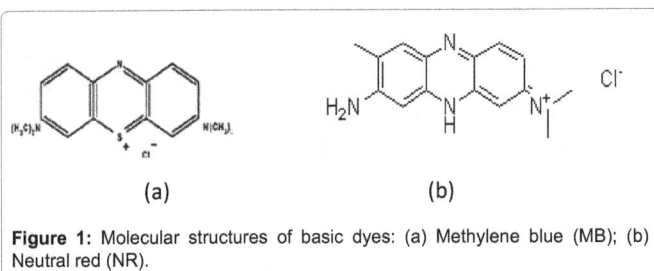

Figure 1: Molecular structures of basic dyes: (a) Methylene blue (MB); (b) Neutral red (NR).

SiO$_2$	Al$_2$O$_3$	Fe$_2$O$_3$	CaO	MgO	Na$_2$O	K$_2$O	SO$_3$	SiO$_2$/Al$_2$O$_3$	LF
14.12	9.84	2.34	26.5	1.76	0.42	0	3.9	1.43	41.93

pH	CEC (mg/g)		CEC (meq/100g)		TSS (m^2/g)	
8.5	75.61		23.66		122.53	

LF: Loss of Fire CEC: Cation Exchange Capacity TSS: Total Specific Surface

Table1: Chemical composition and main properties of the incinerated paper sludge.

suspension by centrifugation. Dye concentration in the supernatant was estimated by UV-visible spectrophotometer.

The removal efficiency η (%) and the amount of dye adsorbed per unit weight of incinerated paper sludge at equilibrium Q (mg/g) were calculated based on following equations Equation (1) and Equation (2):

$$\eta(\%) = \frac{(C_0 - C_e)}{C_e}100 \tag{1}$$

$$Q\left(\frac{mg}{g}\right) = \frac{(C_0 - C_e)}{m}\upsilon \tag{2}$$

where C$_o$ and C$_e$ are the initial dye concentrations and the equilibrium dye concentration (mg/L); V is volume of the solutions (L), and m is the weight of the incinerated paper sludge (g).

Results and Discussion

X-radiation of fluorescence analysis

Table 1 displays the results of chemical composition in the incinerated sludge. The major elements of the incinerated paper sludge sample, expressed in terms of oxide, are primarily of silica, alumina, lime and a strong water content with the moderate presence of the elements such as sodium, magnesium, potassium and iron. The loss of fire raised about 41.93%, sign of the important presence of the organic matter and calcite. The value of pH of the incinerated sludge is estimated at 8.5. This alkalization of the suspension can be attributed to the progressive dissolution of the carbonates (calcite) basically present in incinerated sludge.

Cation exchange capacity and total specific surface

In order to quantify the Cation Exchange Capacity (CEC) and Total Specific Surface (TSS) of MB dye on incinerated paper sludge, the equilibrium adsorption isotherm was carried out using Methylene blue methods [23]. As elucidated in Figure 8, the profile of the MB isotherm is L-3 type, is obtained when the polymoléculaires layers appear only when surface is almost entirely covered with a monomolecular layer [24]. The CEC and SST values were of 75.61 meq /100gr of incinerated sludge and 122.53 m^2/g respectively.

Fourier transformed infrared analysis

As shown in Figure 8, the spectrum of the pure incinerated sludge allowed identifying the characteristic absorption bands of the acid functional groupings of the incinerated sludge such as cellulose (3409.9, 2929.7, 2500, 1427 and 1103.2 cm^{-1}), kaolinite (3693.4 cm^{-1} (OH), 912 cm^{-1} (Al-OH), 700 cm^{-1}, 468.7cm^{-1} (Si-O) and 538.1cm^{-1} (Si-O-AL)), calcite (2515, 1797, 1427.2, 875.6 and 710 cm^{-1}) and talk (675 cm^{-1}). The band at 1630 cm^{-1} is also attributed to physical adsorbed water.

Effect of initial pH solution of NR dye

Qualitative study: The evolution of absorption band maxima of NR in aqueous solution and in NR/ incinerated paper sludge suspension at various pH.

Table 2 illustrates the position and the evolution of the main absorption bands of NR in aqueous solution and upon adsorption on incinerated paper sludge suspension at various pH.

The visible spectra of NR dye, in aqueous solution at various pH values, exhibits two forms: NR^+H and NR, which are in equilibrium with each other. NR, which has absorption bands in the range of 451-453.5 nm under neutral and alkali conditions (pH> 6), whereas NR^+H, is found to lie between 519.5 and 521.5 nm under acidic media (pH ≤6).

As seen in Figure 2, the interactions of NR (NR^+H and conjugate base form, NR) with incinerated sludge differ in certain aspects from those in aqueous solutions. The adsorption process of NR^+H in NR/ incinerated sludge suspension under acidic media (pH= 2.41-6.04) was accompanied by a blue shift of the main absorption bands to shorter wavelengths. The highest shift was observed in pH range of 4.26-6.04. However, under alkali conditions (pH=8.15-12.1) the adsorption profile of NR has a nearly constant absorption band maxima. The shift did not exceed 3nm over the pH range of 8.15-12.1, but is higher for

Initial pH	NR /Aqueous solution Species (nm)		NR /Incinerated sludge suspension Species (nm)		Equilibrium pH
	NR^+H	NR	NR^+H	NR	
2.41	519.5	-	518.5	-	2.62
4.26	520	-	503	-	5.82
5.24	519.5	-	501	-	6.34
6.04	521.5	-	503	-	6.55
8.15	-	452.5	-	453	7.02
10.14	-	451	-	454	7.74
12.01	-	453.5	-	450.5	11.45

NR: Conjugate base form; NR^+H: protonated cation of NR dye

Table 2: The position of the absorption band maxima of NR in aqueous solution and upon adsorption on incinerated sludge as a function of pH for [NR] =50mg/L, [Incinerated sludge]=1g/L, T=298K and contact time of 80 min.

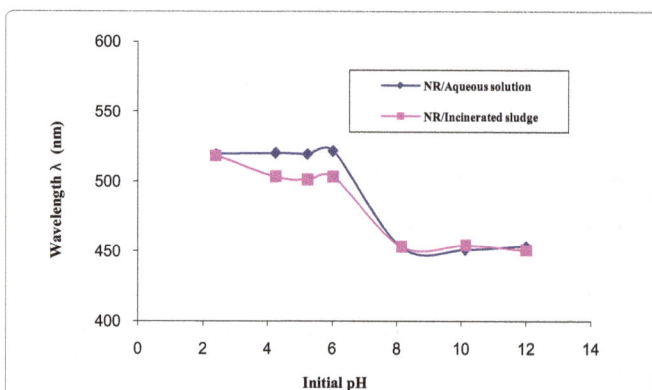

Figure 2: Effect of the pH on the absorption band maxima of NR in aqueous solution and upon adsorption on incinerated sludge for [NR] =50 mg/L, [incinerated sludge]=1g/L, T=298K and contact time of 80 min.

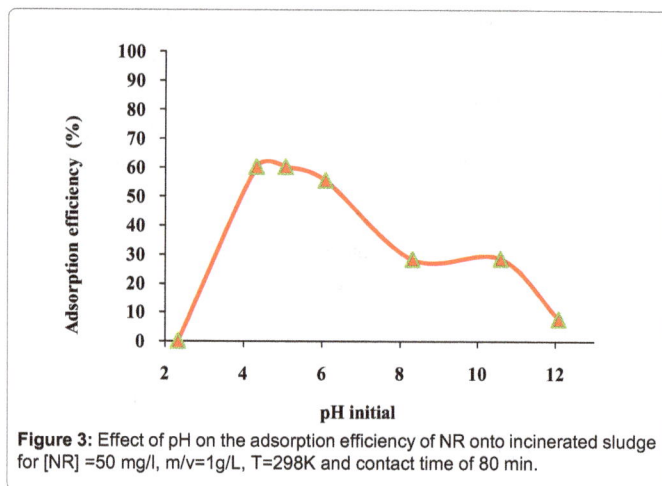

Figure 3: Effect of pH on the adsorption efficiency of NR onto incinerated sludge for [NR] =50 mg/l, m/v=1g/L, T=298K and contact time of 80 min.

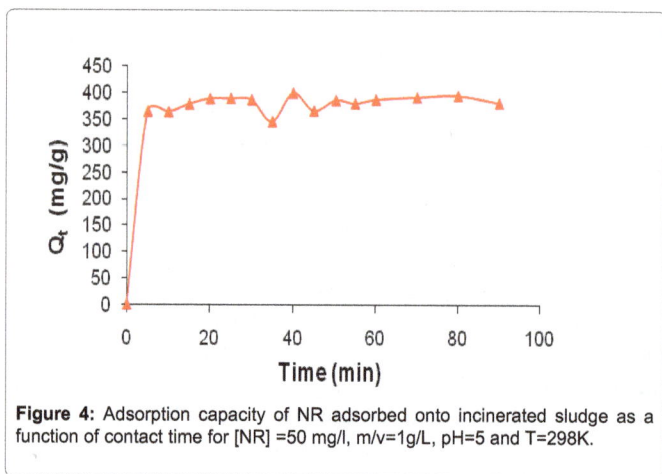

Figure 4: Adsorption capacity of NR adsorbed onto incinerated sludge as a function of contact time for [NR] =50 mg/l, m/v=1g/L, pH=5 and T=298K.

NR^+H over pH ranged between 4.26 and 6.04, being 17, 18.5 and 18.5 nm at pH 4.26, 5.24 and 6.04 respectively. Besides, the results showed a slight increase in pH in NR/incinerated sludge suspension under acid conditions, while these of NR/incinerated sludge were decreased under alkali conditions.

These observations are consistent with chemical reaction between NR^+H monomer cations and dissociated acid functional groupings of incinerated sludge surface and the change in the environmental polarity/acidity of negatively-charged incinerated sludge surface with respect to water surrounding under alkali conditions respectively.

Quantitative study: The evolution of adsorption efficiency of NR in NR/ incinerated paper sludge suspension at various pH: It is commonly accepted that in adsorbate/adsorbent systems, the potential of the surface charge is determined by the activity of ions (e.g. H^+ or pH). The effect of pH solution on the adsorption efficiency of NR onto incinerated sludge was studied over a range of pH values of 2 to12. The pH significantly affected the adsorption rate of NR dye. As shown in Figure 3, the dye adsorption efficiency increased as the initial pH was increased from 2 to 5, and then decreased significantly to reach a percentage of 7. 65% to pH = 12. However no adsorption is observed with pH =2. The maximum dye adsorption of 60.09% of NR was achieved with an initial pH of 5. For this reason, pH 5 is selected for subsequent experiments. The high adsorption percentage observed at pH=5 is due to the presence of dissociated acid functional groupings, which are in general oxygenated functions, on the incinerated sludge surface which boost the attractive interactions between dissociated groupings and NR cations, consequently improving the masse transfer and adsorption rate [25].

Adsorption kinetic

The adsorption kinetics of NR on incinerated paper sludge was measured in the range of 0 to 80 min, by varying the equilibrium time between adsorbate/adsorbent. The adsorption capacity of NR adsorbed on incinerated sludge as a function of contact time was illustrated in Figure 4. The results indicated that the process is found to be very rapid initially, and the equilibrium is achieved within 5 min. The maximum efficiency of 66.78 % was obtained within 80 min at a pH=5. On the other hand, results shown in Figure 5 and 6 indicated that the temperature and pH of system increased as the initial dye concentration was increased within contact time, which can be explained by an endothermic and chemical adsorption as a consequence of an exchange phenomenon between organic cations and incinerated paper sludge respectively. The adsorption data were treated with a pseudo-first order and pseudo-second order kinetic reaction models.

The pseudo-first order equation of Lagergren [26] is generally expressed as follows Equation (3):

$$\frac{dQ_t}{dt} = K_1(Q_e - Q_t) \tag{3}$$

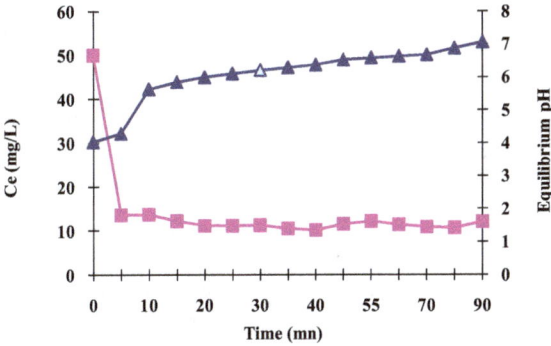

Figure 5: Evolution of initial dye concentration and pH as a function of contact time during the adsorption of NR onto incinerated sludge for [NR] =50 mg/L, pH=5, m/v=1g/L and T=298K.

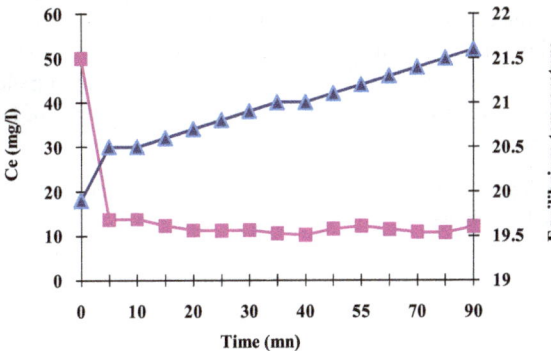

Figure 6: Evolution of initial dye concentration and temperature as a function of contact time during the adsorption of NR onto incinerated sludge for [NR] = 50 mg/L, pH=5, m/v=1g/L and T=298K.

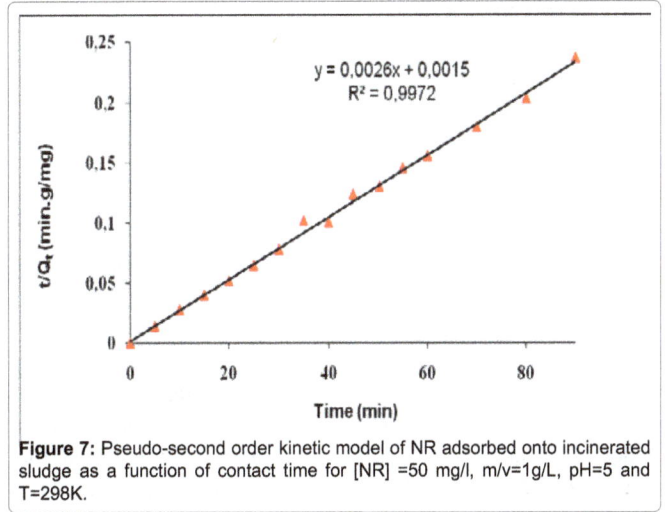

Figure 7: Pseudo-second order kinetic model of NR adsorbed onto incinerated sludge as a function of contact time for [NR] =50 mg/l, m/v=1g/L, pH=5 and T=298K.

System	Experimental result	Pseudo-second order model			Pseudo-first order model		
	Q_{exp} (mg/g)	Q_{max} (mg/g)	K_2 (g/mg,mn)	R^2 (%)	Q_e (mg/g)	K_1 (min^{-1})	R^2 (%)
NR / Incinerated paper sludge suspension	401. 936	384.615	0.0169	0.999	67.050	0.0700	0.899

R^2: Regression coefficient

Table 3: Kinetic parameters of NR dye adsorbed onto incinerated paper sludge for [NR] =50mg/L, [Incinerated sludge] =1g/L, pH =5, T=298K and contact time =80min.

where Q_e and Q_t (both in mg/g) are the amount of dye adsorbed per unit weight of adsorbent at equilibrium and at any time t, respectively and K_1 (L/min) is the rate constant for adsorption of dye. At given boundary conditions for t=0, Q_t=0, the equation Equation (3) can be integrated to give Equation (4):

$$\log_{10}(Q_e - Q_t) = \log_{10}(Q_e) - \frac{K_1}{2.303}t \tag{4}$$

The values of k_1 were calculated from the slopes of the respective linear plots of log (Q_e – Q_t) versus t. The regression coefficients, R^2, (given in Table 3) for the pseudo-first–order model did not exceed the values of 0.90. The calculated Q_e value obtained from pseudo-first-order kinetic model was much different compared with experimental Q_{exp} values. These results suggest that the process does not follow the pseudo-first-order adsorption rate equation of Lagergren.

The sorption kinetic following pseudo-second order model given by Ho [27] is represented in the form Equation (5):

$$\frac{dQ_t}{dt} K_2(Q_e - Q)^2 \tag{5}$$

where Q and Q_e represent the amount of dye adsorbed (mg/g) at any time t; K_2 is the rate of sorption (g/mg.min) and Q_e the amount of dye adsorbed onto incinerated sludge at equilibrium (mg/g). Separating Equation (6), gives:

$$\frac{dQ}{d(Q_e - Q)^2} K_2 dt \tag{6}$$

Integrating Equation (5) with respect to the boundary conditions Q = 0 at t = 0 and Q = Q at t = t, the linearised form of pseudo second order expression can be obtained as Equation (7):

$$\frac{1}{(Q_e - Q)} = \frac{1}{Q_e} + K_2 t$$

(7)

Equation (7) can be further linearised to Equation (8):

$$\frac{t}{Q_t} = \frac{1}{K_2 Q e^2} + \frac{t}{K_2}$$

(8)

The linearity of the plots of t/Q_t versus t for adsorption of NR on incinerated sludge (Figure 7 and Table 3) suggests that all the adsorption data were satisfactorily described by pseudo-second order model ($R^2 = 0.999$), based on the assumption that the rate–limiting step may be chemisorptions involving valency forces through sharing or exchange of electrons between the hydrophilic edge sites of incinerated sludge and polar dye ions [28]. The calculated Q_e value obtained from pseudo-second-order kinetic model was close to the experimental Q_{exp} value.

Adsorption isotherm

Figure 8 shows the adsorption isotherms of the dyes using incinerated sludge. The shape of the NR isotherm does not show the characteristic plateau of a monolayer in the range of the concentration used. This isotherm can be classified as an (S-shape) isotherm according

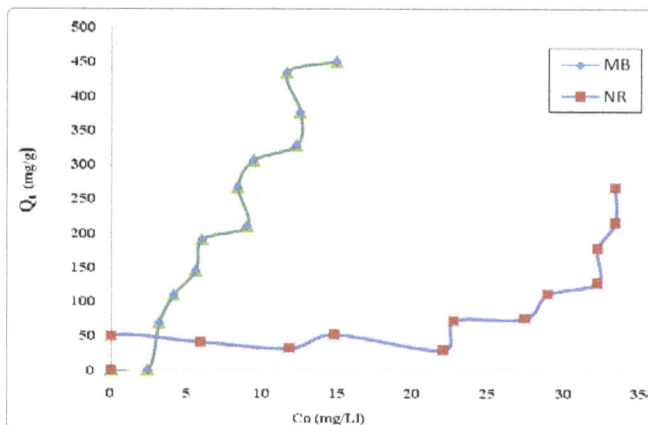

Figure 8: Adsorption isotherms of NR and MB dyes onto incinerated sludge for [NR] =5-60 mg/L, pH=5 (NR) and 7 (MB), m/v=1g/L, T=298K and contact time 80 min.

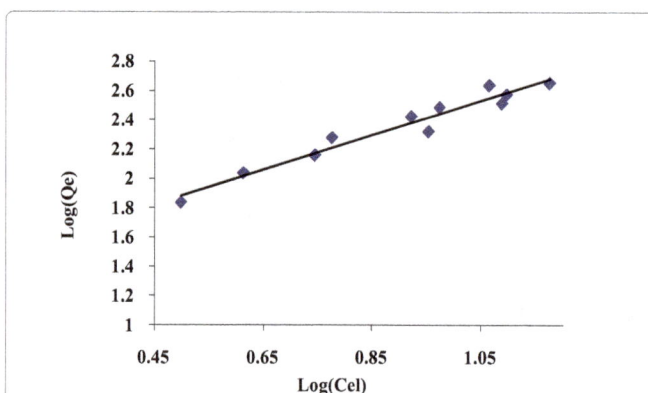

Figure 9: Freundlich isotherm of NR onto incinerated sludge for [NR] =5-60 mg/L, pH=5, m/v=1g/L, T=298K and contact time 80 min.

System	Experimental Results	Langmuir model $C_e/Q_e =1/(K_L Q_{max}) + C_e/Q_{max}$			Freundlich model $lnQ_e = lnK_F + (1/n) lnC_e$		
	Q_{exp}	Q_{max}	K_L	R^2	K_F	1/n	R^2
NR / Incinerated paper sludge suspension	(mg/g)	(mg/g)	(L/g)	(%)	(L/g)		(%)
	374.982	-344.827	-0.0451	35.74	23,4693	1.3746	95.17

R^2: Regression coefficient

Table 4: Adsorption isotherm parameters of NR dye adsorbed onto incinerated sludge for [NR] =5-60mg/L, [Incinerated sludge]=1g/L, pH =5, T=298K and contact time =80min.

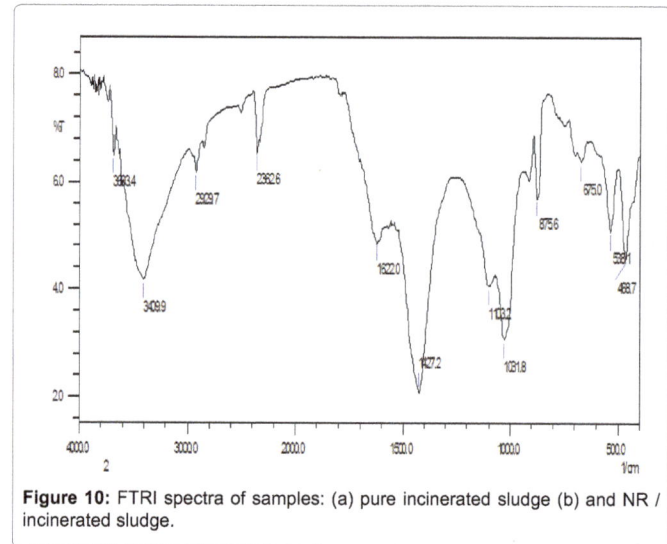

Figure 10: FTRI spectra of samples: (a) pure incinerated sludge (b) and NR / incinerated sludge.

to Giles classification system [29]. The equilibrium adsorption capacity increased more slowly with the increase in initial dye concentration and the maximum adsorption of 374.98 mg/g was obtained for 50 mg/L of NR.

Langmuir [30] and Freundlich [31] isotherms were applied to assess the performance of the adsorption process of NR onto incinerated paper sludge. The linearized equation Equation (9) of Langmuir is given as follows:

$$\frac{Ce}{Qe} = \frac{1}{K_2 Q_{max}} + \frac{Ce}{Q_{max}}$$

(9)

where Q_{max} (mg/g) is the maximum amount of the dye per unit weight of incinerated sludge to form a complete monolayer coverage on the surface bound at high equilibrium dye concentration C_e (mg/L) and K_L (L/g) is the Langmuir constant related to the affinity of binding sites. A plot of C_e/Q_e versus C_e leads to a straight line with the slope of $1/Q_{max}$ and an intercept of $1/Q_{max}K_L$.

The logarithmic form of Freundlich equation Equation (10) is expressed as follows:

$$LnQe = LnK_F + \frac{1}{n} LnCe$$

(10)

where Q_e (mg/g) is roughly an indicator of the adsorption capacity and (1/n) of the adsorption intensity. Values n > 1 represent a favorable adsorption condition. 1/n and K_F (L/g) can be determined from the

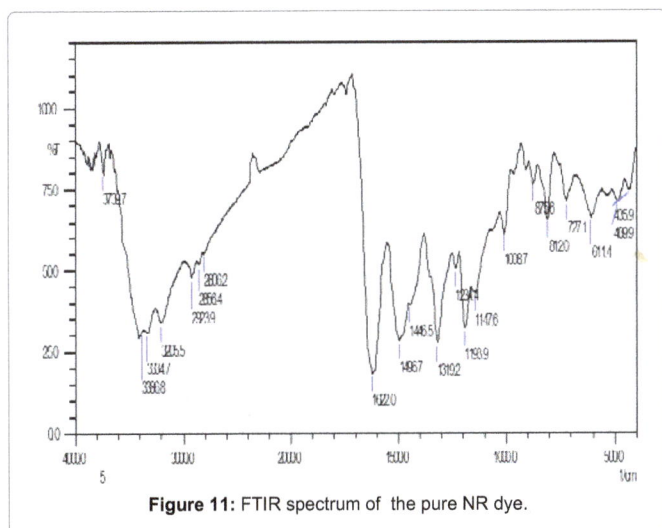

Figure 11: FTIR spectrum of the pure NR dye.

linear plot of $\ln Q_e$ versus $\ln C_e$ (Figure 9). Langmuir and Freundlich isotherms parameters for the adsorption of NR on incinerated sludge were computed in Table 4.

Freundlich model was fitted to the adsorption data with highly significant coefficients of regression ($R^2 > 0.95$). The value of $1/n$ ($1/n > 1$) reveals the nature and strength of adsorptive forces involved indicating the existence of strong adsorption forces operating on heterogeneous surface incinerated sludge.

FTIR analysis

In order to investigate the interactions between NR and incinerated sludge, FTIR analysis was conducted. The FTIR spectra of the incinerated sludge untreated and treated with NR (Figure 10) show a great similarity of some absorption bands with some differences in the intensity. The assignment and interpretation of the bands are the following: a less intense band at 2855 cm^{-1}, assigned to symmetrical aliphatic carbon (-CH$_2$) and a second weak band near 435 cm^{-1} (no identified) corresponding to the NR dye (Figure 11). These results indicated a specific interaction between NR dye molecules and the incinerated sludge surface [32].

Conclusion

Adsorption of neutral red (NR) from aqueous solution using an incinerated paper sludge was investigated using UV visible and FTIR spectrophotometers. Results show that the adsorption efficiency of NR on incinerated sludge was significantly affected by the pH medium. The adsorption process of NR$^+$H on incinerated sludge under acidic media was accompanied by a blue shift of the main absorption bands. The highest shift was observed in pH range of 4.26-6.04. However, under alkali conditions (pH=8.15-12.1) the adsorption profile of NR has a nearly constant main bands. The uptake increased more slowly with increase in initial dye concentration. The adsorption reaction was perfectly described by pseudo-second-order kinetic model. Incinerated sludge exhibited excellent performance for adsorption of NR with a maximum of 374. 98 mg/g. The adsorption data were well fitted with Freundlich adsorptive model. FTIR analysis revealed the strong interaction forces operating on heterogeneous surface of the incinerated sludge between the dissociation of the oxygenated

groupings, which are in general acid functional, and dimethylamine goup [-N+(CH$_3$)$_2$] of the NR dye which could be used to explain the high adsorption capacity of cationic dye onto incinerated sludge.

Acknowledgments

This work was supported by the University of Dr Moulay Tahar, Saida, Algeria

References

1. ADEME Arthur Andersen (1999) Situation du recyclage agricole des boues d'épuration urbaines en Europe et dans divers autres pays du monde.

2. Directive (1999/31/CE) concerning the setting in discharge of waste. The Council of April 26.

3. Dusart O, Serpaud B (1991) La tribune de l'eau. 44:15-22.

4. Linsheng Z, Dobias B (1992) Water Treatment. 7: 221-232.

5. Ciardelli G, Corsi L, Marucci M (2001) Membrane separation for wastewater reuse in the textile industry Resources. Conservation and Recycling 31:2: 189-197.

6. Paprowicz J, Sodczyk S (1988) Application of biologically activated sorptive columns for textile waste water treatment. Environmental Technology. 9: 271-280.

7. Milano JC, Loste-Berdot P, Vernet JL (1995) Photooxydation du Vert de Malachite en Milieu Aqueux en Presence de Peroxyde D'Hydrogene: Cinetique et Mecanisme Photooxidation of Malachite Green in Aqueous Medium in the Presence of Hydrogen Peroxide: Kinetic and Mechanism. Environmental Technology16: 329-341.

8. Yener J, Kopac TDogu Dogu T (2006) Adsorption of basic Yellow 28 from aqueous solutions with clinoptilolite and amberlite. Journal of Colloid and Interface Science 294:257.

9. Aksu Z (2005) Application of biosorption for the removal of organic pollutants: a review. Process Biochemistry 40: 999.

10. Ho YS, Mckay GA (1999) Kinetic study of dye sorption by biosorbent waste product pith. Resources Conservation and Recycling 25: 173.

11. Ho YS, Yuh-Shan C, Tzu-Hsuan H, Yu- Mei (2005) Removal of basic dye from aqueous solution using tree fern as a bioadsorbent. Process Biochemistry 40: 119.

12. Arami M, Yousefi L, Nargess M, Niyaz M, Tabrizi NS (2005) Removal of dyes from colored textile wastewater by orange peel adsorbent: equilibrium and kinetic studies. Journal of Colloid and Interface Science 2: 374.

13. Banat FAl-asheh S, Al-makhadmeh L (2003) Evaluation of the use of raw and activated date pits as potential adsorbents for dye containing waters. Process Biochemistry 39: 195.

14. Garg VK, Kumar R, Gupta R (2004) Removal of malachite green dye from aqueous solution by adsorption using agro-industry waste: a case study of Prosopis cineraria. Dyes and Pigments 62: 3.

15. Gong R, Sun Y, Chen J, Liu H, Yang C (2005) Effect of chemical modification on dye adsorption capacity of peanut hull. Dyes and Pigments 67: 178.

16. Bhattacharyya KG, Sarma A (2003) Adsorption characteristics of the dye, Brilliant green, on Neem leaf powder. Dyes and Pigments 57: 212.

17. Mittal A, Krishnan L, Gupta VK, (2005) Removal and recovery of malachite green from wastewater using an agricultural waste material, de-oiled soya. Separation and Purification Technology 43: 125.

18. Lucia R, Martin P, Miroslav H, Jozef A (2009) Sorption of cationic dyes from aqueous solutions by moss Rhytidiadelphus squarrosus: Kinetic and equilibrium studies. Nova Biotechnologica 9: 1.

19. Arivoli S (2009) Adsorption of malachite green onto carbon, prepared Borassus Bark. The Arabian Journal for Science and Engineering 34: 2A.

20. Guo Tao NNY, Zhang H, Liu YH, Qi J R, Wang ZC, et al. (2003) Adsorption of malachite green and iodine on rice husk-based porous carbon Mater. Chem Phys 82: 110.

21. Safaric I, Safarikova M, Vrchotova N (1995) Study of sorption of triphenyl methan dyes on a magnetic carrier bearing an immobilized copper phtalocyanine dye. Collect. Czech Chem Commun 60: 35.

22. Elaziouti A, Laouedj N, Segheir AA (2010) Preparation and Evaluation of New Adsorbent (Paper Sludge) in the Treatment of Waste Water of Textile Industry. Journal of the Korean Chemical Society 54: 116.

23. Pavan PC, Crepaldi EL, Valim JB (2000) Sorption of Anionic Surfactants on Layered Double Hydroxides. Journal of Colloid and Interface Science 229-347.

24. Chitour SE (1981) Chimie de surfaces. Introduction à la catalyse (2ndedn) O.P.U, Alger.

25. Boudrahe F, Benissad FA (2007) Valorisation d'un déchet solide vue de la récupération des ions de plomb en solution aqueuse. CIGP'07.

26. Lagergren S (1898) About the theory of so-called adsorption of soluble substances. Kungliga Svenska vetenskapsakademiens. Handlingar. 1–39.

27. Ho YS, McKay GA (1998) A comparison of chemisorptions kinetic models applied to pollutant removal on various sorbents. Process Saf Environ 332–340.

28. Allen SJ, Koumanova B (2005) Decolourisation of water/wastewater using adsorption (REVIEW). Journal of the University of Chemical Technology and Metallurgy 40: 177.

29. Giles C H, Mac Ewan TH, Nakhwa S N, Smith D J (1960) Studies in adsorption. Part XI.A system of classification of solution adsorption isotherms, and its use in diagnosis of adsorption mechanisms and measurements of specific areas of solids. Journal of Chemical Society 3973–3993.

30. Langmuir I (1918) The adsorption of gases on plane surfaces of glass, mica, and platinum. Journal Ameican of Chemical society 40: 11361.

31. Chen Z, Xing B, McGill WB (1999) A Unified Sorption Variable for Environmental Applications of the Freundlich Equation. J Environ Qual 28:1422-1428.

32. Arbeloa FL (2002) Spectroscopic characterization of the adsorption of rhodamine 3B dye in smectite-type nanoparticles. Recent Res Devel Nanostructures: 1-23.

8

Estimate of Effective Diffusivity Starting from the Phenol Adsorption Profiles on an Activated Carbon in Discontinuous Suspension

Wahid Djeridi* and Abdelmottaleb Ouederni

Laboratory of Research: Engineering Processes and Industrials Systems (LR11ES54), National School of Engineers of Gabes, University of Gabes, Gabes, Tunisia

Abstract

This study consists on developing a model of internal diffusion with adsorption in a porous particle in order to estimate its effective diffusivity. For this purpose, the particle is put in suspension in an isothermal perfectly agitated reactor in transient state (closed system). The adopted model is based on the adsorption equilibrium on the internal porous surface and assumes that the external concentration varies with time in the external transfer resistance absence. The proposed model equations are numerically solved using the finite differences technique. Experimental concentration profiles of adsorption of dyes either on a commercial activated carbon are smoothed to suit the proposed model. A good agreement between experimental theory profiles is obtained.

Keywords: Kinetic adsorption; Media porous; Numerical solution; Diffusion coefficient

Notation

c: Pore fluid solute concentration, mg/l

C : Fluid phase solute concentration, mg/l

C_0 : Initial concentration, mg/l

K_L : Langmuir adsorption parameter, l/mg

D_e : Diffusion coefficient, cm²/s

K_L : Mass transfer coefficient, cm/s

q_m : Adsorption capacity per unit adsorbent volume, mg/cm³

r : Radial coordinate for particle, cm

R_p : Particle radius, cm

T : Time, mn

V : Solution volume, cm³

V_p : volume of particles, cm³

m : solid mass, mg

γ : intercept of tangent to adsorption isotherm

ε_p : particle porosity

ρ : apparent density particles, g/cm³

Introduction

In literature kinetic adsorption in porous middle has occupied a topic place in researches. Such a phenomen can be caused in solution contains active carbon (porous middle). Many important processes in chemical engineering include modeling of diffusion within particles or fluid spheroids. Accurate, but still computationally simple, solutions to these diffusion equations are, therefore, of great importance. Diffusion in adsorption processes has been subject to many studies since the classical approximate solution by Glueckauf [1]. In that solution, the mass transfer within the spherical particles is described by assuming that the mass- transfer rate depends linearly on the difference between the average concentration within the sphere and the surface concentration, along with the assumption of a constant (time -independent) mass-transfer coefficient. Since then that assumption has been used widely in the adsorber modeling. Furthermore, It has been shown that the linear driving force assumption (LDF) is equivalent to assuming a concentration profile within the particles. Both these approaches are used to ease the complicated solution of the time-dependent material balances in the absorbers [2-6]. In this paper, we research the solution of diffusion equation with the numerical methods and analytic methods.

Diffusion Model

Modeling of diffusion within spherical particles is often an important part of modeling of many processes relevant to chemical engineering such as adsorption. The accurate solution to the differential equations describing diffusion is quite a complicated task involving calculation of infinite series. In view of the highly porous structure of the adsorbent studies and of the relatively low adsorption capacity for different colorant, the pore diffusion model was used to fit the data. We assume the particles are spherical and of radius R_p, that they have an intraparticle porosity ε_p, and that local equilibrium exists at each point within a bead between the pore liquid and the adsorbent surface. The external mass transfer resistance was neglected.

With these assumptions the model is represented by the following conservation equations and boundary conditions. For the particles

$$[\varepsilon_P + (1 - \varepsilon_P) \frac{\partial q}{\partial c}] \frac{\partial c}{\partial t} = \frac{\varepsilon_P}{r^2} D_e \frac{\partial}{\partial r} \left(r^2 \frac{\partial c}{\partial r} \right) \qquad (1)$$

In this equation, D_e and q represents the effective pore diffusivity and the concentration in the adsorbed phase respectively.

*Corresponding author: Wahid Djeridi, Laboratory of Research: Engineering Processes and Industrials Systems (LR11ES54), National School of Engineers of Gabes, University of Gabes, Gabes, Tunisia

With the boundary conditions

$$r = 0, \frac{\partial c}{\partial r} = 0 \tag{1a}$$

$$r = R_p, \ c = C \tag{1b}$$

$$t = 0, c = 0 \tag{1c}$$

'c' represents the concentration in the pore fluid.

At first time we considered that the equilibrium is instantaneous and the concentration in the pore fluid c equal at the initial concentration C_0.

We showed that the adsorption of coloring was well represented by the Freundlich isotherm. Thus, the bracketed term on the left-hand side of eq. (1) can be approximated as

$$\frac{\partial c}{\partial t}\left[1 + \frac{1 - \varepsilon_p}{\varepsilon_p} K_F\right] = \frac{D_e}{r^2} \frac{\partial}{\partial r}\left(r^2 \frac{\partial c}{\partial r}\right) \tag{2}$$

When K_F represent the Freundlich constant.

The implicit method of Crank-Nicolson was used to solve the model equations numerically for different values of D_e and we used Matlab logitiel. If we represent the rapport of concentration with the time, we find the subsequent ensuing for various diffusion constant (Figures 1-4).

When the equilibrium does not happen instantantaneously and the

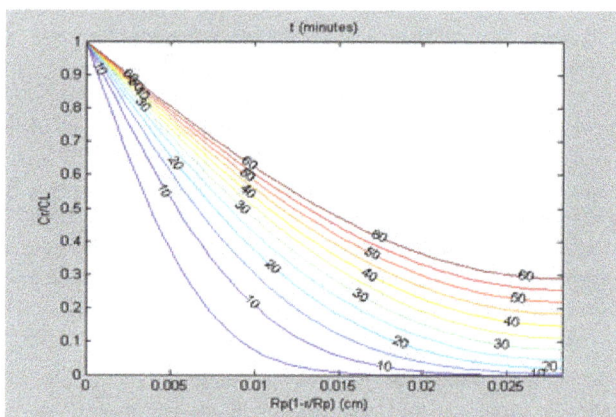

Figure 1: Intrapartical kinetic adsorption with Diffusion coefficient $D_e = 1 * 10^{-7} cm^2/s$.

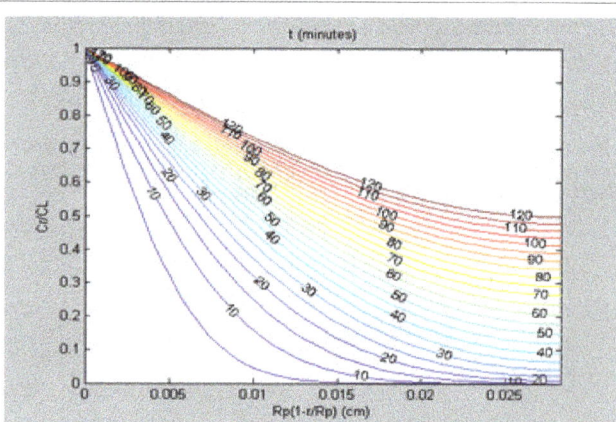

Figure 2: Intrapartical kinetic adsorption with Diffusion coefficient $D_e = 8 * 10^{-8} cm^2/s$.

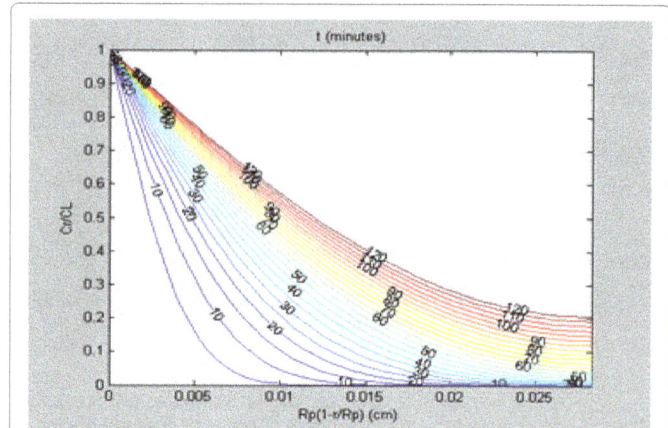

Figure 3: Intrapartical kinetic adsorption with Diffusion coefficient $D_e = 4 * 10^{-8} cm^2/s$.

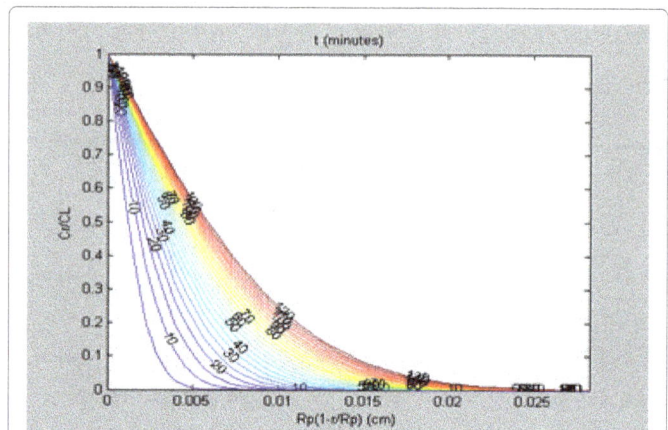

Figure 4: Intrapartical kinetic adsorption with Diffusion coefficient $D_e = 1 * 10^{-8} cm^2/s$.

concentration in the external fluid C variable varies with the time, the model is represented by the conservation equations (1) and boundary conditions (1a), (1b), (1c), and by the equation for the liquid in the vessel

$$\frac{dC}{dt} = -a_p \varepsilon_p D_e \left(\frac{\partial c}{\partial r}\right)_{r=R_p} \tag{3}$$

t=0, C=C$_0$ \hfill (3a)

In these equations, $a_p = 3V_p/VR_p$ is the surface area of the particles per unit volume of liquid and C is the concentration in the external fluid. At equilibrium c=C and, thus, the term in brackets on the left-hand side of eq. (1) can be obtained from the adsorption isotherms. When the adsorption isotherm is linear, (1) and (3) can be solved analytically. The adsorption isotherm is represented by $[\varepsilon_p C + (1 - \varepsilon_p) q] = mC + \gamma$, where m and γ are constants, we obtain

$$\frac{q}{q_\infty} = 1 - 6B(1+b) \sum_{k=1}^{\infty} \frac{\exp(-p_k^2 D_e t / mR_p^2)}{B^2 p_k^2 + 9(1+B)} \tag{4}$$

Where

$$\tan p_k = \frac{3 p_k}{3 + B p_k^2} \tag{4a}$$

In these equations, q$_\infty$ =B(mC$_0$ + γ)/(1+B) and B= V/mV$_p$.

Where the isotherm is non-linear and a finite concentration step is applied, however, a numerical solution is required in general. In this part we showed that the adsorption of coloring was well represented by the langmuir isotherm. Thus, the bracketed term on the left-hand side of eq.(1) can be approximated as

$$\frac{\partial c}{\partial t}\left[1+\frac{1-\varepsilon_p}{\varepsilon_p}\frac{K_L.q_m}{\left(1+K_Lc\right)^2}\right]=\frac{D_e}{r^2}\frac{\partial}{\partial_r}\left(r^2\frac{\partial c}{\partial r}\right) \qquad (5)$$

Where K_L represent Langmuir adsorption parameter.

q_m represent adsorption capacity per unit adsorbant volume.

Validation of the Model

Several work was interested in the study of the kinetics of adsorption of organic molecules or mineral ions in solution by a porous solid. As example work of Grzegorczyk et al. [7], on the adsorption of an amino acid by a porous solid with the use of the operating conditions specified by Table 1 gave very important results. We propose to use these results to test and validate the model developed in our work. Grzegorczyk et al. [7], also used the method of orthogonal collocation based on the use of the finite element to numerically solve the model of the equations of diffusion for various values of the coefficient of diffusion D_e.

In the first simulation, the experimental results published by Grzegorczyk et al. [7], are smoothed by our model for various values of effective diffusivity. It is noticed that this model is in good agreement with the experiments of Grzegorczyk and et al. [7].

Figures 5- 7 represent smoothing by our kinetic model of the experiments of Grzegorczyk by three types of activated carbon.

The values of effective diffusivity thus estimated by our model and the model of Grzegorczyk et al. [7], for the three types of the activated carbon, are represented in Table 2.

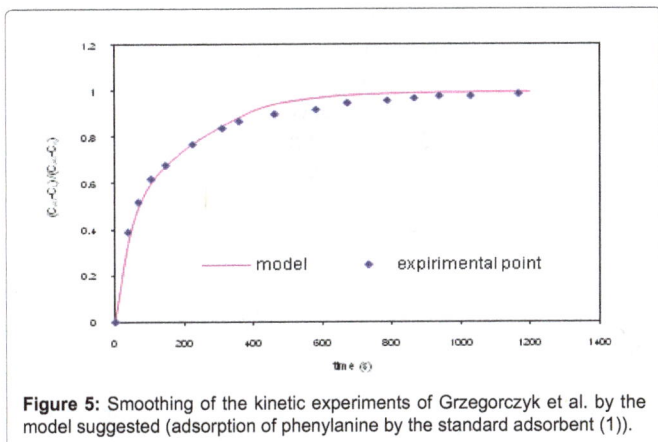
Figure 6: Smoothing of the kinetic experiments of Grzegorczyk et al. by the model suggested (adsorption of phenylanine by the standard adsorbent (2)).

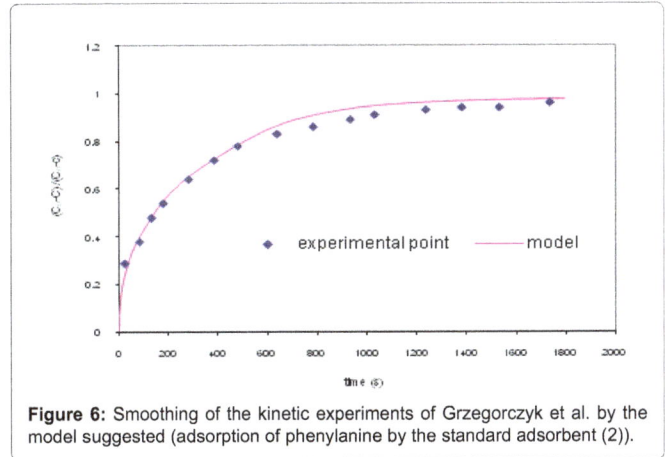
Figure 7: Smoothing of the kinetic experiments of Grzegorczyk et al. by the model suggested (adsorption of phenylanine by the standard adsorbent (3)).

Adsorbent	activated carbon (1)	activated carbon (2)	activated carbon (3)
r_p (mm)	0.283	0.302	0.241
ε_p	0.62	0.51	0.66
C_0 (mol/l)	25×10^{-3}	25×10^{-3}	25×10^{-3}
K_L (l/mol)	34	27	69
W (mg)	10	10	10
V (l)	0.1	0.1	0.1
q (mol/g)	0.77×10^{-3}	1.42×10^{-3}	1.22×10^{-3}

Table 1: Operating conditions of the adsorption of phenylalanine on various adsorbents [7].

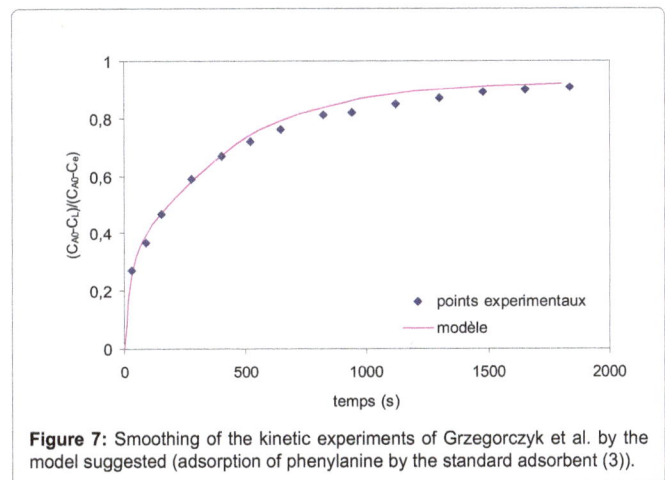
Figure 5: Smoothing of the kinetic experiments of Grzegorczyk et al. by the model suggested (adsorption of phenylanine by the standard adsorbent (1)).

For the three types of the activated carbon, the values of effective diffusivity estimated by the model of Grzegorczyk et al. are of the same order of magnitude as those estimated by our model, which proves in additional the great agreement between our model and also the model of Grzegorczyk et al. [7].

Adsorption of Phenol on Activated Carbon of Norit Type

For elimination by adsorption of phenol in aqueous mediums, one uses in this part an activated carbon of Norit type of average diameter equalizes with 940 µm, the volume of the aqueous phenol solution used is 800 ml to which one adds a mass W=1600 Mg of activated carbon to a temperature fixed θ=40°C. With the same method the implicit method of Crank-Nicolson was used to solve the model equations numerically for different values of D_e and we used Matlab logitiel, the values of R_p, ε_p, q_m, K_L, ρ, V, and V_p are given for the experimental resulting find about Najjar [8], with Norit activated carbon. One represents the evolution of the relative fraction of solution adsorbable of phenol by Norit consistent with time for various concentrations initial C_0, one obtains Figures 8-10.

It is noted that our model is in concord with the experimental points. One notes the values of effective diffusivity estimated by our model for various initial concentrations C_0 of phenol in Table 3.

For the adsorption of phenol on activated carbon of Norit type, when initial concentration C_0 increases the value of the effective

	Standard activated carbon (1)	Standard activated carbon (2)	Standard activated carbon (3)
Effective diffusivity estimated by Grzegorczyk et al. (cm²/s).	1.5×10^{-6}	0.86×10^{-6}	0.52×10^{-6}
Effective diffusivity estimated by our model (cm²/s).	3.5×10^{-6}	1.5×10^{-6}	1×10^{-6}

Table 2: Estimate of the effective diffusivity for the three types of activated carbon.

Initial phenol concentration in solution C_0 (mg/l)	100	50	25
Effective diffusivity D_e (cm²/s)	9×10^{-8}	3×10^{-7}	2×10^{-6}

Table 3: Estimate of the effective phenol diffusivity in activated carbon produced starting from Norit.

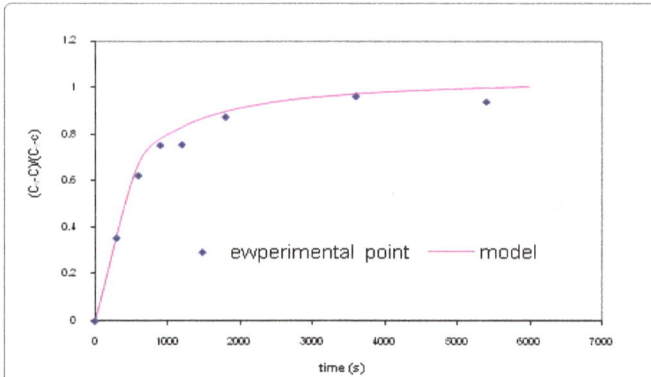

Figure 8: Fraction relating of aqueous solution adsorbed to the balance of phenol by Norit for an initial concentration C_0=25 mg/l.

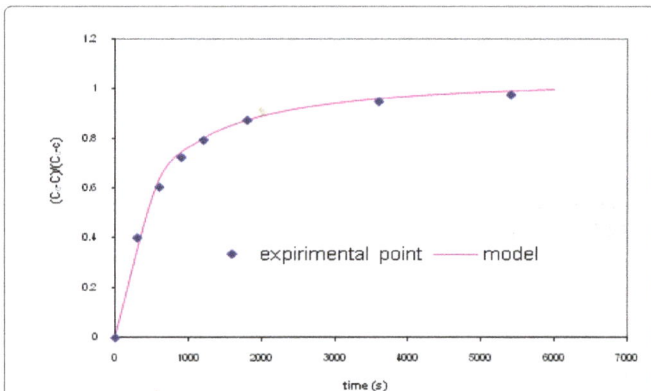

Figure 9: Fraction relating of aqueous solution adsorbed to the balance of phenol by Norit for an initial concentration C_0=50 mg/l.

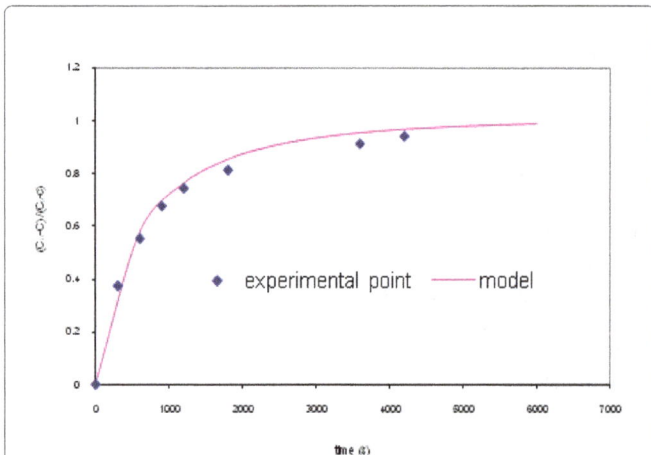

Figure 10: Fraction relating of aqueous solution adsorbed to the balance of phenol by Norit for an initial concentration C_0=100 mg/l.

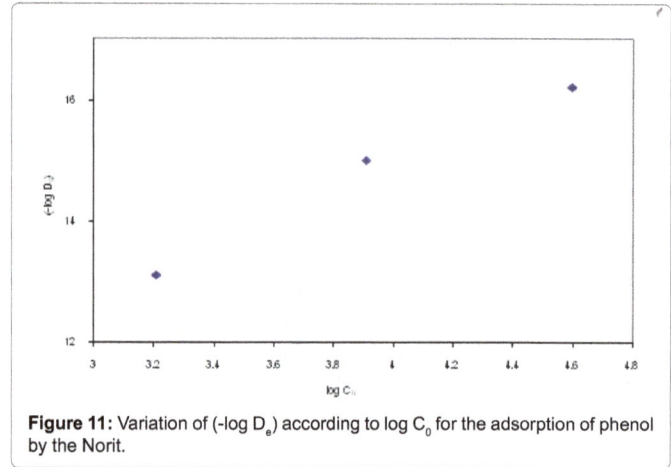

Figure 11: Variation of (-log D_e) according to log C_0 for the adsorption of phenol by the Norit.

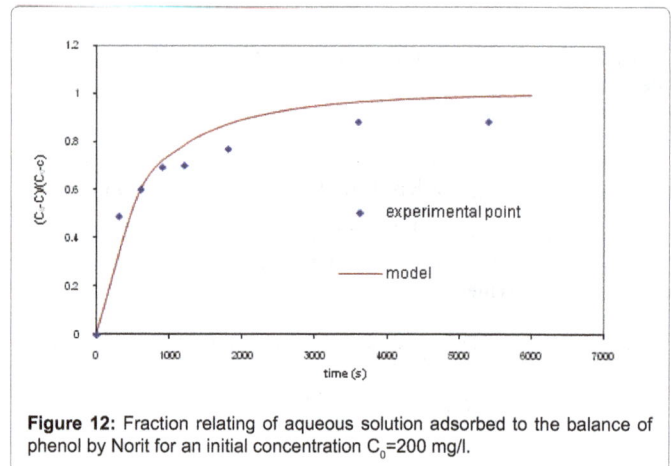

Figure 12: Fraction relating of aqueous solution adsorbed to the balance of phenol by Norit for an initial concentration C_0=200 mg/l.

diffusivity D_e decreases, then one can deduce that effective diffusivity depends on the initial concentration. Figure 11 represents the evolution of ($- \log D_e$) according to log C_0.

For the adsorption of phenol on produced activated carbon of Norit type, the effective diffusivity De varies in opposite direction with initial concentration C_0. One can deduce that effective diffusivity depends on the initial concentration. So now one represents the relative fraction of adsorbed phenol aqueous solution by Norit according to time for great initial concentrations phenol C_0 in the solution, then one obtains Figures 12 and 13. It is noted that for the high initial phenol concentrations in solution C_0= 200 mg/l and C_0= 300 mg/l, the model does not follow the experimental points perfectly.

Conclusion

This work represents a numerical study of kinetic adsorption in porous solid of particles, when the adsorption isotherm is linear the diffusion equations can be solved analytically but when the isotherm is non-linear, a numerical solution is required in general. Several methods are used to obtain the solution, for example the method implicit of Crank-Nicolson. The numerically resulting of diffusion model

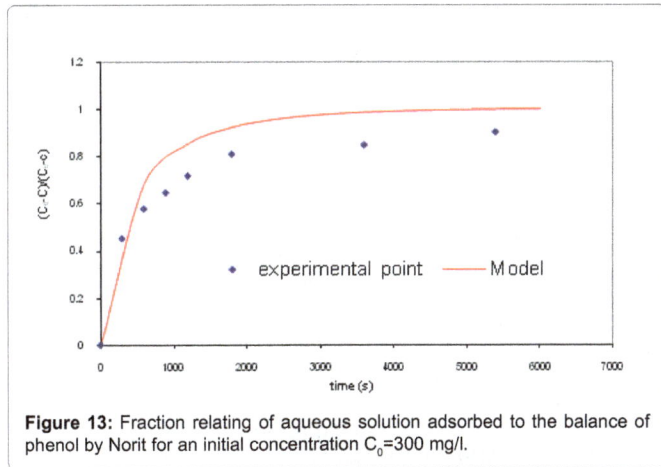

Figure 13: Fraction relating of aqueous solution adsorbed to the balance of phenol by Norit for an initial concentration $C_0=300$ mg/l.

calculated were then compared with the experimental and the diffusion coefficient D_e that smoothed experimental curve.

References

1. Glueckauf E (1955) Theory of chromotography. Part 10-Formulae for Diffusion into Spheres and Their Application to Chromatography. Trans Faraday Soc 51: 1540-1551.

2. Goto M, Smith JM, McCoy BJ (1990) Parabolic Profile Approximation (Linear Driving Force Model) for chemical Reactions. Chem Eng Sci 45: 443-448.

3. Ching CB, Lu ZP (1998) Parabolic Intraparticle Concentration Profile Assumption in Modeling and Simulation of Non linear Simulated Moving-Bed Separation Processes. Chem Eng Sci 53: 1313-1318.

4. Haas OW, Kapoor A, Yang RT (1988) Confirmation of Heavy-Component Rollup in Diffusion-Limited Fixed-Bed Adsorption. AICHE J 34: 1913-1915.

5. Liaw CH, Wang JSP, Greenkorn RA, Chao KC (1979) Kinetics of Fixed-Bed Adsorption : A New Solution. AICHE J 25: 376-381.

6. Li Z, Yang RT (1999) Concentration Profile for Linear Driving Force Model for Diffusion in a Particle. AICHE J 45: 196-200.

7. Grzegorczyk DS, Carta G (1996) Adsorption of amino acids on porous polymeric adsorbents-II. Intraparticle mass transfer. Chem Eng Sci 51: 819-826.

8. Najar S (2004) Fabrication et caractérisation du charbon actif à partir de grignons d'olives, thèse, Ecole Nationale d'Ingénieur de Gabes.

9

Solidification/Fixation of Nickel Ions in Ordinary Portland Cement

Minocha AK and Manish Kumar Goyal*

Environmental Science and Technology Division, Central Building Research Institute, and NIPER, Punjab, India

Abstract

In this research article we represented the results of the studies of addition of varying concentrations of Nickel ions on the physical, chemical and engineering properties of 43 grade ordinary Portland cement. The studies regarding different physical & chemical parameters of OPC like initial and final setting time of cement, bulk density, and compressive strength, microscopic and spectroscopic properties have been carried out and the result of these studies are presented and discussed in the manuscript. Different building products containing varying concentrations of added nickel ions were prepared and subjected to hydrologic environment to investigate the leaching behavior of nickel ions. The efforts have also been made to establish a quantitative co-relation between the concentration of added Nickel ions and the intensity of any effect in above properties of the cement also the possible fixation mechanism of Nickel ions in cement was suggested. Scanning electron microscopy and X- Ray diffraction study of nickel containing cement samples show that nickel have no greater effect on the cement properties. Ni does not show any effect on the hydration of cement & its main compounds but it slightly retards the setting process. Some amount of Nickel is found to distributed throughout to the C-S-H gel phase Nickel get adsorbed the interstitial phases of the cement and formed the insoluble nickel compound. The C-S-H phase of Nickel containing samples found denser than that of control samples.

Keywords: Cement; Heavy metals; Nickel; Setting time; Waste management

Introduction

Industrialization is vital to a nation's socio- economic development as well as its political stature in the international committee of nations. Industries vary according to process technology, sizes, nature of products, characteristics and complexity of wastes discharged. Ideally citing of industries should strike a balance between socio-economic and environmental considerations. Rapid industrial development and the world global growth have led to the recognition and increasing understanding of interrelationship between pollution, public health and environment. While almost industrial activities cause some pollution and produce waste, relatively few industries (without pollution control and waste treatment facilities) are responsible for the bulk of the pollution. It has been reported that industrial effluent has a hazard effect on water quality, habitat quality, and complex effects on flowing waters. Industrial wastes and emission contain toxic and hazardous substances, most of which are detrimental to human health. This type of waste is generated from metallurgical, mining, chemical, leather industry, distillery, sugar, battery, electroplating and pigment industry These include heavy metals such as nickel lead, cadmium and mercury, and toxic organic chemicals such as pesticides, PCBs, dioxins, polyaromatic hydrocarbons (PAHs), petrochemical and phenolic compound. Generally the metal concentration in the waste is too low for economic recovery but high enough to represent toxicity hazards. Therefore, these industries discharge their waste without treatment or with improper treatment in the landfills. Its disposal in hydrologic environment can cause environmental risk due to the mobility of metal ions. These metal ions get mixed in surface water and ground water system, which may be detrimental to human being as well to the environment [1-10]. Various technologies have been developed which purport to render a waste or to reduce the potential for the release of toxic species into the environment. This technique is used to transform potentially hazardous liquid or solid waste into less hazardous and or non-hazardous solid before disposal in a landfill, thus preventing the waste from contaminating the environment. The United State Environmental Protection Agency also recognizes cementitious Solidification/Stabilization as "the Best Demonstrated Available Technology for land disposal of most toxic elements. Many formulations have been developed for the S/S process according to the kinds of wastes. Portland cement can be modified for suitable S/S process using flyash, lime slag soluble silicates etc. One of the most difficult problems in this process is that the hydration of cementitious materials is too retarded to set and harden enough due to the inhibition of hydration reaction of heavy metals in a landfill area. It seems that Solidification/Stabilization (S/S) process would be the best practical technology to treat the nickel containing waste [11-31]. There is very less data reported in the literature about the effects of addition of nickel ions on various properties of cement. Therefore, efforts have been made to fill void in this data. The results of these studies on the effects of addition of nickel on various properties of ordinary Portland cement are presented and discussed in this manuscript. A quantitative co relation between the concentration of nickel added and the intensity of any effect has been established.

Materials, Apparatus and Methods

Materials

To make the standard solutions of different concentrations of nickel metal ions for preparing cement samples, Nickel Nitrate ($Ni(NO_3)_2.6H_2O$) was used from Thomas Baker (Chemicals) Ltd. The metal salt was used as obtained without any further purification. Double distilled water was used to prepare the metal ion solutions throughout the study. Commercial ordinary Portland cement 43 grade was used.

*****Corresponding author:** Manish Kumar Goyal, Environmental Science and Technology Division, Central Building Research Institute, Roorkee-247667 (UA), NIPER, Punjab, India

Constituent	Weight percentage
Silica	20.8
Aluminum oxide	4.4
Iron oxide	3.79
Calcium oxide	66.1
Magnesium oxide	3.3
Anhyd. Sulfuric acid	3
Sodium oxide	0.2
Potassium oxide	0.7

Table 1: Chemical composition of cement.

Physical parameter	Results
Loss on ignition	0.70%
Consistency	31.58%
Soundness	1.0 mm
Bulk density	1.421g/cm³
Initial setting time	175 min
Final setting time	300 min

Table 2: Physical properties of cement.

Apparatus

Atomic Absorption Spectrophotometer (AAS) from Hitachi (Z-7000) was used to determine the metal ion concentrations. Hazardous waste filtration system from Millipore (YT-30142 HW) was used to carry out Toxicity Characteristic Leaching Procedure as recommended by United State Environment Protection Agency (USEPA); UV-Visible spectrophotometer (Aquamate) from Thermo Corporation was used for spectrophotometric studies. Vicat apparatus was used to determine initial and final setting time of control cement as well as metal ion doped cement samples. Compressive strength testing machine from Central Scientific Instruments Company was used to determine the compressive strength of mortar samples. Scanning Electron Microscope (SEM) from LEO 438VP, UK was used for the microstructure visualization of fractured cement surfaces. X-Ray powder diffraction analysis was carried out by using the instrument from Rikagu, Japan to identify the crystalline phases present in the control as well as in metal containing cement samples.

Methods

The chemical analysis of the cement was carried out according to Indian standard specifications IS: 4032:1985 guidelines. The physical properties were tested according to the Indian standard specification IS: 4031:1996 guidelines. The chemical composition and physical properties of Ordinary Portland Cement used are summarized in Tables 1 and 2 respectively.

Preparation of cement pastes and mortars: The Initial and final setting times of blank as well as the cement samples, containing varying concentrations of nickel ions, were determined according to IS: 4031:1996. The results obtained for controls were compared with those obtained for nickel containing samples to know the effect of addition of nickel on the setting time of cement paste. To investigate the effect of addition of nickel on engineering properties like compressive strength, bulk density of mortar samples containing cement and flyash, were cast in 2.78" cubic iron molds. The samples were demolded after 24 hours and were dipped into water for curing the compressive strength of the cubes were determined on 3, 7, 28, 60, 90, 180 and 360 day of curing. Six samples were taken each time and the average value of these results was reported (shown in peak Table 3).

X-ray diffraction analysis: It is possible to find and quantify crystalline components which have over 1% abundance in a sample

if the correct experimental technique is used. XRD can give a semi quantitative or quantitative analysis of the components of the crystalline fraction. Thus XRD gives most information from the crystalline components of the material under investigation. X-ray diffraction analysis was performed with a Rikagu, Japan (Dmax 2200 VK/PC) automated X-ray diffractometer. The samples were crushed to a fine powder with porcelain and sieved through a 45 μm sieve. The XRD scan was made with copper kα radiation from 3°-70° 2θ with 0.02° step width and 1 to 3 s counting time.

Scanning Electron Microscopy: Scanning Electron Microscopy (SEM) is a technique used to know the fractured surface of cementitious material and three dimensional particle level morphology. It is generally used to magnify an image. It can magnify image up to three lack times. There are Three types of rays emitted from the instruments that is back scattered electron rays, secondary rays. Both are used for imaging. And X-rays are used for elemental analysis of the sample. It is used for

2-Theta	d(A)	BG	Height	I%	Area	I%	FWHM
9.134	9.674	227	103	14.8	1419	12.6	0.234
15.759	5.619	174	113	16.2	992	8.8	0.149
18.04	4.913	169	359	51.4	5376	47.9	0.255
18.938	4.682	163	89	12.8	1030	9.2	0.197
22.902	3.88	157	149	21.3	2448	21.8	0.279
26.271	3.39	180	62	8.9	1248	11.1	0.342
26.679	3.339	189	152	21.8	1493	13.3	0.167
29.4	3.036	208	563	80.7	10613	94.6	0.32
30.024	2.974	228	69	9.9	643	5.7	0.158
31.015	2.881	220	88	12.6	1025	9.1	0.198
32.14	2.783	235	474	67.9	11220	100	0.402
32.539	2.749	244	698	100	11108	99	0.271
34.082	2.628	217	517	74.1	10563	94.1	0.347
35.057	2.558	186	64	9.2	1050	9.4	0.279
36.676	2.448	177	83	11.9	648	5.8	0.133
37.256	2.412	168	77	11	1080	9.6	0.238
38.665	2.327	166	56	8	512	4.6	0.155
39.417	2.284	156	105	15	2603	23.2	0.421
40.88	2.206	173	101	14.5	1538	13.7	0.259
41.222	2.188	158	243	34.8	6064	54	0.424
41.578	2.17	161	85	12.2	2479	22.1	0.496
43.183	2.093	166	55	7.9	741	6.6	0.229
44.098	2.052	162	64	9.2	1562	13.9	0.415
44.622	2.029	158	62	8.9	1926	17.2	0.528
45.758	1.981	154	100	14.3	1711	15.2	0.291
47.159	1.926	158	225	32.2	5996	53.4	0.453
49.885	1.827	172	49	7	770	6.9	0.267
50.318	1.812	161	53	7.6	2172	19.4	0.697
50.801	1.796	169	97	13.9	1248	11.1	0.219
51.732	1.766	148	142	20.3	1517	13.5	0.182
54.358	1.686	148	75	10.7	519	4.6	0.118
55.265	1.661	140	57	8.2	631	5.6	0.188
56.401	1.63	142	98	14	1504	13.4	0.261
57.264	1.608	149	44	6.3	271	2.4	0.105
59.984	1.541	146	49	7	903	8	0.313
60.589	1.527	140	50	7.2	1072	9.6	0.364
62.239	1.49	132	134	19.2	2921	26	0.371

Table 3: Peak Search Report (37 Peaks, Max P/N = 11.4) [Sam.7_1.raw] Sam. Nickel1000ppm
PEAK: 21-pts/Parabolic Filter, Threshold=3.0, Cutoff=0.1%, BG=3/1.0, Peak-Top=Summit

S. No.	Concentration of metal ion (ppm)	IST	FST
1	10	175	300
2	50	175	300
3	100	175	300
4	500	180	300
5	1000	190	310
6	2000	190	310
7	5000	205	330
8	7000	205	335
9	8000	205	335
10	10000	205	335

IST- Initial setting time of cement in min
FST- Final setting time of cement in min

Table 4: Effect of fixation of nickel ions on initial and final setting time of cement.

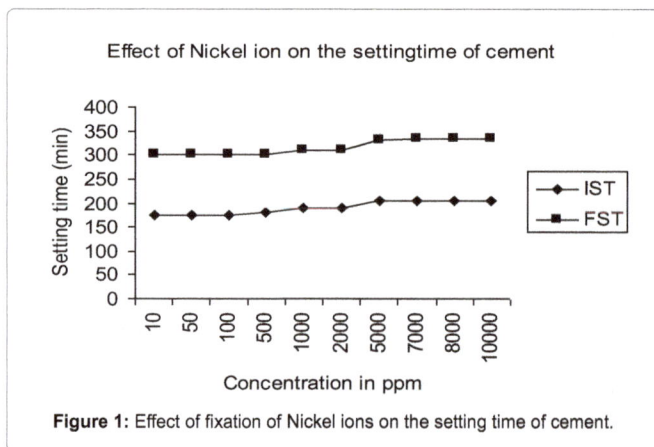

Figure 1: Effect of fixation of Nickel ions on the setting time of cement.

Binder system	No. of days	Average bulk density (± 0.001 g/cm³) nickel		
		Control	500 ppm	1000 ppm
Cement + fly ash	1	2.303	2.303	2.303
	28	2.381	2.359	2.388

Table 5: Average bulk density values of solidified products on 1 and 28 days of curing.

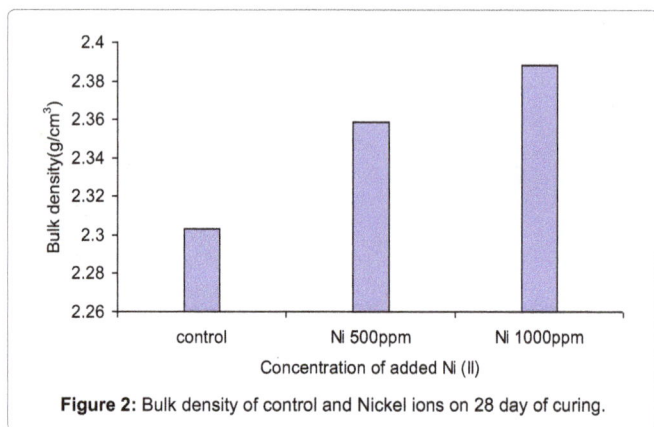

Figure 2: Bulk density of control and Nickel ions on 28 day of curing.

conducting material for non-conducting and biological material like cement coat a conducting material on it. It may be carbon or gold.

Leaching studies: A suitable way to know the effectiveness of the immobilization of contaminants after solidification process is to perform the extraction test. Usually this test is performed under the chosen condition of leaching medium. Standard method No.1311

recommended by United State Environment Protection Agency is followed. Crushed solid material has been taken in hazardous waste filtration system with zero head space extractor. A measured volume of water was added and this assembly has been shaken for 18 hours continuously in agitation assembly. Filtered extract was collected in a closed vessel and analyzed for metal concentration by using Atomic Absorption Spectrophotometer. The results were compared with that of blank samples to know the actual enhancement of the particular metal in the leachate. It was observed that the leachate of the samples doped with 500 ppm and 1000 ppm nickel containing 3.0 ppb and 5.0 ppb nickel in the TCLP leachate in water medium while in acid medium the leaching is somewhat high as 0.25 ppm and 0.65 ppm of nickel is leached out in acid medium. It was found that Nickel mainly incorporated in solid solutions combined with Calcium oxide (CaO) only a little amount of nickel was found in di calcium silicate (C_2S) and tri calcium silicate (C_3S). Nickel has only marginal effects on the formation and hydration of its main compounds. Thus the complete nickel added to the cement mortar is get adsorbed through the matrix and very low leaching is due to the precipitation of nickel.

Results and Discussion

Although hydration of cement begins immediately as water is added into it but there is a period of time-several hours in which cement remains in fluid like state before setting as a rigid, load bearing material. Setting refers the change of state from fluid paste to rigid form. According to IS: 8112:1996. The initial setting time of a 43 grade ordinary Portland cement should be at least 30 min. and final setting time should be within 600 min. The initial and final setting time of control as well as nickel containing samples were determined and presented in Table 4. Nickel was added in the concentrations 10, 50, 100, 500, 1000, 2000, 5000, 7000, 8000 and 10000 ppm respectively the addition of Ni (II) ions upto a concentration of 500 ppm caused no significant effect either on initial setting time or final setting time of cement. Both these parameters increased significantly when the concentration range of Ni (II) ions was increased from 500-7000 ppm as the initial setting time of cement paste bearing 7000 ppm Ni (II) ions was increased to 205 min. and final setting time was increased to 335 min. in comparison with blank sample which shows initial setting time and final setting time as 175 and 300 min. respectively. The retarding effect of Ni (II) ions gets constant when the concentration of Ni (II) is increased. Based on these results it may be stated that Ni (II) ions work as retarder in the concentration range 500-7000 ppm. All the experiments were carried out in triplicate to assure accuracy and reproducibility. All the samples were observed for any visual change before the compressive strength testing each time. There was no significant change found in colour and shape except the surface of nickel containing samples were smoother than that of control samples (Figure 1).

Bulk density of control samples as well as nickel ions containing samples was determined on 28 day of curing. Six samples were taken each time and the average value of these observations was reported. There was no significant change in this parameter for nickel ions containing sample in comparison with control (Table 5). Bulk density of control samples was 2.381 ± 0.001 g/cm³, while it changed to 2.359, and 2.388 ± 0.001 g/cm³ for 500 and 1000 ppm, nickel containing samples respectively at 28th day of curing (Figure 2). The figure shows that bulk density values of nickel containing samples increased with the increase of concentration of nickel although the change is in significant. Compressive strength of control mortar samples as well as samples containing nickel ions was determined on 3, 7, 28, 60, 90, 180 and 360 days of curing. The results of this study are presented in Table 6. It is

Added Nickel ions (ppm)	No. of days							% change in compressive strength in comparison of control						
	3	7	28	60	90	180	360	3	7	28	60	90	180	360
Control	323.4	436	486	570	570	620	620							
500	305	426	445	580	580	600.6	600	94.3	97.7	91.5	101.7	101.7	97.8	96.7
1000	320	425	426	590	600	632	660	98.9	97.5	87.6	103.5	105.2	101.9	106.4

Table 6: Compressive strength (kg/cm^2) of mortar samples.

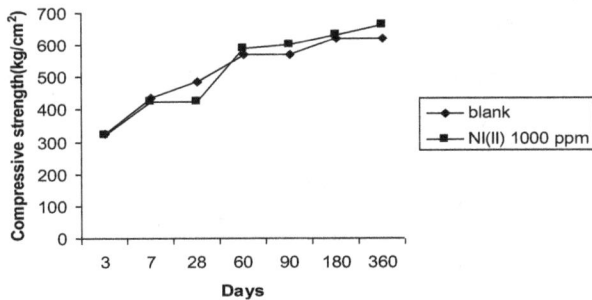

Figure 3: Effect of addition of 1000 ppm Nickel ions on the compressive strength of cement.

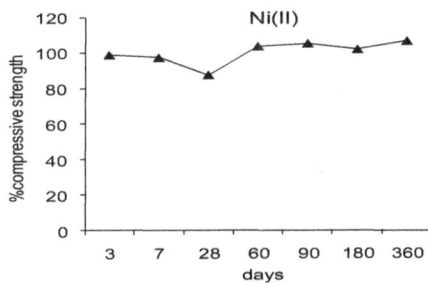

Figure 4: Percentage change in compressive strength of Nickel 1000 ppm containing samples in comparison with control.

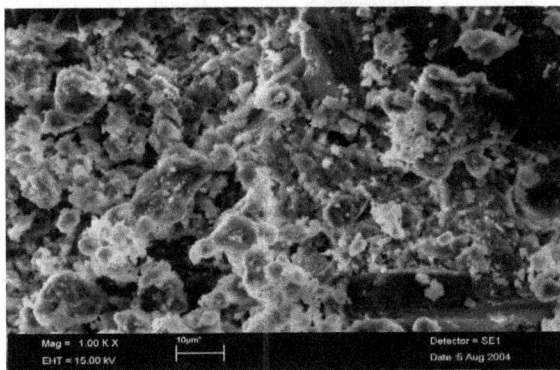

Figure 5: SEM image of pure cement sample on 28 day of curing.

important to mention that according to IS:1882:1989 the compressive strength of 43 grade ordinary Portland cement should be at least 230 kg/cm^2, 330 kg/cm^2 and 430 kg/cm^2 for 3, 7 and 28 days of curing. The addition of nickel ions containing 500 ppm and 1000 ppm of metal ion also caused a decrease in compressive strength in early hydration period as a compressive strength of 3 day is only 305kg/cm^2 and 320 kg/cm^2 for 500 and 1000 ppm and it is only 94.3 and 98.9 percent of the control

sample this decrease is intensified while CS is 445.0 and 426.0 kg/cm^2 on 28 day of curing and the percentage change is 91.4 and 87.5. Utmost care was taken to reduce variability associated with batch preparation step and reagent addition to avoid any substantial variability within a batch. A perusal of data presented in Table 6, exhibits that compressive strength of nickel containing samples decreases significantly in comparison with control sample (Figures 3 and 4).

Microscopic and spectroscopic analysis of the cement paste

The microstructures of hardened control cement paste as well as nickel ions containing cement paste were investigated by Scanning Electron Microscopy (Figures 5 and 6). Scanning electron microscopic analysis was carried out to know the change in the morphology of hardened cement paste on 28 and 180 days of curing.

The graphs obtained by the XRD analysis of the blank as well as the samples containing the varying concentration of heavy metal ions are presented. The qualitative XRD studies correlate well with the findings from the SEM analysis. It is often claimed that stabilization of metal involves the formation of insoluble metal silicate but the SEM and XRD examinations did not reveal any identifiable crystalline zinc silicate, though amorphous gel of calcium silicate was observed in both pure chemfix and metal-bearing chemfixed solids. Scanning electron microscopy and X- Ray diffraction study of nickel containing cement samples (Figures 6 and 7) show that nickel has no greater effect on the cement properties. Nickel does not show any effect on the hydration of cement & its main compounds but it slightly retards the setting process. Some amount of Nickel is found to distributed throughout to the C-S-H gel phase Nickel get adsorbed the interstitial phases of the cement and formed the insoluble nickel compound. The C-S-H phase of Nickel containing samples found denser than that of control samples.

Conclusions

The effect of fixation of nickel on physical, chemical and engineering properties of 43-grade Ordinary Portland Cement was studied. A quantitative correlation between the concentrations of nickel used and the intensity of any change in the properties of 43 grade Ordinary Portland Cement like initial and final setting time, compressive strength, bulk density, leaching studies, microscopic and spectroscopic studies etc. has been established. Nickel was used in the concentration of 500 and 1000 ppm. It may be stated that Ni (II) ions work as retarder in the concentration range 500-7000 ppm. The retarding effect of Ni (II) ions gets constant when the concentration of Ni (II) is increased further.

The addition of nickel ions containing 500 ppm and 1000 ppm of metal ion also caused a decrease in compressive strength in early hydration period. This decrease is intensified while CS is 445.0 and 426.0 kg/cm^2 on 28 day of curing and the percentage change is 91.4 and 87.5. Scanning electron microscopy and X- Ray diffraction study of nickel containing cement samples show that nickel have no greater effect on the cement properties but Some amount of nickel is found to distributed throughout to the C-S-H gel phase, nickel get adsorbed

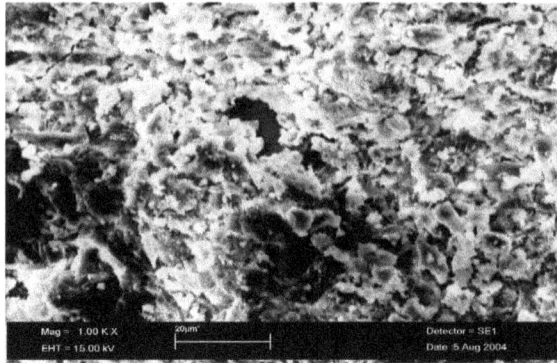

Figure 6: SEM image of nickel 1000 ppm containing cement sample on 28 day of curing.

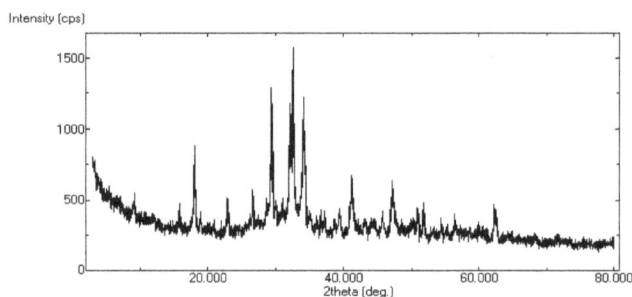

No.		File name	Sample name	Comment	Date
1	☑	Sam.2_1.raw	Sam.2		07-21-04

Figure 7: X-Ray diffractogram of Nickel1000 ppm containing cement sample at 28th day of curing **(Peak).**

the interstitial phases of the cement and formed the insoluble nickel compound. The C-S-H phase of nickel containing samples found denser than that of control samples.

Acknowledgement

The authors are thankful to the Director, Central Building Research Institute, Roorkee. The help of Dr. L.P. Singh, Mr. Jaswinder Singh and Dr. Pankaj Kumar is gratefully acknowledged.

References

1. Park CK (2000) Hydration and solidification of Hazardous waste containing heavy metals using modified cementitious materials. Cem Conc Res 30: 429-435.

2. Cote' PO, Bridle TR, Hamilton DP (1984) Evaluation of Pollutant Release from Solidified Aqueous Waste Using a Dynamic Leaching Test, Hazardous Waste and Environmental Emergencies, Houston TX.

3. Glasser FP (1993) Chemistry of cement solidified waste forms chemistry and microstructure of solid waste forms, Lewis publishers Chelsea, MI-PI.

4. Nocun W, Goteborg JM (1997) Studies on Immobilization of heavy metals in Cement Paste-C-S-H leaching behavior. Proceedings of the 10th international Congress on the chemistry of cement, Sweden.

5. Kindness A, Lachowski EE, Minocha AK, Glasser FP (1994) Immobilization and Fixation of Molybdenum (VI) by Portland cement. Waste Management 14: 97-102.

6. Kasselouri V, Ftikos CH (1997) The effect of MoO3 on the C3S and C3A formation. Cem Conc Res 27: 917-923.

7. Duchesne J, Laforest G (2004) Evaluation of the degree of Cr ions immobilization by different binders. Cem Conc Res 34: 1173-1177.

8. Mayers TE, Eappi EM (1992) Laboratory evaluation of Stabilization/ Solidification   technology for reducing the mobility of heavy metals in New Bedford harbor superfund site sediment stabilization of hazardous radioactive and mixed wastes, 2nd Edn. ASTM publication Philadelphia PA: 304.

9. Adaska WS, Tresouthick SW, West PB (1991) Solidification and Stabilization of wastes using Portland cement. Portland cement Association Skokie.

10. Deja J (2002) Immobilization of Cr6+, Cd2+, Zn2+ and Pb2+ in alkali activated slag binders. Cem Concr Res 32: 1971-1979.

11. Hills CD, Pollard SJT (1997) The influence of interference effects on the mechanical microstructural and fixation Characteristics of cement-solidified hazardous waste forms. Journal of Hazardous Materials 52: 171-191.

12. Wang S, Vipulanandan C (2000) Solidification/Stabilization of Cr (VI) with cement Leachability and XRD analysis. Cem Concr Res 30: 385-389.

13. Macias A, Kindness A, Glasser FP (1997) Impact of carbon dioxide on the immobilization potential of cemented wastes: Chromium. Cem Concr Res 27: 215-225.

14. U.S. Environmental Protection Agency (1986) Test Method for Evaluating Solid Wastes, SW-846, 3rd edn, Office of Solid Waste and Emergency Response, Washington DC.

15. Rossetti VA, Medici F (1995) Inertization of toxic metals in cement matrices: effects on hydration and hardening. Cem Concr Res 25: 1147-1152.

16. Corner JR (1990) Chemical Fixation and Solidification of Hazardous wastes. Van Nostrand- Reinhold, New York.

17. Taylor HFW (1964) The Chemistry of Cement. Academic press, New York.

18. Trezza MA, Ferraiuelo MF (2003) Hydration study of limestone blended cement in the presence of hazardous wastes containing Cr(VI). Cem Concr Res 33: 1039-1045.

19. Minocha AK, Jain N, Verma CL (2003) Effect of Inorganic Materials on the Solidification of Heavy metal Sludge. Cem Concr Res 33: 1695-1701.

20. Halim CE, Amal R, Beydoun D, Scott JA, Low G (2004) Implication of the structure of cementitious wastes containing Pb(II), Cd(II), As(V) and Cr(VI) on the leaching of metals. Cem Concr Res 34: 1093-1102.

21. Kolovas K, Tsivilis S, Kakali G (2002) The effect of foreign ions on the reactivity of the CaO-SiO2-Al2O3–Fe2O3 system Part (II): Cations. Cem Concr Res 32: 463-469.

22. Olmo FI, Chacon E, Irabien A (2001) Influence of Lead, Zinc, Fe(III) and Cr(III) oxide on the setting time and strength development of Portland cement. Cem Concr Res 31: 1213-1219.

23. Roy A, Eaton HC, Cartledge FK, Tittlebaum ME (1991) Solidification/ Stabilization of Heavy metal sludge by a Portland cement/fly ash binding mixture. Hazardous waste and Hazardous materials 8: 33-41.

24. Saygideger S, Gulnaz O, Istifli ES, Yucel N (2005) Adsorption of Cd(II), Cu(II) and Ni(II) ions by Lemna minor L: Effect of Physicochemical environment. Journal of Hazardous Materials 126: 96-104.

25. Goel J, Kadirvelu K, Rajagopal C, Garg VK (2005) Removal of lead(II) by adsorption using treated granular activated carbon : Batch and column studies. Journal of Hazardous Materials 125: 211-220.

26. Kalavrouziotis IK, Koukoulakis PH, Papadopoulos AH (2009) Heavy metal interrelationship in soil in the presence of treated waste water. Global Nest Journal 11: 497-509.

27. Chanakya V1, Jeevan Rao K (2010) Impact of industrial effluents on groundwater quality. J Environ Sci Eng 52: 41-46.

28. Sarma KP, Talukdar B (2008) Sediment Characteristics and Concentration of heavy metals in water and sediment of the effluent discharging water body of Nagaon paper mill, Assam, India. Asian Journal of Water Environment and Pollution 6: 97-102.

29. Ashok K, Bisht BS, Joshi VD, Indus B (2010) Estimation of heavy metals and metalloids from waste water of Bindal River, Dehradun. J Env Bio Sci 24: 195-198.

30. Minocha AK, Goyal MK (2013) Immobilization of molybdenum in Ordinary Portland Cement. J Chem Eng Process Technol 4: 162.

31. Minocha AK, Goyal MK (2013) Effect of immobilization of cadmium(II) ions on the hydration of ordinary Portland cement. J Chem Eng Process Technol 4: 170.

10

Effect of Solution Chemistry on the Nanofiltration of Nickel from Aqueous Solution

Benamar Dahmani and Mustapha Chabane*

Spectrochemistry and Structural Pharmacology Laboratory, Department of Chemistry, Science Faculty, University of Tlemcen, Algeria

Abstract

The rejection of nickel ions on water solutions was studied using aromatic polyamide nanofiltration membrane NF 90 by determination of solution chemistry as the concentration of solution, the pH and ionic strength at 27°C. The experimental results showed that the lower flow solution depends on concentration, solution pH and ionic strength. The solution concentrations showed greater decrease in flux and rejection. Flux decline conducted with a solution of nickel dropped for pH of the solution. At high pH, flux solutions showed higher flux decline than those of low solution pH, while the rejection of ions presented higher rejection. Increased ionic strength had a greater increase in flux decline. The rejection of nickel ions was found to be decreased with decreasing solution pH and increasing ionic strength. Flux and rejection decreased further to the higher ionic strength, which reduces the negative charge repulsion on the surface of the membrane, and thus a decrease of rejection. In addition, comparisons on the decline of flows to co-ions have also been studied in experiments filtration.

Keywords: Flux decline; Membrane; Nickel; Nanofiltration; Rejection

Introduction

The occurrences of metal ions in the aquatic environment have been concern because of their toxicity, while the accumulation may pose various hazards for human health and environment. At the present, the most frequency practiced treatment technology for the removal of metal ions from aqueous solution is chemical precipitation, which only relocated metal ions from aqueous phase, and leave further sludge problems to be solved. Therefore, the removal of metal ions from many industrial wastewaters has stimulated vigorous research activities in the development of appropriate treatment technologies. Membrane separation technologies have been determined to be a feasible option for removal of heavy metal from aqueous solution because of its relative ease of construction and control, and the feasible recovery of valuable metals. Nanofiltration processes is capable of removing heavy metal [1]. They are efficient technologies to remove feed source water in terms of natural organic matter (NOM) [4], inorganic scalants [5-7], salt solution [8-10] and heavy metals [11-15]. Nanofiltration (NF), one of membrane technologies, is a relatively new membrane process, which is considered to be intermediate between ultrafiltration (UF) and reverse osmosis (RO) in terms of operating conditions. NF membrane processes operate at pressures between 50 and 150 psi much lower than RO (200 to 1000 psi), but higher than UF (10 to 70 psi). At the present time, NF is increasingly applied in the field of water treatment. However, membrane fouling caused by organic and inorganic substances can be a major factor for limiting more widespread use of membrane technologies, reducing long-term filtration, and increasing costs for membrane operation through higher labor, cleaning and replacement. Inorganic fouling (i.e. negative and positive ions) can be a significant factor that enhances permeate flux decline during filtration. This may cause an increased concentration polarization that exceeds solubility limit, resulting precipitation (i.e. Ca^{2+}, Mg^{2+}, CO_3^{2-}, SO_4^{2-}, and PO_4^{3-}). This has been recently investigated by Jarusutthirak et al. [7]. Molinari et al. [12] investigated the interactions between membranes (RO and NF) and inorganic pollutants (i.e. SiO_2, NO_3^-, Mn^{+2}, and humic acid). They showed that membrane fouling was caused by the interactions between the membranes and other ions. Other factors, which can cause membrane fouling, are solution pH, ionic strength, concentration, solution composition and operating conditions.

The objective of this work was to investigate the effects of solution chemistry during nanofiltration of nickel solution. The effects of solution chemistry (i.e. concentration, solution pH and ionic strength) were determined on nanofiltration fouling. The discussion of this study was further adapted to improve membrane filtration for long-term operation.

Experimental

Nanofiltration characteristics

An aromatic polyamide thin-film composite NF-90 membrane, produced by Dow-FilmTec., was chosen to determine the effect of solution chemistry on nanofiltration performance. According to the manufacturer, the maximum operating pressure is 600 psi (or 4,137.6 kPa), maximum feed flow rate is 16 gpm (3.6 m^3/hr), and maximum operating temperature 113°F (45°C) and the operating pH is ranged from 1 to 12.

Analytical method

Nickel ion concentration was measured by using atomic absorption (AA) spectrometry (AAnalyst 200 Version 2, Perkin Elmer Corp.). Measurements of solution pH, conductivity and temperature were made using pH meter (shott), and conductivity meter (shott) respectively. Ionic strengths were calculated using a correlation between conductivity and ionic strength of NaCl standard, I.S.[M]=$0.5\Sigma C_i Z_i^2$ (C_i is the ion concentration and Z_i is the number of ions).

***Corresponding author:** Mustapha CHABANE, Spectrochemistry and Structural Pharmacology Laboratory, Department of Chemistry, Science Faculty, University of Tlemcen, Algeria

Flux decline experiments

The experiments were carried out with three liters of solution containing nickel solution ($NiCl_2$ and $Ni(NO_3)_2$) in varied concentration of 10, 20, 50 and 100 mg/L, while varying solution pH from 4 to 6 and ionic strengths (0.01, 0.05 M as NaCl). Flux decline experiments were tested by using a 400-ml dead-end membrane filtration apparatus (Amicon 8400 USA).

A membrane sheet can be fitted to the cell. The membrane active area is about 41.38 cm². The operating pressure was employed via high-pressure regulator of nitrogen cylinder. The permeate flux was kept in a beaker on the electrical balances.

Filtration experiments

Membrane sheets were rinsed with cleaned dionised water, 0.125 M citric acid solution of pH 4, and followed with 10^{-4} M sodium hydroxide solution of pH 10 for 30-min and dried.

Cleaned water flux was determined with a function of transmembrane pressure. Dionised water was subsequently tested for 30-min membrane compaction with velocity rate of 300 rpm. The water flux J_0 was subsequently determined with increased operating

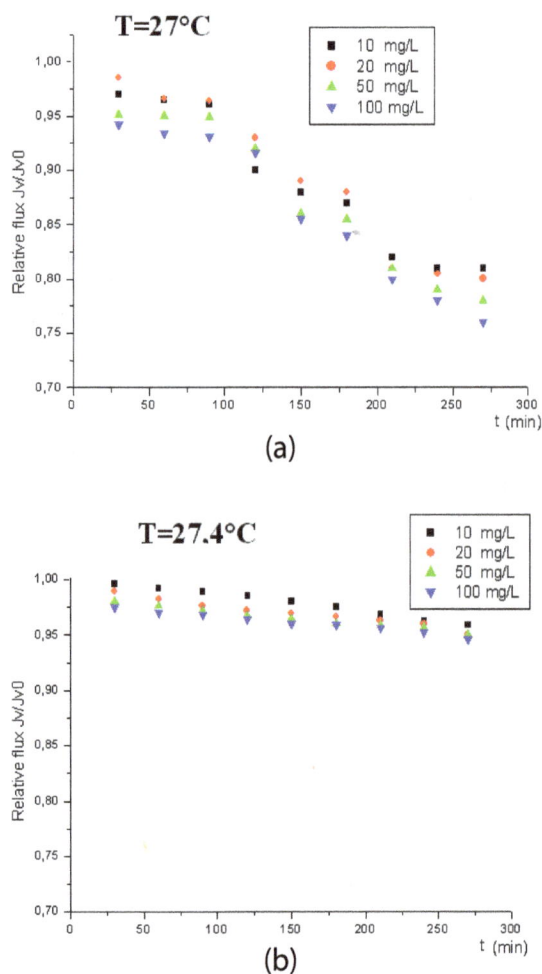

Figure 2: Effect of concentration on rejection; (a) $NiCl_2$ and (b) $Ni(NO_3)_2$ at 27°C.

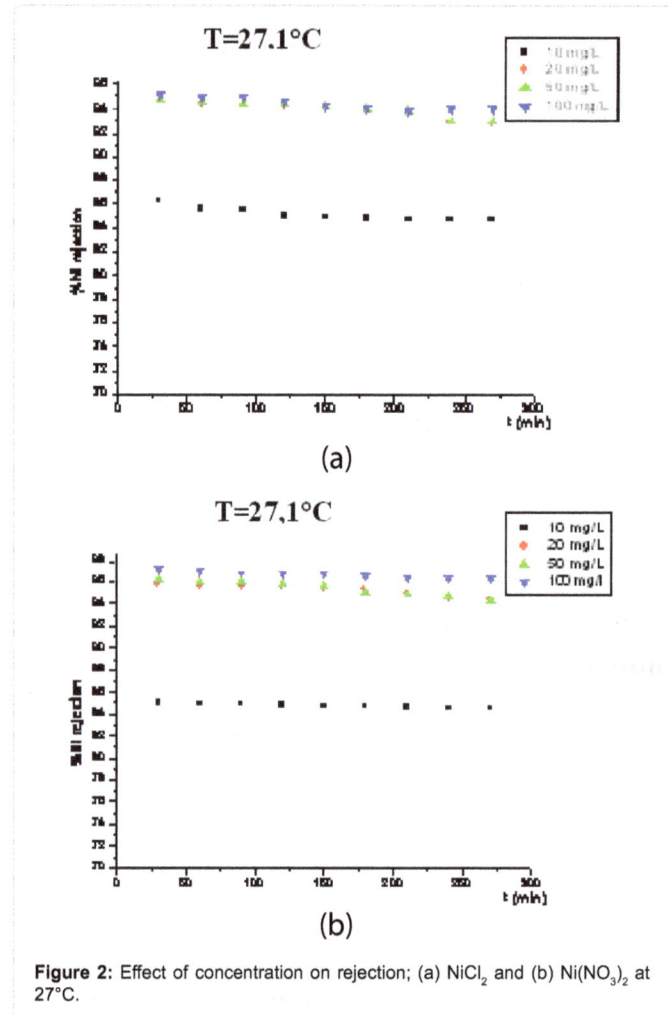

Figure 1: Effect of solution concentration on flux decline; (a) $NiCl_2$ and (b) $Ni(NO_3)_2$.

pressures before nickel solution was used with the system. Feed nickel solutions were prepared for each tested condition. After filtration was terminated, two steps of cleaning, i.e. hydrodynamic followed by chemical cleaning, were performed. For this, the membrane sheet was cleaned with dionised water, then followed with chemical cleaning, acidic solution (using citric acid) with pH of 4 for 30-min each. After each cleaning, water fluxes at different operating pressures were measured to determine water flux recovery.

Analysis of results

The parameters taken into account were: -The volumetric flux J_v (L/m²/h or LMH) was determined by measuring the volume of permeate collected in a given time interval divided with membrane area by the relation:

$$J_v = \frac{Q}{A} \tag{1}$$

Where, Q and A represents flow rate of permeate and the membrane area, respectively. The observed rejection was calculated by the following relation:

$$\%R = (1 - \frac{C_p}{C_i}) \times 100 \tag{2}$$

Where C_p and C_i are the solution concentrations in the permeate and in the initial feed solution, respectively.

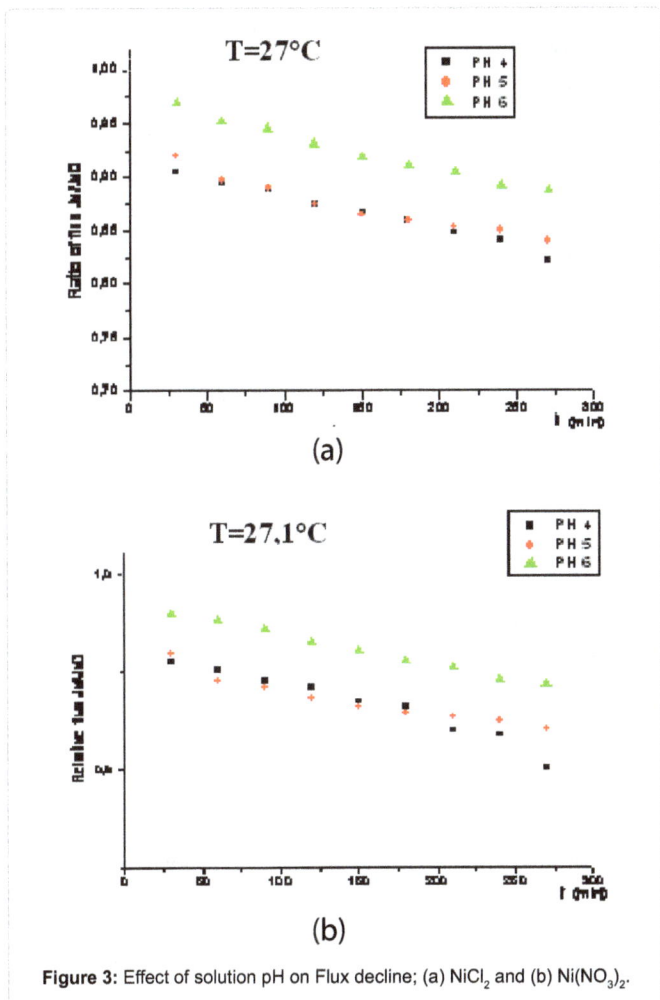

Figure 3: Effect of solution pH on Flux decline; (a) $NiCl_2$ and (b) $Ni(NO_3)_2$.

Results and Discussions

Effect of solution concentration on flux decline and rejection

In order to determine the effect of metal concentration on flux decline and rejection, Nickel ion concentrations were varied for 10, 20, 50 and 100 mg/L at pH 6 and ionic strength of 0.01 M NaCl. As shown in Figure1 and Figure 2, it can be concluded that nickel ion rejection and solution flux decline increased with increasing solution concentrations. The higher solution concentrations for $NiCl_2$ solution had greater ion rejection, about 94%-98 %, while low solution concentration exhibited lower rejection about 85%-97%. For $Ni(NO_3)_2$ solutions having 100 mg/L concentration had the ion rejection about 98-99%, while those of the lower concentrations were 88-95%.

Effects of solution pH on flux

The effects of solution pH on flux decline of $NiCl_2$ and $Ni(NO_3)_2$ solution were carried out at pH 4,5 and 6 with keeping constant ionic strength 0.01 M as NaCl at 60-psi operating pressure. Nickel concentration was about 20 mg/L. Figure 3 showed relative flux with function of operating period for $NiCl_2$ and $Ni(NO_3)_2$ solution. It can be seen that the rate and extent of flux decline increased with increasing solution pH. For the solution of $NiCl_2$ and $Ni(NO_3)_2$ solution at lower pH, flux solutions showed higher flux decline than those of low solution pH. At low pH, it suggested an increased fixed charge of H^+, decreasing electrical double layer thickness within membrane or both, thus decreased the concentration at the membrane surface. At high pH, the

membrane surface and pores become both more negatively charged due to the presence of anion (inorganic). In addition, the osmotic pressure near the membrane surface increase with high salt rejections, thus decreasing the driving pressure. These mechanisms nickel to a decrease in permeate flux and an increase in salt rejection with pH.

Effects of solution pH on rejection

The effects of the solution pH on rejection of $NiCl_2$ and $Ni(NO_3)_2$ solution was carried out with different solution pH from 4 to 6. Solution was maintained constant with Ionic strength of 0.01 M NaCl, 60 psi operating pressure and solution concentration of 20 mg/L during filtration. The obtained results were presented in Figure 4. Nickel ion rejection was found to be decreased with decreasing solution pH level. It was possibly due to higher solution pH, membrane surface take more negative charges, thus attracting greater lead ion. Consequently, solution pH of 5-6 for $NiCl_2$ had greater ion rejection about 96%-98 %, while low solution pH exhibited lower rejection about 88%-91%. For $Ni(NO_3)_2$ solution, the ion rejection percentages of higher solution pH (5-6) and lower solution pH were 91-94% and 76-81%, respectively.

Effects of ionic strength on flux

Figure 5 presents the effect of ionic strength on flux that was carried out at pH 6 with different ionic strengths of 0.01 and 0.05 M as NaCl. It was observed that the extent and rate of solution flux decline increased with increasing ionic strength. In the study, increases in ion

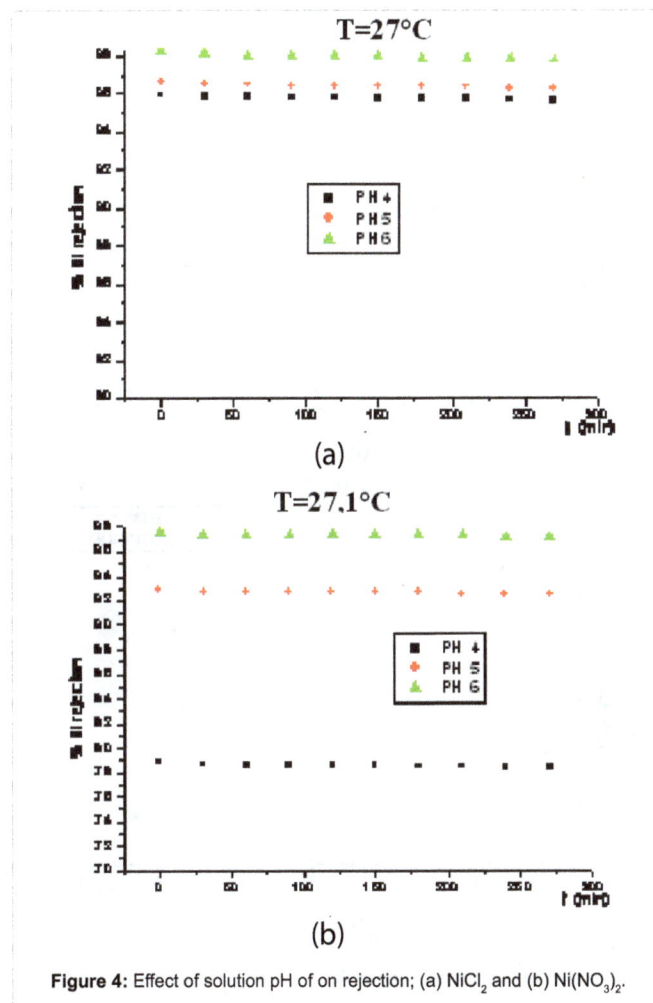

Figure 4: Effect of solution pH of on rejection; (a) $NiCl_2$ and (b) $Ni(NO_3)_2$.

concentration can reduce the permeability of charged membranes [20,21], thus reduced permeate flux. The results showed the similar trend for both of 20 mg/L $NiCl_2$ and $Ni(NO_3)_2$ solution.

Effect of ionic strength on nickel ion rejection

The effect of salt solution on nickel ion rejection were carried out with ionic strengths of 0.01 and 0.05 M as NaCl and the concentration of 20 mg/L, as shown in Figure 6. Solution pH of 6 for $NiCl_2$ and $Ni(NO_3)_2$ solution was kept constant during filtration. It was found that Nickel rejection at ionic strength of 0.05 M showed lower than those at ionic strength of 0.01 M. This was possibly due to increasing salt concentration, reducing membrane permeability, thus allowing nickel ion passage through the membrane surface.

Effect of co-ion on solution flux decline

The effect of the co-ion on Nickel solution flux decline was carried out with two types of Ni^{2+} ($NiCl_2$ and $Ni(NO_3)_2$) solution at the concentration of 20 mg/L. The experiments were performed at pH 6, ionic strength of 0.01 M NaCl and 60 psi operating pressure during filtration. From the experiment, it can be seen that $Ni(NO_3)_2$ solution had a slight lower on flux decline than $NiCl_2$ solution. $NiCl_2$ solution showed higher rejection than $Ni(NO_3)_2$ solution. Since the NF membrane is more negatively charged the monovalent anion of Cl^- is more excluded than NO_3^- resulting in greater rejection. Furthermore, the ion rejection is mainly dependent on its hydration energy in the

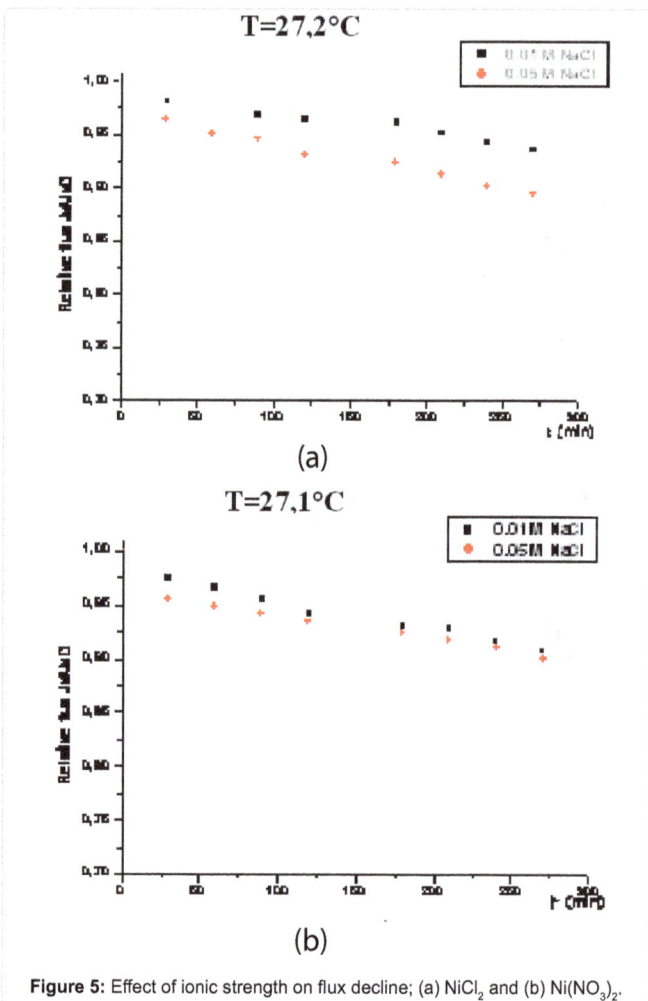

(a)

(b)

Figure 6: Effect of ionic strength on Nickel ion rejection; (a) $NiCl_2$ and (b) $Ni(NO_3)_2$.

solution and it was more retained if it has higher hydration energy, in accordance with the mechanism of solution-diffusion .The hydration energy of Cl^- and NO_3^- are 372 KJ/mol and 328 KJ/mol respectively [21]. However, the higher rejection of chlorides can not be described by the hydration energy owning to the fact that these anions had close hydration energy. It can be explained by the formation of the complexes solution, which involves an electrostatic repulsion with charge membrane and thereafter a strong rejection of the ions.

Conclusion

Nickel ions rejection and flux decline from aqueous solution by nanofiltration was strongly influenced by concentrations, solution pH and ionic strength. Flux decline conducted both in $NiCl_2$ and $Ni(NO_3)_2$ solution decreased for solution pH and concentrations. At higher solution pH, flux solutions showed higher flux decline than those of low solution pH, while le rejection exhibited higher rejection. Increased ionic strength had a greater increase in flux decline. Nickel ion rejection was found to be decreased with decreasing concentrations, solution pH and increasing ionic strength.

(a)

(b)

Figure 5: Effect of ionic strength on flux decline; (a) $NiCl_2$ and (b) $Ni(NO_3)_2$.

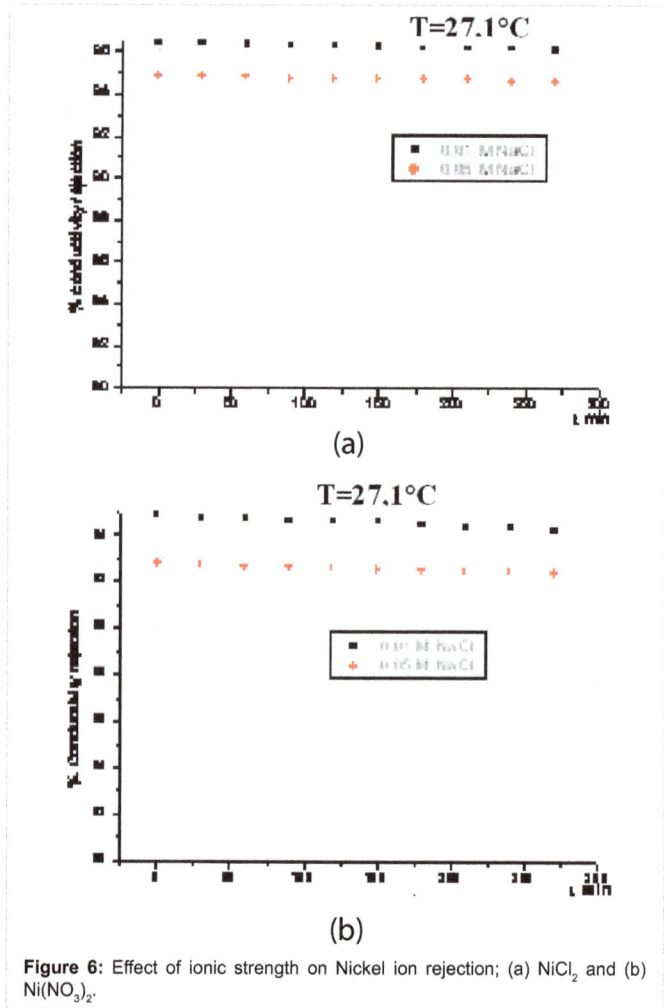

References

1. Thorsen T, Flagstad H (2006) Nanofiltration in drinking water treatment: Literature Review. Techneau D5.3.4B, 31.

2. Cho J, Amy G, Pellegrino J (1999) Membrane filtration of natural organic matter: initial comparison of rejection and flux decline characteristics with ultrafiltration and nanofiltration membranes. Water Res 33: 2517-2526.

3. Schafer AI, Fane AG, Waite TD (2000) fouling effects on rejection in the membrane filtration of natural waters, Desalination 131: 215-224.

4. Kilduff je, mattaraj s, belfort g (2004) flux decline during nanofiltration of naturally-occurring dissolved organic matter: effects of osmotic pressure, membrane permeability, and cake formation. J Memb Sci 239: 39-53.

5. Lisdonk CAC, van Paassen JAM, Schippers JC (2000) Monitoring scaling in nanofiltration and reverse osmosis membrane systems. Desalination 132: 101-108.

6. Lin CJ, Shirazi S, Rao P, Agarwal S (2006) Effects of operational parameters on cake formation of $CaSO_4$ in Nanofiltration. Water Res 40: 806-816.

7. Jarusutthirak C, Mattaraj S, Jiraratananon R (2007) Influence of inorganic scalants and natural organic matter on nanofiltration membrane fouling. J Memb Sci 287: 138-145.

8. Childress AE, Elimelech M (1996) Effect of solution chemistry on the surface charge of polymeric reverse osmosis and nanofiltration membrane. J Memb Sci 119: 253-268.

9. Anne CO, Trebouet D, Jaouen P, Quemeneur F (2001) Nanofiltration of seawater: fractionation of mono-and multi-valent cations. Desalination 140: 67-77.

10. Labbez C, Fievet P, Szymczyk A, Vidonne A, Foissy A, et al. (2003) Retention of mineral salts by a polyamide nanofiltration membrane. Separation and Purification Technology 30: 47-55.

11. Mehiguene K, Garba Y, Taha S, Gondrexon N, Dorange G (1999) Influence of operating conditions on the retention of copper and cadmium in aqueous solutions by nanofiltration: experimental results and modeling. Separation and Purification Technology 15: 181-187.

12. Molinari R, Argurio P, Romeo L (2001) Studies on interactions between membranes (RO and NF) and pollutants (SiO_2, NO_3^-, Mn^{++}, and humic acid) in water. Desalination 138: 271-281.

13. Ku Y, Chen SW, Wang WY (2005) Effect of solution composition on the removal of copper ions by nanofiltration. Separation and Purification Technology 43: 135-142.

14. Ipek U (2005) Removal of Ni(II) and Zn(II) from an aqueous solution by reverse osmosis. Desalination 174: 161-169.

15. Turek M, Dydo P, Trojanowska J, Campen A (2007) Adsorption/Co- precipitation-reverse osmosis system for boron removal. Desalination 205: 192-199.

16. Gaballah I, Kilbertus G (1998) Recovery of heavy metal ions through decontamination of synthetic solutions and industrial effluents using modified barks. J Geochem Explor 63: 241-286.

17. Low KS, Lee CK, Liew SC (2000) Sorption of cadmium and lead from aqueous solutions by spent grain. Process Biochem 36: 59-64.

18. Yaroshchuk A, Staude E (1992) Charged membranes for low pressure reverse osmosis properties and applications. Desalination 86: 115-134.

19. Eriksson P (1988) Water and salt transport through two types polyamide composite membrane. J Membr Sci 36: 297-313.

20. Darbi A, Viraraghavan T, Jin TC, Braul L, Corkal D (2003) Sulfate removal from water. Water Qual Res J 38: 169-182.

21. Novak I, Sipos L, Kunst B (2004) Removal of sulfates and other inorganics from potable water by nanofiltration membranes of characterized porosity. Separation and Purification Technology 37: 177-185.

11

The structure of Isotactic Polypropylene Crystallized from the Melt

Sokainah Rawashdeh[1]* and Ismail Al-Raheil[2]

[1]Applied science department, Faculty of Engineering Technology, Al-Balqaa Applied University, Amman, Jordan
[2]Department of Physics, University College in Lieth, Umm Al-Qura University, Makkah , KSA

Abstract

The melting behavior of isotactic polypropylene was studied in the hot stage mounted on polarizing optical microscope supported by photomonitor. Over a wide range of crystallization temperature there are two main types of spherulites, α and β spherulites for this polymer α spherulites may exist in three forms α_1, α_2 and mixed of α_1 and α_2. The α_2-form can be obtained at high crystallization temperature above 145°C and α_1 can be observed at low crystallization temperatures below 132°C. The mixed α-spherulites can be obtained between 132°C and 145°C. However, the β-phase form small proportion of the total phase not more than 15% and usually exist when the crystallization temperature below 132°C. If the sample crystallized from the melt it shows usually two melting peaks. However, when the temperature of the sample was increased to crystallize in a second step at higher temperature it shows two melting peaks and if the second step of crystallization was very close to melting point it shows only one single peak, since the reorganization of crystals process will be stopped completely.

Keywords: Isotactic polypropylene; Crystallization; Spherulites; Melting point

Introduction

Spherulites develop from crystallites starting from a central nucleus and growing in all directions radially, with small angle of branching in between. The branching provides space filling. The first theory about crystallization in polymers was founded by Padden and Keith [1] and later according to the phenomenological observations that theory was crystallized by Bassett and Vaughan [2]. They found that the small angle fibrillar branching is induced by impurities has low tendency to crystallize and, therefore, segregation will be found in the vicinity of growing front.

According, to Varga revision [3] when isotactic polypropylene is crystallized from the melt, is monitored by polarizing optical microscope, birefrengent spherulites are produced, growing radially at a constant rate under isothermal conditions. Due to impingement of growing spherulitic fronts, the structure formed will consist of polygonal formations confined by straight or curved lines after complete crystallization.

During the crystallization of isotactic polypropylene, Varga [4] found that, two types of spherulites α-and β-modification. Under given thermal conditions the two types develop simultaneously.

Padden and Keith [5] found that three types of α-spherulites might be formed depending on crystallization conditions: positive radial below 133°C, negative radial above with no cross-hatching lamellae above 137°C and a mixed type in between.

In this paper non-isothermal crystallization melt behavior and thermal properties of isotactic polypropylene, ipp, were studied with respect to different crystallization temperatures from the molten state to see the effects on the crystallization behavior and spherulitic structure.

Experimentals

Materials

The materials of this study are a homopolymer of isotactic polypropylene originally supplied by polymer supply and characterization center (PSCC) at Rapra, Shawbury, Shopshire, UK. Its molecular mass has been measured by PSCC to be $M_n=4.7\times10^4$ and $M_w=4.2\times10^5$. The samples were received in the form of pellets.

Optical microscopy

The sample was prepared by squashing ipp sample between slide and cover slip on a hot plate. Then it was inserted inside Mettler (FP82) hot stage connected with a computer reinforced with special software from Mettler Company. The hot stage was mounted on a Nikon microscope type (FX-35DX) with cross polarization condition. The sample inside the hot stage was heated above the melting point of ipp at 200°C, and then it was cooled to different crystallization temperatures below its melting point. The spherulitic growth at crystallization temperature, was monitored in most cases and the resulting spherulites were photographed by Nikon 35 mm camera type (FX-35 DX), or the Nikon microscope was connected to photo monitor to measure the transmitted light intensity of spherulites.

Thermal analysis

The melting and crystallization behavior for isotactic polypropylene samples was studied using Mettler DSC model (FP 85) connected with a computer. Each sample was carefully prepared and measured to have weight between 10-14 mg, by using microbalance with sensitivity 10 micrograms.

Isotactic polypropylene samples were placed inside aluminum crucibles, then the crucibles were heated in the DSC above the melting temperature of ipp at 200°C (the melting point is about 170°C), and the sample kept in the DSC cell for about one minute to melt most of

*Corresponding author: Sokainah Rawashdeh, Applied science department, Faculty of Engineering Technology, Al-Balqaa Applied University, Amman, Jordan

crystalline nuclei. Then the sample was cooled below its melting point to the required crystallization temperature for different times. After that the samples were heated from crystallization temperature at rate $10°C.min^{-1}$ to $200°C$. In sometimes the sample were cooled from the crystallization temperature to lower temperatures, and then heated at rate $10°C.min^{-1}$.

In all measurements the melting behavior was seen in the computer screen, so that the changes of the heat flow with increasing temperature due to structural changes of the samples were recorded.

For optical studies the sample of ipp was squeezed on hot stage glass slide and cover slip and the sample was transferred to Mettler hot stage model FP 82 and crystallized from the melt after heating to 200°C for 1 minute and cooled to the required crystallization temperatures for predetermined times.

Result and Discussion

The various spherulites morphology of isotactic polypropylene was analyzed in details [6]. Two types of spherulitic microstructures are encountered in the α-form. Type $α_1$ spherulites which appear in the temperature lower than 134°C, exhibit an overall positive birefregrance when examined in a polarizing microscope, while the main lamellae are oriented radially, tangential lamellae appear to grow by epitaxial deposition, with a direction of growth which is inclined by approximately 81°C. These tangential lamellae are responsible for the sign of the birefringence that is observed. On the other hand, at a crystallization temperature higher than 140°C, negatively birefrengent spherulites are also identified, these type $α_2$-spherulites, therefore, presumed to contain only a small proportion of tangential lamellae, although the amount of transverse lamellae that is necessary to change the birefringence sign cannot be estimated easily. These are a consequence of in homogeneity of the spherulites where regions of predominantly radial lamellae, giving a negative birefringence, coexist with other regions which contain predominantly tangential lamellae which give a positive birefringence [7]. This assumption is supported by the results in this study when the specimen of isotactic polypropylene was crystallized from the melt at 128°C. In general, the β-spherulites have a larger size than α-spherulites, despite the fact that β-phase occupies, on the whole, a smaller fraction. It can be seen that inter-spherulitic boundary between boundary α and β-types is always curved, with the concavity oriented towards the α-phase. The reason for the concavity is due to the higher crystallization growth of β-spherulites at that temperature.

Upon heating, the β-phase melts at 152°C leaving the α-spherulites alone. The contrast of α-type spherulites can be seen as dark region with positive birefringence due to high density of cross-hatching lamellae. The α-spherulites can be identified as $α_1$-type. With increasing the temperature, the cross-hatching lamellae start to melt gradually. Therefore, the dark contrast of $α_1$-type spherulites changes to bright with negative birefringence due to the melting of tangential lamellae.

According to that there is a transformation from type $α_1$ to $α_2$ spherulites, and the α-spherulites which can be seen after melting of β-spherulites can be considered as $α_2$-type. In fact, the two types of α-spherulites [8] are distinguishable in this study by the different degrees of cross-hatching. Hence, without the cross-hatched lamellae, the spherulites should be a type $α_2$.

The bifurcation of α spherulites was studied by Varga. With increasing crystallization temperature, the relation between the growth rate of β spherulites to α spherulites decreased, and above certain temperature, $T_{βα}$, the growth rate of the α-modification exceeds that of β modifications. The level of $T_{βα}$ as was found by Varga was between 140°C-141°C. He also found above $T_{βα}$ an α-nuclei are found on the surface of growing β spherulites, developing α spherulites is segments which finally overgrow the basic β spherulites completely.

In this study, it was found that the bifurcation of β α-spherulites was not detected. In contrast to the αβ-modification can be observed in many different part of the specimen. For example this can be observed in figure 1 for sample crystallized at 128°C, α spherulites was formed first and in its surface β spherulites was formed later. It is obvious that at 128°C crystallization temperature the growth rate of β spherulites is faster than α spherulites.

Recrystallization of β-spherulites into the α-modification during heated was studied by many authors [9-15]. Studies led to the assumption of thermodynamic instability of β-modification.

In this study it was found that melting of β-modification had a specific feature including a much more complicated process than that outlined above, namely the melting characteristics of β-pp are highly dependent on the thermal post history of the crystalline sample. In contrast to the studies in the above literature, when heating begins from the temperature of crystallization, the β-modification does not recrystallize in α-form; instead they melt separately. However, if the sample containing β-ipp is cooled quickly by immersing it in cold water, upon heating again the partial melting of β-form is accompanied by a recrystallization into α-form (βα-recrystallization) and finally, they melt in α-form. Consequently βα-recrystallization susceptity appears due to recooling.

An optical representation of the above transformation can be observed in figure 2 and 3. Figure 2 shows α and β-modifications at the same time, after isothermal crystallization at 125°C.

When the sample is quenched in cold water, upon heating the βα-recrystallization of β-spherulites transforms into α-spherulites preserving the shape character as shown in figure 3. The appearance of the tendency to βα-recrystallization of β-ipp quenched in cold water can be attributed to the formation of α-phase within the β-spherulites upon quenching, due to a secondary crystallization talking place in the cooling process. According to that, this phase acts as a α-nucleating agent during the partial melting of β-spherulites [16].

The transformation from $α_1$ to $α_2$-spherulites was confirmed from the melting process of ipp crystallized at 125°C by measuring

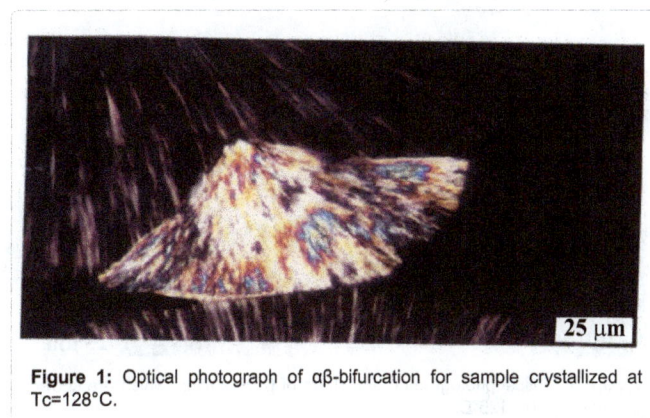

Figure 1: Optical photograph of αβ-bifurcation for sample crystallized at Tc=128°C.

the relative light intensity of α-spherulites between crossed polarizes by applying heating rate of 10°C/min as shown in figure 4. For the α-phase the melting is associated with a continuous decrease of the relative light intensity until 145°C, above that temperature, there is an increase in the relative light intensity associated with transformation of α_1 to α_2-spherulites due to the melting and recrystallization of the cross-hatched lamellae which are thin to form radial perfect lamellae. At crystallization temperature 140°C the form α-spherulites is mixed from α_1 and α_2-spherulites, so that there is a slight increase in the light intensity which is transmitted from the mixed α-spherulites as shown in figure 5. According to that, the transformation from mixed α-spherulites to pure α_2-spherulites is associated with a moderate increase in the light intensity, due to the low contents of cross-hatched lamellae in the mixed α-spherulites.

Pure α_2-spherulites can be obtained if the sample is crystallized in two steps. For example figure 6 shows the relative light intensity transmitted through a mixed α-spherulites grown at 140°C for 18 hours, then the sample was heated in the hot stage until 160°C for 1 hour to form the pure α_2-spherulites, then cooled again to 140°C. Upon heating there was no increase in the transmitted light intensity through the pure α_2-spherulites, at this stage it can be said that, there is no existence to the cross-hatched lamellae in pure α_2-spherulites.

Figure 7 shows the DSC endotherms traces of isotactic polypropylene, curve in figure 7 shows a board peak of endotherm of melt crystallized ipp at 125°C for 1 hour. Annealing the same sample at 155°C for 30 minutes followed by heating to 200°C as shown in curve b changes the melting trace so that there is no longer marked melting before 155°C, but at this temperature the new curve rise to accommodate with the old curve. Evidently, the crystallinity of

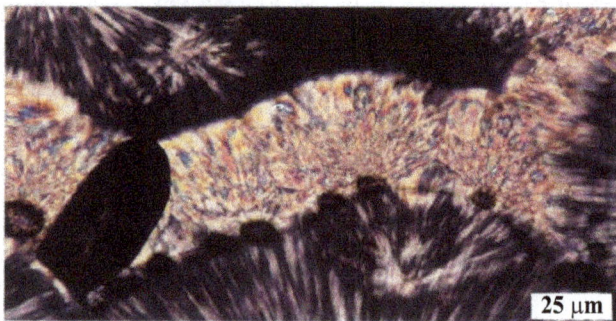

Figure 2: Optical photograph of α and β-modifications after isothermal crystallization at Tc=125°C.

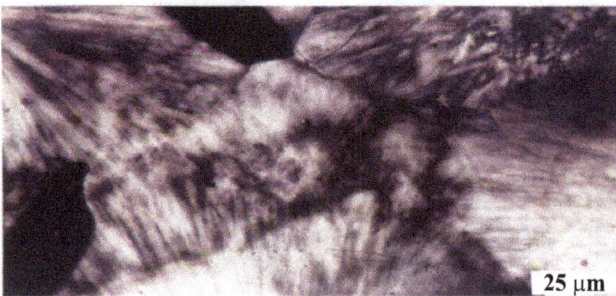

Figure 3: Optical photograph of transform of β-spherulites which are shown in figure (5), Tc=125°C.

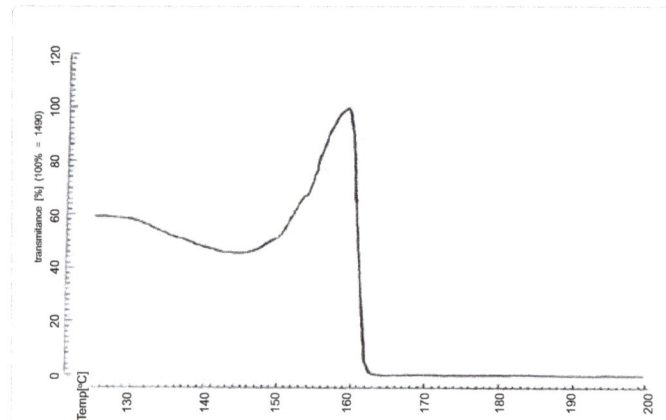

Figure 4: Transmitted light intensity of α-type spherulites of isotactic polypropylene which crystallized at 125°C for 5 hours then heated directly to 200°C at scan rate 10°Cmin⁻¹.

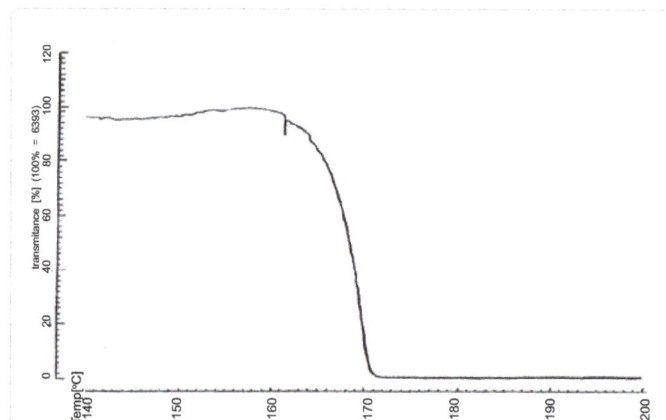

Figure 5: Transmitted light intensity of α-type spherulites of isotactic polypropylene which crystallized at 140°C for 5 days then heated directly to 200°C at scan rate 10°Cmin⁻¹

the cross-hatched lamellae which had melted below the annealing temperature has been vanished from the spherulites structure when the sample crystallized at 125°C and annealed at 155°C.

Comparing the DSC results in figure 7 with the relative light intensity of α-spherulites for sample treated by heat in the same way as shown in figure 4, one can say that the melting behavior as shown in figure 7 belongs only to the melting of α_2-spherulites since α_1-spherulites has been transformed to α_2-spherulites at 145°C as shown from the rising in the light intensity during heating at that temperature. According to that, the melting behavior of the annealed sample at 155°C is due to the melting of the radial lamellae of α_2-spherulites.

This trend applies when the isotactic polypropylene is crystallized at 140°C for 18 hours as shown in figure 8 curve a. once the cross-hatched lamellae are melted after annealing at 155°C, curve b, the double melting behavior still exists, with no broad melting endotherm in the lower side as shown in curve a in the same figure. It is interesting to note that with no β-phase, the double melting behavior is not affected by annealing process at 155°C, therefore, the melting of cross-hatched lamellae does not belong to the lower melting peak as one can expect, it can be suggested that the melting of cross-hatched lamellae belongs only to the endotherm of the broad tail which leads the lower melting peak in curve a in both figures 7 and 8. Annealing the same sample

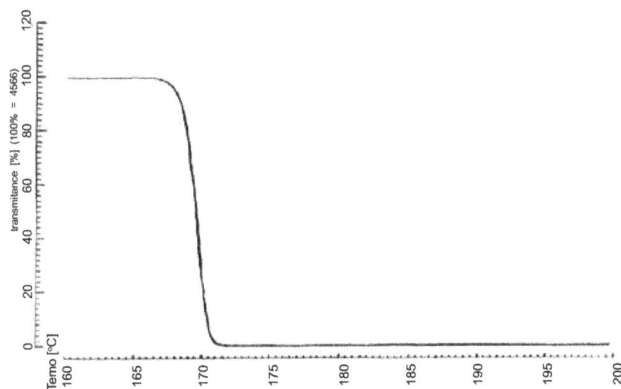

Figure 6: Transmitted light intensity of α-type spherulites of isotactic polypropylene which crystallized at 140°C for 18 hours then heated to 160°C for one hour then heated to 200°C at scan rate 10°Cmin⁻¹.

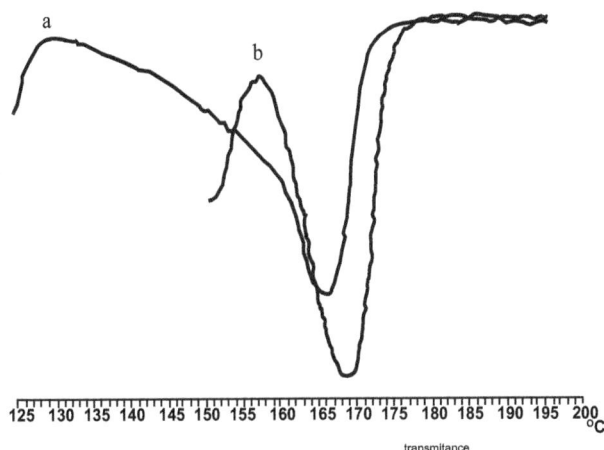

Figure 7: DSC trace for ipp isothermal crystallized from the melt at a- 125°C for one hour then heated to 200°C at scan rate 10°Cmin-1 b-125°C for one hour then annealed at 155°C for one hour then heated to 200°C at scan rate 10°Cmin⁻¹.

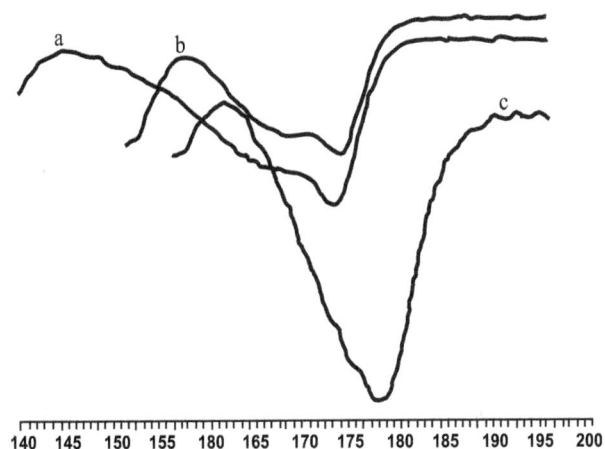

Figure 8: DSC trace for ipp isothermal crystallized from the melt at a- 140°C for 18 hours then heated directly to 200°C at scan rate 10°Cmin-1 b-Tc=140°C for 18 hours then annealed at 155°C for one 10 hour then heated to 200°C at scan rate 10°Cmin-1 c- Tc=140°C for 18 hours then annealed at 160°C for one hour then heated to 200°C at scan rate 10°Cmin⁻¹.

which has been crystallized at 140°C for 18 hours at 160°C for 1 hour then heating directly 200°C as shown in curve c figure 8, changes the double melting behavior to single melting peak. At this high annealing temperature, there is no opportunity to radial lamellae to reorganize themselves again so that the single melting peak represents the true melting point of the existing lamellae in the sample.

Conclusions

On the basis of the results presented above its clear that the α₁-spherulites can be transformed to α₂-spherulites by heating above 145°C, also we can see that β-spherulites can transform to α-type spherulites if the sample of isotactic polypropylene is quenched from the isothermal crystallization temperature in cold water then reheated again, upon reheating β-spherulites transforms into α-form.

For samples crystallized in two steps, it was found that for example if the sample was crystallized at 140°C for a long time it shows two melting peaks. However, if the sample crystallized at 140°C then annealed at 155°C, it still shows double melting peaks upon melting due to reorganization. However if the same sample annealed at 160°C then heated directly it shows only one melting peak due to suppression of reorganization.

References

1. Padden FJ, Keith HD (1966) Crystallization in isotactic polypropylene melts during contraction flow: time-resolved synchrotron WAXD studies. J Appl Phys 37: 4013.

2. Bassett DC, Vaughan AS (1985) On the lamellar morphology of melt-crystallized isotactic polystyrene. Polymer 26: 717-725.

3. Varga (1992) Supermolecular structure of isotactic polypropylene. J Materials Sci 10: 2557-2579.

4. Varga (1995) Crystallization, Melting and Supermolecular Structure of Isotactic Polypropylene. In; Karger-kocsis J (1995) Polypropylene: Structure, Blends and Composites. Structure and Morphology. Kocsis J Karger, Chapman and Hall, London, UK.

5. G. Castelein, G. Coulon, M. Aboulfaraj, C. G'Sell, E. Lepleux (1995) Some Observations of the lamellar Morphology in Isotactic Polypropylene Spherulites by SFM. J Phys III France 5: 547-555.

6. Grein C, Plummer CJG, Kausch HH, Germain Y, Beguelin Ph (2002) Influence of β nucleation on the mechanical properties of isotactic polypropylene and rubber modified isotactic polypropylene. Polymer 43: 3279-3293.

7. Nakamura K, Satoko S, Umemoto S, Thierry A, Lotz B, et al. (2008) Temperature Dependence of Crystal Growth Rate for α and β Forms of Isotactic Polypropylene. Polymer. 40: 915-922.

8. Padden FJ, Keith HD (1959) Spherulitic Crystallization in Polypropylene. J Appl physics. 30: 1479-1484.

9. Ullmann W, Wendorff JH (1979) Studies on the monoclinic and hexagonal modifications of isotactic polypropylene. Polymer science 66: 25.

10. Forgacs BP, Sheromov MA (1981) Polypropylene structure and morphology. Polymer 6: 127.

11. Moos KH, Tigler B (1981) Macromol Chemistry 94: 213.

12. Morrow DR (1969) Journal of Macromol science B3: 53.

13. Guan-Yi S, Huang B, Jung-yun Z (1984) Performance of Plastics. Witold Brostow.

14. Gui-en Z, He ZQ, Jian-min Y, Zwe-wen H (1986) Polypropylene Structure and Morphology. Chapman & Hall.

15. Guan-Yi S, Huang B, Jung-yun Z (1986) Performance of Plastics. Witold Brostow.

16. Varga J (1986) Polypropylene Structure and Morphology. Chapman & Hall.

12

Process Validation of Ceftriaxone and Sulbactam Dry Powder Injection

Shiv Sankar Bhattacharya[1]*, Naveen Bharti[1] and Subham Banerjee[2]

[1]*School of Pharmaceutical Sciences, IFTM University, Moradabad, Uttar Pradesh, India*
[2]*Division of Pharmaceutical Technology, Defence Research Laboratory, Tezpur, Assam, India*

Abstract

In the present work Process validation of Ceftriaxone and Sulbactam as a dry powder injection was carried out. As the manufacturing process of dry powder injection is mainly dependent on blending process. In the present investigation, blending process was validated at different speeds of blender and the % assay was estimated by HPLC method. The octagonal blender was operated at 13, 17 and 20 rpm samples were taken from 10 different locations inside the blender. At 13 rpm there is much variation in assay results, at 17 rpm there is very less variation and also at 20 rpm there is slight variation occurs with reference to the acceptable ranges of assay {i.e. 90-110 %}. The obtained results clearly indicated that the optimum rpm is often necessary for the proper mixing of the drug. Therefore, 17 rpm was considered for the proper mixing at blending stage and it can be successfully employed to manufacture of dry powder injections for further manufacturing. The content uniformity of the net filled content was found to be in ± 5 % of average net content. Hence, it was concluded that process stands validated for the preparation of dry powder injection.

Keywords: Process validation; Ceftriaxone; Sulbactam; Dry powder injection; Good Manufacturing Practices (GMP); Blending

Introduction

Validation is a concept that has been evolving continuously since its formal appearance in the United States in 1978 [1]. According to the FDA's current Good Manufacturing Practices (cGMP) control procedure shall be established to monitor output and to validate performance of the manufacturing processes that may be responsible for causing variability in the characteristics of In-process materials and the drug product [2-4]. Validation is documented evidence that provides a high degree of assurance that a specific process will consistently produce a product that meets its predetermined specifications and quality attribute. Validation study *in evitably* leads to process optimization, better productivity and lower manufacturing cost. The investment made in validation, similar to the investment made in qualified people can only provide an excellent return [5]. The concept of validation has expanded through the years to encompass a wide range of activities from analytical methods used for the quality control of drug substances and drug products, to equipment's, facilities and process for the manufacture of drug substances and drug products to computerized systems for clinical trials, labeling or process control. Validation studies are essential part of Good Manufacturing Practices (GMP) and should be conducted in according with predefined protocols. A written report summarizing results and conclusions should be recorded, prepared and stored. Validation has become one of the pharmaceutical industry's most recognized and discussed subjects. It is a critical success factor in product approval and ongoing commercialization. The objective of the present work is a) to provide documented evidence for the operation sequencing and scheduling of manufacturing processes and to determine the critical parameters of the manufacturing process of dry powder injection. b) to provide assurance that manufacturing process is suitable for intended purpose and consistently meets its predetermined specifications and quality attributes, as per Master Formula Record (MFR) and c) to systematically conduct the validation studies pertaining to the manufacturing activities of Ceftriaxone and Sulbactam dry powder injection and to conclude on a high degree of assurance that manufacturing process, consistently meets the predetermined specifications and quality attributes. Hence the quality product output can be increased, leading to increase in quality, productivity and decrease the need of reprocessing [3,6].

Experimental

List of raw materials

Ceftriaxone Sodium (IP) was from Nectar Lifesciences Ltd, Chandigarh, India and Sulbactam Sodium (USP) was procured from Aurobindo Pharma Ltd, Hyderabad, Telangana, India.

List of packaging materials

Glass Vial 5 ml (Type III) was collected from Neutral Glass & Allied Industries, Surat, Gujrat, India. Grey Bromo Butyl Rubber Bung (20 mm) was purchased from Bharat Rubber Works Pvt. Ltd. Mumbai, Maharashtra, India and Flip Off seal (20 mm, white) was from HBR Packaging Mumbai, Maharashtra, India.

List of equipment's used

All equipment's are perfectly qualified as per Design Qualification, (DQ), Installation Qualification (IQ), Operational Qualification (OQ), and Performance Qualification (PQ) acceptance criteria (Table 1).

Batch operation

The batch operation validation approach means a plan to conduct process validation on different products manufactured with the same processes using the same equipment. The validation process using these approaches must include batches of different strengths or products which should be selected to represent the worst case conditions or scenarios to demonstrate that the process is consistent for all strengths or products involved. In process validation three consecutive batches

*Corresponding author: Shiv Sankar Bhattacharya, School of Pharmaceutical Sciences, IFTM University, Delhi Road, Moradabad-244 001, Uttar Pradesh, India

Name of the Equipment's	DQ	IQ	OQ	PQ
Bung Processor & Autoclave	Complies	Complies	Complies	Complies
Vial Washing Machine	Complies	Complies	Complies	Complies
Sterilization & Depyrogenation Tunnel	Complies	Complies	Complies	Complies
Vial Filling & Bunging Machine	Complies	Complies	Complies	Complies
Blender	Complies	Complies	Complies	Complies
LAF (Filling)	Complies	Complies	Complies	Complies
LAF (Cooling zone)	Complies	Complies	Complies	Complies
LAF (Blending area)	Complies	Complies	Complies	Complies
LAF (Mount stand)	Complies	Complies	Complies	Complies

Table 1: Equipment qualification details.

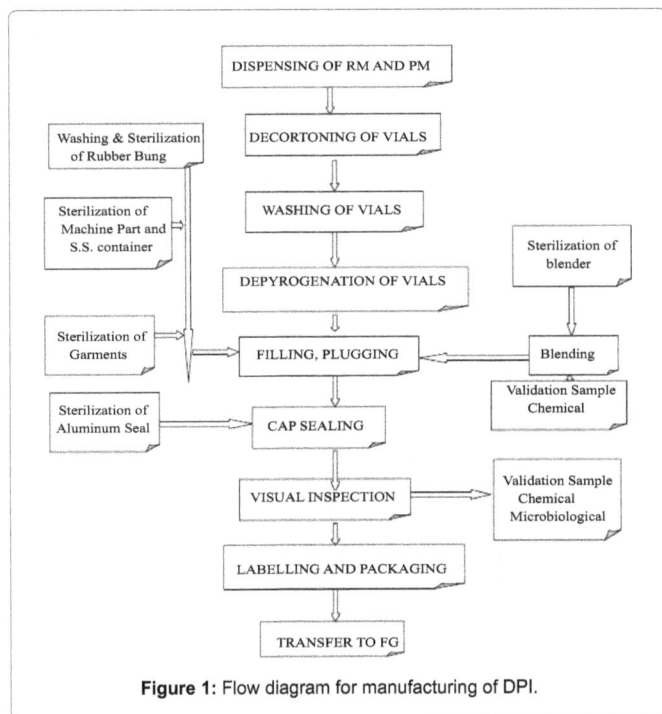

Figure 1: Flow diagram for manufacturing of DPI.

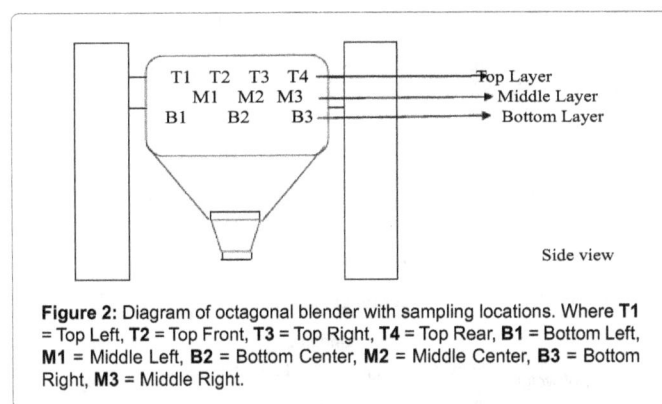

Figure 2: Diagram of octagonal blender with sampling locations. Where **T1** = Top Left, **T2** = Top Front, **T3** = Top Right, **T4** = Top Rear, **B1** = Bottom Left, **M1** = Middle Left, **B2** = Bottom Center, **M2** = Middle Center, **B3** = Bottom Right, **M3** = Middle Right.

are used for manufacturing operation these batches are of the size which will be produced during the routine marketing of the product. The given process flow diagram is for manufacturing DPI, to be performed in various stages (Figure 1) and points indicated different sampling locations (Figure 2). Sampling analysis report was reported in terms of assay, uniformity of weight, pH, particulate matter, uniformity of content by weight at various intervals at different operations as per the sampling plan at blending stage. The following rotational variations

were investigated at 13, 17 and 20 rpm for a fixed blending time of 60 min according to the master production control record.

Results and Discussion

The process validation was started at the qualification of instrument all the instrument was qualified at the time of process validation. Environmental condition monitoring of manufacturing area is critical process parameter for process validation. In environmental monitoring critical parameter like, temperature, relative humidity, and differential pressure, viable or non-viable particles are generally monitored. The maximum and minimum temperature was found to be 26°C and 24°C respectively in different processing area. The maximum and minimum % relative humidity was found 45% and 30%, respectively in different processing area. The differential pressure was found to be not more than (NMT) 10 Pascal. The viable particles were not found during observation. The maximum non-viable particles of ≥ 0.5 μ were found NMT 3520 per m³ in sterile filling area. Similarly, the maximum non-viable particles of ≥ 5.0 μ were found to be NMT 29 per m³ in sterile filling area. The visible and non-visible particulate matter was checked during vial washing, sterilization and filling stages, the particulate matter was found to be as per acceptance criteria. During vial filling and stoppering the weight variation and content uniformity of dosage unit was also calculated/checked. The result was found under acceptance criteria (Table 2). Sealing integrity test was performed after vial sealing with the help of sealing integrity test apparatus no defects was observed in this test. Analytical test and sterility test of finished product was performed by quality control and microbiology department both test were complies. In the process validation of dry powder for injection, the main focus was done on Blending stage. The dry powder for both the drugs i.e. Ceftriaxone and Sulbactam was blended in the octagonal blender at various revolutions per minute (13 or 17 or 20 rpm) for 60 minutes according to the master production control record (Figures 3-5). The comparison of % assay of Ceftriaxone and sulbactam at different rpm were represented in Figures 6 and 7 respectively. The above results and graph showed that more consistent % assay values were found at 17 rpm. The results obtained at 13 rpm were not complying with the acceptance limits of 90.0-110.0% because the values were found to be less than 90.0%. It may be due to the incomplete or improper blending occurrence at the 13 rpm. The results at 20 rpm were also inconsistent in their % assay values and the values were found to be greater than 100.0% either due to segregation or improper mixing. The results at

Parameters	Area	Acceptance criteria	Observation comply / not comply
Temperature (°C)	Vial filling area Cooling zone Vial washing room Vial sealing room	NMT 24°C NMT 24°C NMT 26°C NMT 26°C	comply
Relative Humidity (%)	Vial filling area Cooling zone	NMT 30% NMT 45%	comply
Differential Pressure (mm)	Vial filling vs Vial washing Vial filling vs Cooling zone Vial filling vs Air lock	NLT 10 pascal NLT 10 pascal NLT 10 pascal	comply
Sterile Filling Area Particle Count	Viable particle count Non-viable particle coun	1 CFU/m³ ≥ 0.5 μ=NMT 3520/m³ ≥ 5 μ= NMT 29/m³	comply
Area Adjacent to Sterile Area Particle Count	Viable particle count Non-viable particle count	2 CFU/m³ ≥ 0.5 μ=NMT 352000/m³ ≥ 5 μ= NMT 2900/m³	comply

Table 2: Environmental Condition of Manufacturing Area.

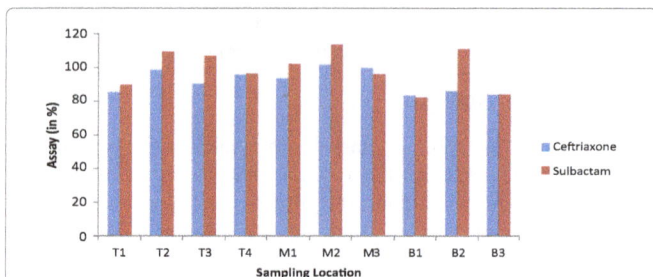

Figure 3: Comparative assay (in %) of Ceftriaxone and Sulbactam at 13 RPM.

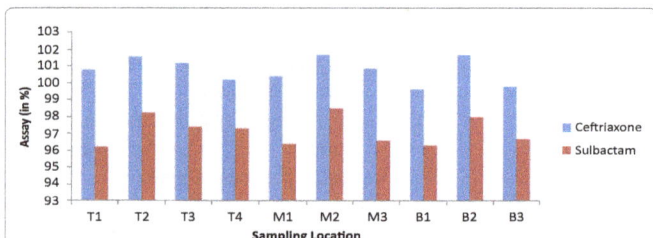

Figure 4: Comparative assay (in %) of Ceftriaxone and Sulbactam at 17 RPM.

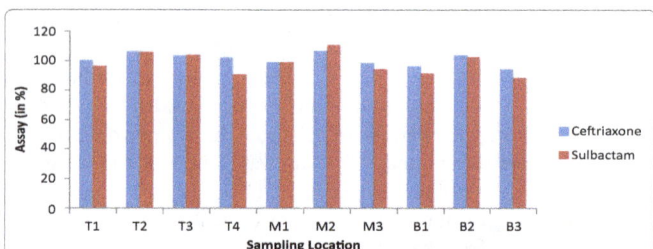

Figure 5: Comparative assay (in %) of Ceftriaxone and Sulbactam at 20 RPM.

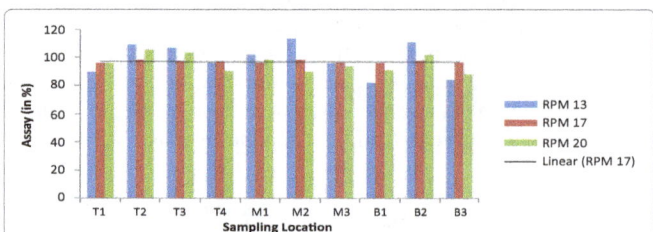

Figure 6: Showing comparison of % assay of Ceftriaxone at different rpm. The graph also indicates that assay of Ceftriaxone at RPM 17 show consistent result (due to linearity).

criteria. Hence the product can be successfully manufactured at the commercial scale and the sterile manufacturing process is validated.

Conclusion

From the present study, following conclusion can be drawn i) blending stage of the Process Validation play key role in the manufacturing of dry powder injections, ii) rpm was a critical parameter at blending stage of process validation which governs proper & uniform mixing and thus responsible for good assay results, iii) drug content in all the vials was within the limit and iv) attempts were made in present study to prepare a stable composition of dry powder injection of Ceftriaxone & Sulbactam combination. These results clearly reflect that the prepared dry powder injections of Ceftriaxone & Sulbactam offers good assay results and within limit. Thus, Optimum blending speed i.e. 17 rpm at blending stage can be successfully employed to manufacture Dry Powder Injections for further scale up. Finally, all the test result was found to be as per acceptance criteria or compiled. Based on observation of three batches it was concluded that the product can be successfully manufactured and the sterile manufacturing process is validated.

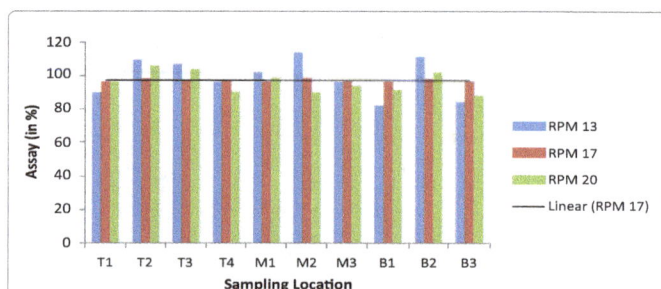

Figure 7: Showing comparison of assay of Sulbactam at different RPM. The graph also indicates that assay of Sulbactam at RPM 17 show consistent result (due to linearity).

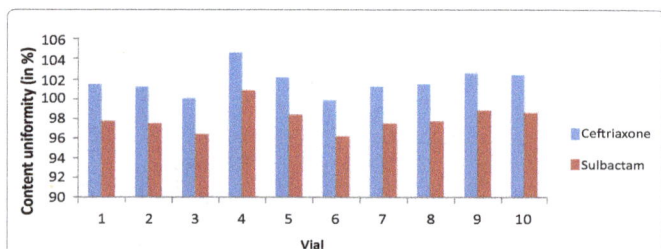

Figure 8: Comparison of % content uniformity of Ceftriaxone and Sulbactam.

17 rpm were found to be consistent and also graphically showed the linearity indicating that at this rpm the process stands validated and the results are reproducible. The results for uniformity of content by weight were observed in Figure 8. The above results of % content uniformity by weight for all the 10 vials sampled were found to be in acceptance criteria range of 85-115% which indicates that filling of powder process was producing reproducible results of acceptance limits. The uniformity of weight was also found to comply with the acceptable range limits i.e. ± 5% of average net content. The pH of the finished sample was done and found to be 6.45 i.e. in limits of 4.5-9.0. The particulate matter test was also complying with the acceptance criteria (Table 3).

So the data of all three batches were complying with its acceptance

Test parameter	Acceptance criteria	Observation comply / Not comply
Assay	As per monograph	comply
Uniformity of weight	Individual weight ± 5% of target fill weight	comply
pH	4.5 to 9.0	comply
Particulate matter	Vials should be essentially free from visible particulate matter. Sub-visible particulate matter: ≥ 10 μ: NMT 3000/vial ≥ 25 μ: NMT 300/vial	comply
Uniformity of dosage units (By weight variation)	Meets the requirement, (NMT ±15.0%)	comply
Sealing of vials	No. defects should be observed.	comply

Table 3: Observation Report.

References

1. US FDA (1987) Guideline on General principles of process validation. Center for Drug Evaluation and Research (CDER), Rockville, MD.

2. Chows SC (1997) Pharmaceutical validation and process control in drug development. Drug Information Journal 31: 1195-1201.

3. WHO (1992) Good Manufacturing Practices for Pharmaceutical Products. WHO Expert Committee on Specifications for Pharmaceutical Preparations. 32nd Report, WHO Technical Report Series no.823, Geneva 14-96.

4. Potdar MA (2007) Pharmaceutical Quality Assurance. 2nd edition, Nirali Prakashan.

5. Chaloner-Larsson G, Anderson R, Egan A (1997) A WHO guide to GMP requirements: Part 2: Validation, guide to Good Manufacturing Practice. World Health Organization, Geneva.

6. Haider SI (2006) Pharmaceutical Master Validation Plan. 1st edition, CRC Press LLC, 2-6.

13

Use Alloy Quasicrystalline $Al_{62,2}Cu_{25,3}Fe_{12,5}$ for Steam Reforming of Methanol

Jamshidi ALCL[1], Nascimento L[1]*, Rodbari RJ[2], Barbosa GF[3], Machado FLA[4], Pacheco JGA[1], Barbosa CMBM[1]

[1]Center for Technology and Geosciences -CTG/UFPE, Av. Moraes Rego, 1235 – Cidade Universitária, CEP: 50670-901, Recife, PE, Brazil
[2]Department of Sociology-DS, Islamic Azad University, Tehran-Iran
[3]Department of Physics, Department of Exact and Natural Sciences-DENS/UFERSA, Rua da Harmonia Alto de São Manoel, CEP: 59625210-Mossoró, RN-Brazil
[4]Department of Physics -DP, Center of Exact and Natural Sciences-CENS/UFPE, Av. Prof. Moraes Rego, 1235-Cidade Universitária, CEP: 50670901, Recife- PE, Brazil

Abstract

This study shows the good performance of quasicrystal $Al_{62,2}Cu_{25,3}Fe_{12,5}$ as catalyst in catalytic reactions. The metal catalyst without being leached with acid or base with the stoichiometric composition of dry $Al_{62,2}Cu_{25,3}Fe_{12,5}$ among the reactions shown to be a partial oxidation occurred , which formation of the products was methanol , methanal + methanoic acid, water and dimethyl ether. For this research used experimental techniques as X-Ray Diffraction-XRD to follow the evolution of the alloy phase, the Scanning Electron Microscopy-SEM allowing the study of surface microstructure, and Transmission Electron Microscopy-TEM studies the morphology of internal phase, and defects quasicrystalline nuclei; tests for the catalytic conversion of methanol and selectivity and products formed from this material used as catalyst. The activity and stability of catalyst quasicrystal for steam reforming of methanol showed sufficient performance compared to other catalysts. The Fe and Cu species highly dispersed in the homogeneous layer quasicrystal catalyst increases the catalytic activity and suppresses the aggregation of Cu particles. We propose that the quasicrystal can be a good catalyst to be used in catalytic steam reforming, with high catalytic activity and excellent thermal stability.

Keywords: Catalytic activity; Quasicrystal; Methanol

Introduction

Quasicrystals are solid materials that exhibit a unique form of matter with long range order without periodicity and symmetries of non crystallographic rotation (To symmetries of order 5, 8, 10 and 12). Since were studied for the first time in 1984 by Shechtman et al. [1], more than 100 binary alloys with metal composition, ternary and quaternary these systems were found.

These alloys containing stable quasicrystalline phases. Moreover, they have good physical properties, high electrical resistance. The low chemical reactivity of quasicrystals is attributed to the presence of pseudo gaps [2], which is a reduction of the electronic density of states at the Fermi surface. This can be interpreted as a low reactivity with oxygen when comparing the reactivity of quasicrystalline league Al-Cu-Fe with their crystalline analogs. In quasicrystalline materials, heterogeneities in the form of second phases, precipitates, segregates, grain boundaries, dislocations and stacking faults exist. In contrast, amorphous alloys are composed of a single phase of homogenous solid solution without any physical and chemical heterogeneity. Thus, alloys quasicrystalline and amorphous with sufficient concentrations of corrosion-resistant elements show superior corrosion resistance that has never been found in any crystalline metallic alloys.

The oxidation of alloys subjected to high temperatures is quite complex, since various technical processes are involved activation. The potential application of interest and quasicrystals for catalysis was investigated by Agostinho [3]. For these alloys have stable equilibrium phases even at high temperatures and can be used in high temperature catalytic reactions. In this study on catalytic activity of quasicrystals did analysis of various alloys Al-Pd-Mn and Al-Cu-Fe in the decomposition reactions in methanol. In the reaction of steam reforming of methanol was studied by Tsai et al. [4]. In the composition of the system quasicrystalline alloy Al-Cu-Fe in icosahedral phase. Some combinations of alloys such as were observed (Al-Pd-Mn, Al-Cu- Fe, Al-Co-Ni and Ti-Zr-Ni) and these show the adsorption and/

or reactivity of simple molecules such as CO, CH_3OH and H_2. In catalytic reaction of interest for increased production of hydrogen gas using a catalyst quasicrystalline be noted that even a good performance achieved by obtaining a greater amount hydrogen gas at a temperature lower than the beginning of the reaction. Several quasicrystalline alloy containing Pd were tested and were highly active for the decomposition of methanol.

According to the studies of Hao et al. [5] reported that the alloy with the chemical composition of Ti-Zr-Co contain a quasicrystalline icosahedral phase showing a high catalytic activity and selectivity for the oxidation of cyclohexane.

However, quasicrystals based on Al have been considered as the most promising candidate for the surface catalysis because of their heat stability and ease of production, because this metal is quite abundant and inexpensive cost. The oxidation of aluminum in the alloy has quasicrystalline a change of concentration in the area near the surface and may induce a transformation of icosahedral phase to the crystalline phase , but emphasizing that the icosahedral phase is predominantly the presence of aluminum.

In the icosahedral phase and in the presence of oxygen atoms aluminum move to the surface, assuming that it is due to the driving

*Corresponding author: Nascimento L, Center for Technology and Geosciences-CTG/UFPE, Av. Moraes Rego, 1235-Cidade Universitária, CEP: 50670-901, Recife-PE, Brazil

force provided by the exothermicity of the oxide is greater than that of the other constituent of the alloy [6]. In this case forms a thin layer of amorphous Al_2O_3 oxide layer called passivating. This layer is alumina in catalytic reactions is favored due to their structural and chemical characteristics, its intrinsic acidity resulting from the combination of two hydroxyl groups neighbors forming water during thermal dehydration, leaving exposed Al^{3+} ions which by their electron deficiency, act on acid sites Lewis or through the hydroxyl groups retained act as proton donors, or Bronsted acid sites, and the alumina these types of reactions. The copper-based catalyst has a tendency to favor the catalysis of reactions, can be thermodynamically stable, and it is this feature provides better conditions for the steam reforming reaction. The quasicrystalline alloy is used as a precursor and/or a support associated with it, in the case of a zeolite. The research already done with the quasicrystalline alloys, to leached or dry, always obtained good results in catalytic reactions. It is interesting to say in general, the catalytic activity depends on feature the surface, the atomic structure of the surface, the electronic nature of the surface and the surface energy. The reactions involving the conduction electrons, where magnetic for Fe and Cu moments both interact with their electric fields in translational periodicity of quasiperiodic structure and exchange interaction between the conduction electrons in sublevels s-d and the magnetic moments localized on the sublevel Fe, facilitating organic reactions on its surface. But it must be said, the mode of preparation of the catalyst can affect the structure and composition of the catalyst, as well as its activity, but have due care in preparation. So, you want to obtain a good quasicrystalline catalyst for certain purposes reactions in heterogeneous catalysis, with the intention of having some products such as methanol, dimethyl ether, hydrogen gas and olefins.

This paper studies the behavior of catalytic quasicrystalline alloy $Al_{62,2}Cu_{25,3}Fe_{12,5}$ in the reaction of steam reforming of methanol in the review that this league supports high temperatures and maintaining its stability. Whereas, good catalytic activity depends characteristics of the electronic configuration $3d^{n-2}4s^2$, first complete orbital later if the d orbitals. Therefore quasicristal is very promising for reaction catalysis.

Reaction of Methanol in Quasicrystals

The reactions involving alcohols are widely studied due to the high importance of these reagents either as products or even as many catalytic reactions. The high industrial production of methanol and simplicity of their structure makes one of the most studied molecules in terms of the surface [7]. The main industrial reactions involving methanol in metallic and quasicrystalline surfaces. These reactions are usually activated metal catalysts based on Cu, Fe or Pd dispersed in a thin bed oxide matrix.

Given that the OH group is the most reactive part of an alcohol, methanol serves as a model for chemical reactions which involved most carbonated chain alcohols are. The steam reforming reaction is considered promising and has the following advantages over other methods: it does not require oxidation of the fuel, has increased efficiency, reduced size of the system as a whole and produces high hydrogen concentration. The steam reforming of methanol (SRM) is potentially a good process for on board production of hydrogen for mobile fuel cells yielding the maximum amount of hydrogen. Iron and copper based catalysts have been identified as outstandingly effective for the SRM and therefore are subject of intensive research.

The formal reaction network of SRM over quasicrystals based catalysts mainly consists of three reactions. Steam reforming of methanol (Eq.1) is an endothermic reaction which is as good as irreversible at temperatures above 200°C and ambient pressure [8].

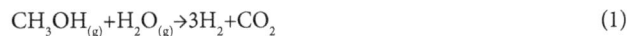

$$CH_3OH_{(g)} + H_2O_{(g)} \rightarrow 3H_2 + CO_2 \tag{1}$$

$\Delta H = 49.6\ KJ.mol^{-1}$

Since this reaction is endothermic, the reactor needs to be heated. This is usually done by catalytic methanol combustion. A side reaction of less importance is the decomposition of methanol (Eq.2), also endothermic and nearly irreversible at temperatures above 200°C and ambient pressure.

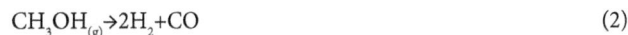

$$CH_3OH_{(g)} \rightarrow 2H_2 + CO \tag{2}$$

$\Delta H = 90.2\ KJ.mol^{-1}$

The reaction products of SRM suffer the consecutive endothermic reverse water-gas shift reaction (rWGS, Eq.3), which is also known to be catalyzed by amorphous alloys by iron and even the nanostructured zeolite associated with quasicrystalline alloys catalysts.

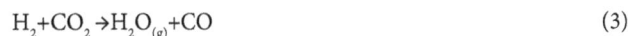

$$H_2 + CO_2 \rightarrow H_2O_{(g)} + CO \tag{3}$$

$\Delta H = 41.1\ KJ.mol^{-1}$

Although this reaction is thermodynamically strongly disfavored in the typical temperature range of SRM and due to the presence of water in the reactant mixture, it be-comes important as it was found to be the main reaction pathway for carbon monoxide formation. The reduction of carbon monoxide, due to its poisoning effect on fuel cell electrodes, is an aim of new catalyst development with higher selectivity for SRM. Alternative routes, adding oxygen to the feed (combined reforming of methanol) or using a molar excess of water, lower the carbon monoxide concentration but still not sufficiently for the direct use of the product gas mixture in a fuel cell. Up to now, carbon monoxide needs to be eliminated in an expensive clean-up unit, where carbon monoxide is selectively oxidized. Further by-products re-ported by several research groups are methane, formaldehyde, dimethyl ether and methyl formate as trace components, respectively.

The composition of the product gas from the steam reforming of methane is much less complicated than that from the partial oxidation due to oxygen in fewer chemical reactions. With this causes a decrease and there is less risk of the active surface oxidation of copper oxide (CuO) can be maintained. Although different reaction mechanisms that are proposed, it is believed that the dissociative adsorption of methanol is an essential step of the reaction. This step probably needs a partially oxidized surface CuO, because methanol is adsorbed very weak on a surface of pure copper. Generally two types of adsorbents sites are considered: one absorbs hydrogen, and the other absorbs all the intermediates through oxygen linkages. The first probably corresponds to a reduction of copper; the latter is probably related species of Cu oxidized. After adsorption, the methoxy groups are gradually dehydrogenated, but there is general agreement on the route that leads to $CO_2 + H_2$. Some authors assumed the formation of methyl ethanoate $HCOOCH_3$ as an intermediate which hydrolyzes and HCOOH CH_3OH: HCOOH then decomposes into $CO_2 + H_2$. There formation of the adsorbed methanol is dehydrogenated to formaldehyde, formic acid, and finally CO_2, passing through intermediate dioxometilene. The products formed in the reform of methanol gas are H_2, CO_2, H_2O and CH_3OH not converted. Small amounts of carbon monoxide (byproduct) are observed when methanol is almost completely converted. From the foregoing, it is the most favorable for the production of gas with a high hydrogen content (reaches 75%) process compared with previous processes and also has a high selectivity for carbon dioxide. Another way of producing hydrogen from methanol is a combination of partial oxidation with steam reforming. Its main advantage is that the heat

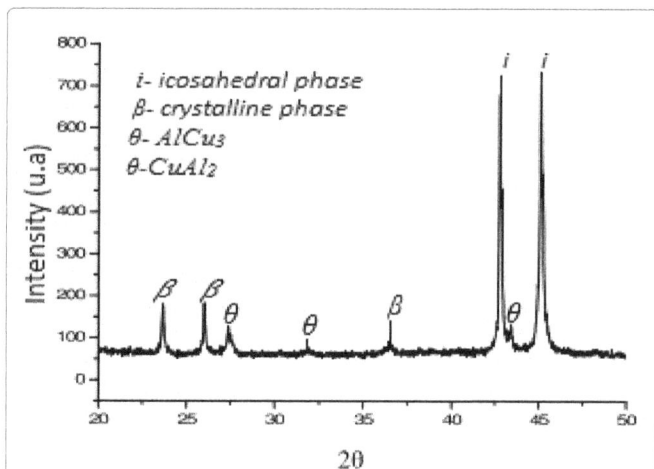

Figure 1: X-Ray diffraction (XRD) of the sample $Al_{62,2}Cu_{25,3}Fe_{12,5}$ heat treated in time of 8 h.

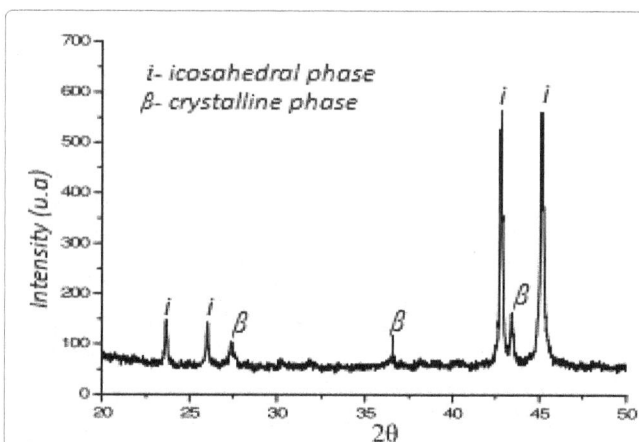

Figure 2: X-Ray diffraction (XRD) of the sample $Al_{62,2}Cu_{25,3}Fe_{12,5}$ heat treated at 24 h.

application is as a catalyst, since this alloy has a thermal stability in the temperature range 1073K.

Materials and Methods

The preparation of quasicrystalline alloys consists of a nominal composition $Al_{62,2}Cu_{25,3}Fe_{12,5}$ according to their grain size , having a purity of 99,9 % . The alloy was obtained by melting the pure elements of air. The induction furnace was used in controlled atmosphere Argon 5.0, in order to obtain a good homogenization of quasicrystalline phase.

The sample preparation was performed at the Laboratory of Physics of Materials-LPM, Federal University of Paraiba Center of Exact Sciences and Earth/UFPB where a high-frequency generator (40 kVA) manufacturing POLITRON was used . Each element that makes up the league quasicrystalline contains 10g, this measurement procedure was performed on a Shimadzu balance. The training method was through solidification in the cold hearth furnace for generating a heterogeneous league, being a common procedure a mixture of quasicrystalline phase to the crystalline phase. To be a proportional increase in quasicrystalline phase in the sample, it is necessary to heat treatment, so that will further the transformation of peritectic phases. This was done using the heat treatment in a furnace Nabertherm resistance mark, in which the samples remained during the period of 8h and 24h at a temperature of 750 °C.

The Ray Diffraction-(XRD) was used to monitor the evolution of the phases and sample stability during casting. Was used for both a Diffractogram SIEMENS D5000 being used the CuKα radiation whose wavelength is Å. To analyze the morphology of the quasicrystalline placed a LEO Scanning Electron Microscope; Model 1430, coupled with an OXFORD probe was used. The samples after casting and were placed in catalytic tests dispersion in isopropyl alcohol solution. The energy dispersive spectroscopy (EDS) of the sample quasicrystalline alloy $Al_{62,2}Cu_{25,3}Fe_{12,5}$ micrograph was obtained at an energy dispersive analysis after 8 hours heat treatment coupled with the SEM sample which has previously been metallized with gold (average thickness 12 nm). In the description of the structure and morphology of the samples to verify surface defects, dislocations and size of particles , made using a Transmission Electron Microscope (TEM) Tecnai 20 with the tension between 20 kV to 200 kV, and 1.9 Å resolution of point 0 2 nm. The catalytic tests were carried out on the alloy catalytic evaluation unit, Model TCAT -1 at atmospheric pressure, 300mg weighed sample which was introduced in Pyrex reactor heated at room temperature to 450°C at a heating rate of air in an atmosphere (argon) to flow.

After reaching a temperature of 450°C in 2h the sample remained under these conditions to remove the physically firssor life water.

The waste products of the reactor were successively injected "on line" by a 10-way valve on a Varian CP3800 gas chromatograph with thermal conductivity detector at 15 minute intervals until you reach the pseudo-steady state".

Results and Discussion

Diffractogram X-ray

The spectra of X-ray diffraction of the sample stoichiometry, $Al_{62,2}Cu_{25,3}Fe_{12,5}$ are shown in Figures 1 and 2, respectively, for heat treatment of 8 hours and 24 hours. These diffraction peaks can identify phases; icosahedral phase i-, β- crystalline phase composition Al_{50-x} $(Cu, Fe)_{50+x}$ and θ-AlCu$_3$ tetragonal phases and θ-CuAl$_2$. These results also suggest that the β phase is formed directly from the liquid alloy. In addition, the β phase transforms below 600°C to form the λ and

required for the process can be supplied by the reaction itself (auto thermal reaction). However, the production of hydrogen in the product gas and the conversion of methanol are lower than that of steam reforming of methanol [9].

A metal composition binary or tertiary alloy, quasicrystals in the case of dispersion will improve or alter the electronic structure of a catalyst. More importantly, the presence of a promoter can change the adsorption characteristics of the surface of the catalyst, changing the reducibility of the catalyst or, in some cases, alter the catalytic performance [10]. Generally burning methanol (CH_3OH) initially occurs in very high temperature with a catalyst thermal stability for no deactivation of the same at the beginning of the catalytic reaction [11] is necessary. In an early stage of this study it was observed that the Cu/Cr$_2$O$_3$/Fe$_2$O$_3$ high temperature shift catalyst exhibits remarkably low activity and poor selectivity for methanol steam reforming, and this combination will definitely never reach technical application for this reaction. However, this catalyst was not suspended from further experiments since there is a pool of information about a reaction system that cannot be discovered by the analysis and investigation of good and optimized catalysts.

This is the case of the alloy quasicrystalline $Al_{62,2}Cu_{25,3}Fe_{12,5}$ of the

Figure 3a: Image quasicrystalline the surface of the alloy $Al_{62,2}Cu_{25,3}Fe_{12.5}$ showing the icosahedral phase, at 8 h.

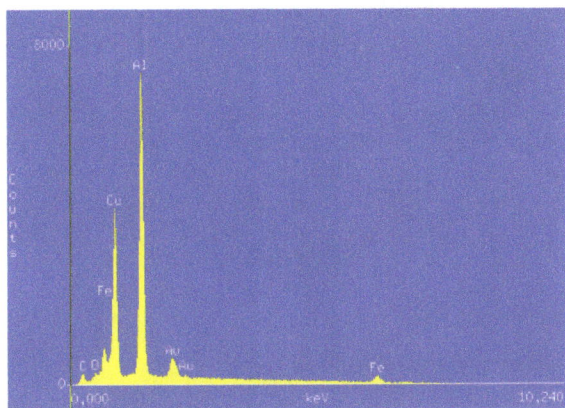

Figure 3b: EDS elemental analysis of quasicrystalline alloy $Al_{62,2}Cu_{25,3}Fe_{12.5}$ showing the icosahedral phase after heat treatment at 8 h.

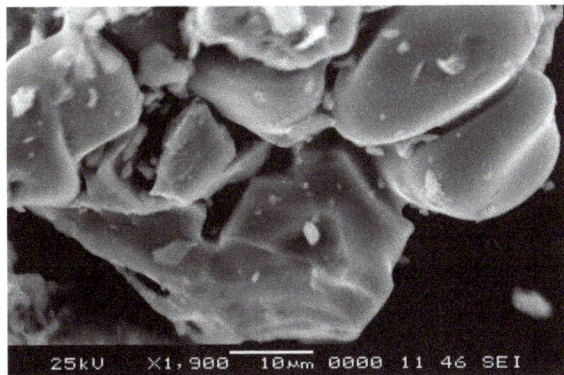

Figure 4: Alloy quasicrystalline $Al_{62,2}Cu_{25,3}Fe_{12,5}$ to 8 h heat treatment and subjected to a catalytic reaction.

θ - phases which are induced by the solubility of Cu and Fe. The $Al_{62,2}Cu_{25,3}Fe_{12,5}$, treatment of the alloy 5 at 850°C solid solution is shown in Figure 1. In Al-Cu-Fe system, icosahedron quasicrystal (i-$Al_{62,2}Cu_{25,3}Fe_{12,5}$) and ($\beta$-$Al_{10}Cu_{10}Fe_{10}$) have been proposed to have a close structural relationship from the viewpoint of valence electron correlation in the concentration of electrons free driving in their atomic structure. These two phases are distinguished in the XRD pattern (Figures 1 and 2), namely icosahedral quasicrystalline (i-phase) and tetragonal phases (θ-$CuAl_2$ and θ-$AlCu_3$ and β-Al_3Cu) phase. Intensity

peaks corresponding to the i-icosahedral layer is higher than the peaks are specifically related to the β-phase.

However, the alloy $Al_{62,2}Cu_{25,3}Fe_{12,5}$ fusion, i- phase co-existed with a small amount of AlFe (Cu) solid solution [12].

The crystalline β-phase crystalline phase is a solid solution solubility of copper (Cu) and iron (Fe). The θ-$AlCu_3$ tetragonal phases and θ-$CuAl_2$, they make a sequence corresponding to the gradual increase of temperature in the constitution of the league. The formation of quasicrystalline phase depends on the composition and shape as the transformation of the crystalline phases occurs, that is, with increasing temperature within the range of composition suitable for aesthetic transformation to occur. However the results of the two diffraction patterns of X-ray indicates that the peak intensity is greater quasicrystalline the icosahedral phase. It should be noted when comparing Figure 1 with Figure 2 shows the sample that underwent a heat treatment of 24 h, it is found that the quasicrystalline league became homogeneous, which is almost single phase because the peak of the crystalline β-phase is low intensity almost imperceptible.

Scanning electron microscopy and EDS

Figures 3a, 3b, 4 and 5 respectively show the sample results of scanning electron microscopy in samples of alloy composition quasicrystalline $Al_{62,2}Cu_{25,3}Fe_{12,5}$ catalytic reaction before and after the samples were subjected to catalysis. For catalysts trimetal Al-Cu-Fe in continuing training (one of the promising reactions in this area is the direct synthesis of dimethyl carbonate CO_2 and methane) from CO_2 and methanol.

The Scanning Electron Microscopy (SEM) and Energy Dispersive Spectroscopy (EDS) applied to the sample type $Al_{62,2}Cu_{25,3}Fe_{12,5\ 5}$ produced Figures 4a and 4b respectively for heat treatment 8pm. Composition analysis was performed with SDS showed that regions were composed mainly of Cu, Fe, O and a small amount of Al that can complex with Cu, Al_2O_3, $AlFe_3$, Al_3Cu, Cu_2O and Fe_3O_4 or $CuFeO_2$ [13]. The image of the microstructure of the alloy quasicrystalline icosahedral phase of Figure 4a corresponds to a crystal polyhedron formed by a combination of pentagonal and hexagonal phases, with geometric uniformity. In Figure 4b it is observed in the EDS elemental analysis of the spectrum, there is a greater predominance of aluminum than other elements (copper and iron) that makes up the quasicrystalline league. Figures 4 and 5, respectively, show the results of the analysis of scanning electron microscopy in the sample after the catalytic reaction.

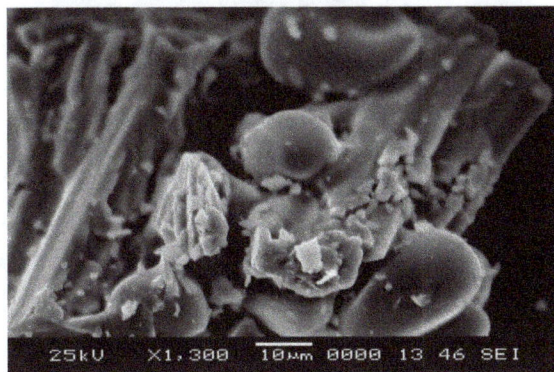

Figure 5: Alloy quasicrystalline $Al_{62,2}Cu_{25,3}Fe_{12,5}$ to 24 hours of heat treatment and subjected to a catalytic reaction.

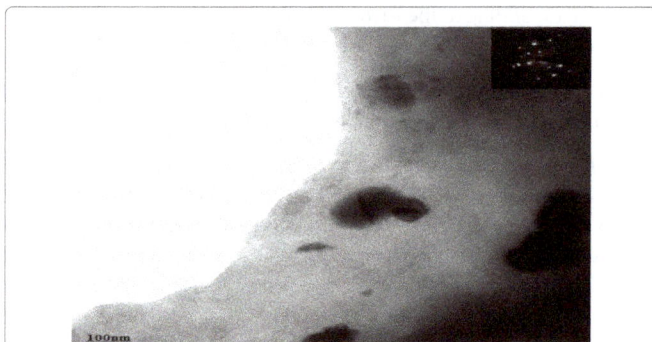

Figure 6: Image Transmission Electron Microscopy - MET league of quasicrystalline $Al_{62,2}Cu_{25,3}Fe_{12,5}$ after a catalytic reaction.

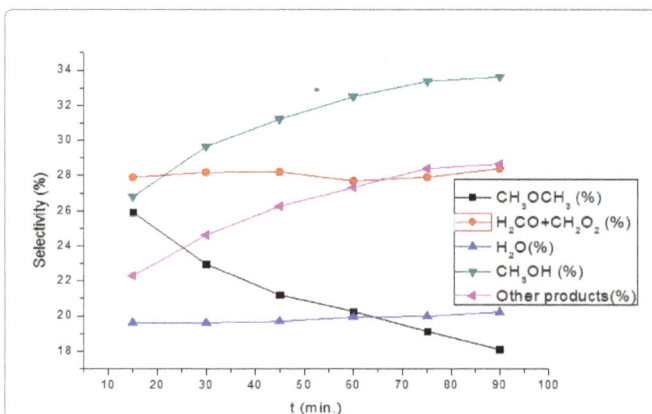

Figure 7: Catalytic behavior of the sample quasicrystal $Al_{62,2}Cu_{25,3}Fe_{12,5}$ on the basis of selectivity (%) of the product and time in minutes.

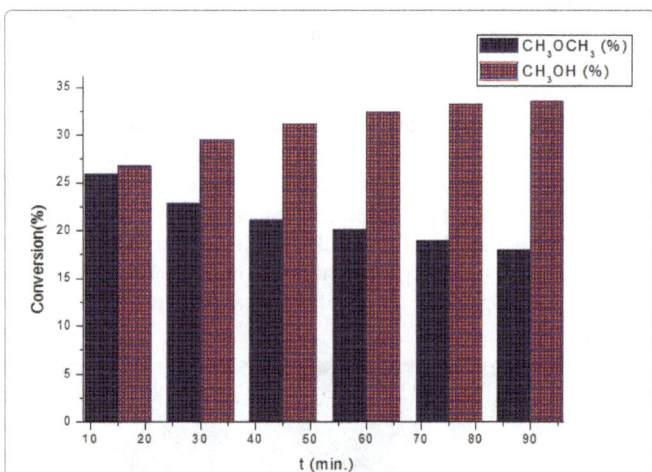

Figure 8: Comparison of conversion rate (%) of the product methanol and methanoic acid + methanol obtained in the use of quasicrystal $Al_{62,2}Cu_{25,3}Fe_{12,5}$ as a catalyst versus time (minutes).

The microstructure of the alloy samples quasicrystalline Figures 4 and 5 by which underwent the heat treatment of 8 hours and 24 hours process, and also by a catalytic reaction can analyze these figures there quasicrystalline occurrence of decomposition of crystalline phases with strong presence of the intermetallic phase. Of course, the reduction occurred in quasicrystalline phase and the intermetallic phase oxidation, which in turn acted as nuclei or catalytic sites.

Transmission Electron Microscopy (TEM)

The Transmission Electron Microscopy (TEM) sample showed 6 points of diffraction profiles obtained from the dry process, showed a dark layer as shown in Figure 6 that this can be attributed to Cu, Cu_2O and Fe_3O_4. The dry regions generated in quasicristal had uniform composition and consisted of homogeneous mixture of Cu, Fe, Al, and their oxides. Within the area of particle quasicrystal (Al-Cu-Fe) presented symmetry with double reflections with a quasi periodic structure, featuring quasicrystalline surface with standard well defined symmetry. Transverse TEM observations with clear evidence that the microstructure of the layer passivating Dry is heavily dominated by quasicrystalline precursor alloy $Al_{62,2}Cu_{25,3}Fe_{12,5}$. TEM observations showed taken in a cross-section of the sample revealed a cubic intermetallic phase in the form of spinel $Cu_xFe_{3-x-y}Al_yO_4$ formed on the outermost layer of quasicrystal after dry heat treatment. The orientation can stabilize the nanoparticles through a binding Cu Cu-Fe-O at quasicrystal for the production of olefins, dimethyl ether and methanol itself. This drastic increase in catalytic activity is responsible for the fine Cu nanoparticles in the composite.

The image in Figure 6 shows a low magnification bright field image of $Al_{13}Fe_4$ and defective dendritic shape, and the dark field of small particles of copper (Cu) and iron (Fe) in which elements are alloying. On the other hand, the higher selectivity for methanol reforming and decomposition products and the same is related to the transition metal in this case should tell Cu and Fe, the oxidation reaction of methanol on the surface of copper (Cu) yields the formation of formaldehyde as the main product [14].

Evaluation of catalytic processes

Evaluation of catalytic activity in terms of selectivity (%) of quasicrystalline alloy $Al_{62,2}Cu_{25,3}Fe_{12,5}$ investigated, and the results obtained from the two tables, it was possible to built the two graphs of Figures 7 and 8 with the curves of selectivity (%) versus time (minutes). Table 2 presents the data conversion (%) of methanol and methanoic acid + methanol obtained as a function of time (minutes).

The results obtained with the catalyst formed by connecting quasicrystalline $Al_{62,2}Cu_{25,3}Fe_{12,5}$ had its performance tested in a catalytic evaluation unit, which shows the percent yield of methanol plus the product of the reaction. Considering the selectivity of the products obtained, it was observed that there was good selectivity progressed according to length, for products methanol + methanoic acid, water and methanol, except for dimethyl ether.

In the first 40 minutes, the dimethyl ether showed a good

t(min.)	15	30	45	60	75	90
CH_3OCH_3 (%)	25,91	22,97	21,2	20,25	19,12	18,09
$H_2CO+CH_2O_2$ (%)	27,89	28,18	28,19	27,68	27,89	28,38
H_2O(%)	19,61	19,61	19,68	19,93	19,97	20,21
CH_3OH (%)	26,79	29,64	31,22	32,49	33,37	33,62
Other products. (%)	22,28	24,61	26,23	27,30	28,38	28,64

Table 1: Selectivity (%) of methanol in the reaction products as a function of injection time (min).

t(min.)	15	30	45	60	75	90
CH_3OCH_3 (%)	25,91	22,97	21,2	20,25	19,12	18,09
CH_3OH (%)	26,79	29,64	31,22	32,49	33,37	33,62

Table 2: Conversion (%) Products of methanol and methanoic acid methanol + versus time (min).

percentage of selectivity. However, on 50 minutes following a trend of decay was observed. The ethanol + methanoic acid showed a selectivity curve practically constant over the time intervals. The analysis of the selectivity curve of the percentage of methanol showed an increase of nearly 10%. Such growth can be attributed to a complete oxidation of methanol verified later in the reaction of formation of the methoxy radical (-O -CH$_3$). This radical is considered the most stable intermediate after adsorption on the surface of methanol. In reaction that occurs due to the increase of the methanol temperature there is a decomposition of formaldehyde. Of course, this next stage of formaldehyde yields in varying degree the carbon monoxide CO and carbon dioxide CO$_2$. In general the reactions occur in sequential steps. We observe, in the first step of the reaction of methanol dehydrogenation one.

Conclusions

The main conclusions of the research are as follows:

1. The league quasicrystalline Al$_{62,2}$Cu$_{25,3}$Fe$_{12,5}$ and are thermodynamically stable at high temperatures, one of the favorable prerequisite for the use of catalysis, and their electronic and surface of quasicrystal provides a catalytic activity properties;

2. The microstructure of the sample quasicrystal, Al$_{62,2}$Cu$_{25,3}$Fe$_{12,5}$ after the catalytic reaction, it was observed that there was a breakdown of the crystalline phase quasicrystalline with a strong presence of intermetallic phase;

3. The selectivity and conversion of quasicrystalline alloy Al$_{62,2}$Cu$_{25,3}$Fe$_{12,5}$ demonstrated the adsorption of the metal surface of methanol quasicrystal influencing the percentage of products obtained from the reaction medium. Particles of copper (Cu) are present on the surface of quasicrystal being and is the metallic transition element that favors the oxidation of methanol;

4. The low cost of alloying encourages their use in catalytic reactions, and is one of the most favored indices showing the possibility of quasicrystal be used as an industrial catalyst;

5. The species dispersed in the Fe homogeneous dry layer increases the catalytic activity and suppresses the aggregation of Cu, producing other products such as formaldehyde, methanoic acid and other olefins produce methanol.

6. This is probably the formation of other olefins formed from the decomposition of methyl formate, the primary intermediate surface quasicrystal due to their electronic properties during the course of heterogeneous catalysis;

7. The quasicrystalline surfaces in these catalysts were dominated by methoxy and formate groups, the intermediate formaldehyde and methyl formate dioxomethylene can be observed during the steam reforming of methanol.

Acknowledgment

The authors thank PRH 28/MCT/ANP for financial support of this work and the Laboratory Group of Magnetism and Magnetic Materials MMM - Department of Physics, Center of Exact and Natural Sciences of UFPE.

References

1. Shechtman D, Blech I, Gratias D, Cahn JW (1984) Metallic Phase With Long Range Orientational Order and No Translational Symmetry. Physical Review Letters 53.

2. Nascimento L, Agostinho LCL, Cavalcanti FB (2012) Grouping Model in Fermi Surface Applied to Quasicrystals. Revista Colombiana de Materiales N 3: 55-62.

3. Agostinho LCL (2009) Estudo da Aplicabilidade dos Quasicristais AlCuFe em Reações Catalíticas na Oxidação do Metanol,Dissertação (Mestrado em Ciências de Materiais). Universidade Federal da Paraíba,João Pessoa-Paraíba.

4. Tsai AP, Yoshimura M (2001) Highly active quasicrystalline AlCuFe catalyst for steam reforming of methanol, Applied Catalysis A 214: 237-224.

5. Hao J, Wang J, Wang Q, Yu Y, Cai S, et al. (2009) Catalytic oxidation of cyclohexane over Ti-Zr-Co catalysts. Applied Catalysis A, General 368: 29-34.

6. Nascimento L, Agostinho LCL, Cavalcanti BF (2012) Oxidação na Fase Icosaedral do Quasicristal Al62,2Cu 25,3Fe12,5. Ciência & Tecnologia dos Materiais 24: 73-79.

7. Yoshimura M, Tsai AP (2002) Quasicrystal application on catalyst. Journal of Alloys and Compound 342: 451-454.

8. Shen WJ, Jun KW, Choi HS, Choi HS, Lee KW (2000) Thermodynamic investigation of methanol and dimethyl ether synthesis from CO2 hydrogenation. Korean J Chem Eng 17: 210-216.

9. Rosen MA (1991) Thermodynamics Investigation of Hydrogen Production by Steam Methane reforming. International Journal of Hydrogen 16: 207-217.

10. Lenarda M, Storaro L, Frattini R, Casagrande M, Marchiori M, et al. (2007) Oxidative methanol steam reforming (OSRM) on PdZnAl hydrotalcite derived catalyst. Catalysis Communications 8: 467-470.

11. Lindstrom B, Pettersson LJ (2001) Hydrogen generation by steam reforming of methanol over copper-based catalysts for fuel cell applications. International Journal of Hydrogen Energy 26: 923-933.

12. Nascimento L, Agostinho LCL, Cavalcanti BF (2009) Comportamento da Oxidação na Fase Icosaedral do Quasicristal Al62,2Cu25,3Fe12,5. Acta Microscópica l. 18: 295-303.

13. Estrella M, Barrio L, Zhou G, Wang X, Wang Q, et al. (2009) In Situ Characterization of CuFe2O4 and Cu/Fe3O4 Water-Gas Shift Catalysts. J Phys Chem C 113: 14411–14417.

14. Agostinho LCL, Barbosa CMBM, Nascimento L, Rodbari JR (2013) Catalytic Dehydration of Methanol to Dimethyl Ether (DME) Using The Al62,2Cu25,3Fe12,5 Quasicrystalline Alloy. J Chem Eng Process Technol 4: 2-8.

14

Droplet Size Distribution in a Kenics Static Mixer: CFD Simulation and Experimental Investigation of Emulsions

Farzi GA[1]*, Reza-Zadeh N[2] and Parsian Nejad A[2]

[1]Material and Polymer Engineering Department, Hakim Sabzevari University, Sabzevar, Iran
[2]Mechanical Engineering Department, Hakim Sabzevari University, Sabzevar, Iran

Abstract

The minimum achievable droplet sizes created by a simple in-line Kenics Static Mixer (KSM) under various flow rates and mixing time in oil in water (O/W) emulsion were investigated through turbulent flow system. First, a Computational Fluid Dynamics (CFD) method is utilized to predict final droplet sizes in different Reynolds number. Then, an experimental setup was used in order to validate CFD results. The droplet size was monitored using Dynamic Light Scattering (DLS) technique by means of a Malvern zetasizer machine. Breakup/coalescence of droplets under constant volume fractions of oil was studied when flow rate was varied from 36.7 to 85 ml/s. Results showed that droplet size distribution highly depends on flow rate and mixing time. Droplets break more easily and faster at higher flow rates. The results proved that the obtaining small enough droplets using static mixer in less than 40 minutes at the flow rates above 36.7 ml/s at moderate concentration of oil volume fraction.

Keywords: Two phase flow; Emulsion; Droplet sizes; Kenics Static Mixer; CFD

Introduction

Over recent years a great deal of attention has been paid to the formation and stability of micro/nano scale emulsions and precise control of droplet size and size distribution [1]. Two phase liquid dispersion is one of the most complex processes among mixing operations. Agitating two immiscible liquids results in the dispersion of one phase in the other in the form of small droplets whose characteristics depend on the equipment and the operating conditions [2]. It is practically impossible to make stable dispersions of uniform droplet size distribution, because of the wide range of properties and flow conditions [3]. A large amount of work can be found in the literature concerning the prediction of drop size distributions in turbulent liquid-liquid dispersions in static mixers (SM). Most of them use the concept of a turbulent energy cascade to predict the maximum stable droplet diameter, referring to the Hinze-Kolmogorov theory [2-8].

Static mixers are introduced as an alternative device believed to have a significant industrial potential to produce stable emulsions [9]. While SMs are widely used in other agro and petrochemical processes, they have not been studied in depth for the generation of mini-emulsion droplets. It is clear that they can be economically practical, safely used and can be utilized on larger scales. However, in terms of dispersion systems, their role in droplet breakage is an area of ongoing research. In our previous work [10], we reported a successful experimental production of mini-emulsions which was produced by making the mixture circulate two immiscible liquids (oil and aqueous phases) through the pipe in which the SM was inserted.

In this investigation, a CFD code is used to calculate the flow in the KSM and results are validated by means of DLS measurements of a final droplets diameter in a specific time. In the first section, the numerical methods and governing equations are proposed and the model and simulation properties are described. In the second section, two cases in the same manner of experimental setup but different in order of flow rate are defined and material properties are described. Finally, the CFD results are compared with the experimental results to evaluate the results, then numerical results and the dynamic behavior of the KSM is discussed in more detail.

Theoretical Model-Numerical Methodology

Breakup of bubbles and droplets, has been the subject of investigation for several decades starting with the pioneering work by two researchers, Kolmogorov [11] and Hinze [12], who proposed a formula for the maximum drop size, independently. Thereafter, Luo base on spherical assumption of droplet shapes proposed a model for breakup of fluid drop, description of the stability of mono-dispersed colloids, Population Balance Equations (PBEs) have found diverse applications in areas involving particulate systems [13]. Recently, Solsvik proposed an algebra of the high-order least-squares method, which linked to the implementation issues of a problem describing the drop size distribution within a liquid-liquid emulsion [14]. The high-accuracy, low numerical diffusion of the least-squares method for these types of solution has been proved, regarding the published literatures [15-18].

In this work, a 3D CFD model of the two-phase flow in continuous KSM is developed. Based on an Eulerian-Eulerian two-fluid model, the high-order least-squares method (HOLS) is used to solve the PBE [19]. The PBE and CFD models are both solved by the non-commercial CFD code developed by our researcher team.

A general form of the population balance equation can be expressed as follows:

$$\frac{\partial_n(L;x,t)}{\partial_t} + \nabla \cdot [\bar{u}n(L;x,t)] = -\frac{\partial}{\partial L}[G(L)n(L;x,t)] + B_{ag}(L;x,t) - D_{ag}(L;x,t) + B_{br}(L;x,t) - D_{br}(L;x,t) \quad (1)$$

Where, $n(L;x,t)$ is the number density function with droplet diameter (L) as the internal coordinate, $G(L)n(L;x,t)$ is the droplet flux due to molecular growth rate, $B_{ag}(L;x,t)$ and $D_{ag}(L;x,t)$ are the birth and

*Corresponding author: Farzi GA, Material and Polymer Engineering Department, Hakim Sabzevari University, Sabzevar, Iran

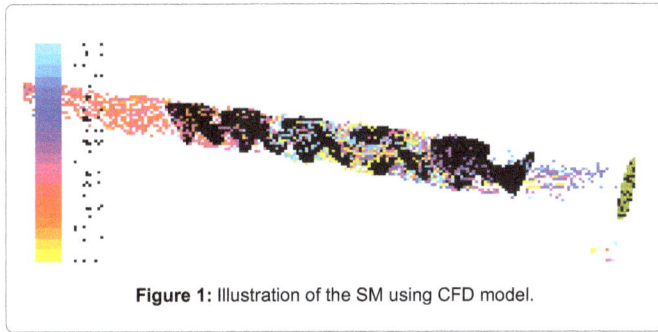

Figure 1: Illustration of the SM using CFD model.

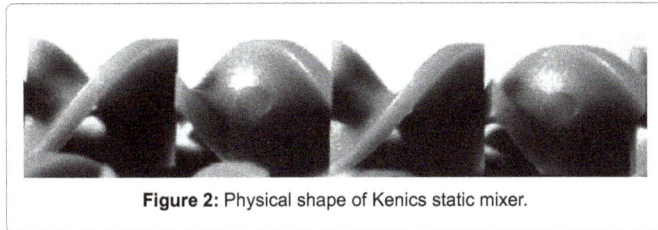

Figure 2: Physical shape of Kenics static mixer.

name	First Case		Second Case	
	material	amount	material	amount
Continues phase	de-ionized water	60% (211 g)	de-ionized water	85% (418 g)
Dispersed phase	methyl methacrylate (MMA)	40% (84.4 g)	sunflower oil	15% (72 g)
surfactant	sodium dodecyl sulfate (SDS)	1 g/L	sodium dodecyl sulfate (SDS)	0.4 g/L

Table 1: Emulsion Recipes.

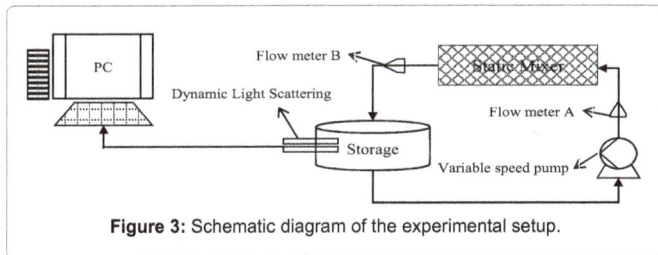

Figure 3: Schematic diagram of the experimental setup.

death rate of droplets diameter (L) due to aggregation, respectively, and $B_{br}(L;x,t)$ and $D_{br}(L;x,t)$ are the birth and death rate of droplets diameter (L) due to breakage, respectively. In eqn 1, the first term on the left hand is the transient term, the second term is the convective term, and the terms on the right hand are the source term describing droplet growth, aggregation, and breakage dynamics, respectively.

Regarding to its actual properties, the simulated SM has an inner diameter of 25 mm, a height of 25 mm, and 10 standard static elements fabricated from polyacetal plastic, arranged alternatively at 90° (Figure 2). In addition, Grid sensitivity was carried out initially, and the results indicated that a total amount of 325 K cells was adequate to conserve the mass of each phase in the dynamics model.

In order to obtain suitable mesh size in our CFD model at initial step, fluid velocity was varied from 0.11 to 10 m/s which provides Reynolds number from 3 K to 280 K and mesh size was adapted for minimal error based on numerical and experimental Re number. Significant differences was seen for course mesh size, however, after re-meshed the model with super fine mesh, ($\sim 10^{-6}$) when looking at the overall flow characteristics, which are shown below, one can see that the differences

are not too large and in general they agree well. Detailed information of mesh independency study and prediction errors is presented in Table 2.

It should be note that, although the physical geometries of SM is adapted with model (Figure 1), there may exist some differences between results of our model and experiments due to small deviation of SM geometry and other assumptions.

The phase-coupled SIMPLE (Semi-Implicit Method for Pressure Linked Equations) algorithm was used to couple pressure and velocity [20]. A one stage calculation and two cases with different Reynolds numbers were implemented. The flow field was simulated with bulk velocity started from 36.7 ml/s, 60.6 ml/s and 62 ml/s in the first case and 36.7 ml/s up to 82 ml/s in the second case. Then results were compared with those obtained experimentally from DLS technique. The breakage and coalescence process was simulated by utilizing the energy and PBE model regarding droplet tracing technique until the average nano-size Sauter mean diameter (d_{32}) reached. It should be noted that since sufficient amount of literatures proved there is negligible ratio of breakage occurs in storage tank compared to those in SM, droplet breakage in storage can be disregarded [9,10,21-23].

Experimental

To carry out the experimental studies two cases were considered for oil phases; methyl methacrylate (MMA) and sunflower oil. In both systems de-ionized water was used as continuous phase. General formulations of mini-emulsions are shown in Table 1.

In the second case, Sunflower oil as dispersed has a density of 902.4 kg/m³, Refractive index of 1.4646, and viscosity of 47.11 g/m.s, all measured at 25°C.Geometry and dimensions of SM were modeled using Solid Works 3D CAD software, and exported into commercial software GAMBIT 2.1 and an appropriate mesh is generated.

A schematic diagram of the experimental setup is provided in Figure 3. A circulator pump was made to function with variable electrical current to ensure a series of known flow-rates. The mixture of two immiscible fluids was pumped from a 2 liter capacity reservoir to the SM. The fluid flow unit consists of piping section (with inner diameter of 25 mm and total piping lengths of 1571 mm) preceded by an inlet section where two phases are co-axially introduced into the piping section without any pre-mixing process. However, immediately after entrance into the pipe they mixed due to the turbulence fluid flow system. As mentioned previously, the oil droplet size was measured by DLS after certain time achieving steady state condition. Furthermore, a feedback system is used to measure the flow rate.

Results and Discussion

At the initial phase of the validation process, the results of experimental emulsification using the KSMs according to first formulation were compared with those obtained from CFD model. These experimental results were previously published elsewhere [10]. Figure 4 shows droplet size as emulsification time. In this figure the points to the graph are experimentally captured for different flow rates, whereas the lines indicate calculated values using CFD code.

Regarding to the Figure 4, droplet diameter decreasing asymptotically with increasing homogenization time and smaller droplet obtained at higher flow rates. In other respects, an increase in mechanical energy can help overcome the limit imposed by interfacial tension, thereby inducing more breakage. One might theorize that at higher flow rates, more energy is input into the system allowing breaking up large droplets. By means of that, intensifies the distribution

Re. No.	100		200		400		800		1.2 k		12 k		50 k		80 k		120 k	
Mesh Size	Pre. Err. %	Vel. Pro. Err. %	Pre. Err. %	Vel. Pro. Err. %	Pre. Err. %	Vel. Pro. Err. %	Pre. Err. %	Vel. Pro. Err. %	Pre. Err. %	Vel. Pro. Err. %	Pre. Err. %	Vel. Pro. Err. %	Pre. Err. %	Vel. Pro. Err. %	Pre. Err. %	Vel. Pro. Err. %	Pre. Err. %	Vel. Pro. Err. %
10^{-1}	725	911	608	552	-	-	-	-	-	-	-	-	-	-	-	-	1332	1592
10^{-2}	408	391	418	382	427	371	-	-	-	-	-	-	-	-	-	-	528	717
10^{-3}	244	212	238	203	217	193	225	201	-	-	-	-	-	-	-	-	-	-
10^{-4}	68	118	72	128	59	113	73	141	68	140	-	-	-	-	-	-	-	-
10^{-5}	16	33	19	25	12	16	14	22	11	16	9	13	13	15	17	26	14	18
10^{-6}	16	31	17	21	13	15	14	21	10	14	9	14	12	15	14	23	16	21
10^{-7}	17	38	17	20	11	10	12	20	7	11	10	12	12	14	14	23	14	20

Table 2: Mesh Independency Results. Pre. Err=Prediction Error, Vel. Pro. Err=Velocity Profile Error.

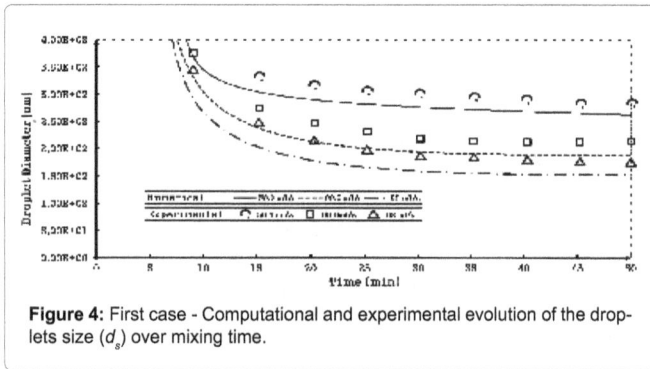

Figure 4: First case - Computational and experimental evolution of the droplets size (d_s) over mixing time.

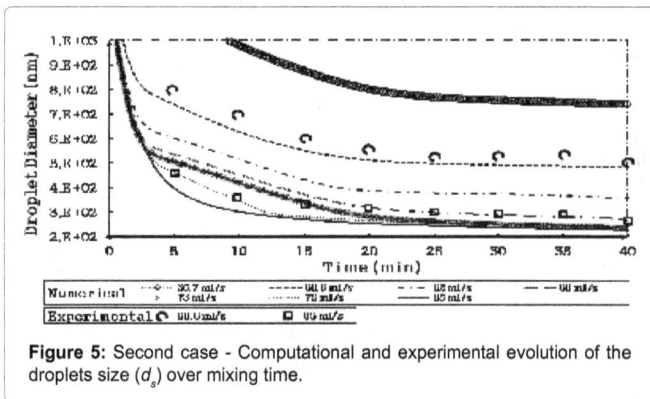

Figure 5: Second case - Computational and experimental evolution of the droplets size (d_s) over mixing time.

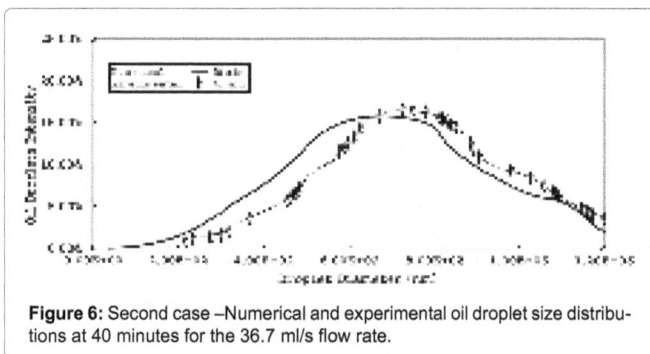

Figure 6: Second case –Numerical and experimental oil droplet size distributions at 40 minutes for the 36.7 ml/s flow rate.

and decreases the average size. This leads to narrow distribution droplet size.

Considering Figure 4, discrepancy between the CFD and experimental results for 36.7 ml/s flow rates in 6 times interval are, 28, 31, 27, 35, 32 and 29 nm respectively, with total computational discrepancy about 14%. These results have been 35, 27, 41, 36, 32 and 27 with total computational discrepancy about 18% for 60.6 ml/s flow rate. Similarly, the results shows about 45, 42, 37, 31, 29 and 30 nm mismatch in droplet size at 62 ml/s flow rate and 21% of total computational discrepancy.

Figure 5 shows the effect of different flow rates on the mean droplet size at fixed values of the sunflower oil and surfactant concentrations. The flow rate was set to 60.6 and 85 ml/s for experiments and varied from 36.7 to 85 ml/s for numerical studies. As it can be seen, at 60.6 ml/s flow rate, the experimental data shows larger droplets than numerical results. When flow rate is higher than 68 ml/s from the beginning of process for few minutes droplet diameter of emulsion is in good accordance with those of CFD results. However, in general there is a meaningful difference between experimental and CFD results. This is due to the fact of the problem of "lost" droplets, saying, droplet trajectories are trapped near a solid wall accentuate in lower flow rates [21].

The experimental results of 85 ml/s flow rates shows relatively lower difference of droplet size between numerical and experimental results in compared with those obtained for lower flow rates.

With increasing flow rate, the Non-linear relationship between the flow rate and the average droplets size appears even at first stage of emulsification.

In order to evaluate the validity of our CFD model results for a given homogenization time, droplets size of emulsion prepared within 40 minutes at 36.7 ml/s flow rate is compared with those obtained from numerical data in Figure 6. This figure clearly displays similar trends for numerical and experimental results.

It is possible to determine the frequency of coalescence and breakup for numerical results in Figure 6. This may help us to have an idea for experimentally coalescence and breakup of droplets. The brakeage of droplet has been studied extensively, the incorporation of two different breakage behavior that accounted for large droplets to break easier due to turbulent shear [24,25] and on the other hand, small droplets break due to collisions between droplets and turbulent eddies [26,27].

However, theoretical and experimental results are not ideally matched, but in order to gain an insight to the results of previous investigations it is worthy to discuss the frequency of coalescence and break up of droplets based on Figure 6 in three different group of smaller than 400 nm, between 400-800 nm and larger than 800 nm. Where,

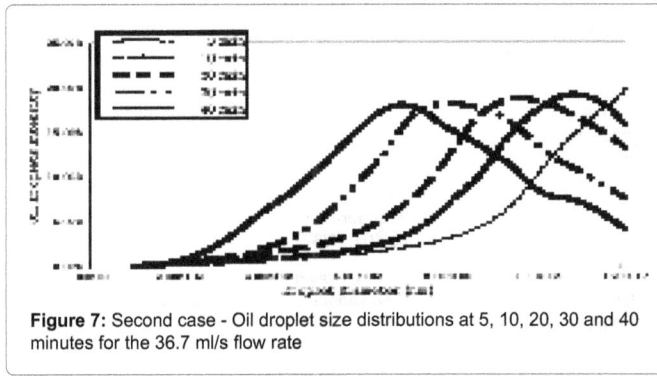

Figure 7: Second case - Oil droplet size distributions at 5, 10, 20, 30 and 40 minutes for the 36.7 ml/s flow rate

Flow rate (ml/s)	Homogenization time (min)	Average Num. Err. (%)		
		Under 400 nm	400 to 800 nm	Over 800 nm
36.7	5	12.51	19.28	13.25
	10	18.92	17.05	12.08
	20	18.26	21.59	23.24
	30	13.85	16.12	21.25
	40	20.82	23.20	28.56
60.6	5	14.28	21.02	14.70
	10	13.25	19.66	14.86
	20	16.02	23.74	24.00
	30	15.11	16.83	21.63
	40	16.28	17.47	17.68
62	5	13.80	15.84	19.83
	10	12.52	15.36	16.80
	20	25.53	17.70	19.18
	30	19.28	18.63	18.92
	40	17.28	18.95	24.84
68	5	14.50	17.67	15.19
	10	17.08	16.92	18.22
	20	24.53	17.51	17.16
	30	16.24	22.93	22.16
	40	15.23	23.56	13.12
73	5	17.52	14.14	16.90
	10	23.89	16.47	19.29
	20	24.57	23.72	17.54
	30	21.99	15.08	21.67
	40	16.29	16.48	24.06
75	5	17.24	16.93	16.02
	10	19.82	22.62	17.70
	20	23.87	19.79	15.93
	30	16.55	13.47	22.03
	40	13.83	13.60	22.86
85	5	17.91	15.53	17.81
	10	22.88	12.51	19.11
	20	18.01	15.39	21.40
	30	18.63	21.90	18.21
	40	24.21	15.77	15.09

Table 3: Discrepancies in the computational and experimental results.

numerical results for the average frequency of droplets under 400 nm showed average coalescence of $1.243 \times E10^6$ and $8.315 \times E10^2$ of breakage per droplet. Whereas, these values are $6.218 \times E10^4$ and $2.386 \times E10^4$ for second group and also $9.624 \times E10^2$, $9.582 \times E10^8$ for third group, respectively. These results only calculated during the time a droplet was a member of the groups. These comparison shows that smaller droplets tend to more coalescence and larger droplets should break up more frequently than smaller ones. These results are in agreement with those

of other researchers. If the previously described breakage frequency is valid, then our experimental data supports the dependency of the breakage efficiency to the droplet size. Regarding the numerical data there is negligible breakage rates predicted for small droplet sizes.

It is reported that the coalescence of droplets, depends on the evolution of overall surface area and shape of drops [28,29] and/or on the diameter of drops [30] and/or on the volume of the droplet [30,31]. Now it is interesting to turn our attention to check whether or not these well-founded phenomena may satisfy with our CFD results. Since we use PBEs, the exact number of droplets is available for each of previously mentioned groups of droplets. It was determined that the number of droplets under 400 nm is only 2.89% of total number of droplets, while those between 400 to 800 nm are 39.24% and larger than 800 nm are 57.87%. Based on these results total surface area of droplets are $5.292 \times E10^9$ nm^2, $2.634 \times E10^{14}$ nm^2 and $1.109 \times E10^{14}$ nm^2 for droplets groups less than 400 nm, between 400-800 nm and larger than 800 nm, respectively.

Thus, considering the total surface area of droplets in compare with coalescence and breakage rate reveals a logical conformity with respect to previews judgment; this means that the coalescence rate strongly depends indirectly on the droplet size and with decreasing droplet sizes, harmonically increasing total droplets surface area, coalescence frequency increased in agreement.

Figure 7 shows numerical results of oil droplet size distributions after 5, 10, 20, 30 and 40 minutes of homogenization for the 36.7 ml/s flow rate. One can see as the slope of the 5 min indicator curve increased dramatically after 900 nm droplet size, the ratio is express limitation of droplet breakage to the 900 nm.

It also reproduced the positive trend that the mean diameter decreased with increasing homogenization time. However, the numerical results show some difference in droplet diameter, especially for the lowest flow rates. Taken collectively, these results suggested that the functional dependencies of the mixing time and breakage rate was reasonable but that quantitative predictions with the base case model parameters may be difficult. Below also provided further numerical details of the full drop size distribution (See Table 3).

Conclusions

Droplet breakage using KSM has been simulated by means of CFD technique. In the preliminary validation stage, the simulation has captured the droplet changes successfully and reasonable difference between computed and measured results was shown. Fluid flow rate, mean droplet size and homogenization time were considered as important parameters. The CFD results were evaluated for two experimental systems with different oil phase. Droplet size was measured for these systems using Dynamic Light Scattering method. In theoretical model mesh size was adapted for our system using mesh dependency studies. Comparing theoretical results and experimental results may pursuit one that population balance equations can be suitable technique to simulate droplet creation at homogenization process. A more in depth study considering much more parameters is required to gain better understanding of homogenization process using SM.

References

1. Chang-Hyung C, Weitz DA, Lee CS (2013) One step formation of controllable complex emulsions: from functional particles to simultaneous encapsulation of hydrophilic and hydrophobic agents into desired position. Advanced Materials 25: 2536-2541.

2. Musa SH, Basri M, Masoumi HRF, Karjiban RA, Malek EA, et al. (2013) Formulation Optimization of Palm Kernel Oil Esters Nano emulsion-Loaded with Chloramphenicol suitable for Meningitis Treatment. Colloids and Surfaces B: Biointerfaces 112: 113-119.

3. MeeRyung L,Ha-Neul C, Ho-Kyung H, Won-Jae L (2013) ARTICLE: Production and Characterization of Beta-lactoglobulin/Alginate Nanoemulsion Containing Coenzyme Q_10: Impact of Heat Treatment and Alginate Concentrate. Korea food science of animal resources 33: 67-74.

4. Sheeran PS, Matsunaga TO, Dayton PA (2013) Phase-transition thresholds and vaporization phenomena for ultrasound phase-change nanoemulsions assessed via high-speed optical microscopy. Physics in medicine and biology 58: 4513.

5. Sanela DM, Nebojsa CD, Tanja IM, Jela RM, Milic, Gordana VM, et al. (2013) Nanoemulsions produced with varied type of emulsifier and oil content: An influence of formulation and process parameters on the characteristics and physical stability. Hemijskaindustrija 67: 795-809.

6. Schmidt J, Damm C, Romeis S, Peukert W (2013) Formation of nanoemulsions in stirred media mills. Chemical Engineering Science 102: 300-308.

7. Mat Hadzir N, Basri M, Abdul Rahman MB, Salleh AB, Raja Abdul Rahman RN, et al. (2013) Phase behaviour and formation of fatty acid esters nanoemulsions containing piroxicam. AAPS PharmSciTech 14: 456-463.

8. Campardelli R, Cherain M, Perfetti C, Iorio C, Scognamiglio M, et al. (2013) Lipid Nanoparticles production by supercritical fluid assisted emulsion-diffusion. The Journal of Supercritical Fluids 82: 34-40.

9. El-Jaby U, Farzi G, Lami EB, Cunningham M, Timothy MF (2009) Emulsification for latex production using static mixers. In Macromolecular Symposia 281: 77-84.

10. Farzi G, Mortezaei M, Badiei A (2011) Relationship between droplet size and fluid flow characteristics in miniemulsion polymerization of methyl methacrylate. Journal of Applied Polymer Science 120: 1591-1596.

11. Tikhomirov VM (1991) On the breakage of drops in a turbulent flow. Selected Works of A. N. Kolmogorov Mathematics and Its Applications (Soviet Series) Volume 25: 339-343.

12. Hinze JO (1955) Fundamentals of the hydrodynamic mechanism of splitting in dispersion processes. AIChE J 1: 289-295.

13. Luo H, Svendsen HF (1996) Theoretical model for drop and bubble breakup in turbulent dispersions. AIChE J 42: 1225-1233.

14. Solsvik J, Jakobsen HA (2014) Solution of the dynamic population balance equation describing breakage–coalescence systems in agitated vessels: The least squares method. The Canadian Journal of Chemical Engineering 92: 266-287.

15. Solsvik J, Jakobsen HA (2013) On the solution of the population balance equation for bubbly flows using the high-order least squares method: implementation issues. Reviews in Chemical Engineering 29: 63-98.

16. Solsvik J, Borka Z, Becker PJ, Sheibat-Othman N, Jakobsen HA (2014) Evaluation of breakage kernels for liquid–liquid systems: Solution of the population balance equation by the least-squares method. The Canadian Journal of Chemical Engineering 92: 234-249.

17. Bork Z, Jakobsen HA (2012) Evaluation of breakage and coalescence kernels for vertical bubbly flows using a combined multifluid-population balance model solved by least squares method. Procedia Engineering 42: 623-633.

18. Solsvik J, Jakobsen HA (2013) Evaluation of weighted residual methods for the solution of a population balance model describing bubbly flows: The least-squares, galerkin, tau, and orthogonal collocation methods. Industrial & Engineering Chemistry Research 52: 15988-16013.

19. Santos FP, Favero JL, Lage PLC (2013) Solution of the population balance equation by the direct dual quadrature method of generalized moments. Chemical Engineering Science 101: 663-673.

20. Reddy NS, Rajagopal K, Veena PH, Pravin VK (2013) A Pressure Based Solver for an Incompressible Laminar Newtonian Fluids. International Journal of Fluids Engineering 5: 21-28.

21. Talansier E, Dellavalle D, Loisel C, Desrumaux A, Legrand J (2013) Elaboration of controlled structure foams with the SMX static mixer. AIChE Journal 59: 132-145.

22. Baumann A, Jeelani SAK, Holenstein B, Stössel P, Windhab EJ (2012) Flow regimes and drop break-up in SMX and packed bed static mixers. Chemical Engineering Science 73: 354-365.

23. Meijer HEH, Singh MK, Anderson PD (2012) On the performance of static mixers: A quantitative comparison. Progress in Polymer Science 37: 1333-1349.

24. Rafiee M, Simmons MJH, Ingram A, Stitt EH (2013) Development of positron emission particle tracking for studying laminar mixing in Kenics static mixer. Chemical Engineering Research and Design 91: 2106-2113.

25. Lee L, Hancocks R, Noble I, Norton IT (2014) Production of water-in-oil nanoemulsions using high pressure homogenisation: A study on droplet break-up. Journal of Food Engineering 131: 33-37.

26. Focke C, Kuschel M, Sommerfeld M, Bothe D (2013) Collision between high and low viscosity droplets: Direct Numerical Simulations and experiments. International Journal of Multiphase Flow 56: 81-92.

27. Kunnen RPJ, Siewert C, Meinke M, Schröder W, Beheng KD (2013) Numerically determined geometric collision kernels in spatially evolving isotropic turbulence relevant for droplets in clouds. Atmospheric Research 127: 8-21.

28. Munz M, Mills T (2014) Size dependence of shape and stiffness of single sessile oil nanodroplets as measured by atomic force microscopy. Langmuir 30: 4243-4252.

29. Ata S, Pugh RJ, Jameson GJ (2011) The influence of interfacial ageing and temperature on the coalescence of oil droplets in water. Colloids and Surfaces A: Physicochemical and Engineering Aspects 374: 96-101.

30. Mertaniemi H, Forchheimer R, Ikkala O, Ras RH (2012) Rebounding droplet-droplet collisions on superhydrophobic surfaces: from the phenomenon to droplet logic. Adv Mater 24: 5738-5743.

31. Heng F, Striolo A (2012) Mechanistic study of droplets coalescence in Pickering emulsions. Soft Matter 8: 9533-9538.

15

Adsorption on Activated Carbon from Olive Stones: Kinetics and Equilibrium of Phenol Removal from Aqueous Solution

Thouraya Bohli[1]*, Nuria Fiol[2], Isabel Villaescusa[2] and Abdelmottaleb Ouederni[1]

[1]Laboratory of Research: Engineering Processes and Industrials Systems (LR11ES54), National School of Engineers of Gabes, University of Gabes, St Omar Ibn Elkhattab, 6029 Gabes, Tunisia
[2]Department d'Enginyeria Quimica, Agraria i Tecnologia Agroalimentaria,Universitat de Girona, Avda Lluis Santolo, 17003 Girona, Spain

Abstract

Activated carbon is prepared with chemical activation of olive stones, by using H_3PO_4. Batch adsorption of phenol from aqueous solution was investigated. The adsorptive properties were studied in terms of pH, equilibrium time, initial concentration (C_0: 25-300 mg/L) and particle sizes (0.125-1.6mm) effects. The experimental kinetic data fitted well the pseudo second order model and the equilibrium isotherm data the Langmuir model. The results indicate that chemical olive stones activated carbon is suitable to be used as an adsorbent material for adsorption of phenol from aqueous solution.

Keywords: Adsorption; Olive stones; Activated carbon; Phenol

Introduction

Phenol and its derivatives are toxic and carcinogenic in nature and are among the priority pollutants of the European Union and US Environmental Protection Agency [1]. Several methods, such as microbial degradation, adsorption, chemical oxidation, solvent extraction and reverse osmosis are being used for removing phenols from wastewater [2].

Activated carbons, the most important commercial adsorbents, are materials with large specific surface areas, high porosity, adequate pore size distributions and high mechanical strength [3,4]. Activated carbons can be produced from different carbonaceous materials such as coal, wood, peat and agricultural wastes especially lignocellulosic by-products. They are widely employed in water and wastewater treatment processes for removing organic compounds such as phenol and its derivatives [5,6].

Tunisia is classified in the fourth ranking of Mediterranean country in the production of olive oil. Therefore, olive trees cultivation constitutes a strategic position within the Tunisian agricultural sector, representing 1.6 million hectares under cultivation (30% of agricultural land area) and counting an estimated 60 million olive trees [7]. This high production implies a big quantity of olive stone waste. In a previous work [8], olive stones were used to prepare an activated carbon by chemical processes using H_3PO_4 (COSAC). This precursor is a by-product of oleic factories producing olive oil in Tunisian country. The produced activated carbon is characterized by a high surface area, developed micropores and heterogeneous functional groups.

The aim of this study was to investigate kinetics and equilibrium aspects of the adsorption of phenol onto olive stones activated carbon. Three kinetics models including pseudo-first order, pseudo-second order and intra-particle diffusion models were used to discuss adsorption mechanisms. Experimental equilibrium isotherms were fitted with Langmuir and Freundlich equations to analyze the adsorption process.

Materials and Experimental Methods

Activated carbon

Preparation: Activated carbon was prepared from olive stones by chemical activation with orthphosphoric acid (H_3PO_4) according to the method developed by Gharib et al. [8]. Initially, the precursor was washed thoroughly with water, dried and then impregnated with a dilute phosphoric acid solution for 9 hours at 110°C. The resulting materials were washed with distilled water to remove any leachable impurities. The dried solid was carbonized in a vertical tubular fixed bed reactor heated at controlled temperature by electric furnace and fed with a continuous nitrogen flow. The carbonization time and temperature were optimized to be 2 h 30 min and 410°C respectively. The producer of carbon was washed with distilled water and then dried to be ready for using.

Characterization of the activated carbon:

a. Specific surface area and textural proprieties: Specific surface area and pores characteristics of activated carbon were determined by nitrogen adsorption and desorption isotherms at 77.7 K with an automatic Sorptiometer Autosorbe-1C Quantachrome apparatus (Common Services Research Unit of ENIG). The activated carbon sample is previously out gassed at 250°C and under vacuum. The micropores volume is estimated from nitrogen isotherm by analyzing the nitrogen adsoprtion isotherm at 77 K by Dubinin and Radushkevich micropore analysis method. The total pore volume is estimated from the adsorption quantity of nitrogen at relative pressure near unity. The measured values of textural characteristics are reported in Table1, we note the important micropore volume and a high specific surface area.

b. Determination of pH_{pzc}: The point of zero charge (pHPZC), pH at which the adsorbent is neutral in aqueous suspension, was determined following the procedure of Lopez-Ramon [9]. In this method 50 ml of 0.01M NaCl solutions were filled in closed Erlenmeyer

*Corresponding author: Thouraya Bohli, Laboratory of Research: Engineering Processes and Industrials Systems (LR11ES54), National School of Engineers of Gabes, University of Gabes, St Omar Ibn Elkhattab, 6029 Gabes, Tunisia

PARAMETERS	VALUE
Apparent specific weight (g/cm³)	0.55
Specific surface area BET (m²/g)	1040
Micropores volume (cm³/g)	0.463
Mesopores volume (cm³/g)	0.014
Total pore volume (cm³/g)	0.477
Average pores diameter (nm)	1.835
pH_{pzc}	3.00

Table 1: Physical and chemical characteristic of olive stones activated carbon.

flasks under agitation at room temperature of about 25°C. The pH of each solution is initially fixed at value lying from 2 to 12 by adding 0.1M HCl or 0.1M NaOH solutions. Then 0.1g of solid adsorbent was added to each flask and the final pH was measured after 48 h. pH_{PZC} is localized at the point where the curve pH_{final} versus $pH_{initial}$ intersects the first bisector. The fined value is pHpzc=3, showing that the activated carbon is anionic.

Phenol

Analytical-reagent grade phenol (purity>99%), was used as the adsorbate. A stock solution was prepared by dissolving required amount of phenol in double distilled water. Different initial concentrations (C_0) of phenol, in the range of 25-300 mg/l, were obtained by successive dilutions.

Phenol concentration was determined by UV absorption at 270 nm wave-length using a calibrated UV-Visible spectrophotometer.

Isotherm equilibrium adsorption construction

Experiments were carried out by dispersing known quantities of adsorbent (0.1- 0.5 g) within 200 ml of 400 mg/l phenol solution in 250 ml flasks. Adsorption equilibrium isotherms were measured via batch mode adsorption technique by placing volumetric flasks in a shaking mixer and using a thermostatic bath to control and to fix the temperature at 40°C. After 48 h of mixing at fixed instant speed of 500 rpm, samples were taken out from each flask and filtered. Initial and residual concentrations of phenol were measured using previously calibrated UV absorption at 270 nm analytical method.

For pH effect experimental study, the adsorption of phenol by COSAC was investigated over a pH range of 2 to 9 at 40°C with an initial solute concentration of 100 mg/l and an adsorbent carbon dose of 0.1 g. Experiments were carried out for 48 h at constant agitation speed of 500 rpm. Initial solution pH was adjusted by adding a diluted 0.1 M HCl or 0.1 M NaOH.

Kinetics adsorption

Kinetic adsorption was studied at various initial phenol concentrations (25-300 mg/l) and for different activated carbon particle sizes (0.125-1.600 mm) and the same quantity: 1.6 g in batch mixed suspension with 800 ml of phenol solution at given initial concentration. The temperature of the suspension was maintained constant at 40 °C by using a thermostatic bath. Samples of 5 ml were withdrawn at regular times and the residual phenol in the filtrates was measured.

The amount of phenol adsorbed on activated carbon q_t (mg/l) was followed versus time and was calculated using the following equation:

$$q_t = \frac{(C_O - C_t)V}{m}, \text{ (mg/g)} \qquad (1)$$

Where C_0 and C_t are the concentration of phenol solution (mg/l) at initial and at sampling time (t) respectively; V the volume of solution

(l) and m is the weight (mg) of activated carbon. Each experiment was carried out in duplicated way and the average results are presented in this work.

Results and Discussion

Effect of initial pH on the adsorption

The pH of solution is one of the most important parameter affecting phenol adsorption processes because it affects the surface charge of the adsorbent as well as the degree of ionization and speciation of phenol [10]. Effect of initial pH on adsorption of phenol was studied with initial concentration of 100 mg/l and optimum carbon dose of 1 g/l and at a temperature of 40°C. Figure 1 shows the influence of solution pH on phenol removal by COSAC in the pH range of 2.0 to 9.0. The phenol adsorption decreases slowly with the increase of pH for values up than pH=4.0 and the adsorption sharply decreases for pH>6. The maximum phenol uptake obtained at pH lower than pHpzc (pHpzc=3) can be explained by the fact that at this pH range the COSAC surface is charged positively and phenol was protonated [11] this create a strong electrostatic interaction between phenol molecule carbon surface. However, with the increase of pH solution behind pHpzc, phenol become more and more dissociated and COSAC surface is charged more negatively leading to increased electrostatic repulsion force between the anionic phenol form and OH- groups on COSAC surface and between phenolate-phenolate anions in solution [12] resulting in a decrease of phenol uptake. Taking into account the obtained results further experiments were carried out at pH 2.3 without adjustment.

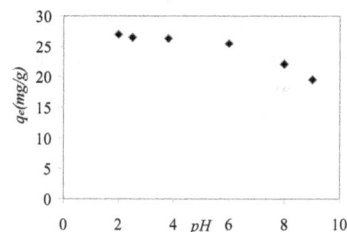

Figure 1: Effect of pH on the uptake of phenol onto COSAC (Temperature 313 K, initial concentration100 mg/l).

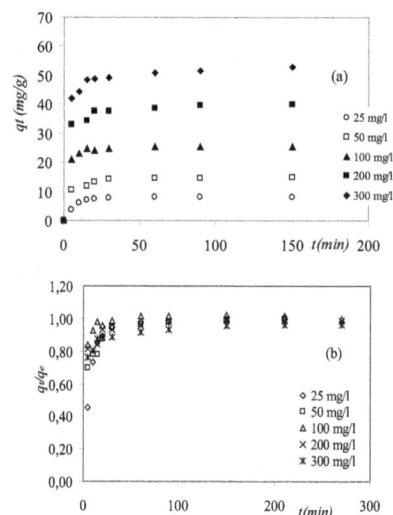

Figure 2: Initial concentration effect on process kinetics for the adsorption of phenol onto COSAC (Particle sizes: 0.630-1.000 mm, temperature 313 K, initial pH: 2.3).

Effect of contact time and initial concentration

Initial concentration effect on the removal of phenol is reported in Figure 2a. When initial phenol concentration increased from 25 to 300 mg/l, the adsorption capacity of COSAC increased from 8 to 53 mg/g. The time evolution of the amount adsorbed phenol indicates that the equilibrium time was reached at about 60 min for all the initial concentrations (Figure 2b). We observe two kinetics regions: the first one is characterized by a high adsorption rate and this is due to that initially the number of sites of activated carbon available is higher and the driving force for the mass transfer is greater. Therefore, phenol reaches the adsorption site easily. As time progress the number of free site of COSAC decreases and the non adsorbate molecules are assembled at the surface thus limiting the capacity of adsorption. The increase of loading capacities of COSAC with increasing phenol concentration may be due to higher π-π interaction between phenol and the surface function of activated carbon. π-π interaction is usually the mean involved mechanism of phenol adsorption [13,14].

Effect of the particle size on the adsorption kinetic of phenol

The amount adsorbed phenol extent versus the agitation time for different particles sizes (0.125 - 1.6 mm) at the same phenol initial concentration of 100 mg/l, natural pH (around 2.3) and an stirrer speed of 500 rpm are shown in Figure 3. This high mixing speed was considered to be sufficient to overcome the external film diffusion resistance. Results show that: the smaller the particle size, the faster was the adsorption kinetic. We can also observe, for all particle sizes, that the adsorption processes could be subdivided in two steps: rapid first one followed by a slow step, this is become evident for small particle size. By else, with increasing particle size the phenol adsorption quantity decreases and the kinetic is slower. This confirms the hinting effect of the internal mass transfer process in the dynamic of the adsorption process.

Figure 3 suggests that there are two linear sections in each plot. The first section corresponds to the diffusion in the micropores. At the beginning of the phenol diffusion into COSAC, there is a fast initial uptake (part 1) with high intraparticle diffusion rate constant. It accounts for approximately 91% of the overall adsorption capacity for a particle size between 0.63 and 1 mm.

The first rapid step is due to a high interaction between adsorbent surface and phenol (sites with high affinity). The second step is relative to a less energetic adsorption with carbon sites having low affinity and formation of multilayer.

Adsorption kinetics

Pseudo-first-order, pseudo-second-order and intra particular

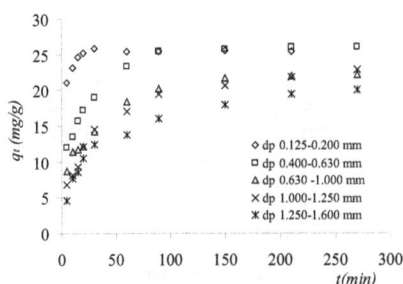

Figure 3: Effect of contact time on the amount of phenol onto COSAC for different particle size, (initial concentration of 100 mg/L, temperature 313 K, initial pH of 2.3).

diffusion models are applied to describe the kinetic data. The effect of initial concentration and particle sizes was investigated to find the best kinetic model.

Pseudo-first order model: The pseudo-first-order of Lagergren [15] is expressed in a linear form:

$$Ln(q_e - q_t) = k_1 t + Ln q_e \qquad (2)$$

Where k_1 is the rate constant of pseudo-first order adsorption (mn^{-1}), q_t and q_e are respectively the instantaneous and equilibrium amount of phenol adsorption per unit weight of activated carbon (mg/g) and t is the time (min). The plot of Ln $(q_e - q_t)$ versus t should give a linear relationship with the slope equal to: k_1.

Results given in Table 2 indicate that first order kinetic model does not correctly fit the experimental data for different initial concentration and particle sizes; coefficients regression are between 0.598 and 0.920 and the phenol uptakes are less then these given by experiment.

Pseudo-second order model: The kinetic data are fitted with pseudo-second order model resulting in the following linear equation [16]:

$$\frac{t}{q} = \frac{1}{K_2 q_e} + \frac{1}{q_e} t \qquad (3)$$

Where K_2 is the rate constant of pseudo-second order.

The linear plots of t/q_t versus t show that the experimental data agree with the pseudo-second order kinetic model. The calculated q_e (mg/g) (Table 2) values agree very well with the experimental data and the correlation coefficients close to the unit in all cases. These indicate that the adsorption of phenol from aqueous solution on COSAC obey the pseudo-second order kinetic model. It was observed also that the constant kinetic, k_2, decreased with the increasing of the initial phenol concentration and particles size. This shows the importance of transport process, particularly the internal diffusion in overall adsorption process kinetics some observations are reported by Srihari et al. [16].

Intra-particle diffusion: Intra-particle diffusion was often considered as the limiting step which limits the kinetics in the most process of adsorption. The possibility of a limitation by the diffusion in the pores is explored by plotting phenol uptake against the square root of time, using the Weber and Morris model:

$$q_t = k_d \sqrt{t} + C \qquad (4)$$

Where q_t is the instantaneous amount of phenol adsorption per unit gram of activated carbon (mg/g), C (mg/g) is a constant giving an idea about the thickness of the boundary layer: the larger the value of C is more important the lay effect limitation.

The intraparticle diffusion model plots for the adsorption of phenol onto COSAC, under the effect of initial phenol concentration and carbon particle size, suggest that there are two linear parts in each plot. The initial parts are attributed to boundary layer diffusion effects or external mass transfer effects. These lines do not pass through the origin indicating that the intra-particle diffusion is not the only process that can control kinetics of adsorption [16,17]. Whereas, the second parts may be attribute to intra-particle diffusion effects.

The rate constants: k_{d1} and k_{d2}, of intraparticle diffusion (mg/g), was calculated respectively from the slope of the first and second linear portion of the plots. It is obvious from Table 2 that k_{d1} is higher than K_{d2}. This can be related to the fact that firstly the number of pores available are very high and after there is possibility of pore blockage or steric hindrance exerted by the adsorbed molecules on the carbon surface, it will eventually slow down the adsorption process and give rise to other linear sections with smaller intraparticle rate constants.

The validity of the order of adsorption processes is based on the regression coefficients and on predicted q_e values. Table 2 shows that correlation coefficients of the pseudo-second order kinetic model are the highest as compared to these given by the pseudo-first order and interparticle diffusion model. Either the q_e (mg/g) calculated by the pseudo-second order, agree very well with the experimental data. Thus indicate that the adsorption of phenol from aqueous solution on COSAC obey the pseudo-second order kinetic model.

Isotherm of adsorption

The adsorption isotherm of phenol on COSAC was studied at 30°C and an initial solution pH of 2.3. As can be seen from Figure 4 [18] and according to the classification of Brunauer, Emett, and Teller the adsorption isotherm of phenol on CACOS is of type IV shows the formation of two successive layers of phenol on the surface of activated carbon when the interactions between the molecules of phenol and the surface of adsorbent are stronger than the interactions between the adsorbed molecules. In this case, the sites of adsorption of the second layer begin to fill perform only when the first layer is complete.

The same shape of isotherm was found by Calace et al. [19], when they study the adsorption 4-nitrophenol (4-NP) on papermill sludges.

The first part of the equilibrium experimental isotherm was modeled by Langmuir and Freundlich equations (Figure 5). According to the theory of adsorption, the model of Langmuir is based on the fixation of a monolayer of adsorbate molecules on the surface of pores. The model assumes uniform adsorption on the surface and no transmigration in the plane of the surface. Langmuir's equation is mathematical expressed as follows.

$$q_e = \frac{q_{max} K_L C_e}{1 + K_L C_e} \tag{5}$$

Where: K_L is the equilibrium adsorption constant related to the free energy of the adsorption (l/mg) and q_{max}: the maximum adsorption capacity (mg/g), C_e the equilibrium concentration (mg/l), is the q_e

amount adsorbed at equilibrium (mg/g). The linear form of equation (5) is given equation (6):

$$\frac{C_e}{q_e} = \left(\frac{1}{q_{max} K_L}\right) + \left(\frac{1}{q_{max}}\right) C_e \tag{6}$$

The essential characteristics of Langmuir model can be expressed by dimensionless constant called separation factor, R_L, with is given by this equation:

$$R_L = \frac{1}{1 + K_L C_0} \tag{7}$$

The adsorption is considered as irreversible $R_L=0$, favourable when $0<R_L<1$, linear when $R_L=1$ and unfavourable when $R_L>1$.

The Freundlich equation is an empirical model that considers heterogeneous adsorptive energies on the adsorbent surface [15].

$$q_e = K_F C^{1/n} \tag{8}$$

The linear form of equation (8) is given equation (9):

$$Log q_e = Log K_F + \frac{1}{n} Log C_e \tag{9}$$

Where, K_F (l/g) and $1/n$ are Freundlich constants. The parameters of both models are calculated and summarized in Table 3.

Figure 5 shows that Langmuir equation is more reasonably applicable than Freundlich equation with correlation coefficients R^2 equal to 0.99 and an R_L of 0.061 indicating favourable adsorption of phenol on COSAC.

Ozkaya [20] studied the removal of phenol using a commercial activated carbon, Langmuir adsorption model give a maximum uptake of 49.7 mg/g. Srivastava et al. [15], were studied the adsorptive removal of phenol by bagasse fly ash (BFA) and two activated carbons ACC and ACL, maximums amount given by Langmuir model are respectively 23.83, 30.22 and 24.65 mg/g. Kilic et al. [21], investigated phenol adsorption from aqueous solutions by activated carbon prepared from tobacco residues, results show that maximums removal are 17.83 and 0.55 mg/g respectively for ACK1 and ACK2. In the present work, Table 2 shows a maximum uptake related to the first part of phenol isotherm of 58.82 mg/g and Figure 2 give an experimental value of the amount

	First order k_1(min⁻¹) q_e (mg/g) R^2			second order k_2(min⁻¹) q_e R^2			intra-particle diffusion k_{d1}(mg/g) R^2 kd_2(mg/g) R^2			
Initial concentration (mg/L)										
25	0.0627	4.525	0.889	1.9847	25.575	1	1.6359	0.936	0.0675	0.7864
50	0.0544	9.387	0.734	0.1045	27.100	0.999	1.1192	0.929	0.0937	0.9512
100	0.1743	13.538	0.784	0.0698	23.419	0.998	1.4511	0.813	0.0194	α
200	0.0423	13.279	0.641	0.0493	24.038	0.998	1.8901	0.747	0.1822	0.7535
300	0.0384	15.825	0.598	0.0468	21.276	0.997	3.2479	0.952	0.3152	0.8573
Particle size (mm)										
0.125-0.2	0.1401	10.479	0.814	0.2104	8.803	0.996	1.464	0.915	0.0426	α
0.4-0.63	0.0341	18.541	0.935	0.3257	15.240	0.999	2.091	0.992	0.2557	α
0.63-1	0.0253	16.629	0.920	0.5509	25.773	1	1.535	0.926	0.4246	0.867
1-1.25	0.0226	17.101	0.923	0.5560	40.322	0.999	2.481	0.961	0.0613	0.957
1.25-1.6	0.0275	17.995	0.920	0.6428	52.631	1	2.386	0.984	0.7051	0.960

α: for values less than 0.7

Table 2: Kinetic parameters for the adsorption of phenol onto COSAC.

Langmuir parameters				Freundlich parameters		
q_{max} (mg/g)	K_L (l/mg)	R_L	R^2	K_F (L/g)	n	R^2
58.823	0.038	0.061	0.99	3.935	1.865	0.964

Table 3: Langmuir and Freundlich sorption parameters.

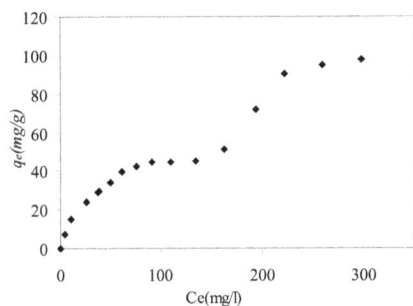

Figure 4: Experimental adsorption isotherms of phenol onto COSAC (temperature: 313K, initial pH: 2.3, particle size: 0.630 -1.000 mm).

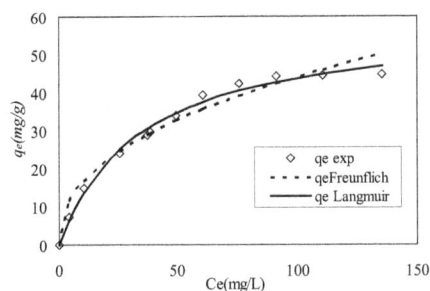

Figure 5: Comparison of experimental and predicted adsorption isotherms of phenol onto COSAC according to Langmuir and Freundlich models for the first part (temperature 313K, initial pH 2.3, particle size: 0.630-1.000 mm).

uptake related to the second part of this isotherm of about 97.6 mg/g. These results indicate that COSAC can be consider as a good adsorbent of phenol as compared by other materials for removing phenol, COSAC shows a high capacity to sorbs this pollute from water.

Conclusion

In this study, the adsorption of phenol from aqueous solution using olive stones activated carbon (COSAC) was investigated. Results indicate that adsorption capacity of COSAC was considerably affected by pH, initial concentration, and particle size. Phenol adsorbent increased with increasing initial concentration and decrease with the increase of particle size of activated carbon.

Kinetic experiments carried out with different particle sizes of COSAC showed that the smaller the particle size, the faster was the diffusion of phenol into adsorption sites of adsorbent. Kinetics results of phenol adsorption were examined using the pseudo-first-order, pseudo-second-order and intra particular diffusion kinetics models. The results obtained show that the pseudo-second order model was the best to describe adsorption kinetic data for phenol. Equilibrium data were fitted to Langmuir and Freundlich isotherms and the best fit was given by the Langmuir isotherm model, with a maximum monolayer adsorption capacity of about 58.8 mg/g. thus indicates that COSAC is suitable to be used as an adsorbent material for adsorption of phenol from aqueous solution it may also be effective in removing other harmful species such as heavy metals.

References

1. Santana CM, Ferrera ZS, Torres Padrón ME, Santana Rodríguez JJ (2009) Methodologies for the Extraction of Phenolic Compounds from Environmental Samples: New Approaches. Molecules 14: 298-320.

2. Liu QS, Zheng T, Wang P, Jiang JP, Liu N (2010) Adsorption isotherm, kinetic and mechanism studies of some substituted phenols on activated carbon fibers. Chemical Engineering Journal 157: 348-356.

3. Ugurlu M, Gurses A, Acikyildiz M (2008) Comparison of textile dyeing effluent adsorption on commercial activated carbon and activated carbon prepared from olive stones by ZnCl2 activation. Microporous and Mesoporous Materials 111: 228-235.

4. Mohanty K, Das D, Biswas MN (2005) Adsorption of phenol from aqueous solutions using activated carbons prepared from Tectona grandis sawdust by ZnCl2 activation. Chemical Engineering Journal 115: 121-131.

5. Baudu M, Guibaud G, Raveau D, Lafrance P (2001) Prediction of adsorption from aqueous phase of organic molecules as a function of some physicochemical characteristics of activated carbons. Water Quality Research Journal of Canada 36: 631-657.

6. Ania CO, Parra JB, Pis JJ (2008) Effect of texture and surface chemistry on adsorptive capacities of activated carbons from phenolic compounds removal. Fuel Processing Technology 78: 337-343.

7. Hannachi H, Msallem M, Elhadj SB, El Gazzah M (2007) Influence du site géographique sur les potentialités agronomiques et technologiques de l'olivier (Olea europaea L.) en Tunisie. Comptes Renduus Biologies 330: 135-142.

8. Ouederni A, Gharib H (2005) Processing olive pomace Tunisian activated carbon chemically with phosphoric acid. Recent Advances in Process Engineering, SFGP, Paris, France.

9. Lopez-Ramon MV, Stoeckli F, Moreno-Castilla C, Carrasco-Marin F (1999) On the characterization of acidic and basic surface sites on carbons by various techniques. Carbon 37: 1215-1221.

10. Maleki A, Mahvi AH, Ebrahimi R, Khan J (2010) Evolution of Barley straw and its Ash in Removal of Phenol from Aqueous System. Word Applied Sciences Journal 8: 369-373.

11. Lü G, Hao J, Liu L, Ma H, Fang Q, et al. (2011) The Adsorption of Phenol by Lignite Activated Carbon. Chinese Journal of Chemical Engineering 19: 380-385.

12. Mareno-Castilla C (2004) Adsorption of organic molecules from aqueous solutions on carbon materials. Carbon 42: 83-94.

13. Li D, Wu Y, Feng L, Zhang L (2012) Surface properties of SAC and its adsorption mechanisms for phenol and nitrobenzene. Bioresource Technology 113: 121-126.

14. Li Y, Du Q, Liu T, Peng X, Wang J, et al. (2013) Comparative study of methylene blue dye adsorption onto activated carbon, graphene oxide, and carbon nanotubes. Chemical Engineering Research and Design 91: 361-368. 15.

15. Srivastava VC, Swamy MM, Mall ID, Prasad B, Mishra IM (2006) Adsorptive removal of phenol by bagasse fly ash and activated carbon: Equilibrium, kinetics and thermodynamics. Colloids and Surfaces A: Physicochemical Engineering Aspects 272: 89-104.

16. Srihari V, Das A (2008) The kinetic and thermodynamic studies of phenolsorption onto three agro-based carbons. Desalination 255: 220-234.

17. Hall KR, Eagleton LC, Acrivos A, Vermeulen T (1996) Pore-and solid-diffusion kinetics in fixed-bed adsorption under constant-pattern conditions. Industrial & Engineering Chemistry Fundamentals 5: 212-223.

18. Khalfaoui M, Knani S, Hachicha MA, Ben Lamine A (2003) New theoretical expressions for the five adsorption type isotherms classified by BET based on statistical physics treatment. J Colloid Interface Sci 263: 350-356.

19. Calace N, Nardi E, Petronio BM, Pietroletti M (2002) Adsorption of phenols by papermill sludges. Environmental Pollution 118: 315-319.

20. Ozkaya B (2006) Adsorption and desorption of phenol on activated carbon and a comparison of isotherm models. Journal of Hazardous Materials 129: 158-163.

21. Kilic M, Apaydin-Varol E, Pütün AE (2011) Adsorptive removal of phenol from aqueous solutions on activated carbon prepared from tobacco residues: Equilibrium, kinetics and thermodynamics. Journal of Hazardous Materials 189: 397-403.

16

Kinetic Model for the Sorption of Ni(II), Cu(II) and Zn(II) onto Cocos Mucifera Fibre Waste Biomass from Aqueous Solution

Augustine K. Asiagwu[1], Hilary I. Owamah[2]* and **Izinyon O. Christopher[3]**

[1]Chemistry Department, Delta State University, Abraka, P.M.B.1, Delta State, Nigeria
[2]Civil Engineering Department, Landmark University, P.M.B1001, Omu-Aran, Kwara State Nigeria
[3]Civil Engineering Department, University of Benin, Benin, Nigeria

Abstract

Sorption of divalent metals ions Ni(II), Cu(II) and Zn(II) unto coconut cocos mucifera fibre waste biomass over a wide range of operation conditions and equilibrium-sorption kinetics were studied. The batch experiment showed that pH 2-3, was the best range for the sorption of the metal ions onto the biomass. The time-dependent experiments showed that the binding of the metal ions onto the adsorbent, was quite rapid and occurred within 25 minutes and completed within 50 minutes. The sorption process was examined by means of the Langmuir and the Freundlich isotherm models. The monolayer sorption capacity obtained using the Langmuir equation was 0.09 mg/g Ni (II), 0.08 mg/g Cu(II) and 0.09 mg/g Zn(II). The Freundlich isotherm model was not too appropriate for the sorption process, since R^2 for all the three metals are less than 0.90. However the K_F value of Zn(II) (0.880), is greater than that of Ni(II) (0.077) and Cu(II) (0.075), suggesting that Zn(II) has greater adsorption tendency towards the biomass. The kinetics of the sorption mechanism was evaluated using the pseudo-first order rate model and pseudo-second rate model. The results indicated that the pseudo-second order model provides a more appropriate description of the single and mixed metal-ion sorption process of Ni(II) Cu(II) and Zn(II) onto coconut fibre biomass.

Keywords: Metal ions adsorption process; Adsorption conditions and sorption kinetics

Introduction

The increasing levels of metals in the environment from various anthropogenic sources have become a source of concern for environmentalists and scientists alike. Unlike the toxic organics that in many cases can be degraded, metals deposited into the environment tend to persist indefinitely, accumulating in living tissues through food chain. These effects and many others have made it necessary to adapt measures that will help remove these heavy metals from wastewater, surface and groundwater supplies, using any available cost effective means. The problems of our ecosystem are increasing with advancement in technology. Heavy metal pollution is one of these problems. Toxic heavy metal release into the environment has been increasing continuously as a result of man's industrial activities and technological development.

The release of these heavy metals posses a significant threat to the environment and public health because of their toxicity, bioaccumulation in food chain and persistence in nature [1]. Lead is a heavy metal that affects the functioning of the blood, liver, kidney and brains of human beings. Lead is a component of most industrial and domestic paints. Nickel which causes gastrointestinal irritation and lung cancer is often obtained from Ni/Fe storage batteries. Due to the magnitude of the problem of heavy metal pollution, research into new and cheap methods of removal has been on the increase recently. Several workers have reported on the potential use of agricultural by-products as good adsorbents for the removal of metal ions from aqueous solutions and waste water [2].

This process attempts to put into use the principle of using waste to treat waste and becomes even more efficient because these agricultural by-products are readily available and often pose some waste disposal problems. Hence, they are available at little or no cost, since they are waste products. This makes the process of treating waste waters with agricultural by-products adsorbents more cost effective than the use of conventional adsorbents like commercial activated carbon.

In addition, there is no need for a complicated regeneration process when using agricultural by-products for waste water treatment [3]. The ability of some agricultural by-products to adsorb heavy metals from waste water and aqueous solutions has been reported in literatures and these include: Cotton seed hulls, rice straw and sugarcane bagasse, maize cob [3,4]. Physio-chemical methods such as chemical precipitation, lime coagulation, ion exchange, solvent extraction, reverse osmosis, chemical oxidation or reduction, electrochemical treatment, evaporation recovery, filtration and membrane technologies have been widely used to remove heavy metal ions from industrial wastewater [5-8]. These processes may be ineffective or expensive and non- environment friendly, especially when the heavy metal ions in solutions are contained in the order of 1-100 mg of the dissolved heavy metal ions [9].

Moreover, the disadvantage like incomplete metal removal, high reagent and energy requirements, generation of toxic sludge or other waste products that require careful disposal has made it imperative for a cost effective treatment method that is capable of removing heavy metals from aqueous effluents. In recent times, a great deal of interest has been given to the utilization of agricultural by-products as adsorbents for removal of trace amounts of toxic and valuable heavy metals from municipal and industrial wastewater, particularly because of low cost, high availability of these materials, absence of complicated

*Corresponding author: Hilary I. Owamah, Civil Engineering Department, Landmark University, P.M.B1001, Omu Aran, Kwara State Nigeria

regeneration process and their capability of binding to heavy metals by adsorption, chelation and ion-exchange [10,11,3].

This present study is aimed at determining the optimum conditions for the sorption of three divalent metals Ni(II), Cu(II) and Zn(II) onto cocos mucifera fibre, via equilibrium adsorption isotherms and kinetic studies. The equilibrium of the adsorption process is often described by fitting the experimental points with models usually used for the representation of equilibrium adsorption isotherm [12]. The isotherm models for single-solute system are the Langmuir and Freundlich isotherm models. Evaluation of equilibrium sorption performance needs to supplemented by process-oriented studies of the kinetics and eventually by dynamic continuous flow tests.

Presently, we are not yet aware of any proven documented work on the efficient reuse and recycling of cocos mucifera waste. It is on this premise that this study was undertaken to investigate the possibility of using cocos mucifera waste as a cost effective adsorbent for heavy metals removal from wastewater.

Experimental

Sample collection

20 fresh coconuts (cocos mucifera) were collected from a farm in Eku in Ethiope East Local Government Area of Delta State, Nigeria. These were all hand plucked from the trees thoroughly washed and the fibre carefully carved out of the shell. The fibres, while still fresh, were washed with deionized water and allowed to air-dry. The dried fibres were torn into shreds and cut into small chips; these were then sun-dry for five days. The dried samples were ground using mechanical grinder and then severed through a B.S. standard screen to obtain a particle size of 100μm and stored in a plastic container for further analysis.

Activation and purification of the biomass

It is often necessary to activate a solid before using it as an adsorbent for sorption studies. The purpose of activation is to increase the surface area of the solid by introducing a suitable degree of porosity into the solid matrix [13]. Again, activation may also produce structural defects in solids, which may be favorable to sorption processes [14]. The experiments on activation and purification of the biomasses were carried out according to the previous works of [15]. 500 g of finely divided biomass was activated and at the same time, purified by soaking in excess 0.3 M HNO_3 for 24 hours to remove any metals and debris that might be in the biomass prior to experimental metal ion exposure, followed by washing thoroughly with deionized water until a pH of 7.1 ± 0.1 was attained and then air-dried. The air dried biomass was then washed with deionized water and re-suspended in 1.0M hydroxylamine to remove all O-acetyl groups. To remove all other soluble materials, the biomass was washed with deionized water and centrifuged at 300rpm for five minutes, and the supernatants obtained were dried at room temperature. The physical property of the modified adsorbent is as shown in table 1.

Effect of metal ion concentration on sorption

The experiment on the effect of metal ion concentration on sorption was performed according to the pervious works of [13]. Several standard solutions of 10, 20, 30, 40, 50, 60,70 and 80 mg/1 were prepared from spectroscopic grade standards for Ni(II) (from $NiSO_4$),for Cu(II) from $CuSO_4$, and for Zn(II) from $ZnSO_4.6H_2O$ the metal solutions made separately were adjusted to pH 5.0 (since most metals are soluble at this pH). Using conc. HC1, 50ml of each metal ion solution was added

S/N	Parameter	Value
1	Ash Content	5.8
2	Moisture content	4.6
3	Total pore volume	0.71
4	Iodine number	600
5	Surface Area(m²/g)	630
6	Volatile matter	1.9
7	Mean particle size(mm)	0.11
8	Apparent density(g/ml)	0.60
9	Fixed carbon	69.5
10	pH	6.5
11	Methylene blue number	20
12	Micropore volume(ml/g)	0.14 ± 0.02

Table 1: Physical properties of the different types of adsorbents.

to accurately weighed (2.50±0.01mg) activated/purified biomass in different flasks and agitated for 50 minutes to ensure that equilibrium was reached. At the end of the time, the suspension was filtered through Whiteman No. 45 filter paper and centrifuged at 2800rpm for 5 minutes (to remove all traces of cloudiness). The supernatants were analyzed for metal ions by AAS.

Effect of contact time on sorption

The experiment on the effect of contact time on the metal ions binding was performed according to the pervious works of [13]. Activated/purified biomasses (2.50 ± 0.01mg) were weighed into several flasks. Ni(II), Cu(II) and Zn(II) solution (2.50mg in 50ml of water) were added to the biomasses. The flasks were then labeled for time intervals of 5,10,15,20,25,30,40 and 50 minutes. The pH's of these suspensions were adjusted to 5.0. The flasks were tightly covered and shaken at the appropriate time intervals. At the end of each interval, the suspensions were filtered using Whiteman No. 45 filer paper and then centrifuged at 2800 rpm for 5 minutes. The metal ion concentration was determined using Atomic Adsorption Spectrometer (AAS).

Effect of pH on sorption

Activated/purified biomass (2.50 ± 0.01 mg) was weighed into several flasks, Ni(II), Cu(II) and Zn(II) solutions (2.50 mg in 50 ml of water) were added to the biomass. These suspensions were then adjusted to pH 2.0, 3.0, 4.0, 5.0, 6.0, 7.0, 8.0, 9.0 by adding a solution of either HCl or NaOH. The flasks were then tightly covered with cellophane and shaken for 50 minutes. The suspensions were filtered using Whiteman No. 45 filter paper and then centrifuged at 2800rpm for 5 minutes in order to remove all traces of cloudiness. The metal contents were determined using AAS after recording the final pH of the solutions in the test tubes.

Analysis of metal content

The Ni(II), Cu(II) and Zn(II) contents in all experiments were determined with an Atomic Absorption Spectrometer (AAS). Spectroscopic grade standards were used to calibrate the instruments, which were checked throughout the analysis for instrument's response. The batch experiments were performed in triplicates and the means were computed for each set of values for quality assurance.

Data evaluation

The amount of Ni(II),Cu(II) and Zn(II) ions removed by the biomass during the series of the batch investigations were determined using a mass balance equation as shown in Equation (1):

$$q_e = \frac{V}{M}(C_O - C_e) \tag{1}$$

Where;

q_e = metal-ion concentration on the biomass (mg/g) at equilibrium.

C_e = metal-ion concentration in solution (mg/I) at equilibrium.

C_o = initial metal-ion concentration in solution (mg/l)

V = volume of solution (L)

M = Mass of biomass used (g)

Kinetic treatment of experimental data

In order to comprehensively investigate the mechanism of adsorption, the pseudo – first order and the pseudo-second order equations as have been used in [16,17] were applied to the experimental data. The linear from of pseudo-first order model is given by Equation (2):

$$\ln\left(q_e - q_t\right) = \ln q_e - Kt \tag{2}$$

Where;

q_e = mass of metal adsorbed at equilibrium (mg/g)

q_t = mass of metal adsorbed at time t (mg/g)

K = equilibrium constant

A linear plot of $\ln(q_e\text{-}q_t)$ versus t confirms the model

The pseudo – second order equation earlier used for the sorption system of divalent metal-ions using sphagnum moss plant was adopted [16]. The linear from of the pseudo-second order model is generally expressed as given in Equation (3):

$$\frac{t}{q_t} = \frac{1}{h_o} + \frac{t}{q_e} \tag{3}$$

Results and Discussions

Effect of metal-ion concentration

The experimental results of the uptake of Ni(II), Cu(II)and Zn(II) ions onto the coconut fiber waste biomass at various initial metal ion concentrations are as shown in table 2 and figure 1. The sorption capacity increased from 0.08-0.085 mg/g Ni(II); 0.077-0.079 mg/g Cu(II); and 0.081 -0.091 mg/g Zn(II), with increase in the metal ion concentration from 10-80 mg/1 at biomass dose of 2.5 mg. The three metals in this study were adsorbed in this order: Zn(II) > Cu (II)> Ni(II). However,

Initia lmetal ion concentration (mg/1)	Amount of metal ion adsorbed (mg/g), q_e	Amount of metal ion adsorbed (mg/g),	Amount of metal ion adsorbed (mg/g), q_e
	Ni	Cu	Zn
10	0.078	0.077	0.081
20	0.082	0.074	0.088
30	0.082	0.078	0.089
40	0.084	0.077	0.089
50	0.081	0.076	0.092
60	0.084	0.077	0.091
70	0.086	0.077	0.091
80	0.085	0.079	0.091

(Mass of biomass 2.5mg, temperature 25 °C, pH 5.0, time 50 minutes, volume 50ml).

Table 2: Effect of initial metal ion concentration on metal ion removal.

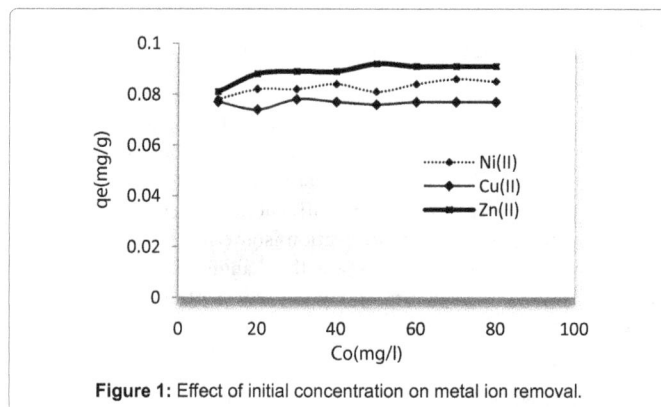

Figure 1: Effect of initial concentration on metal ion removal.

Time (minutes)	Amount of metal ion adsorbed (mg/g), q_e Ni(II)	Amount of metal ion adsorbed (mg/g), Cu(II)	Amount of metal ion adsorbed (mg/g), q_e Zn(II)
5	0.077	0.069	0.069
10	0.078	0.071	0.071
15	0.078	0.071	0.071
20	0.079	0.072	0.071
25	0.080	0.073	0.072
30	0.080	0.073	0.073
40	0.081	0.074	0.074
50	0.082	0.074	0.074

(Mass of biomass 5g, temperature 25°C, pH 5.0, concentration 10mg/1, volume 50ml)

Table 3: Effect of contact time on metal ion removal.

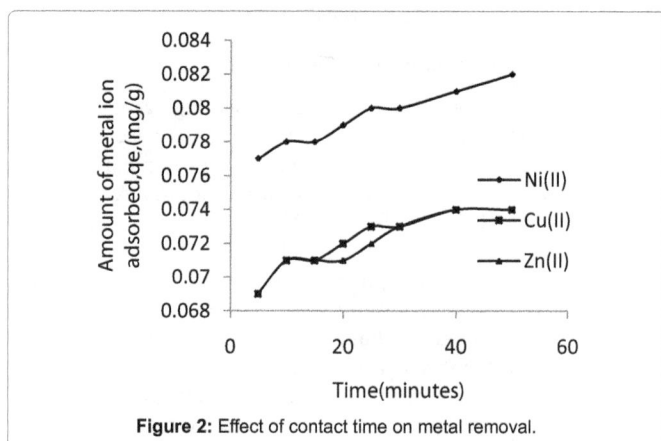

Figure 2: Effect of contact time on metal removal.

the actual percent removal of the metal ions from solutions was found to increase with increase in initial metal-ion concentration (Figure 1) this may be due to the fact that at lower concentrations, adsorption of the metal-ions occurred slowly, and further increase in initial metal-ion concentration led to a competition for available bonding sites on the biomass surface by the metal-ions and thus increased adsorption. Similar adsorption patterns have also been reported by other researchers [9,18-22].

Effect of contact time

Time dependence experiments were conducted in order to obtain how long the coconut fiber waste biomass would take to absorb the metal-ions at optimum pH. The data from the time dependent experiments for the removal of the trace metals is as presented in table

3 and figure 2. As the contact time was increased from 5-50 minutes, the amount of metal-ions removed by the biomass was observed to also increase rapidly. These data suggest that metal ion removal by the biomass increased with increase in contact time for the three metal ions investigated. Within the first 5-25 minutes, the biomass was capable of removing over 70% of each metal ion. The rapid adsorption of the metal ions by the biomass indicates that adsorption might have taken place on the cell wall of the biomass, since most, if not all soluble components were removed during washing. This relatively fast and early removal of metal ions also indicates that physi-sorption as well as chemisorption processes were involved in the reaction between the metal-ions and the coconut fibre waste biomass. Using other different biosorpents, similar adsorption trends have been observed by other researchers [19,13,22].

Effect of pH

Biosorbent materials are made up of complex, organic residues such as lignin and cellulose, which contain different types of polar functional groups. These groups can be actively involved in chemical bonding, and may be responsible for the typical cation exchange characteristics evident in biomaterials. The pH dependence data for the sorption of the three metals under investigation are presented in table 4 and figure 3.

The data showed that the sorption of Ni(II), Cu (II) and Zn(II) increased as pH increased from 2-5, with optimum adsorption of 82-86%,occurring between pH 2 and 3. Above pH 5, a gradual decrease in the amount of metal-ions removed by the biomass was observed. In general, this pH-dependent study showed that the sorption of the three metal-ions were similar; being highest at pH 5 and then began to decrease with further increase pH. This indicates that the adsorption mechanism for the metals investigated might be an ion exchange process. Again, the adsorption mechanism for these metals were stable and rapid, implying that adsorption was taking place on the cell surface

pH	Amount of metal ion adsorbed (mg/g), q_e	Amount of metal ion adsorbed (mg/g),	Amount of metal ion adsorbed (mg/g), q_e
	Ni(II)	Cu(II)	Zn(II)
2	0.082	0.078	0.085
3	0.082	0.080	0.085
4	0.084	0.082	0.086
5	0.084	0.083	0.083
6	0.079	0.076	0.083
7	0.079	0.076	0.082
8	0.078	0.075	0.082
9	0.078	0.074	0.081

(Mass of biomass 2.5mg, time 50 minutes, temperature 25°C, pH5.0, concentration, 10mg/1, volume 50ml)

Table 4: Effect of ph on metal-ion removal.

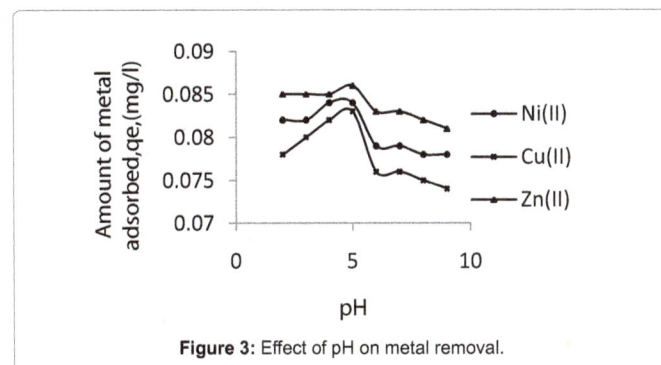

Figure 3: Effect of pH on metal removal.

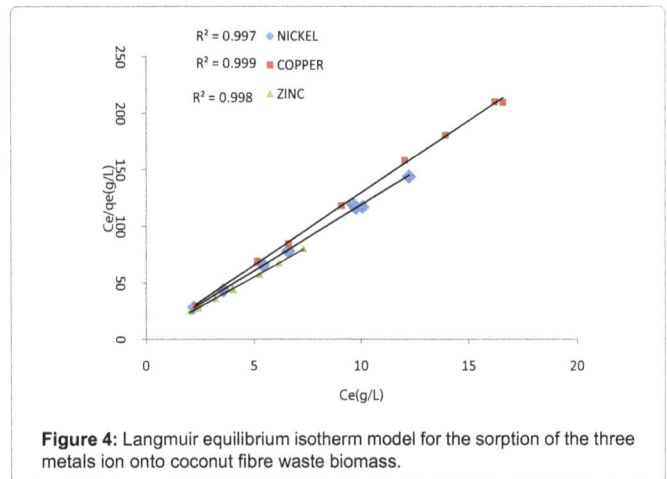

Figure 4: Langmuir equilibrium isotherm model for the sorption of the three metals ion onto coconut fibre waste biomass.

Metal ions	q_m(mgg-1)	K_1(Lg-1)	R^2	S_f
Ni(II)	0.09	4.303	0.997	0.023
Cu(II)	0.08	6.617	0.999	0.015
Zn(II)	0.09	4.914	0.998	0.020

Table 5: Linear Langmuir Isotherm Parameters.

of the coco -nut fibre waste biomass. The trend of this pH-dependent adsorption suggests that by reducing the pH, the bound metal-ions could be desorbed, and the spent biomass regenerated. Once the metals are recovered, the biomass material, which is biodegradable, will cause no environmental damage and may be utilized as natural soil conditioners or fertilizers. Hence, the dry coconut fibre waste biomass could be used for cleaning the environment and industrial effluents. Similar findings using other biomaterials have been observed by other researchers [3,10,11,20,21].

Langmuir isotherm

The Langmuir isotherm was adopted for the estimation of the maximum adsorption capacity corresponding to complete monolayer coverage on the biomass surface. The plots of specific sorption (C_e/q_e) against the equilibrium concentration (C_e) for Ni(II), Cu (II), and Zn(II) ions are as shown in figure 4 and the linear isotherm parameters, q_m, k_L and the coefficient of determination R^2 are presented in Table 5. Langmuir equilibrium isotherm model was used to analyze the sorption of the three metals ion onto coconut fibre waste biomass. The R^2 values of the three metal ions indicate that the Langmuir isotherm provides a good model for the sorption system. The sorption capacity, q_m, which is a measure for the maximum adsorption capacity, corresponding to a complete monolayer coverage showed that the coconut fiber waste biomass has a higher capacity for Ni(II) (0.09g/g) and Zn(II) (0.09g/g) than Cu(II) (0.08g/g). The adsorption coefficient, K_L, which is related to the apparent energy of sorption was greater for Cu(II) (6.617Lg-1) than Ni(II) (4.30Lg-1) and Zn(II) b(4.914Lg-1). This observation showed that the energy of adsorption is not very favorable for Cu(II), probably due to its large ionic radius hence, not all binding sites would have been available to Cu(II). Similar sorption capacity orders have also been reported by [19] for different metal-ions onto fluted pumpkin waste biomass. This adsorption capacity is slightly higher than the adsorption capacities obtained for the three metals ions, using other biosorbents [23-25].

Furthermore, the favorability of adsorption of the three metal-ions onto the coconut fibre waste biomass was tested using the essential features of the Langmuir isotherm model, expressed in terms of a dimensionless constant called the "Separation factor" [26]. The separation factor, S_f is defined as expressed in Equation (4):

$$S_F = \frac{1}{1 + K_L C_O} \qquad (4)$$

C_o =10mg/1

The S_F indicates the shape of isotherm as follows:

S_F	>	1	Unfavourable isotherm
S_F	=	1	Linear Isotherm
S_F	=	0	Irreversible isotherm
$0 > S_F < 1$			Favourable isotherm

The separation parameters for the three metals are less than unity, indicating that the coconut fibre waste biomass is an excellent adsorbent for the three metal-ions. The observed separation factor (Table 6) and indicates that high concentration of Cu(II), Zn(II) and Ni(II) ions in an effluent will not be a limiting factor in the ability of coconut fibre waste biomass to sorb these metal-ions. Similar separation parameter favourability has been observed for the sorption of the metal ions using other biosorbent [19,26]. Based on the R^2 values, the linear form of the Langmuir isotherm appears to produce a reasonable model for the sorption of the three metals, since their R^2 values are all greater than 0.990, thus, showing that coconut fibre waste biomass is an excellent biomaterial for the removal of metal-ions from aqueous solution.

Freundlich isotherm

The Freundlich isotherm model was chosen to estimate the adsorption intensity of the solute (metal-ion) onto the sorbent surface. The linear Freundlich isotherm parameters for the sorption of the three divalent metals onto coconut fibre waste biomass are presented in table 6. An examination of the plot (Figure 5) reveals that the Freundlich isotherm was not a very appropriate model for the sorption study of the

Metal ions	1/n	K_f	R^2
Ni(II)	0.042	0.077	0.533
Cu(II)	0.009	0.075	0.591
Zn(II)	0.071	0.080	0.153

Table 6: Linear Freudlich Isotherm Parameters.

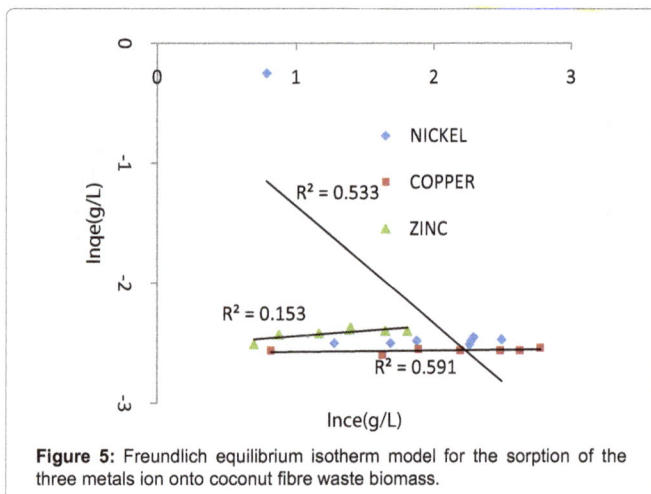

Figure 5: Freundlich equilibrium isotherm model for the sorption of the three metals ion onto coconut fibre waste biomass.

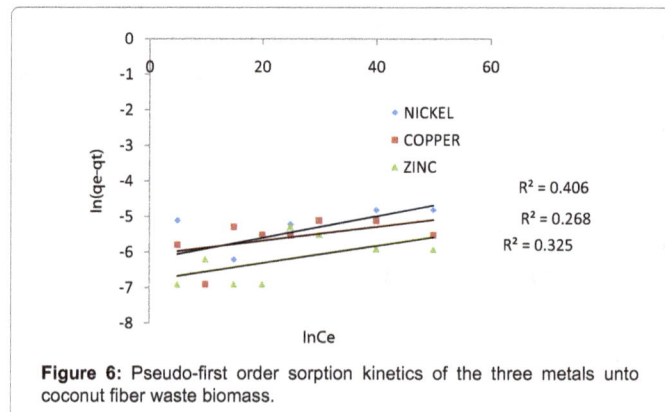

Figure 6: Pseudo-first order sorption kinetics of the three metals unto coconut fiber waste biomass.

Metal ions	K_1	$q_e(mgg^{-1})$	R^2
Ni(II)	0.031	0.002	0.406
Cu(II)	0.019	0.002	0.268
Zn(II)	0.024	0.001	0.325

Table 7: Kinetic parameter values for the pseudo-first order rate for the sorption of the three metal-ions.

three metal-ions since the R^2 values of the three metals were all less than 0.900. The K_f value of Zn(II) (0.880) is greater than that of Ni(II)(0.077) and Cu(II) (0.075), suggesting that Zn(II) has the greatest adsorption tendency towards the waste biomass than the other two metal-ions. The Freundlich equation parameter, 1/n, which is a measure of the absorption intensity for Zn(II) (0.071) is higher than those of Ni(II) (0.042) and Cu(II) (0.009), indicating a preferential sorption of Zn(II) by the waste biomass. Similar absorption intensities have been observed for the metal ions [26].

Kinetics of Sorption

This is probably the most important factor for determining the rate at which sorption takes place for a given system and is also very essential in understanding sorbent design, sorbate residence time and reactor dimension [19]. However according to [18], sorption kinetics shows a large range dependence on the physical and/or chemical characteristics of the sorbent materials, which also influences the sorption process and the mechanism.

Pseudo-first order model

A plot of $\ln (q_e-q_t)$ versus $\ln C_e$ (Figure 6) gives the pseudo-first order kinetics. From the plot, it is observed that the relationship between the metal ion diffusivity, $\ln (q_e-q_t)$ and time, t, is non-linear, indicating that the diffusivity of the metal ion onto the biomass surface is film-diffusion controlled. The non linearity of the diffusivity plot showed that the first -order equation was not adequate for describing the adsorption of the three divalent metal ions onto the biomass surface. This trend has also been reported in [15] for the kinetic study of different ions onto caladium bicolor biomass. Also, it was observed that the pseudo-first order equation did not provide a very good description for the sorption of the three metal ions onto cocos mucifera biomass as, their R^2 were all less than 0.990 for all the three metal ions. Hence, no further consideration of this model was attempted. Table 7 shows the various determined kinetic parameters for the pseudo first order model.

Pseudo-second order model

The initial sorption rate, h_o, the equilibrium sorption capacity,

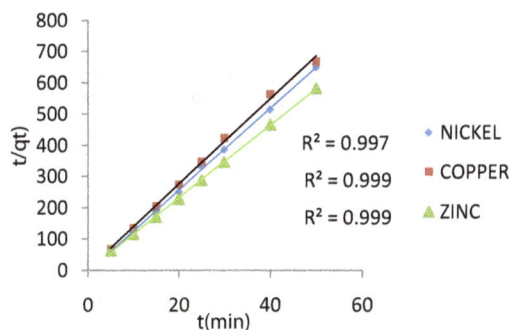

Figure 7: Pseudo-Second order sorption kinetics of the three metals unto coconut fiber waste biomass.

Metal ions	H_o(mgg^{-1}min^{-1})	K_2(mgg^{-1}min^{-1})	q_e(mgg^{-1})	R^2
Ni(ii)	0.153	26.49	0.076	0.999
Cu(ii)	0.331	62.11	0.07	0.998
Zn(ii)	1.293	175.50	0.086	0.999

Table 8: Values of kinetic parameters for the pseudo-second order rate for the sorption of the three metal ions on the biomass.

Metal ions	R_2^2	R_1^1
Ni(II)	0.998	0.406
Cu(II)	0.999	0.268
Zn(II)	0.999	0.325

Table 9: Comparison of coefficients of determination, R^2 for the pseudo-first (R_1^2) and pseudo-second (R_2^2) order rate models.

q_e the pseudo-second order rate constant, K_2 and the coefficients of determination, R^2 were determined experimentally from the slopes and interceptions of the plot of t/q_t against t (Figure 7) and are shown in table 8. The data showed that Zn(II) had a higher sorption rate than Cu(II) and Ni(II). This implies that in a mixed metal-ion system of the three metals, Zn(II) will be adsorbed better. The second pseudo order equation was also found suitable for describing the sorption of the three metal ions using other biosorbents [19].The R^2 values for the two rate constants are as listed in table 9. The data shows a good compliance with the pseudo-second order equation, as the R^2 for the three metal ions adsorption onto the biomass were all > 0.990. The coefficients of determination for the pseudo-first order kinetic model were all smaller than those of the pseudo-second order, indicating that the pseudo-second model was more appropriate in describing the sorption kinetics. Similar pattern of coefficients of determination have been reported for the kinetics of sorption of different metal ion on caladium bicolor biomass [15].

Conclusions

The sorption of Ni(II),Cu(II) and Zn(II) onto coconut fiber waste biomass was found favorable. The kinetic data has provided information on the suitability of coconut fiber waste biomass as an excellent biosorbents for Ni(II),Cu(II),and Zn(II) from aqueous solution. The actual percent removal, of the three metal ions from solution increased with increase in initial metal ion concentration. Optimum pH for the sorption of the three metal ions was 5.The sorption of the three metal ions was found to increase rapidly with contact time. The equilibrium data fit both the Langmuir and the Freundlich isotherms well with the Langmuir model showing a better fit. On the whole, the coconut fiber waste biomass has been proven to be a cost effective biosorbents for treating heavy metal-contaminated wastewater as an alternative to the available conventional methods.

References

1. Ceribasi HI, Yetis U (2001) Water SA 27: 15-20.

2. Eromosele IC, Otilolaye OO (1994) Binding of iron, zinc, and lead ions from aqueous solution by shea butter (Butyrospermun parkii) seed husks. Bull Environ Contam Toxicol 52: 530-537.

3. Abia AA, Igwe JC (2005) Sorbent Kinetics and Intraparticulate Diffusivities of Cd, Pb and Zn ions on Maize Cobs. African Journal of Bio-Technology 4: 509-512.

4. Nale BY, Kagbu JA, Uzairu A, Nwankwere ET, Saidu S, et al. (2012) Kinetic and Equilibrium Studies of the Adsorption of Lead(II) and Nickel(II) ion from Aqueous Solutions on Activated Carbon Prepared from Maize Cob. Der Chemica Sinica 3: 302-312.

5. Suh JH, Kim DS, Song SK (2001) Inhibition Effect of Initial Pb(II) Concentration on Pb(II) Accumulation by Saccharomyces Cereviseae and Aurebaridium Pullulans. Bioresource technology 79: 99-102.

6. Chong KH, Volesky B (1995) Description of Two-Metal Desorption Equilibria by Langmuir-Type Models. Biotechol Bioeng 47: 451-460.

7. Asiagwu AK, Okoye PAC, Eboatu AN (2006) Simulation and Modeling-an Option for the Control of Heavy Metals In Surface Water. Journal of Chemical Society of Nigeria 31: 114-117.

8. Horsfall M, Spiff AI (2005) Effect of Metal Ions Concentration on the Biosorption of Pb(II) and Cd(II) by Caladium Bicolor (wild cocoyam). African Journal of Biotechnology 4: 191-196.

9. Abia AA, Jnr Horsfall M, Didi O (2003) Studies on the Uses of Agricultural By-Products for the Removal of Trace Metals from Aqueous Solution. Journal of Applied Science and Environmental Management 6: 89-95.

10. Gardea-Torresday JL, Tiemam KJ, Gonzalez JH, Henning JA, Townsend MS (1996) Ability of Silica-Immobilized Medicago Sativa (Alfalfa) to Remove Copper Ion from Solution. J Hazard Mater 48: 181-190.

11. Gang S, Weixing S (1998) Sunflower as Adsorbent for the Removal of Trace Metals ion from Waste water. Industrial Engineering and Chemistry Research 37: 1324-1328.

12. Jnr Horsfall M, Abia AA, Spiff AI (2003) Removal of Cd(II), Zn(II) ions from waste water by Cassava (Manihot Esculenta Cranz) Waste Biomass. African Journal of Biotechnology 2: 360-364.

13. Jr Horsfall M, Abia AA, Spiff AI (2005d) Sorption of Lead, Cadmium, and Zinc on Sulfur-Containing Chemically Modified Wastes of Fluted Pumpkin (Telfairia occidentalis Hook f.). Chemistry and Biodiversity 2: 373-385.

14. Anusiem ACI (1999) Applied Surface and Colloid Chemistry (Owerri Varsity Print).

15. Jnr Horsfall M, Spiff AI (2005) Kinetic Studies On The Sorption Of Lead And Cadmium Ions From Aqueous Solutions By Caladium Bicolor (Wild Cocoyam) Biomass. J Chemical Society of Ethiopia 19: 89-102.

16. Ho YS, Wase DAJ, Forster CF (1995) Bata Nickel Removal from Aqueous Solution by Sphagnum Moss Peat. Wat res 29: 1327-1332.

17. Fourest E, Roux CJ (1992) Heavy Metals Biosorption by Fungal Mycihal By-products: Mechanism and Influence of pH. Applied Microbiology and Biotechnology 37: 610-617.

18. Ho YS, Wase DAJ, Forster CF (1995) Kinetic Studies of Competitive Heavy Metal Adsorption by Sphagnum Moss Peat. Environ Techno 17: 71-77.

19. Vaishnav V, Kailash D, Suresh C, Madan L (ISC-2011) Adsorption Studies of Zn(II) ions from Wastewater Using Calotropis Procera as an Adsorbent. Res J Recent Sci 1: 160-165.

20. Gardea-Torresday JL, Gonzalez JH, Tiemam KJ, Rodriguez O, Gamez G (1998) Phytofiltration of hazardous cadmium, chromium, lead and zinc ions by biomass of Medicago sativa (Alfalfa). J Hazard Mater 57: 29-39.

21. Pankaj P, Sambi SS, Sharma SK, Singh S (2009) Batch Adsorption Studies for the Removal of Cu (II) Ions by ZeoliteNaX from Aqueous Stream. Proceedings of the World Congress on Engineering and Computer Sciences, San Francisco, USA.

22. Jnr Horsfall M, Spiff AI (2004) Equilibrium Sorption Study of Al3+, Co2+ and Ag+ in Aqueous Solutions by Fluted Pumpkin (Telfairia Occidentalis HOOK f) Waste Biomass. Acta Chim Solv 52: 174-181.

23. Asrari A, Hossein T, Manoosh H (2010) Removal of Zn(II) and Pb (II) ions Using Rice Husk in Food Industrial Wastewater. J Appl Sci Environ Mange 14: 159-162.

24. Zhao G, Wu X, Tan X, Wang X (2011) Sorption of Heavy Metal Ions from Aqueous Solutions: A Review. The Open Colloid Science Journal 4: 19-31.

25. El-Sayed GO, Dessouki HA, Ibrahim SS (2010) Biosorption Of Ni (II) And Cd (II) Ions From Aqueous Solutions Onto Rice Straw. Chemical Sciences Journal 9.

26. Radenovic A, Malina J, Strkalj A (2011) Removal of Ni(II) from Aqueous Solution by Low-Cost Adsorbents. The Holistic Approach to Environment 1.

17

Phosphogypsum Conversion to Calcium Carbonate and Utilization for Remediation of Acid Mine Drainage

J Mulopo[1]* and D Ikhu-Omoregbe[2]

[1]Council for Scientific and Industrial Research, Natural Resources and the Environment, Pretoria, South Africa
[2]Chemical Engineering Department, Cape Peninsula University of Technology, Cape Town, South Africa

Abstract

The conversion of phosphogypsum waste (a waste product of phosphoric acid production in Richard Bay, South Africa), using sodium carbonate was tested. Response surface methodology (RSM) was used to investigate the combined effect of relevant process variables to maximize the production of calcium carbonate in a batch reactor. The process variables include time (0, 60, and 120 min); slurry content (5, 10, and 15%); agitation speed (100, 300, and 500 rpm); and sodium carbonate/gypsum molar ratio (0.8, 1.4, and 2). In the optimum conditions of the process (the slurry concentration of 5%, molar ratio sodium carbonate/gypsum of 2, stirring rate of 500 rpm) the conversion of waste gypsum to calcium carbonate after 105 minutes reaches over 98.5%. The calcium carbonate produced in this work compares favorably to the commercial calcium carbonate (laboratory grade) during Acid Mine Drainage (AMD) neutralisation.

Keywords: Optimization; Waste gypsum; Calcium carbonate; Acid Mine Drainage

Introduction

South Africa has a highly-developed domestic and export oriented phosphate industry with the largest igneous phosphate deposit located in Phalaborwa in the low veld of the Northern Province of the country. Waste gypsum is produced as a by-product during treatment of phosphate concentrate with sulphuric acid to produce phosphoric acid according to the following simplified reaction:

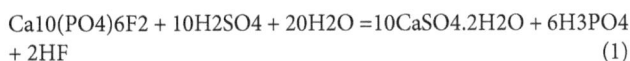

$$Ca10(PO4)6F2 + 10H2SO4 + 20H2O = 10CaSO4.2H2O + 6H3PO4 + 2HF \tag{1}$$

The Richard Bay Foskor phosphoric acid plant is a complex import/export orientated industry with raw materials including large quantities of phosphate rock from Phalaborwa and approximately 350 000 tonnes of sulphur which are imported annually. The bulk of the phosphoric acid produced in Richard Bay is exported whilst approximately 10% is used in South Africa for the manufacture of phosphatic fertilizers and animal feeds. The effluent from this plant consists of about 10 000 tonnes/day of waste gypsum, containing 20-30% free water and which is pumped into the sea or stockpiled. The readily apparent disadvantages of these disposal options is that it results in loss of all commercial value or poses environmental threat by the leaching action of rainwater and/or storage water producing an acidic effluent that may enter the nearby surface and/or groundwater regime, creating an environmental hazard [1].

The beneficiation of the waste gypsum generated during production of phosphoric acid has been for a long time an important challenge for the phosphoric acid industry, ecologists and all those concerned with disposal of waste products such as government and civil society. In view of the chemical characteristics of waste gypsum from the phosphoric acid industry (more than 93% is CaSO4.2H2O) and its economic potential, and of the prevailing environmental pollution concerns there is a clear value in promoting the use of waste gypsum as an alternative raw material for many applications. Hitherto a number of options for waste gypsum beneficiation have been proposed in South Africa:

Fertilizer and soil conditioner

Around 40.000 tonnes per year of waste gypsum are sold by the Omnia group to the agriculture sector. Waste gypsum has been also used worldwide as an agricultural fertiliser or for soil stabilisation amendments [2,3].

Cement and wallboard

The Omnia group upgrades around 200.000 tonnes per year of waste gypsum for use in the cement industry. The use of waste gypsum has been also elsewhere reported in the cement industry as a setting regulator in place of natural gypsum [4].

Road construction materials

Unfortunately, these solutions have not proven reliable in terms of large-scale volume reduction of waste gypsum.

Security of water supply has become a key strategic issue for the sustained economic growth of South Africa as it is a water-stressed country. Although the South African mining sector is one of the critical pillars and drivers of the South African economy, mining activities are also associated with environmental pollution such as acid mine drainage (AMD). AMD is highly acidic water, usually containing high concentrations of metals, sulphides, and salts as a consequence of mining activity. The major sources of AMD include drainage from underground mine shafts, runoff and discharge from open pits and mine waste dumps, tailings and ore stockpiles, which make up nearly 88% of all waste produced in South Africa. The gold mining industry in South Africa (principally the Witwatersrand Goldfield) is in decline, but the post-closure decant of AMD represents an enormous threat. For example, the potential volume of AMD for the Witwatersrand

*Corresponding author: Jean Mulopo, Council for Scientific and Industrial Research, Natural Resources and the Environment, Pretoria, South Africa

Goldfield alone amounts to an estimated 350ML/day [5-8]. While some mines have established ad hoc treatment processes to treat localized water pollution, South Africa urgently needs a regional, consolidated approach to AMD. It is against this background that the government of South Africa has suggested that the pumping and partial treatment of mine water is critical and should be implemented in the short term in the western, central and eastern basins through neutralisation and metals removal. The partial treatment of mine water to neutralise acidity and remove metals will require considerable amount of calcium carbonate which will put extra pressure on the existing local market. It is therefore imperative that alternatives options of calcium carbonate production be considered. In this paper we consider the use of sodium carbonate for waste gypsum treatment and recovery of calcium carbonate. We have completed batch laboratory tests required to gather process data and determine the best process conditions for optimized production of calcium carbonate from waste gypsum using sodium carbonate. Response Surface Methodology (RSM) was used to optimize the waste gypsum conversion.

Materials and Methods

Feed stock material

Identification of mineralogical characteristics of phosphogypsum waste is important from processing viewpoint as the beneficiation of waste gypsum is in some extent controlled by mineralogical and textural characteristics. Samples of phosphogypsum from the Foskor plant in Richards Bay (South Africa) were collected for this study. Splits of these samples were subjected to chemical and physical analyses. The data on particle size distribution were obtained by MICROTRAC S3500 Particle Size Analyser. The waste gypsum used in this work has a particle size ranging between 0.003 and 0.055 mm in diameter with an average particle size of 0.017 mm in diameter. X-Ray Fluorescence (XRF) Oxford Instruments TWIN-X and HITACHI S-7000 Scanning Electron Microscope (SEM) were used to determine the elemental composition

and morphological structure of the waste gypsum. X-Ray Fluorescence (XRF) results in Table 1 show the collected phosphogypsum generally contains traces of Co, Cu, Zn, Ga, Sr, Y, Zr, Ba, Ce, W, Pb and Th and some rare earth elements like Y. The morphological structure of untreated phosphogypsum samples, as determined by SEM, is illustrated in Figure 1 and reveals a homogeneous, prismatic piling arrangement. Radioactivity analysis was also conducted on the Foskor phosphogypsum for gross alpha/beta-activity and for selected radio nuclides in the uranium and thorium decay series by the South African Nuclear Energy Corporation and the result shows that the collected phosphogypsum contains low levels of U-series radionuclide. The most important source of radioactivity is 226Ra and 228Ra as shown in Table 2. Other researchers have reported a wide variety of radionuclide concentrations in different phosphogypsum sources [9,10]. The concentrations of Ra-226 and Ra-228 are respectively 316 and 424 Bq/kg in Foskor phosphogypsum.

Equipment

Waste Gypsum treatment experiments were carried out batch-wise using a 3 litre Perspex reactor as shown in the experimental setup (Figure 2). The reactor had four equally spaced baffles and was equipped with an outer flow jacket. An overhead stirrer equipped with a radial turbine impeller was used for mixing combined with an external recirculation pump. A Hanna HI2829 multi-parameter meter was used to log electrical conductivity, pH and temperature data. A hypodermic syringe was used to draw out sample aliquots from the gypsum reactor. A PERKIN ELMER ANALYST 700 Atomic Absorption Spectrometer (AAS) and a VARIAN Inductively Coupled Plasma (ICP) spectrometry were used to analyse for Ca and Na during experiments. A LECO CS200 carbon/sulphur analyser was used to determine the inorganic Carbon (C) content of the calcium carbonate produced.

Experimental procedure

Waste gypsum treatment experiments was carried out by mixing

Cr (%)	Mn (%)	Fe (%)	Co (ppm)	Ni (ppm)	Cu (ppm)	Zn (ppm)	Ga (ppm)	Ge (ppm)	As (ppm)	Se (ppm)
0.025	0.01	0.07	51.90	12.40	12.90	22.31	7.81	<1.20	0.71	1.31
Br(ppm)	**Rb(ppm)**	**Sr(ppm)**	**Y(ppm)**	**Zr(ppm)**	**Ag(ppm)**	**Cd(ppm)**	**Sn(ppm)**	**Sb(ppm)**	**Te(ppm)**	**I(ppm)**
3.100	2.60	2183	105.61	11.11	<0.71	1.71	<0.91	1.30	0.51	<2.51
Cs(ppm)	**Ba(ppm)**	**La(ppm)**	**Ce(ppm)**	**Hf(ppm)**	**Ta(ppm)**	**W(ppm)**	**Tl(ppm)**	**Pb(ppm)**	**Th(ppm)**	**U(ppm)**
<5.1	216.90	594.30	1435	11.71	10.90	34.00	2.40	18.90	20.21	4.41
Al (%)	**Na (%)**	**Mg (%)**	**Si (%)**	**P (%)**	**S (%)**	**K (%)**	**Ca (%)**	**Ti (%)**	**V (%)**	
<0.018	<0.45	<0.08	0.34	0.80	19.66	<0.01	25.52	<0.01	0.02	

Table 1: Chemical composition (wt. %) of the Foskor phosphogypsum as determined by XRF analysis.

Phosphogypsum raw material				Calcium carbonate produced			
Nuclide	**Value(Bq/kg)**	**Unc.**	**MDA**	**Nuclide**	**Value (Bq/kg)**	**Unc.**	**MDA**
238 U	14.71	0.51	0.59	238 U	8.65	0.43	0.60
234U	14.81	0.51	0.60	234U	8.72	0.43	0.61
226 R a	316	32	90	226 R a	413	37	99
210Pb	171	20	48	210Pb	237	24	51
235U	0.68	0.03	0.03	235U	0.40	0.02	0.03
228 R a	424	18	33	228 R a	442	20	42
228Th	114	6	8.81	228Th	123	6	11
40K	< MDA	-----	140	40K	< MDA		160

Table 2: Analysis of Foskor phosphogypsum and produced calcium carbonate for selected radionuclide (Bq/kg).

The uncertainty (Unc. column) is quoted at 1 sigma (or coverage factor k = 1). The uncertainty is calculated mainly from counting statistics and it is not the standard deviation obtained from replicate measurements. No uncertainty value is reported of a less than MDA ("< MDA") is indicated in the Value column. The minimum detectable activity concentration (MDA column) is calculated with a 95% confidence level.

Figure 1: SEM micrograph of the Foskor phosphogypsum.

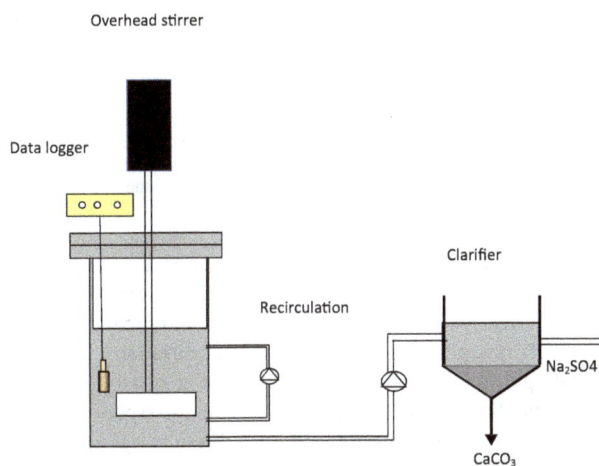

Figure 2: Experimental Setup.

100 g of waste gypsum with a pre-calculated amount of water and placed in the reactor. The slurry was then mixed for 20 minutes for ambient experiments or heated up to a predetermined temperature under continuous agitation for experiments above ambient temperatures until stabilization. A certain amount of sodium carbonate corresponding to different molar ratios was introduced into the reactor and reacted for defined time period. Sample aliquots were collected at predetermined time intervals over the entire reaction period. The collected samples were immediately filtered; the sulphate content determined by standard analytical procedures as described in Standard Methods for the Examination of Water and Wastewater (APHA, 1985). The calculation of the reaction conversion applied to all the experimental data is as follows:

Reaction conversion (%) = (Sulphate in filtrate /sulphate in feed sample) × 100   (2)

Results and Discussion

Statistical analysis

Optimization of process conditions is usually one of the most important factors to reduce the production cost. The conventional method, which involved varying one variable at a time while keeping the other variables constant is lengthy and often does not produce

the effect of interaction of different variables. To this end, Response Surface Methodology (RSM) was used to optimize the waste gypsum conversion. Laboratory scale confirmation of the products was then used to validate the feasibility of the derived optimum conditions. The tests were performed to investigate mainly the effect of the investigated factors on the waste gypsum conversion to calcium carbonate. Four factors were taken into consideration in the experimental planning: time (min) (A); slurry content (%) (B); Molar ratio of sodium carbonate to gypsum (C) and stirring speed (rpm) (D) as shown in Table 3.

The efficiency of fit of the model was checked by the determination coefficient (R^2). In this case, the value of the determination coefficient (R^2 = 0.96) indicates that only 4% of the total variations are not explained by the model (Table 5). The R^2 value for phosphogypsum conversion indicates that the proposed model is adequate with no significant lack of fit as 96 % of the variability in the response could be explained by the model. The value of the adjusted determination coefficient (Adj. R^2 = 0.91) is also very high, which indicates a high significance of the model. R^2 values increase as more variables are

Variable	Parameter	Low Value	Median Value	High Value
Time (min)	A	0	60	120
Slurry (%)	B	5	10	15
Molar ratio (Na_2CO_3/ Gypsum)	C	0.8	1.4	2
Stirring speed (rpm)	D	100	300	500

Table 3: Levels of factors chosen for the experimental design.

Experiment No.	t (min), A 30,45, 60	Slurry (%), B 5, 10, 15	Molar ratio, C (moles Na_2CO_3/ moles gypsum) 0.8, 1.4, 2	v(rpm) D 100, 300; 500	Phospho-gypsum Conversion (%)
1	0	5	0.80	100	7
2	120	5	0.80	100	54
3	0	15	0.80	100	2.61
4	120	15	0.80	100	52
5	0	5	2	100	12.60
6	120	5	2	100	86
7	0	15	2	100	2.40
8	120	15	2	100	37
9	0	5	0.80	500	11
10	120	5	0.80	500	72
11	0	15	0.80	500	4.43
12	120	15	0.80	500	58
13	0	5	2	500	4.8
14	120	5	2	500	97
15	0	15	2	500	5.60
16	120	15	2	500	61.71
17	0	10	1.40	300	4.32
18	120	10	1.40	300	58.81
19	60	5	1.40	300	69
20	60	15	1.40	300	51
21	60	10	0.80	300	53
22	60	10	2	300	62
23	60	10	1.40	100	42
24	60	10	1.40	500	83
25	60	10	1.40	300	58.21
26	60	10	1.40	300	57.91
27	60	10	1.40	300	59.61

Table 4: Experimental results for treatment of Foskor phosphogypsum with Sodium Carbonate using central composite design.

added to a model; the adjusted R2 is often used to summarize the fit as it takes into account the number of variables in the model. For multiple regression, although R2 is still the percent of the total variation that can be explained by the regression equation, the largest value of R2 will always occur when all of the predictor variables are included, even if those predictor variables don't significantly contribute to the model. R2 will only decrease or stay the same as variables are removed, but never increase. The Adjusted R2 uses the variances instead of the variations. That means that it takes into consideration the sample size and the number of predictor variables. The value of the adjusted R2 can actually increase with fewer variables or smaller sample sizes.

This correlation is also proven by the plot of predicted versus experimental values of phosphogypsum conversion in Figure 3. Other elements, such as Ca and Na, were not considered as response factors because the only important parameter for following the reaction process evolution was considered to be the sulphate removal. However for the purpose of this study the solubilization of Ca was also monitored and was always very low (0.1-0.5 mg/1). In Table 5 the coefficients A, B, D and A2 are all significant model terms.

The effect of the independent variables, slurry (%) and time on the phosphogypsum conversion can be accessed from Table 4. Phosphogypsum conversion mainly depend on the reaction time since both its linear and quadratic effects are significant (Prob>|t| less than 0.05). The waste gypsum conversion is decreased when the phosphogypsum slurry content (%) is increased. Table 4 shows that the linear effect of the slurry content has a negative sign. However the quadratic effect of the slurry content on the phosphogypsum conversion has a positive sign, which suggests that the phosphogypsum conversion will have a curvilinear shape in the time-slurry (%) space as the negative value of the linear slurry coefficient will cause an initial downward slope, and the positive value of the coefficient for the slurry quadratic terms will cause the curve linearity (upward due to the positive sign). The effect of the independent variables, time and sodium carbonate/gypsum molar ratio on the waste gypsum conversion can also be assessed from Table 4. The phosphogypsum conversion does not depend significantly on the sodium carbonate/gypsum molar ratio as both its linear and quadratic effects are not significant (Prob>|t| more than 0.05). However the phosphogypsum conversion to calcium carbonate depends on the stirring speed as its linear effect is significant.

There are a number of combinations of variables that could give maximum levels of phosphogypsum conversion. The optimum experimental conditions were obtained by the differentiation of

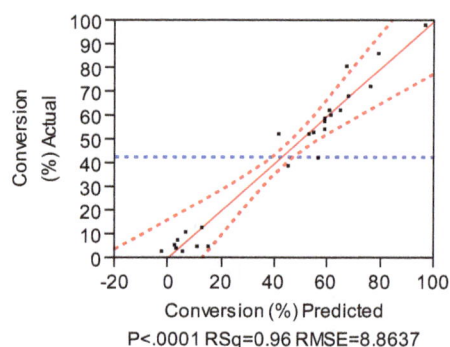

Term		Estimate	Std Error	t Ratio	Prob>|t|
Intercept		58.34	3.51	16.60	<.0001*
Time (min)(0,120)	A	29.30	2.09	14.03	<.0001*
slurry (%)(5,15)	B	-7.45	2.09	-3.57	0.0044*
Molar Ratio(0.8,2)	C	3.10	2.09	1.49	0.1654
Stirring Rate (rpm)(100,500)	D	5.59	2.09	2.67	0.0217*
Time (min)*slurry (%)	AB	-4.80	2.22	-2.16	0.0533
Time (min)*Molar Ratio	AC	2.92	2.22	1.32	0.2143
slurry (%)*Molar Ratio	BC	-4.21	2.22	-1.90	0.0841
Time (min)*Stirring Rate (rpm)	AD	3.73	2.22	1.68	0.1205
slurry (%)*Stirring Rate (rpm)	BD	0.61	2.22	0.27	0.7888
Molar Ratio*Stirring Rate (rpm)	CD	-0.08	2.22	-0.04	0.9707
Time (min)*Time (min)	A²	-26.45	5.54	-4.78	0.0006*
Slurry (%)*slurry (%)	B²	1.65	5.54	0.30	0.7714
Molar Ratio*Molar Ratio	C²	-0.85	5.54	-0.15	0.8807
Stirring Rate (rpm)*Stirring Rate (rpm)	D²	3.15	5.54	0.57	0.5811

*significant

Table 5: Analysis of the response surface quadratic model obtained from experimental designs.Prob>|t| less than 0.05 indicate model terms that are significant, R² = 0.961226, Adjusted R² = 0.911878, t Ratio are equal to the coefficients/std errors.

the quadratic model using an algorithm carried out on the Maple 9.5 program (Waterloo Maple, Inc., Canada) to achieve maximum phosphogypsum production. The derived optimum conditions were: Time = 105 min, phosphogypsum slurry concentration = 5%, sodium carbonate/phosphogypsum molar ratio = 2 and speeding rate = 500 rpm. The predicted waste gypsum conversion corresponding to these levels was 98.6%.Validation of the predicted results was accomplished by performing additional experiments in duplicate with the parameters suggested by the numerical modeling. The two experiments yielded an average waste gypsum conversion of 98.4%. Good agreement between the predicted and experimental results confirmed the experimental adequacy of the model and the existence of the optimal conditions.

Application

The quality and the neutralisation potential of the precipitated calcium carbonate produced by conversion of waste gypsum using sodium carbonate were assessed. Partial treatment, as normally applied to Acid Mine Drainage, involves chemical neutralization of the acidity followed by precipitation of iron and other suspended solids and usually the chemicals used for Acid Mine Drainage treatment include limestone and hydrated lime [11-14]. The use of limestone requires that its calcium content be as high as possible to reduce the drawbacks for using limestone which include slow reaction time. In the case of limestone, the neutralization process can be described as the replacement of the undesirable cation components (H^+, Fe^{2+}, Fe^{3+}) in AMD by a more acceptable cation, which is Ca^{2+} and the overall reaction can be written as follows:

$$CaCO_3(s) + 2H^+ = Ca^{2+} + H_2O + CO_2$$

XRF analyses were conducted to identify elemental composition of the calcium carbonate produced at the optimum conditions derived above.

The main objective of the product evaluation was to compare the

Figure 3: Actual (experimental) response conversion (%) versus predicted (model) conversion (%).

Parameter	pH	Acidity	SO_4^{2-}	Ca	Cl^-	Fe	Mg	Mn	Na	Al
Units		mg $CaCO_3$	mg/L	mg/L	mg/L	mg/L	mg/L	mg/L	mg/L	mg/L
Values	3.01	125	2700	468	45	246	135	70	82	2.62

Table 6: Typical composition of Acid Mine Drainage neutralized using calcium carbonate.

Sample	Inorganic C	SiO_2	SO_4^{2-}	Al_2O_3	Fe_2O_3	MgO	CaO	Na_2O	K_2O	Cr_2O_3
Commercial $CaCO_3$	54.31	0.49	0.07	0.26	0.07	0.19	58.13	< 0.16	< 0.01	0.04
Produced $CaCO_3$	48.93	1.63	2.41	0.36	0.01	< 0.05	46.84	2.49	0.01	0.04

Table 7: Produced and commercial composition (wt. %) for major oxides, inorganic carbon as determined by XRF and LECO analysis.

neutralizing potential of the produced calcium carbonate vis-à-vis the commercial calcium carbonate. The test consists of simple stirred beaker filled with 100 ml AMD of composition shown in Table 6 and 200 mg of the neutralizing material (calcium carbonate). The pH was followed through the entire testing period.

The main results of the batch tests performed with the two types of calcium carbonate are presented in Table 7. This test allows evaluating the influence of the composition of the limestone. There is a rapid increase of pH during the first 2 minutes from 3.2 to 5.6 for commercial calcium carbonate and to 4.5 for calcium carbonate produced by treatment of phosphogypsum with sodium carbonate. Then a difference in the neutralization potential between these two calcium carbonates is monitored. However both solutions reach a pH of 6.9 after 60 minutes Figure 4. These results confirm that the neutralizing potential of the produced calcium carbonate is comparable in the long run to that of commercial calcium carbonate although the calcium and inorganic carbon content of the produced calcium carbonate is slightly lower than that of commercial calcium carbonate as shown in Table 7. The radionuclide in the calcium carbonate produced from phosphogypsum have been also analysed as shown in Table 2. The results show that Radium-226, Radium-228 and Pb-210 are the most significant radionuclide and that their concentrations have increased slightly compared to the initial values in the feed phosphogypsum although concentrations are still much less than 1Bq/g. This result is in agreement with previous reported results that most radio nuclides follow the calcium carbonate during treatment of waste gypsum [10].

Conclusion

This paper constitutes the first step in an attempt to develop economical processes that could exploit and consume the phosphogypsum from the phosphate industry in South Africa for calcium carbonate production. Environmental concerns are related with the large stockpiles of phosphogypsum and their negative impact on surrounding land, water and air as most of the phosphogypsum generated each year is dumped on land or in the sea. In this study, selected parametersnamely the reaction time, the phosphogypsum slurry concentration (%), the stirring rate and the molar ratio of sodium carbonate to phosphogypsum were subjected to second order polynomial models to determine their significance on the phosphogypsum conversion to calcium carbonate using sodium carbonate. The optimum conditions for the four independent variables were derived from the experimental results and new experiments were carried out to cross-validate their liability of the optimum conditions. The conversion of Foskor phosphogypsum to calcium carbonate using sodium carbonate is estimated to be optimum if the treatment process maintains the slurry concentration at 5%, the sodium carbonate/gypsum molar ratio at 2, the stirring rate at 500 rpm and the time at 105 minutes. This study further demonstrated that the regenerated CaCO3 exhibits effective neutralisation ability during AMD pre-treatment compared to commercial laboratory grade CaCO3. The recovery of calcium carbonate from waste gypsum represents an opportunity to alleviate the high cost of the AMD treatment technologies.

References

1. Szlauer B, Szwanenfeld M, Jakubiec HW, Kolasa K (1990) Hydrobiological characteristics of ponds collecting effluents from a phosphogypsum tip of the police chemical works near Szczecin. Acta Hydrobiologica 32: 27-34.

2. Degirmenci N, Okucu A, Turabi A (2007) Application of phosphogypsum in soil stabilization. Building and Environment 42: 3393-3398.

3. Reijnders L (2007) Cleaner phosphogypsum, coal combustion ashes and waste incineration ashes for application in building materials, A review. Building and Environment 42: 1036-1042.

4. Akin AI, Yesim S (2004) Utilization of weathered phosphogypsum as set retarder in Portland cement. Cement and Concrete Research 34: 677-680.

5. Naicker K, Cukrowska E, McCarthy TS (2003) Acid mine drainage arising from gold mining activity in Johannesburg, South Africa and environs. Environ Pollut 122: 29-40.

6. Duane MJ, Pigozzi G, Harris C (1997) Geochemistry of some deep gold mine waters from the western portion of the Witwatersrand Basin, South Africa. Journal of African Earth Sciences 24: 105-123.

7. Roychoudhury AN, Starke MF (2006) Partitioning and mobility of trace metals

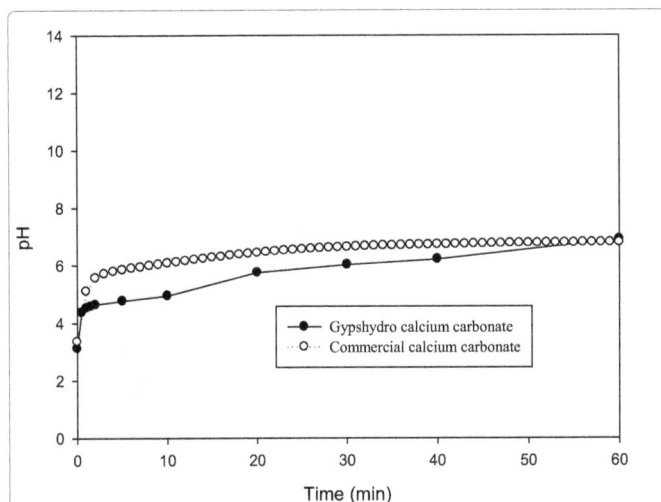

Figure 4: pH profiles during neutralization of AMD with produced calcium carbonate and commercial calcium carbonate. (Conditions: 2g calcium carbonate/1000ml AMD-315 rpm).

in the Blesbokspruit: Impact assessment of dewatering of mine waters in the East Rand, South Africa. Applied Geochemistry 21:1044-1063.

8. Tutu H, McCarthy TS, Cukrowska E (2008) The chemical characteristics of acid mine drainage with particular reference to sources, distribution and remediation: The Witwatersrand Basin, South Africa as a case study. Applied Geochemistry 23: 3666-3684.

9. Burnett WC, Schultz MK, Carter DH (1996) Radionuclide flow during the conversion of phosphogypsum to ammonium sulphate. Journal of Environmental Radioactivity 32: 33-51.

10. Rutherford PM, Dudas MJ, Samek RA (1994) Environmental impacts of phosphogypsum. Science of the Total Environment 149: 1-38.

11. Roessler CE (1984) the radiological aspects of phosphogypsum, In:

Proceedings of the Seminar on Phosphogypsum, Miami, Florida Institute of Phosphate Research 11-36.

12. Hedin RJ, Watzlaf GR, Nairn RW (1994) Passive treatment of coal mine drainage with limestone. Journal of Environmental Quality 23:1338-1345.

13. Maree JP, du Plessis P, van der Walt CJ (1992) Treatment of acid effluents with limestone instead of lime. Water Science and Technology 26: 345-355.

14. Maree JP, du Plessis P (1994) Neutralization of acid mine water with calcium carbonate. Water Science and Technology 29: 285-296.

15. Maree JP, van Tonder GJ, Millard P, Erasmus TC (1996) Pilot- scale neutralisation of underground mine water. Water Science and Technology 34: 141-149.

18

Natural Fiber Reinforced Composite: A Concise Review Article

Ashish Chauhan*, Priyanka Chauhan and Balbir Kaith

Department of Chemistry, Dr. B. R. Ambedkar National Institute of Technology, Jalandhar 144 011 (Pb) India

Abstract

Fiber reinforced composites were in use since ancient times. Due to the disadvantage of the synthetic and fiber glass as reinforcement, the use of fiber reinforced composite gained the attention of the young scientists. The review article explores the use of variable fiber as reinforcement in composites. With the advancement of science and technology the new means of characterization and evaluation of physico-chemico-thermo-mechanical properties of the composite have been used that have explored the new horizon of utilizing them for various applications.

The term 'composite' in material science refers to a material made up of a matrix containing reinforcing agents. Reinforcement is the part of the composite that provides strength, stiffness, and the ability to carry a load. Wood is a natural composite of cellulose fibers in a matrix of lignin. In manufacturing, fibers are the most commonly used reinforcement that yields Fiber Reinforced Composite (FRC). The reinforcement is embedded into the matrix. Common matrixes include mud (wattle and daub), cement (concrete), polymers (fiber reinforced plastics), metals and ceramics. The most common polymer-based composite materials include fiberglass, carbon fiber and kevlar. Fiberglass is probably one of the most familiar reinforcing composite materials that were introduced in 1940, consisting of glass fiber reinforcement of unsaturated polyester matrix [1-3]. This glass fiber had numerous drawbacks that led to search for alternate substitute as reinforcement. Fiber as reinforcement to the composite had outstanding physio-chemico-thermo-mechanical performance, durability and eco-friendly nature that highlighted and promoted its scope.

The beginning of composite materials may have been the bricks fashioned by the ancient Egyptians from mud and straw. The ancient brick-making process can still be seen on Egyptian tomb paintings in the Metropolitan Museum of Art. Commercialization of the composites could be traced to early century when the cellulose fibers were used to reinforce phenolics, urea and melamine resins. Composites in the world of today have wide range of applications, wherever high strength-to-weight ratio remains and important consideration for use. Its principal use is found in automotive, marine and construction industries. In majority of cases, requiring high performance in the automotive and aerospace industries, the discontinuous phase or filler is in the form of a fiber. In most cases, composite matrices are the thermosets having carbon and ceramics for high temperature applications. Thermosets (epoxy, polysulfones) and thermoplastics (polyetherether ketone, polyimide) due to high strength and performance are pioneer for research and industrial applications. Nano-composites in the latest advances have high aspect ratio and improved electrical, mechanical and thermal properties that could be fabricated for various purposes [4-6].

The use of natural fiber as reinforcement in composite was a challenging task. Ferreria et al. [7] improved the fatigue strength by using hybrid fiber composites with a polypropylene hemp layer adjacent to the bond interface which was expected to produce more uniform stress in transient regions. Richardson and Zhang [8] applied flow visualization experiments using resin transfer molding for developing a better understanding of the mold filling process for hemp mat reinforced phenolic composites. *Eucalyptus urograndis* pulp used as the reinforcement for thermoplastic starch showed an increase of 100% in tensile strength and more than 50% in modulus with respect to non-reinforced thermoplastic starch [9]. Fiber reinforced composite materials offered a combination of strength and modulus that are either comparable to or better than many traditional metallic materials. Increase in the flax and jute fiber content in polyurethane based composites increased the shear modulus and impact strength. However, increasing the micro void content in the matrix decreased its strength [10]. Jayaraman and Bhattacharya [11] reported the mechanical performance of wood fiber waste based plastic composites and observed that tensile strength does not generally change with fiber content. Zulkifli et al. [12] prepared Natural Rubber (NR)-Polypropylene (PP) composites by increasing the amount of NR in PP by 5-20% increase in its composition, inter-laminar fracture properties as well as the resistance of material to delaminate crack propagation. With increase in NR amount of this inter-laminar fracture the toughness of composite material decreased. Thwe and Liao [13] studied the resistance of bamboo fiber-PP hybrid composites to hygrothermal aging and their fatigue behavior under cyclic tensile load. The use of maleic anhydride polypropylene as a coupling agent suppressed the moisture absorption and degradation in such composites. Herrera-Franco and Valadez-Gonzalez [14] reported that fiber matrix adhesion promoted the fiber surface modification on alkaline treatment and matrix pre-impregnation e.g. use of silane as coupling agent in case of henequen fiber-HDPE composite. The increase in mechanical strength was found to be raised between 3–43% for longitudinal tensile and flexural properties whereas in transverse direction, the increase was greater than 50% with respect to the properties of composites made of untreated fiber. Increase in stiffness was approximately 80% of the calculated values.

*Corresponding author: Ashish Chauhan, Department of Chemistry, Dr. B. R. Ambedkar National Institute of Technology, Jalandhar 144 011 (Pb), India

Electrical properties of the wood polymer composites from agro-based materials such as banana, hemp and agave fiber using novolac resins have been reported by Naik and Mishra [15]. Mitra et al. [16] treated the unwoven jute fiber with precondensate like formaldehyde, melamine formaldehyde and polymerized cashew nut shell liquid-formaldehyde, prior to its use as reinforcing material for the preparation of composites. The treatment reduced the moisture absorbance of the jute. Kandola et al. [17] reported the fabrication of novel glass reinforced epoxy composites containing phosphate. Eichhorn and Young [18] studied the deformation in micro mechanics of natural cellulose fiber networks and composites. Kaith et al. [19,20], Singha AS et al. [21] prepared polymer matrix based composites using flax-g-copolymers, flax fiber and mercerized flax as reinforcing agent. It was observed that the reinforcement increased the endurance of the composite to higher loads as compared to pure polystyrene. Mercerized fiber was found to be more effective reinforcing agent for wear resistance, tensile strength and compressive strength as compared to the grafted fibers. However, reduction in moisture absorbance and increase in the chemical resistance on graft copolymerization was observed.

Chauhan A [22] utilized the *Hibiscus sabdariffa* (Roselle) stem as reinforcing agent in phenol formaldehyde matrix based composites. Roselle fiber was graft copolymerized with monomers like methyl acrylate, ethyl acrylate, butyl acrylate and acrylonitrile and used these graft copolymer as reinforcement in the composite. The grafted fiber and composites were characterized by SEM, XRD, TGA, DTA and evaluated for physico-chemico-thermal properties. It was observed that the modified grafted fiber incorporated into the composite enhanced the physico-chemico-thermo-mechanical competence. Since, the grafted monomer acted as a coupling agent. The mechanical evaluation was done on the basis of wear, tensile, compressive strength test, flexural strength, young's modulus and hardness. However, some variation was seen in few cases but in most of the cases the strength improved. The better mechanical behavior could be accounted due to compatible fiber-matrix interaction and orientation of the fiber. However, some deviation in the results could be justified by other governing factors for overall mechanical performance like nature and amount of matrix and fiber, orientation, distribution of the fiber with respect to the matrix axis, form of reinforcement used (woven or non-woven, grafted or ungrafted), strength of the interfacial bond between the fiber and matrix, length of the fiber (continuous or discontinuous), aspect ratio that on mere imbalance may lead to de bonding and cracking [22-32].

So, we have seen above that various researchers have utilized the low weight and high strength of fibers like hemp, flax, jute as reinforcement to form fiber reinforced composite. These reinforcements have improved the strength and properties of composites if used after the graft copolymerization of the fiber like Roselle. We are blessed with variable natural resources and fiber but very less has been explored and utilized as yet. Fiber reinforced composites are one of the means to utilize the natural resources. But, with time these renewable resources and fiber will soon deplete. So, there is a great need to sustain and procure them for the future use. We should seek more fruitful means to explore the maximum potential and utilize the natural fiber for the development of science and technology.

References

1. Tsai SW, Hahn HT (1980) Introduction to Composite Materials. Technomic Pub., West Post.

2. Nielsen LE (1974) Mechanical properties of Polymers and Composites. Marcell Dekker Inc, New York.

3. Nicolais L (1975) Mechanics of Composites. Polym Eng Sci 15: 137-149.

4. http://144.206.159.178/ft/862/46270/822929.pdf

5. Krishnamoorti R, Vaia R (2002) Polymer Composites: Synthesis Characterization and Modeling. American Chemical Society Symposium Series 804, Washington DC.

6. Vaia RA, Giannelis EP (2001) Polymer Nanocomposites: Status and Opportunities. MRS Bulletin 26: 394-401.

7. Ferreira JM, Silva H, Costa JD, Richardson M (2005) Stress analysis of lap joints involving natural fibre reinforced interface layers. Composites Part B: Engineering 36: 1-7.

8. http://www.ingentaconnect.com/content/els/135983 5x/2000/00000031/00000012/art00008

9. Curvelo AAS, de Carvalho AJF, Agnelli JAM (2001) Thermoplastic starch-cellulosic fibers composites: preliminary results. Carbohydrate Polymers 45: 183-188.

10. Bledzki AK, Zhang W, Chate A (2001) Natural fiber reinforced polyurethane microfoams. Compos Sci Technol 61: 2405-2411.

11. Jayaraman K, Bhattacharya D (2004) Mechanical performance of woodfibre-waste plastic composite materials. Resources Conservation and Recycling 41: 307-319.

12. Zulkifli R, Fatt LK, Azhari CH, Sahari J (2002) Interlaminar fracture properties of fiber reinforced natural rubber/polypropylene composites. J Mater Process Technol 128: 33-37.

13. Thwe MM, Liao K (2003) Durability of bamboo-glass fiber reinforced polymer matrix hybrid composites. Compos Sci Technol 63: 375-387.

14. Herrera-Franco PJ, Valadez-Gonzalez A (2004) Mechanical properties of continuous natural fibre-reinforced polymer composites. Compos Part A Appl Sci Manuf 35: 339-345.

15. http://cat.inist.fr/?aModele=afficheN&cpsidt=16182404

16. Mitra BC, Basak RK, Sarkar M (1998) Studies on jute-reinforced composites, its limitations, and some solutions through chemical modifications of fibers. J Appl Polym Sci 67: 1093-1100.

17. Kandola BK, Horrocks AR, Myler P, Blair D (2003) Mechanical performance of heat/fire damaged novel flame retardant glass-reinforced epoxy composites. Compos Part A Appl Sci Manuf 34: 863-873.

18. Eichhorn SJ, Young R J (2003) Composite micromechanics of hemp fibres and epoxy resin microdroplets. Compos Sci Technol 63: 1225-1233.

19. http://www.autexrj.com/cms/zalaczone_pliki/6-07-2.pdf

20. Kaith BS, Singha AS, Susheel K (2006) Mechanical Properties of raw flax and Flax-g-poly(MMA) reinforced Phenol-Formaldehyde Composites. International Journal of Plastics Technology 10: 572-577.

21. Singha AS, Susheel Kumar, Kaith BS (2005) Preparation of flax-g- copolymer reinforced phenol-formaldehyde composites and evaluation of their physical and mechanical properties. International Journal of Plastics Technology 9: 427-435.

22. Chauhan A (2009) Synthesis and Evaluation of Physico-Chemico-Mechanical properties of polymer matrix based Composites using Graft copolymers of Hibiscus sabdariffa as reinforcing agents. PhD Thesis, Punjab Technical University, India.

23. Kaith BS, Chauhan A (2008) Synthesis, Characterization and Mechanical Evaluation of the Phenol-Formaldehyde Composites. E Journal of Chemistry 5: S1015-S1020.

24. Chawla Shashi (2002) A Text book of Engineering Chemistry, Dhanpat Rai and Co. (Pvt.) Ltd., Educational and Technical Publishers Delhi.

25. Kaith BS, Singha AS, Chauhan Ashish (2006) X-Ray Diffraction Studies and Thermogravimetric/Differential Thermal Analysis of Graft Co-polymers of Methylacrylate onto Hibiscus sabdariffa Fiber, Journal of Polymer Materials 26: 349-356.

26. Kaith BS, Chauhan Ashish, Mishra BN (2008) Studying the morphological transformation in graft copolymers of Binary Mixture of Methyl acrylate and Acrylonitrile onto Hibiscus sabdariffa fiber by XRD and TGA/DTA. Journal of Polymer Materials 25: 69-76.

27. Kaith BS, Chauhan Ashish (2008) Synthesis, Characterization and Evaluation of the Transformations in Hibiscus sabdariffa-graft-poly(butyl acrylate). E Journal of Chemistry 5: S980-S986.

28. Kaith BS, Ashish C, Singha AS, Pathania D (2009) Induction of the morphological changes in Hibiscus sabdariffa fiber on graft copolymerization with Binary vinyl monomer mixtures. International Journal of Polymer Analysis and Characterization 14: 246-258.

29. Chauhan A, Kaith BS, Singha AS, Pathania D (2010) Induction of the morphological changes in Hibiscus sabdariffa on graft copolymerization with acrylonitrile and co-vinyl monomers in binary mixture. Malaysian Polymer Journal 5: 140-150.

30. Chauhan A, Kaith B (2011) Synthesis, Characterization and Chemical studies of Hibiscus sabdariffa-g-copolymers. Fibers and Polymers.

31. Chauhan Ashish, Kaith Balbir (2010) Thermo-Chemical Evaluation of the Roselle Graft Copolymers. Polymer from Renewable Resources 1: 173-187.

32. Chauhan A, Kaith B (2011) Thermal and Chemical studies of Hibiscus sabdariffa-graft-(Vinyl monomers). International Journal of Polymeric Materials 60: 837-851.

19

Synthesis, Characterization and Chelation Ion-Exchange Studies of a Resin Copolymer Derived From 8-Hydroxyquinoline-Formaldehyde-Pyrogallol

Soumaya Gharbi[1]*, Jameleddine Khiari[2] and Bassem Jamoussi[1]

[1]Research Laboratory Analytical Chemistry, Macromolecular and Heterocyclic, Ipest, Tunisia
[2]Preparatory Institute for Engineering Studies of Bizerte, Tunisia

Abstract

An ion exchange resin chelator was synthesized from 8-hydroxyquinoline and pyrogallol using formaldehyde as a crosslinking agent to 120°C in a solution of DMF. The resin was characterized by elemental analysis and FTIR. The morphology of the synthesis resin was examined by optical photograph and scanning electron microscopy (SEM). The physicochemical properties of the resin were studied. The cation exchange capacity was measured and the effect of pH and metal ion concentration on the ability of the ion exchange were studied. The ratio of cation exchange reaction and the distribution coefficient in tartaric acid medium at different pH were also studied using the method of batch equilibration.

Keywords: Chelating resin; Batch equilibration; Physico-chemical properties; Distribution coefficient

Introduction

Over the past years, there has been a growing concern for the immobilization of metal ions introduced into bodies of water and wastewater by increasing human technological activities. It has been established beyond doubt that the ions of heavy metals in the environment (air, soil and water) pose a serious threat to human health. With the exponential growth of the population, it is necessary to control the release of toxic heavy metal ions before entering the complex ecosystem.

Separation, removal, and the enrichment of metals in trace amounts in aqueous solutions, have an important role in wastewater, industrial or geological sample analysis. The solid phase extraction on the metal ion was granted fast acceptance due to its various advantages over other the invention also provides methods. The extraction of metal ions using ion exchange resin is a chelating power green analytical method, since it does not involve the use of toxic organic chlorine compounds, which are very frequently used in conventional techniques liquid-liquid extraction.

The main objective of much of the research on chelating resins was the preparation of the insoluble part functionalized polymers that can provide in conjunction with more flexible working conditions good stability and high capacity for metal ions. The use of the modified clay minerals for the adsorption of metal ions from aqueous solutions for purification of industrial water or waste water, the treatment has been widely studied.

These clay minerals, when used as colloids or powders were found to be effective as ion equally exchange resin, but it is difficult to retrieve from these adsorbents filters after use. This regeneration makes very difficult to reuse clay adsorbents. Chelating ion exchange resins have also been prepared by copolycondensing or 8-hydroxyquinoline phenol derivatives such as o-aminophenol, resorcinol or resorcylic acid with formaldehyde [1].

Antico and al synthesized from a gel-type ion exchange resin, Glycol 8-hydroxyquinoline methacrylate and used to investigate separating Pb (II) and Cu (II) in the chloride solution [2]. Chelating ion exchange resins also have been synthesized by the Friedel-Crafts condensation 8-hydroxyquinoline [3] and substituted 8-hydroxyquinoline with [4]

1,2-dichloroethylene. The synthetic resins have been found selective for certain metal ions over a wide pH range. The chelating behavior poly (8-Quinoline-5, 7-dimethylene) [5] and its cross-linked polymer [6] from the reaction with different amounts of bisphenol-a to certain trivalent ions of lanthanides such that La (II) and Gd (II) was investigated by static a load balancing Method. Vernon and al. prepared and studied chelating properties of toxin resin to the transition metal ions [7,8] stability test, they suggested that the gel polymers must never be allowed to dry; otherwise their favorable properties are destroyed.

Shah et al. [9,10] and Warshwsky et al. [11] reported certain resins based on 8-hydroxyquinoline and substituted 8-hydroxyquinoline, respectively, and chelating properties to transition and post-transition metal ions. A chelating ion exchange resin was synthesized from 8-hydroxyquinoline and catechol using formaldehyde have been synthesized by Shah et al. [12].

Until now, no resin based on 8-hydroxyquinoline pyrogallol-formaldehyde in DMF has been inserting for quantitative removal and separation transition metal ions and post transition. As industrial effluents are often rich in transition and transition metal ions after, removing the metals a large industrial task. The study described in the present communication processes of synthesis and characterization of absolvers together with the systematic studies of various properties of ion exchange resin.

Materials and Methods

Materials

8-Hydroxyquinoline (Prolabo, IGT Paris) and formaldehyde were purified by recrystallization method.

*Corresponding author: Soumaya Gharbi, Research Laboratory Analytical Chemistry, Macromolecular and Heterocyclic. Ipest, Tunisia

Pyrogallol (Glaxo extra pure) was purified by rectified spirit. Metal ion solutions (KANTO CHEMICAL CO, INC.) were used as received.

Synthesis method

8-hydroxyquinoline (14.5 g, 0.1 moles) has been ground into fine powder and taken in a (250 ml) round bottom balloon and dissolved in DMF (25 ml) to give a clear yellow solution. Formaldehyde (7.5 ml, 0.25 mol as 37%) added and stirred until a solution of red color.

A pyrogallol solution (11 g, 0.1 mole) in (10 ml) DMF has been added to above the solution and stirring for 3 h. Then, the mixture has been heated to reflux on a water bath at 90°C under constant stirring for 3 to 5 h.

Condensation reagents was carried out in the presence of an acid HCl 2M on sand bath by heating to 120°C during 7 -8 h until a viscous solution with formation of a hard mass of brown colored resin has been obtained.

The synthetic resin has been removed from the reaction vessel and cured in an oven at 70-80°C during 12 h. The resulting resins having been washed with DMF and deionized water to remove unreacted monomers and impurities. After the complete washing cycle, the reaction yield production of the resin synthesized was 50% (15 g).

The resin sample has been purified and dried finely crushed and sieved to obtained uniform particles of 50-70 mesh and stored in polyethylene bottle. The resin has been then screened characterized using different instrumental analysis techniques and was used for the entire experiments during the search period. Testing of solubility of the resin in a different solvent was performed at room temperature and pressure with intermittent agitation. The resin has been found being insoluble in all common organic solvents such as acetone, ethanol, benzene, chloroform etc. and all acids and alkalis of higher strengths.

Infra-Red spectra of the synthesized resin sample have been carried out without solvent on a Thermo Scientific Nicolet spectrometer IR-200 FT-IR, which is the measurement accuracy of 4 cm^{-1} in the area 400-4000cm^{-1}. The allocation FTIR spectra peaks are presented in Figure 1. Elemental analysis was performed on Carlo ZAF EDAX quantification (Standard Less). The results of the elemental analysis are in good accordance with the calculated values of %C, % H and % N which is presented in Table 1.

To convert the resin sample in its H$^+$ form, it has been equilibrated with a solution of HCl 1M for 24 h and washed with deionized water until it is free from chloride by testing with silver nitrate solution. The H$^+$ form of the resin have been used for subsequent studies. The batch equilibrium method has been adopted for studying the ion-exchange properties. Physico-chemical properties such as moisture and solid contents, apparent density, true density and void volume fraction have been studied by the methods of the literature [13].

The sorption properties such as exchange rate, pH effect on the exchange capacity, effect of the cation concentration on the exchange capacity and distribution coefficient (K$_d$) values for various metal ions as a function of pH and concentration of the electrolyte have been studied with the literature methods [13,14]. The value of the void volume fraction has been found 0.36, while other resins signaled 8-hydroxyquinoline formaldehyde [8], 8-hydroxyquinoline formaldehyde resorcinol [9] and 8-hydroxyquinoline formaldehyde catechol [12] exhibited void volume fraction values of 0.54, 0.59 and 0.36, respectively.

Rate of exchange of metal ions

For the experiment, an accurately weighed (0.250 ± 0.001 g) dry resin in the H$^+$ form was taken into different glass bottles stoppered and equilibrated with a desired pH values with buffer solution for 24 h. After decanting of the buffer solution, 50 ml (0.2 M) of metal ion solution was added with the same pH. The amount of non-chelated metal ion has been determined by Atomic Absorption Spectroscopy (AAS) at fixed time intervals. The results are shown in Figure 2.

Effect of pH on metal ion exchange capacity

To study the pH effect on absorbing metal ions, it is necessary to buffer the resin and solutions used. For this purpose, buffer solutions of

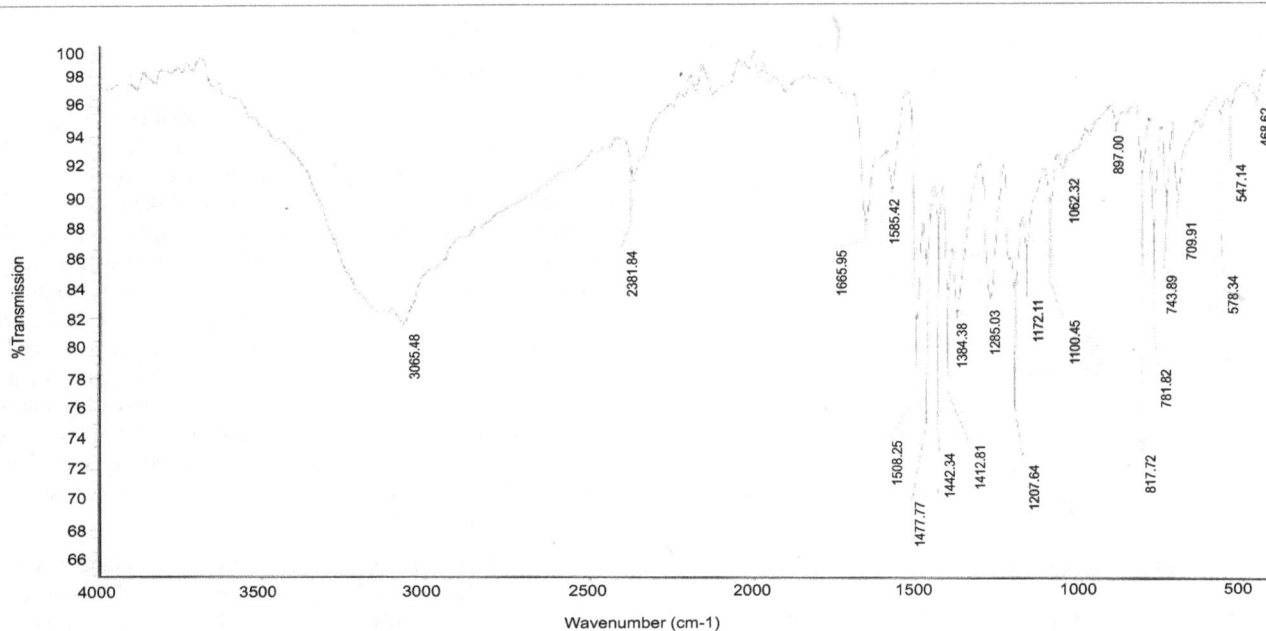

Figure 1: FTIR spectrum of resin.

% Calculated (% Found)		
C	H	N
69.15	4.04	4.74
68.15	3.98	4.65

Table1: Element Analysis of the Resin.

Figure 2: Optical photograph of resin 250X Magnifications.

Property	Values (S.D)
% Moisture content	5.25 (±0.05)
% Solid	93 (±0.05)
True density	0.95 (±0.01) g.cm^{-3}
Apparent density	0.51 (±0.01) g.cm^{-3}
Void volume fraction	0.38 (±0.025)

Table 2: Physicochemical Properties of the synthesized Resin.

pH 3-6 have been prepared from 0.2 M acetic acid and 0.2 M sodium acetate solutions [15].

A pH-meter (Elico, model CL-44) has been used for measuring the pH. Different weighing assemblies (0.250 ± 0.001 g) dry resin have been equilibrated with buffer in different bottles closed during 24 h, so that the resin attained the desired pH value.

After 24 h, buffer solutions have been decanted and (50 ml) of 0.2 M metal ion solutions pH ranging from 3 to 6 have been added. Metal ion solutions have been equilibrated at room temperature for 24 h with intermittent agitation.

After 24 h, the solutions were decanted and metal ion concentration in the supernatant was measured by a method Atomic absorption spectroscopy (AAS) which is a spectral procedure for the quantitative determination of chemical elements employing the absorption of optical radiation (light) by free atoms in the gaseous state.

The same method was followed throughout the study to calculate the ion exchange capacity of the resin as:

$$Exchange\ capacity\ (mmol.g\text{-}1) = \frac{\big([Initial\ molarity\ of\ the\ metal\ ion] - [Remaining\ molarity\ of\ the\ metal\ ion]\big)}{(Atomic\ molar\ mass\ of\ the\ metal * Weight\ of\ the\ resin\ sample)}$$

Effect of metal ion concentration on exchange capacity

To study the metal ion concentration effect on uptake of different metal ions by the resin, the resin has been equilibrated with an acetate buffer to desired pH (pH value of highest exchange) for 24 h and buffer

solutions were then decanted. An accurately weighed (0.250 ± 0.001 g) dry resin has been equilibrated with metal ion solutions (50 ml) with variant mass concentration (e.g., 1 ppm, 2 ppm, 3 ppm, 4 ppm to 5 ppm) at the same pH value at room temperature during 24 h with intermittent agitation.

After 24 h, the metal ion solutions were decanted and non-chelated metal ions have been estimated by atomic absorption spectroscopy (AAS).

K_d Values for metal ions in presence of electrolyte (tartaric acid) solution

Measurement of distribution coefficient of metal ions over a wide range of condition is the best method to avoid choosing elution conditions for separation columns by a strictly trial and error process. The batch distribution coefficient, K_d is defined as:

$$K_d = \frac{mmole\ of\ metal\ ion\ on\ resin * volume\ of\ metal\ ion\ solution}{mmomle\ of\ metal\ in\ solution * weight\ of\ dry\ resin}$$

Although this distribution coefficient is measured on a base discontinuously, it can be used to predict elution behavior metal ions eluted from an ion-exchange column. For separating two substances, the conditions should be chosen such that the distribution coefficient of one of them is small (preferably 1 or less) so that it eluting the column is fast, while the distribution coefficient of the other substance, under the same conditions should be as large as possible (more than 10 times) such that the substance is firmly held by the resin [16].

Effect of different concentrations and pH of the electrolyte (tartaric acid) to adsorbing metal ion the synthetic resin was investigated. A dry resin sample weighed exactly (0.250 ± 0.001 g) has been suspended in the electrolyte solution (tartaric acid) of 50 ml different known concentrations (e.g. 0.1 M, 0.2 M, 0.3 M, 0.5 M and 1 M). The pH of the suspension was adjusted to the desired value using acetate buffer and the resin has been equilibrated for 24 h. To the suspension, 2.0 ml of different metal ion solutions under study (5 mg.ml^{-1}) was added to be equilibrated for 24 h with intermittent agitation. After 24 h, the solutions were decanted and unabsorbed metal ions were estimated.

Results and Discussion

Physico-chemical properties

Physicochemical properties of the synthesized resin are presented in Table 2. Moisture content of a resin providing a measure of its water swelling capacity or its loading capacity. The water content depends on numerous factors, such that the composition of the resin matrix, the cross-linking degree or the nature of the active groups and the ionic resin form. The degree of cross-linking a resin has an effect on the moisture content and moisture content of the resin, and therefore, has an effect on the selectivity. In a high moisture content of the resin, the active groups are more spaced apart, e.g. strong acid cation resins contain about 50% moisture. The water content percentage of synthetic resin, as shown in Table 2 is 5.25%. Therefore, the resin has a low moisture percentage range relative to the commercial resins. Resins synthesized from salicylic acid and furfural-benzidine p-hydroxybenzoic acid-furfural-benzidine [17] has lower moisture contents (4.01 and 4.9% respectively) than the synthesized resin.

The resin synthesized from 8-hydroxyquinoline-pyrogallol-formaldehyde [12] and salicylic acid formaldehyde-m-cresol [18] had slightly higher moisture content (5.64 to 9.4%, respectively).The moisture differential may be caused by different experimental conditions, such as carriers, wherein the resins have been synthesized,

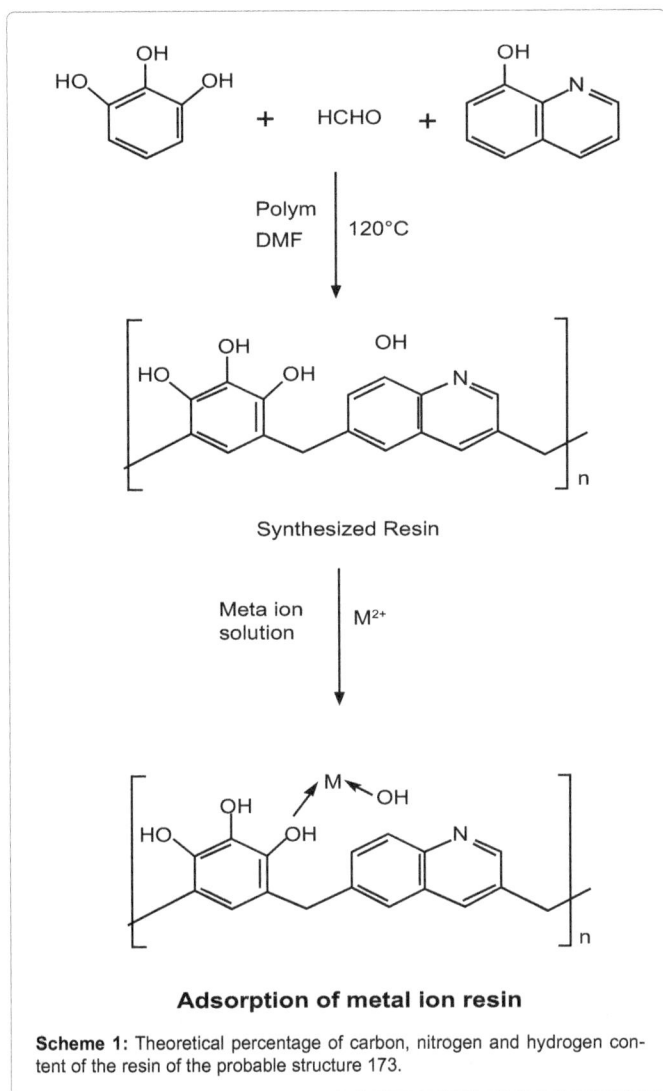

Scheme 1: Theoretical percentage of carbon, nitrogen and hydrogen content of the resin of the probable structure 173.

the polymer backbones, the cross-linking degree and the functional groups involved.

The true density of the synthetic resin is 0.95 g.cm^{-3}, which is given in Table 2. The actual density commercial resins are generally between 1.1 to 1.7 g.cm^{-3}. To prevent flutter of resin particles, the actual density floating must be greater than a resin particles is not desirable in chromatographic studies, as disturbs forming a compact column. The optimum density and size uniform particles allow the perfect column packing and column performance.

Measuring the density of the column or the apparent density is necessary that the resins are commercially available on a volume basis and packed on a weight basis. The bulk density of the polymer synthesized is given in Table 2, from which to see that the apparent density of the resin is in the range of 0.51 g cm^{-3}, which is comparable to the density commodity resins. This can be due to changes in the polymer matrix, different functional groups and the synthesis process. The apparent density parameter gives an indication of the required length of a column packed with a study of the ideal chromatographic column.

The value of the void volume fraction resin is also given in table a vacuum volume fraction of the synthetic resin was the order of 0.38.

The sensitive value the void volume fraction is broadcast on the resin and exchangeable ion exchange rate increases leaving ions. The essential minimum void volume has an improved diffusion exchangeable ion and gives the feasibility of the operation of the column thereby.

Spectral characterization of resin

The FTIR spectrum of the resin is shown in Figure 1. A strong band at 3300 cm^{-1} is due to the υ(O-H) stretching of phenolic group, a medium band at 3066 cm^{-1} is due to the υ(C-H) stretching of aromatic ring and the presence of a medium-strong band at 2900 cm^{-1} is due to the υ(C-H) stretching of methylene group. The bands at 1666, 1585 and 1447 cm^{-1} can be assigned to υ(C=N) heterocyclic ring and υ(C=C) aromatic ring stretching, respectively. A band at 1442 cm^{-1} is also due to δ(C-H) deformation of methylene group [9]. The presence of a medium-strong band at 1384 cm^{-1} can be assigned to in plane δ(O-H) bending of aromatic and a band at 1285 cm^{-1} is due to aromatic υ(C-O) stretching. A sharp single band at 781 cm^{-1}, which can be assigned to the presence of 1,2,3,4,5-penta substituted benzene ring [10], confirms the polymerization of monomers.

Elemental analysis

The theoretical percent of carbon, nitrogen and hydrogen content of the resin have been calculated from the general formula ($C_{17}H_{13}NO_4$) of the repeating unit of the structure likely (Scheme 1). Table 1 show that the results of the elemental analysis are in good accordance with the calculated values. The elemental analysis results are supported the proposed structure of the resin presented in Scheme 1.

Optical and SEM photographs

The morphology of the insert resin sample was investigated by scanning electron micrographs, which are represented in Figure 2 and Figure 3, respectively. The synthetic resin optical photographing showed that it is of a brown color. The morphology of the resin shows a fringed model of the, crystalline-amorphous structure.

The fringes represent transition state between the crystalline and amorphous phase [19]. The resin exhibits more amorphous character with closed packed surface having deep pits (Figures 3a and 3b) as it compared with anthranilic acid-formaldehyde-resorcinol resin and 8-hydroxyquinoline-formaldehyde-pyrogallol resin reported earlier [12-19].

Surface analysis has been found to be useful in the understanding surface features of the material. The morphology of the resin parts crystal growth from polymer solutions corresponding to the largest organization large scale in polymers, e.g. in size spherulites of a few millimeters. Ideally, the spherulites are aggregates of size sub-

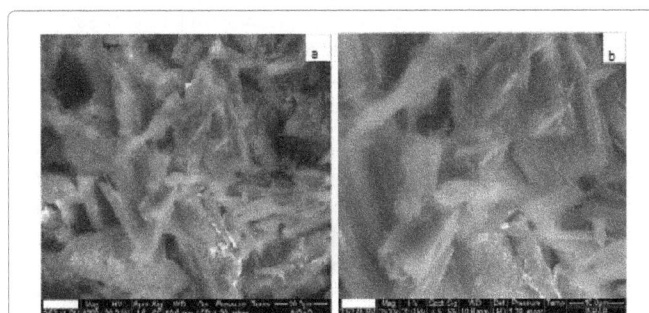

Figure 3: SEM photograph of resin at (a) 1200X and (b) 2500X Magnifications.

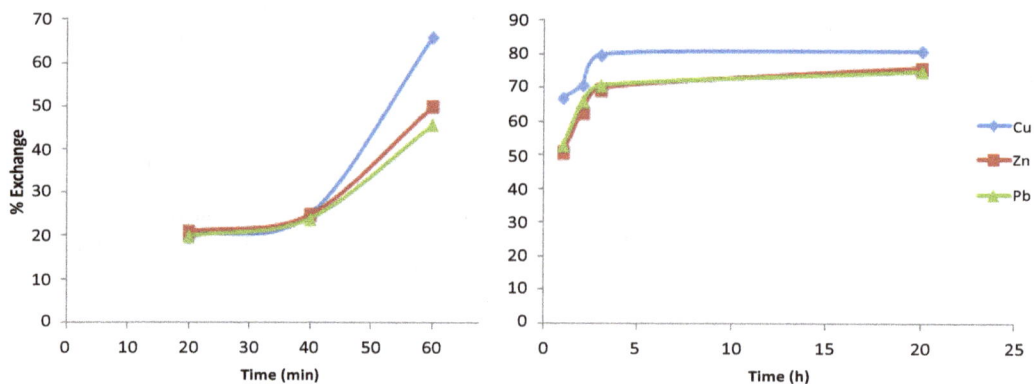

Figure 4: Rate of exchange of cations on resin.

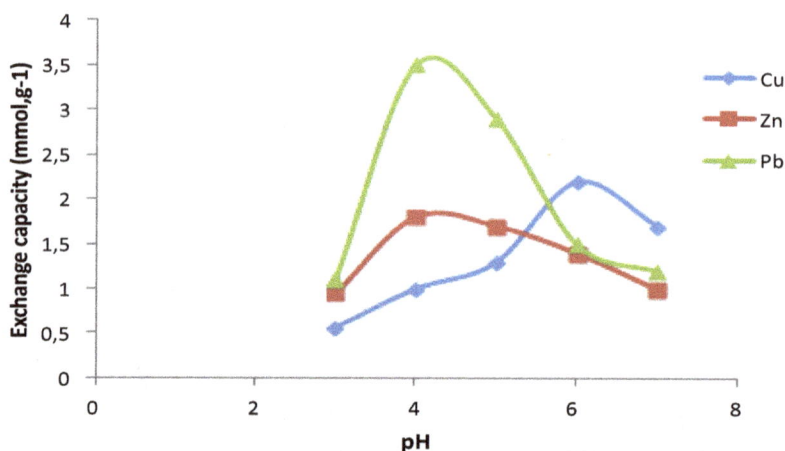

Figure 5: Effect of pH on cation exchange capacity.

microscopic particles. Spherulites are characterized by the secondary structure features, such as faint corrugations. The higher magnification 500 'shows the coexistence of an appreciable amorphous fraction with a small amount of shallow pits. The surface of the resin (Figure 3) contributes greater segments of crystalline regions.

Rate of exchange for metal ions

As the base resin structure is important in physical absorption metal ions by the different copolymers of the resin. From the results, viewed that the metal ion exchange rate is faster at the beginning and then slow. These occur due to the removal or exchange of ions of the solution the surface of the resin and after the entire available site (group) were occupied resin. It results in slower rate of the exchange. Figure 4, shows the rate of exchange of the metal ion of the synthetic resin.

Metal ion exchange on the resin has a time dependent phenomenon. The rate of different metal ion exchange was to determine the shortest time interval for which equilibrium can be performed [20]. The graph represented on Figure 4 indicates that time required for the exchange of 50% ($t_{1/2}$) for Cu (II) is 30 min and Zn (II) and Pb (II) are 50 and 56 min, respectively. This is assigned to the Cu (II) hydrated have radii smaller (0.419 nm) that hydrated Zn (II) (0.430 nm) and thus have greater access to the resin surface. The order of the exchange capacity is: Cu (II)>Zn (II)>Pb (II).

Rapid exchange rate in the beginning can be explained on the basis of mass action of the law and the equilibrium state. The exchange rate greater facilitates the chromatographic separation column. Metal ion exchange kinetics mainly depends on various physical properties, including the particle size distribution, the size pores, the physical base structure and diffusion of counterion [12,13].

Effect of pH on exchange capacity

Metal removal ions from an aqueous solution by sorption are strongly dependent on pH of the solution which influences the surface charge of the sorbent [19]. Chelating ligands form complexes with various metal ions in specific pH conditions [20]. Therefore, the synthesis resin has been used to study the effect of varying pH on its chelating ability to various metal ions. The results of the exchange capacity depending on the pH for different metal ions are presented in Figure 5. The results show that the sorption metal ions is increased with increasing pH to a maximum value and thereafter decreased. Maximum sorption occurred Cu (II) at pH 6, Zn (II) at pH 4 and Pb (II) at pH 4. The order of selectivity for the metal ions is: Cu (II)>Zn (II)>Pb (II). An increase in pH increases the negatively charged nature of the sorbent surface. This leads to an increase in the electrostatic attraction between positively charged metal ions and negatively charged sorbent and results in increased sorption of metal ions. At lower pH, the removal

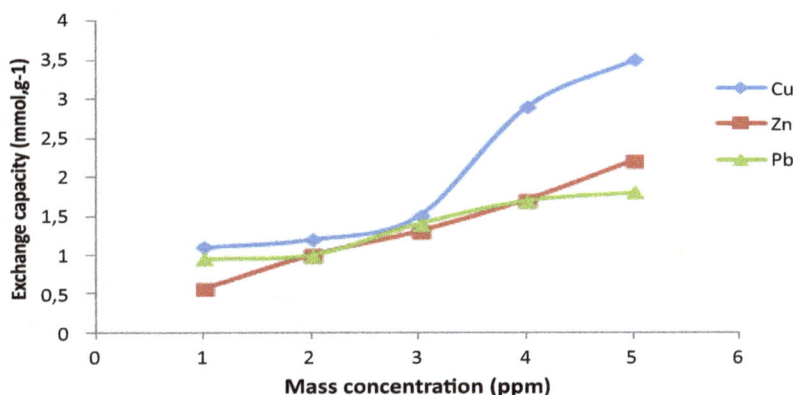

Figure 6: Effect of cation concentration on exchange capacity.

Metal ion	[Tartric acid] (M)	K_d values at different pH				
		3	3.5	4	5	6
Cu(II)	0.1	1101	671	527	202	155
	0.2	698.3	320	130	110	79.7
	0.3	250.1	106	80.5	50.8	38.9
	0.5	55.6	55.6	39.7	52.3	55
	1	14	19.3	11.8	8.4	6.5
Zn(II)	0.1	1800	1320	432	402	200
	0.2	640.7	618	220	111	131
	0.3	588.2	300	95	71.8	59
	0.5	134	115	74	28.1	36.8
	1	81	35.5	25.2	11.8	5.3
Pb(II)	0.1	2221	980	610	421	261
	0.2	668.4	640	231	116	142
	0.3	435.9	341	95.3	82.2	30.4
	0.5	130.2	82.5	80	40.3	4.7
	1	10.8	45.4	34.9	17	2.8

Table 3: K_d Values of Metal Ions at Various Tartaric Acid Concentrations and pH Values.

of metal ions is decreased due to the higher concentration of H^+ ions present in the reaction mixture which compete with the metal ions for the sorption sites at the surface.

Meanwhile, the observed decrease in sorption capacity is due to the formation of insoluble metal ion hydroxides [21].

In the case of Pb(II) purely electrostatic factors are responsible. Due to the less deep pits, resin exhibits lower ion exchange capacity for Pb(II). Pb(II) has bigger hydrated radii, so it cannot easily penetrate to originate in more crystalline region of the polymeric network. Therefore, the cleaner separation can be obtained from the browser binary mixture Pb (II) with transition metal ions such as Cu (II) or Zn (II).

Effect of metal ion concentration on exchange capacity

Examination data presented in Figure 6 shows that the amount of adsorbed metal ions increases with increasing concentration of metal ions in the solution to a maximum value, and will remain constant while new increase of metal concentration. A low concentration of metal ions, the available number of metal ions in the solution is low relative to the arrangement sites on the sorbent [21]. However, at higher concentrations, the sorption available sites remain same as more metal ions are available for sorption and subsequently sorption becomes almost constant then after [9,10].

Effect of electrolyte concentration and pH on distribution coefficient (k_d) values

Batch process of equilibrium is useful to determine the distribution coefficient (K_d) for the metal ions depending on the concentration of tartaric acid. Tartaric acid is a potent chelating elution agent. It contains six oxygen atoms with unshared electron pairs, which from complexes of chelating metal ions more stable with resin-metal complexes. Tartaric acid can easily eluting resin metal ions and give clear separation.

Distribution coefficient values (K_d) for different metal ions have been determined by the batch equilibration method. The K_d values have been investigated metal ions depending on pH and the concentration of the electrolyte solution and the results are shown in Table 3. The present survey limit distribution studies to a certain pH to each metal ion to avoid hydrolysis metal ions at a pH above [9]. As is evident from the following Table 3, any electrolyte concentrations and the entire pH range, the values of K_d decrease in order Pb (II)>Zn (II)>Cu (II).

Meanwhile, in all cases, the K_d values decrease with increasing electrolyte concentration and increase with increasing in pH. It is expected that distribution coefficients metal ions can vary depending on the stability metal complexes with chelating groups of the resin. Best results for stabilizing higher distribution coefficients. This study limit distribution studies to a certain pH to each metal ion to avoid hydrolysis metal ions at a higher pH.

It is evident that to obtain sharper separation metal ions, large values of ΔK_d should selected from the same experimental conditions [16].

Analytical Application

Removal of toxic metal from industrial effluents

The chelated lead has been eluted with a tartaric acid solution to 1 M to pH 3. It has been found that removal and recovery of copper from the effluent were quantitative.

The pH of the industrial effluent containing Zn (100 cm³) has been adjusted to 3 using a buffer solution and passed through the resin column at a rate of 1 cm³ min⁻¹ stream, followed by washing with water carefully.

The results showed 92% Pb (II) and 89% Pb (II) recovery of industrial effluents using the resin. The synthetic resin may be adopted for the industrial processing (treatment) of wastewater.

Conclusions

In conclusion, the chelating resin derived from Pyrogallol and 8-Hydroxyquinoline with formaldehyde was useful cation exchanger for divalent metal ions. A SEM image of the resin establishes the amorphous nature of the resin which helps for the higher metal ion uptake. The synthesized resin can be used for the removal of toxic heavy metals from aqueous media and industrial wastewater containing Cu(II) and Pb(II). The recovery of the metals from Industrial effluents gives an indication of the utilization potential of the synthesized resin for wastewater treatment.

Acknowledgement

We are thankful to Physicochemical National Institute of Research and Analysis INRAP (Technopole Sidi Thabet, Tunisia) for providing SEM and optical photograph facility.

References

1. Pennington L, Williams M (1959) Chelating ion exchange resins. Industrial and Engineering Chemistry 15: 759-762.

2. Antico A, Masana A, Salvado V, Hidalgo M, Valiente M (1995) Separation of Pd(II) and Cu(II) in chloride solutions on a glycol methacrylate gel derivatized with 8-hydroxyquinoline. Journal of chromatography A 706: 159-166.

3. Patel BS, Patel SR (1979) Chelation ion-exchange properties of poly(8-hydroxyquinolinediylethylene), Macromolecular Chemistry and physics 180: 1159- 1163.

4. Patel BS, Choxi GS, Patel SR (1979) Synthesis and study of poly(8-hydroxyquinoline-7,5-diylethylene). Macromolecular Chemistry and physics 180: 897-904.

5. Ebraheem K, Mubarak M, Yassien Z, Khalili F (1998) Chelation properties of poly(8-hydroxyquinoline 5,7-diylmethylene) towards some trivalent lanthanide metal ions. Solvent Extraction and Ion Exchange 16: 637-649.

6. Ebraheem K, Mubarak M, Yassien Z, Khalili F (2000) Chelation properties of poly(8-hydroxyquinoline 5,7-diylmethylene) crosslinked with bisphenol-A toward lanthanum(III), cerium(III), neodimium(III), samarium(III), and gadolinium(III) ions. Sep Sci Technol 35: 2115-2125.

7. Vernon F, Nyo KN (1977) Synthesis optimization and the properties of 8-hydroxyquinoline ion-exchange resins. Analytica Chimica Acta 93: 203-210.

8. Vernon F, Eceles H (1973) Chelating ion-exchangers containing 8-hydroxyquinoline as the functional group. Analytica Chimica Acta 63: 403-414.

9. Shah BA, Shah AV, Bhandari BN (2004) Recovery of transition metal ions from binary mixtures by ion exchange column chromatography using synthesized chelating ion resin derived from m-cresol. Asian Journal Chemistry 16: 801-1810.

10. Shah BA, Shah AV, Bhandari BN (2003) Selective elution metal ions on a new chelating ion exchange resin derived from substituted 8-hydroxyquinoline. Asian Journal of Chemistry 15: 117-125.

11. Warshawsky A, Wang Y, Berkowitz B (2003) 8-Hydroxyquinoline-5-Sulfonic acid (HQS) impregnated on Lewatit MP 600 for cadmium complexation: Implication of solvent impregnated resins for water remediation. Separation Science and Technology 38: 149-163.

12. Shah BA, Shah AV, Bhandari BN, Bhatt RR (2008) Synthesis, Characterization and Chelation Ion-Exchange Studies of a Resin Copolymer Derived from 8-Hydroxyquinoline-Formaldehyde-Catechol. J Iran Chem Soc 5: 252-261.

13. Helfferich F (1962) Ion Exchange, McGraw-Hills, New York.

14. Kunnin R (1958) Ion exchange resins, 2nd edn., Wiley, London.

15. Vogel AI (1989) Qualitative inorganic analysis, 5th edn., Longaman, London.

16. Fritz JS, Pietrzyk DJ (1961) Non-aqueous solvents in anion-exchange separations. Talanta 8: 143-162.

17. Kapadia RN, Vyas MV (1980) Synthesis and Physicochemical Studies of Some New Amphoteric Ion Exchangers. Journal of Applied Polymer Science 27: 3793-3807.

18. Shah A1, Devi S (1987) A new chelating ion-exchanger containing p-bromophenylhydroxamic acid as functional group-IV: column separations on a hydroxamic acid resin. Talanta 34: 547-550.

19. Devi S, Shah A (1987) A new chelating ion-exchanger containing p-bromophenylhydroxamic acid as functional group-IV: column separations on a hydroxamic acid resin. Talanta 34: 547-50.

20. Prabhakar LD, Umarani C (1994) Coordination polymers derived from poly(2-acryloxybenzaldheyde thiosemicarbazole)-divinylbenzene. Journal of Polymer Materials 11: 147-156.

21. Prasad HH, Popat KM, Anand PS (2002) Synthesis of crosslinked methacrylic acid-coethyleneglycol dimethacrylate polymers for the removal of copper and nickel from water. Indian Journal of Chemical Technology 9: 385-393.

20

ZnO-Assisted Photocatalytic Degradation of Congo Red and Benzopurpurine 4B in Aqueous Solution

Elaziouti[1]*, N. Laouedj[2] and Bekka Ahmed[1]

[1]LCPCE Laboratory, Faculty of sciences, Department of industrial Chemistry, University of the Science and the technology of Oran (USTO M.B). BP 1505 El M'naouar 31000 Oran, Algeria
[2]Dr. Moulay Tahar University, Saida, Algeria

Abstract

The photocatalytic degradation of two commercial azo dyes Congo red (CR) and Benzopurpurine 4B (BP4B) in aqueous solution was investigated under UV-A light at different operating conditions, including irradiation time, pH solution, initial dye concentration, amount of catalyst, light intensity as well as band gap of other semiconductor groups by UV-spectrophotometric monitoring. The highest decomposition were obtained at pH 8 as a result of 95.02 and 97.24 % degradation efficiencies of CR and BP4B for 60 and 80min of irradiation time respectively. Photodecomposition reactions of both dyes were correlated with pseudo-first-order kinetic model. For BP4B, the degradation data were satisfactory described by Langmuir–Hinshelwood (L-H) mechanism, whereas those of CR were not sufficient to conclude that the L-H mechanism is the most suitable one to describe the photocatalytic process of CR. These findings can support the design of remediation processes and also assist in predict their fate in the environment.

Keywords: Photocatalysis; Zinc oxide; Congo red; Benzopurpurine 4B; Langmuir–Hinshelwood (L-H)

Introduction

Heterogeneous photocatalysis oxidations performed with light irradiated semiconductors dispersions has been extensively investigated owing to their highly efficiency to completely mineralize the harmful organic and inorganic ions species to CO_2 and water [1]. Most researches has been focalized on the heterogenic systems based on high dispersion TiO_2 with a crystalline modification of anatase (Degussa P25, Hombriat UV-100, Aldrich, etc.) as a result of their high photocatalytic activity and widespread uses for large-scale water treatment. However, the relatively elevated intrinsic band gap of anatase TiO_2 (3.2 eV), limited their efficiencies under solar light, so that the effective utilization of solar energy is limited to about 4% of total solar spectrum. In order to meet the requirement of future environment and energy technologies, it is necessary to develop highly efficient, non toxic and chemically stable photocatalyst. Various semiconductor catalysts such as MO_2 (M= CeO_2, ZrO_2, SnO_2), M'_2O_3 (M'=α-Fe_2O_3, Bi_2O_3, Al_2O_3, Sb_2O_3 ect..) metal oxide and DS (D=Zn, Cd, Bp) metal chalcogenide groups were investigated, but their practical uses have been constrained by their low photocatalytic activity under solar light, short-term stability against photo- and chemical corrosion as well as potential toxicity [2]. Many attempts have been made to study ZnO-mediated photocatalytic degradation of organic compounds [3-6]. Semiconductor, on irradiation with photon of sufficient energy, greater than or equal to the band gap energy of the semiconductor (hv ≥ Eg), a free electron (e⁻) and electronic vacancy-a hole (h⁺) are generated and recombine or migrate in the semiconductor surface being partially localized on structural defective centers of its crystalline lattice Equation (1). The photogenerated electrons take part in the reduction reaction with dissolved oxygen, producing superoxide anion (O^-_{2ads}), hydroperoxide (HO_{2ads}) radicals and hydrogen peroxide (H_2O_{2ads}) Equation (2-4), while the photogenerated holes can oxidize either the organic compound directly Equation (5) or both hydroxylic ions and water molecules adsorbed on the photocatalyst surface Equation (6-7) forming the organic cation-radicals ($R^{·+}_{ads}$) and hydroxylic radicals

($HO^·_{ads}$). The stepwise photocatalytic mechanism is illustrated below:

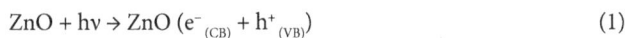

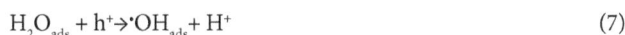

$$ZnO + hv \rightarrow ZnO\ (e^-_{(CB)} + h^+_{(VB)}) \tag{1}$$

$$O_{2ads} + e^- \rightarrow O^{·-}_{2ads} \tag{2}$$

$$O^{·-}_{2ads} + H^+ \rightarrow HO_{2\ ads} \tag{3}$$

$$O^{·-}_{2ads} + 2H^+ + e^- \rightarrow H_2O_{2ads} \tag{4}$$

$$R_{ads} + h^+ \rightarrow R^·_{ads}{}^+ \tag{5}$$

$$HO^-_{ads} + h^+ \rightarrow {}^·OH_{ads}{}^+ \tag{6}$$

$$H_2O_{ads} + h^+ \rightarrow {}^·OH_{ads} + H^+ \tag{7}$$

The hydroxylic, peroxide and hydroperoxide radicals are the main oxidizing agents in the heterogeneous photocatalytic systems used in the water treatment technologies. The heterogeneous photocatalytic processes substantially depend on a variety of environment conditions such as surface charge and electronic structure of catalyst, the nature of surface-active center, the localization degree of photogenerated charges, the amphoteric properties of photocatalyst, pH, temperature, nature of pollutant, photocongeners, crystalline structure, synthesis method and photorector dimension [7].

Aggregation is one of the features of dyes in solution. Based on literature, Congo red dye tends to aggregate in aqueous and organic

---

*Corresponding author: Elaziouti, LCPCE Laboratory, Faculty of sciences, Department of industrial Chemistry, University of the Science and the technology of Oran (USTO M.B) BP 1505 El M'naouar 31000 Oran, Algeria

solutions leading to dimer formation and sometimes even higher order aggregates due to hydrophobic interactions between aromatic rings of dye molecules. This aggregation phenomenon is more noticed for high Congo red concentrations, at high salinity and/or low pH. The formed aggregates (micelles) separate and precipitate onto solid surfaces.

In the present work, the potential ability of ZnO-assisted photocatalytic degradation of Congo red (CR) and benzopurpurine 4B (BP4B) was assessed in terms of evolution of the photodecomposition efficiency at different operating parameters such as, irradiation time, pH solution, initial dye concentration, amount of catalyst, light intensity and as well as band gap of other semiconductor groups. The experimental data were quantified by applying the pseudo-first order kinetic and Langmuir-Hinshelwood (L-H) model to accommodate reactions occurring at a solid-liquid interface.

Materials and Methods

Nanoparticle semiconductors ZnO (BET surface area, S= 10 m^2/g and particle size D=60 nm, 99.99%), TiO_2 (anatase 99.99%), Al_2O_3(99.99%), CeO_2 (99.99%), and Fe_2O_3 (99.99%), were obtained from Merck, and were used without further purification. Congo red (C.I. 22 120, MW = 696.67 g mol^{-1}, $C_{32}H_{24}N_6O_6S_2.2Na$, λ max =497 nm and pKa=4) and benzopurpurine 4B (C.I.23500, MW =724.74 g mol^{-1}, $C_{34}H_{26}N_6O_6S_2Na_2$ λ max=500 nm and pKa=6.8). The molecular structure of the dye is illustrated in Figure 1. Distilled water was used for preparation of various solutions.

Photocatalytic reactions were carried out inside a (BLX-E365) photoreactor equipped with 6UV-A lamps with an emission maximum at λ of 365 nm. The suspension was irradiated perpendicularly to the surface of solution, and the distance between the UV source and vessel containing reaction mixture was fixed at 15 cm. The experiments were performed at 298K. In all photocatalytic degradation experiments, 300mL CR solution of appropriate concentration was taken in photocatalytic reactor vessel of 600 ml capacity. A known quantity of semiconductor was added and mixture was stirred to obtain uniform suspension. The suspension pH values were previous adjusted using $NaOH/H_2SO_4$ solutions via pH meter (HANNA HI 83141).

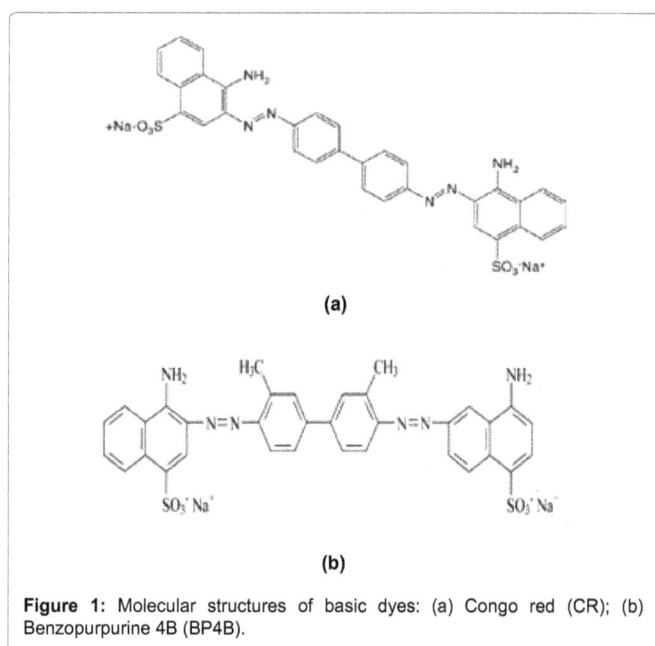

Figure 1: Molecular structures of basic dyes: (a) Congo red (CR); (b) Benzopurpurine 4B (BP4B).

Before irradiation, photocatalyst/substrate suspension was stirred in the dark for 30 minutes at 298K to ensure the adsorption equilibrium was established. Next, the lamp was switched on to initiate the photocatalytic degradation reaction. During irradiation, agitation was maintained by a magnetic stirrer to keep the suspension homogeneous. The suspension was sampled at regular intervals of time and immediately centrifuged using (EBA-Hetlich) at 3500 rpm for 15 min to completely remove photocatalyst particles. The residual concentration of the solution samples was monitored using UV-Vis Spectrophotometer (Shimadzu UV mini-1240) at λ = 497 and 500 nm for CR and BP4B respectively, as a function of irradiation time.

The effect of initial pH on the photocatalytic degradation of Benzopurpurine 4B was researched over a range of pH values from 2 to 10. But for Congo red, the experiments were only conducted from pH 6 to 10 for avoiding dye aggregation. The experiments were also performed by varying the initial dye concentration from 20 to 60 mg/L, amount of photocatalyst from 0.25 to 3 g/L and light intensity from 50 to 90 j/ cm^2 as well as band gap of semiconductors by replacing ZnO with TiO_2, Al_2O_3, CeO_2 and Fe_2O_3 nanoparticles.

The data obtained from the photocatalytic degradation experiments were then used to calculate the degradation efficiency η (%) of the substrate Equation (8):

$$\eta(\%) = [\frac{(C_i - C_f)}{C_i}]100 \qquad (8)$$

where C_i: dye initial concentration (mg·L^{-1}) and C_f : dye residual concentration after certain intervals (mg·L^{-1}).

To calculate the corresponding energy at UV-A wavelength. The energy of an electro-volt, E (eV), at a given wavelength, λ (nm), is given by Equation. (9):

$$E(eV) = \frac{hc}{\lambda j} \qquad (9)$$

where h is Planck's constant (6.626 10^{-34} J s); c is the speed of light (3 10^8 m/s); and j is the number of electro-volt per joule (joule=1.6 10^{-19} electro-volt). The corresponding light energy at UV-A (365 nm) wavelength was estimated to E_{UV-A} = 3.4 eV.

The photocatalytic degradation efficiency of ZnO catalyst for the degradation of CR and BP4B was quantified by measurement of dyes apparent first order rate constants under operating parameters and Langmuir–Hinshelwood modified kinetic analysis to accommodate reactions occurring at a solid-liquid interface. Surface catalyzed reactions can often be adequately described by a monomolecular Langmuir–Hinshelwood mechanism, in which an adsorbed substrate with fractional surface coverage θ is consumed at an initial rate given as follow Equation (10) [8]:

$$-[\frac{dC}{dt}] = r_0 = K_{app}\theta = \frac{K_1K_2C_0}{1 + K_1C_0} \qquad (10)$$

where K_1 is a specific rate constant that changes with photocatalytic activity, K_2 the adsorption equilibrium constant, and C_0 is the initial concentration of the substrate (Congo Red and Benzopurpurine 4B in our cases). Inversion of the above rate equation is given by Equation (11):

$$\frac{1}{K_{app}C_0} = \frac{1}{K_1K_2} + \frac{C_0}{K_1} \qquad (11)$$

Thus, a plot of reciprocal of the apparent first order rate constant $1/K_{app}$ against initial concentration of the dye C_0 should be a straight line with a slope of $1/K_1$ and an intercept of $1/K_1K_2$. Such analysis allows one to quantify the photocatalytic activity of ZnO catalyst through the specific rate constant K_1 (with larger K_1 values corresponding to higher photocatalytic activity) and adsorption equilibrium constant K_2 (K_2 expresses the equilibrium constant for fast adsorption-desorption processes between surface of catalyst and substrates). The integrated form of the above equation (Equation 10) yields to the following Equation (12):

$$t = \frac{1}{K_1K_2}\ln\frac{C_0}{C} + \frac{1}{K_2}(C_0 - C) \qquad (12)$$

where t is the time in minutes required for the initial concentration of the dye C_0 to decrease to C. Since the dye concentration is very low, the second term of the expression becomes small when compared with the first one and under these conditions the above equation reduces to Equation (13):

$$\ln\frac{C_0}{C} \approx K_1K_2t = K_{app}t \qquad (13)$$

where k_{app} is the apparent pseudo-first order rate constant, C and C_0 are the concentration at time 't' and 't=0', respectively. The plot of ln C_0/C against irradiation time t should give straight lines, whose slope is equal to K_{app}. The half-life of dye degradation at various process parameters was raised from Equation (14):

$$t_{1/2} = \frac{0.5C_0}{K_2} + \frac{0.693}{K_1K_2} \qquad (14)$$

where half-life time, $t_{1/2}$, is defined as the amount of time required for the photocatalytic degradation of 50% of the RC and BP4B dyes in aqueous solution by ZnO catalyst.

Results

Effect of UV light and catalyst

Figures 2 and 3 illustrate the photocatalytic degradation kinetics of 20 mg/L of dyes in aqueous solution under three different experimental conditions through UV-A alone, dark/ ZnO and UV-A/ZnO. The degradation rate was found to increase with increase in irradiation time

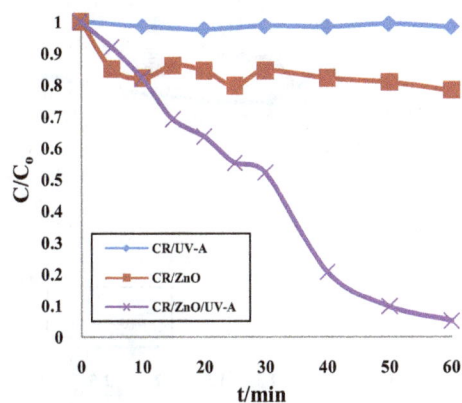

Figure 3: Photocatalytic degradation of BP4B under different experimental conditions ([ZnO]=1/L, [BP4B]=20mg/l, natural pH =8, T=298K, λ_{max}=365 nm and I=90j/cm²).

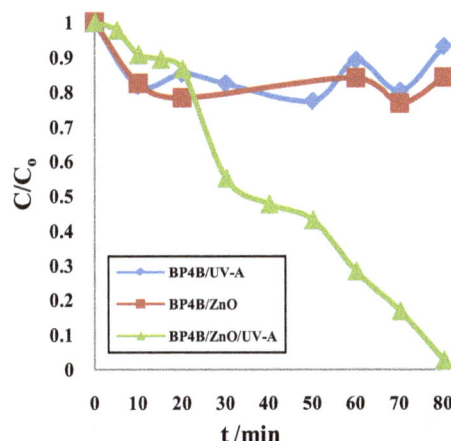

Figure 4: Effect of pH on photocatalytic degradation of dyes ([ZnO]= 0.5g/L(for CR) and 1g/L (for BP4B), [CR]= [BP4B]= 20mg/l, T=298K, λ_{max}=365 nm and I=90j/cm²).

and 95.02 and 97.24% of degradation were achieved within 60 and 80 min for CR and BP4B respectively (curve Dye/ZnO/UV-A). When 20 mg/L of both dyes along with ZnO were magnetically stirred for the same optimum irradiation times in the absence of light, lower (20.78 and 13.17 %) degradation were observed (curve Dye/ZnO) for CR and BP4B respectively, whereas, disappearance of dyes was negligible (0.49% for CR and 4.45% for BP4B) in the direct photolysis (curve Dye/UV-A) indicating that the observed high decomposition of both dyes in the UV/ZnO process is exclusively attributed to the photocatalytic reaction of the semiconductor particles. Similar results have been reported for ZnO-assisted photocatalytic degradation of azo dyes such as Congo red [9] and Reactive Black 5 [10].

Effect of pH solution

In order to study the effect of initial pH on the degradation efficiency of ZnO catalyst on photodecomposition of both dyes, experiments were carried out at various pH, ranging from 2 to10, except for CR, where tests were done from 6 to 10 for avoiding dye aggregation. The results showed that the pH significantly affected the degradation efficiency for both dyes. As shown in Figure 4 and Table

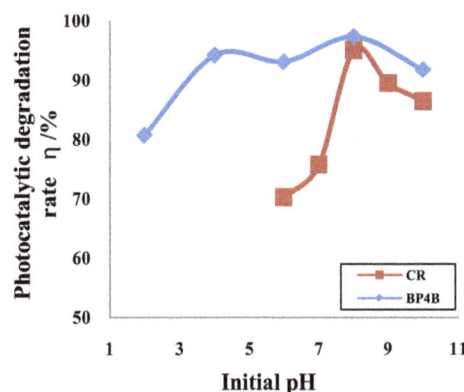

Figure 2: Photocatalytic degradation of CR under different experimental conditions ([ZnO]=0.5g/L, [CR]= 20mg/l, natural pH =8, T=298K, λ_{max}=365 nm and I=90j/cm²).

Experimental parameters			Experimental results		Pseudo-first order model			Experimental results		Pseudo-first order model		
					CR					BP4B		
	RC	BP4B	ADS / %	PCD /%	k_{app} /min^{-1}	$t_{1/2}$ /min	R^2/%	ADS/%	PCD/%	k_{app}/min^{-1}	$t_{1/2}$/min	R^2/%
Initial pH solution (pH)	6	2	18.19	70.25	0.014	49.511	79.8	86.09	80.70	0.023	9.57	32.2
	7	4	16.13	75.68	0.025	27.726	96.8	41.51	94.19	0.037	14.29	98.3
	8	6	10.50	95.02	0.041	16.906	88.1	32.84	93.07	0.035	128.36	98.0
	9	8	14.94	89.38	0.038	18.241	79.0	3.89	97.24	0.026	26.45	95.6
	10	10	11.81	86.339	0.035	19.804	74.7	24.84	91.69	0.036	144.40	88.0

R^2: Regression coefficient, ADS: Adsorption and PCD: Photocatalytic degradation

Table 1: Kinetic parameters of photocatalytic degradation of dyes in aqueous solution as a function of pH ([ZnO]=0.5g/L(for CR) and 1g/L (for BP4B), [CR]=[BP4B]=20mg/L, T=298K, λ_{max}=365 nm and I=90j/cm^2).

1, for CR, the degradation rate increased from 70.25 to 95.02% as the pH value was increased from 6 to 8, and then decreased to 86.34% at pH 10, whereas, for BP4B, the degradation activity raised from 80.70 to 97.24% when the pH was increased from 2 to 8 and then decreased to 91.69 % at pH 10. The maximum degradation rate of CR (95.02%) and BP4B (97.24%) were achieved at pH 8. For this reason, for both dyes, the pH 8 was selected for subsequent experiments.

It is commonly accepted that in photocatalyst/aqueous systems, the potential of the surface charge is determined by the activity of ions (e.g. H$^+$ or pH). A convenient index of the tendency of a surface to become either positively or negatively charged as a function of pH is the value of the pH required to give zero net charge (pH zpc) [11,12]. pH zpc is a critical value for determining the sign and magnitude of the net charge carried on the photocatalyst surface during adsorption and photocatalytic degradation process. Most of the semiconductor oxides are amphoteric in nature, can associate (Equation 15) or dissociate (Equation 16) proton. To explain the relationship between the layer charge density and the adsorption, so-called Models of Surface Complexation (SCM) was developed [13], which consequently affects the sorption–desorption processes and the separation and transfer of the photogenerated electron–hole pairs at the surface of the semiconductor particles. In the 2-pK approach we assume two reactions for surface protonation.

The zero point charge pH zpc for ZnO is 9.0. For pH values lower than the pH zpc of ZnO, the surface becomes positively charged, according to the following reaction Equation (15):

$$\text{pH} < \text{pH zpc} \quad \text{Zn-OH} + \text{H}^+ \rightarrow \text{ZnOH}_2 \tag{15}$$

ZnO surface becomes negatively charged for pH values higher than pH$_{pzc}$, given by the following reaction Equation (16):

$$\text{pH} > \text{pH zpc} \quad \text{Zn-OH} + \text{OH}^- \rightarrow \text{ZnO}^- + \text{H}_2\text{O} \tag{16}$$

The experimental data revealed that higher degradation rate of BP4B was observed in acidic medium. Since BP4B is an anionic dye, its adsorption mainly performed via an electrostatic interactions between the positive ZnO surface and BP4B anions, leading to a maximum extent at pH 2. Thus, a strong adsorption can lead to a drastic decrease in the active centers on the catalyst surface, which results in decrease in the absorption of the light quanta by the catalyst and consequently to a reducing of the kinetic reaction. As a result, the high degradation efficiency cannot be ascribed to the photocatalytic oxidation of the BP4B dye, but to catalyst behavior under strong acid pH. ZnO nanoparticles can undergo photo-corrosion through self-oxidation

at pH lower than 4 Equation (17) In particular, ZnO can be photo-oxidized with decreasing the pH Equation (18)

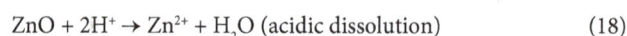

$$\text{ZnO} + 2\text{h}^+_{(VB)} \rightarrow \text{Zn}^{2+} + 1/2\text{O}_2 \tag{17}$$

$$\text{ZnO} + 2\text{H}^+ \rightarrow \text{Zn}^{2+} + \text{H}_2\text{O} \text{ (acidic dissolution)} \tag{18}$$

Photocatalytic activity of anionic dyes (mainly sulfonated groups) such as CR and BP4B reaches a maximum value in lower pH zpc (i.e. pH =8). At alkaline mediums, excess of hydroxyl anions facilitate photogeneration of ·OH radicals which is accepted as primary oxidizing species responsible for photocatalytic degradation, resulting in enhancement of the efficiency of the process. Furthermore we found that, where the adsorption of dyes was weak, degradation scarcely occurred. The adsorption affects strongly the accessibility of the surface reducing species to the CR and BP4B reduction kinetics. However, adsorption is not the only factor that controls the photocatalytic degradation of dyes. Although the adsorption extents of both dyes were lower, the degradation rates were in the reverse order.

At pH higher than pH zpc value (i.e. pH =10), a dramatically decrease in the degradation efficiency could be explained on the basis of amphoteric behaviors of ZnO catalyst. The negatively surface of ZnO catalyst (highly concentration of hydroxide ions) and the great negatively charged RC and BP4B dye anions results in electrostatic repulsion electrostatic.

Moreover, the stability of ZnO may not be guaranteed at this high

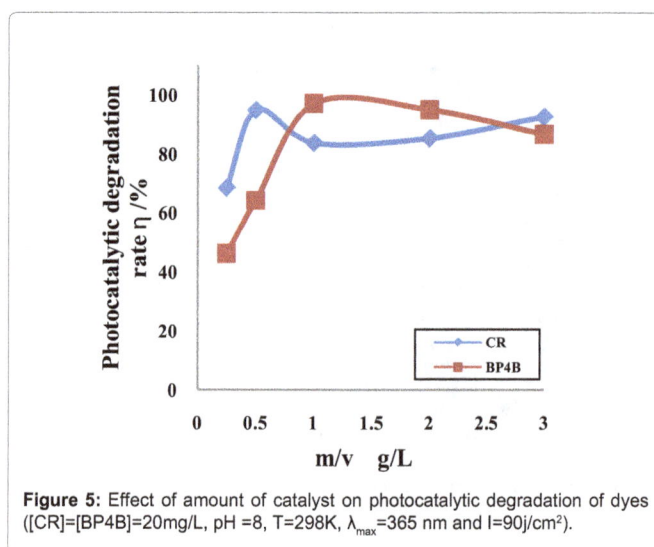

Figure 5: Effect of amount of catalyst on photocatalytic degradation of dyes ([CR]=[BP4B]=20mg/L, pH =8, T=298K, λ_{max}=365 nm and I=90j/cm^2).

Experimental parameters		Experimental results		Pseudo-first order model			Experimental results		Pseudo-first order model		
		CR					BP4B				
		ADS / %	PCD /%	k_{app} /min^{-1}	$t_{1/2}$ /min	R^2/%	ADS/%	PCD/%	k_{app}/min^{-1}	$t_{1/2}$/min	R^2/%
Catalyst amount ZnO/g/L	0.25	8.93	68.73	0.030	23.105	82.6	38.31	46.29	0.014	49.511	46.1
	0.5	10.50	95.02	0.041	16.906	88.1	2.34	64.21	0.018	38.508	68.4
	1	33.59	83.80	0.037	18.734	93.0	3.89	97.24	0.026	26.456	95.6
	2	41.98	85.45	0.040	17.329	79.0	35.54	95.26	0.037	18.734	96.7
	3	53.40	92.68	0.048	14.441	93.0	60.85	86.97	0.026	26.660	96.2

R^2: Regression coefficient, ADS: Adsorption and PCD: Photocatalytic degradation

Table 2: Kinetic parameters of photocatalytic degradation of dyes in aqueous solution as a function of amount of photocatalyst ([CR]=[BP4B]=20mg/L, pH =8, T=298K, λ_{max}=365 nm and I=90j/cm^2).

pH due to possibility of alkaline dissolution of ZnO Equation (19):

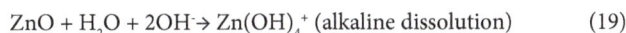

$$ZnO + H_2O + 2OH^- \rightarrow Zn(OH)_4^+ \text{ (alkaline dissolution)} \qquad (19)$$

Effect of the amount of catalyst

The effect of the amount of catalyst (m/v) on photocatalytic degradation of dyes was conducted over a range of catalyst amount from 0.25 to 3g/L. As observed in Figure 5 and Table 2, the degradation rate of CR increased from 68.73 to 95.02 % in CR and from 46.29 to 97.24% in BP4B when the ZnO amount was raised from 0.25 to 0.5 g/L for CR and from 0.25 to 1g/L for BP4B respectively. This increase in degradation rate with the photocatalyst amount can be explained in terms of availability of active sites on the catalyst surface and the penetration of UV light into the suspension as a result of increased screening effect and scattering of light. Further increase in the catalyst amount beyond of 0.5 g/L for CR and 1g/L for BP4B, the rate of degradation remains nearly constant for CR, but it slightly decreased for BP4B due to overlapping of adsorption sites as a result of overcrowding of adsorbent and deactivation of activated catalyst particles owing to collision with ground state catalyst as shown below Eq. (20) [14]:

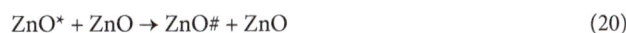

$$ZnO^* + ZnO \rightarrow ZnO\# + ZnO \qquad (20)$$

where ZnO* is ZnO with active species adsorbed on its surface; ZnO# deactivated form of ZnO* shielding by ZnO may also take place. The adsorbent dose of 0.5 and 1 g/L for CR and BP4B were used in all other parameter experiments.

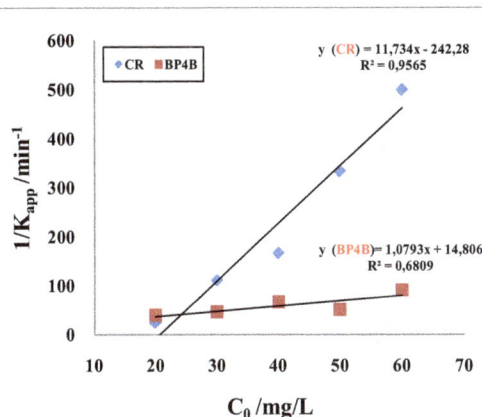

Figure 6: Effect of initial dye concentration on photo catalytic degradation of dyes ([ZnO]=0.5g/L (for CR) and 1g/L (for BP4B) pH =8, T=298K, λ_{max}=365 nm and I=90j/cm^2.

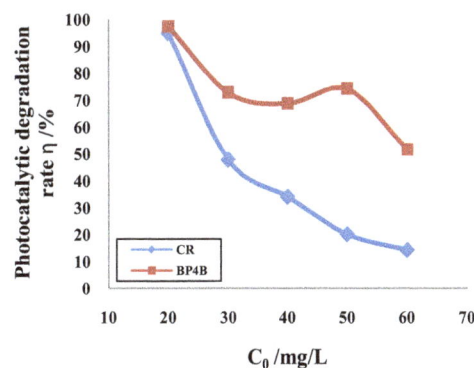

Figure 7: Langmuir–Hinshelwood analysis for photo catalytic degradation of CR and BP4B ([ZnO]=0.5g/L (for CR) and 1g/L (for BP4B) pH =8, T=298K, λ_{max}=365 nm and I=90j/cm^2.

Effect of initial dye concentration

Figure 6 illustrates the effect of initial dye concentration on the photo catalytic degradation rate of dyes in the range of 20 to 60 mg/L. As it can be observed, disappearance rate was found to be inversely affected by initial concentration of dyes. The drastic decrease in the degradation activity with dyes concentration is ascribed to the increase in the local concentration of CR as well as BP4B on the ZnO surface, while the UV light irradiation time and photocatalyst amount are kept constant, leading to the formation of dimer and higher order aggregates owing to hydrophobic interactions between aromatic rings and hence the rate formation of hydroxylic and superoxide anion radicals are dramatically reduced thereby decreasing rate of degradation. The maximum concentration of both dyes that could be degraded by 0.5 and 1g/L of ZnO for CR and BP4B respectively is found to be 20mg/L. Similar trend was observed in the photocatalytic degradation of Reactive Black 5 and Reactive Orange 4 dyes using ZnO and TiO_2 as photocatalysts [15]. Thus 20 mg/L CR and BP4B was selected as optimum concentration for the study of other experiments.

The photocatalytic degradation process profiles of both dyes by ZnO catalyst at low dye concentrations and under pH solution, amount of catalyst and initial dye concentration follow apparently pseudo-first-order kinetics. The linear plot of lnC_0/C against irradiation time t (Table 1, 2 and 3) should give a straight line with relatively high regression coefficients, whose slope is equal to the apparent first order rate constant K_{app}.

On the other hand, the effect of initial concentrations of both dyes

Experimental parameters		Experimental results		Pseudo-first order model			Experimental results		Pseudo-first order model		
		CR					BP4B				
		ADS / %	PCD /%	k_{app} /min^{-1}	$t_{1/2}$ /min	R^2/%	ADS/%	PCD/%	k_{app}/min^{-1}	$t_{1/2}$/min	R^2 /%
Initial dye concentration Dye/mg/L	20	10.50	95.02	0.041	16.906	88.1	3.89	97.24	0.026	26.456	95.6
	30	20.93	47.87	0.009	77.016	91.0	11.25	72.85	0.022	31.651	82.6
	40	15.26	34.10	0.006	115.525	91.0	2.12	68.87	0.015	45.904	98.2
	50	1.95	20.15	0.003	231.049	95.0	1.18	74.24	0.020	34.485	81.9
	60	3.86	14.25	0.002	346.574	86.00	33.11	51.71	0.011	62.446	95.9

R^2: Regression coefficient, ADS: Adsorption and PCD: Photo catalytic degradation

Table 3: Kinetic parameters of photo catalytic degradation of dyes in aqueous solution as a function of initial dye concentrations.([ZnO]=0.5g/L (for CR) and 1g/L (for BP4B), pH =8, [CR]=[BP4B]=20-60mg/L, P^H =8 T=298K, λ_{max}=365 nm and I=90j/cm^2).

in the photocatalytic degradation rate can be assessed in terms of the Langmuir–Hinshelwood (LH) kinetic model modified.

The plot of $1/K_{app}$ against C_0 (Figure 7) should yield a straight line with high regression coefficients (R^2=0.96 and 0.68 for CR and BP4B respectively). The K_1 and K_2 values were calculated from the slope (1/K_1) and the intercept (1/K_1K_2) respectively. The values of K_1 and K_2 were found to be 0.048 L/g and 0.085g/L.min for CR and 0.073L/g and 0.927g/L.min for BP4B respectively. The product of K_1K_2 = 0.004 and 0.068 min^{-1} for CR and BP4B respectively which represent the apparent rate constant K_{app} for low initial concentrations of dye and is in agreement with the experimental results obtained from (Equation 13) for BP4B.This results suggest that the photocatalytic degradation data were satisfactorily described by Langmuir–Hinshelwood (LH) kinetic model in a wide range of dye concentration. However, that of CR are not sufficient to conclude that the L-H mechanism is the most suitable one to describe the photocatalytic process of CR, since the calculated K_{app} (K_{app}=K_1K_2) value deduced from pseudo-first-order kinetic equation was much different compared with experimental K_{app}, despite its regression coefficient is higher than of BP4B.

Effect of the UV-A light intensity

Intensity of the irradiation has been reported to be an important parameter influencing the degradation of organic chemicals by photo catalytic activity. The effect of light intensity on the rate of photo catalytic degradation of dyes was investigated by varying the light intensity of UV-A between 50 and 90 j/ cm^2. Results reported in Figure

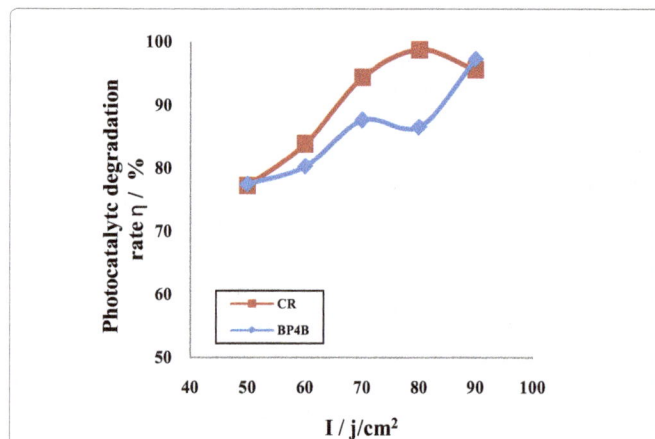

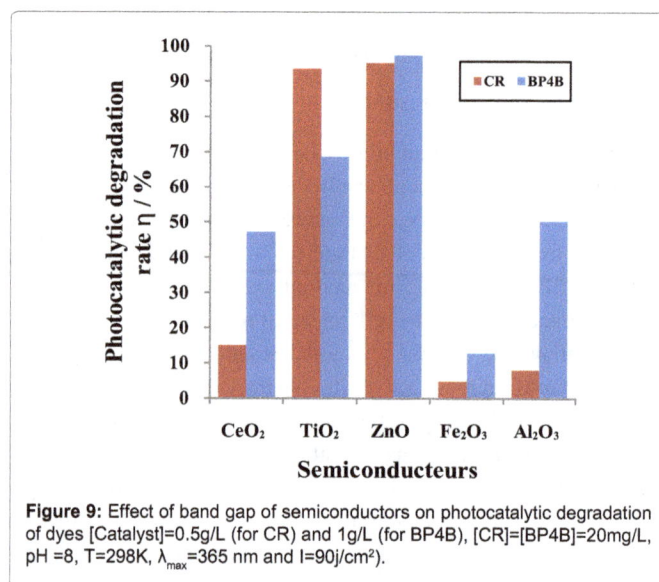

Figure 9: Effect of band gap of semiconductors on photocatalytic degradation of dyes [Catalyst]=0.5g/L (for CR) and 1g/L (for BP4B), [CR]=[BP4B]=20mg/L, pH =8, T=298K, λ_{max}=365 nm and I=90j/cm^2).

8 and Table 4, depicts that the degradation efficiency of BP4B increased linearly with the light intensity, whereas that of CR it increased up to 70j/cm^2 and after no changes are observed. The linear increase of the degradation efficiency for CR and BP4B at light intensity, ranging from 50 to70 j/cm^2 and from of 50 to90 j/cm^2 for CR and BP4B respectively is assigned by more photons would be available for excitation at the semiconductor surface and in turn more electron hole pairs will be generated. Thus this resulted in enhanced rate of degradation. The value of k_1 was found to increase with the increase in light intensity, a typical characteristic of a photo catalytic reaction. At a higher light intensity than 70 j/cm^2 for CR, there is no effect on degradation activity on further increase in light intensity. The results indicate that maximum numbers of photons which are required for excitation are available in specific constant range of irradiating light intensity. Further increase in light intensity no significant changes are observed in photocatalytic degradation efficiency because all photo catalyst particles are exited, so, the rate of degradation remains unchanged [16].

Effect of the band gap of semiconductors

The photocatalytic degradation reactions were further performed in four other semiconductors having different band gap values (Figure 9 and Table 5). It is evident that the photocatalytic degradation of dyes greatly depends on the electronic structure and properties of semiconductor surface/solvent. Generally, semiconductor on irradiation with light energy greater than or equal to band gap

Figure 8: Effect of light intensity on photocatalytic degradation of dyes [Catalyst]=0.5g/L (for CR) and 1g/L (for BP4B), [CR]=[BP4B]=20mg/L, pH =8, T=298K and λ_{max}=365 nm).

energy of the semiconductor (hv ≥ Eg), a free electron (e⁻) and electronic vacancy-a hole (h⁺) are formed and recombine or migrate in the semiconductor surface. Probability of electron transfer in the semiconductor/adsorbate system is determined by a relative position of the valence band, the photocatalyst conductance band and the value of the oxidation-reduction potential (ORP) of the oxidant and the substrate. The photogeneration of electrical charge is in dynamic equilibrium with their recombination substantially reducing the quantum yield of the photocatalytic process. The ORP of water oxidation, hydroxylic ions, and most of organic compounds below of reducing photogenerated holes within a wide interval of the pH due to which the formation of hydroxylic radicals and organic cation-radicals of photocatalyst surface are thermodynamically possible processes. It has already reported that semiconductors such as ZnO and TiO_2 having band gaps larger than 3 eV are excellent photocatalysts. Obviously, ZnO and TiO_2 semiconductors exhibit a higher degradation activity than that of the other systems because their band gaps (Eg = 3.2 and 3.3 eV for ZnO and TiO_2 respectively) are slightly equal to that of UV-A irradiation source (E_{UV-A} =3.4 eV). The photogenerated electron (e⁻)-hole (h⁺) pairs can be easily separated and transferred to the semiconductor/adsorbate interface efficiently, consequently improving the photocatalytic activity [17].

On the other hand, Al_2O_3 and Fe_2O_3 catalysts showed lower activity for the photocatalytic activity of dyes than ZnO and TiO_2, since their conductance bands of 5.6 and 3.7 eV respectively, are much higher than of the E_{UV-A} light irradiation source as a result of low light energy conversion efficiency, so the photogenerated electrical charge in semiconductor cannot efficiently transfer in the surface and are lost due to recombination.

The minimum energy required for excitation of an electron from the valence band to conductance band for the semiconductor such as CeO_2 is 2.7 eV. The photodecomposition process of both dyes in CeO_2

system was much lower than ZnO and TiO_2 catalysts, although it band gap being smaller than of E_{UV-A} light energy. The CeO_2 system might reduce the life of electron-hole pairs, and enhance the opportunities of their recombination. The order of the photocatalytic degradation efficiency is ZnO > TiO_2 > CeO_2 > Al_2O_3 > Fe_2O_3. Consequently, the photocatalytic ability of semiconductor is significantly dependent on their band gap values.

Table 4 and 5 record the kinetic parameters of photocatalytic degradation of CR and BP4B in aqueous solution.

The results show that the photocatalytic decolorization of both the dyes in aqueous solutions under light intensity and band gap of catalysts can be described by the pseudo-first-order kinetic model. The semi-logarithmic plots of the experimental data under optimized conditions ($\ln C_0/C$ against t) yielding to a straight line. The regression coefficients for the fitted lines were calculated to be R^2 = 0.881 and 0.956 for CR and BP4B respectively. The apparent rate constants, K_{app} and the half-life time, $t_{1/2}$ were calculated to be 0.041 min⁻¹ and 16.906 min for CR and 0.026 min⁻¹ and 26.46 min for BP4B.

Table 4 and 5 report the kinetic parameters of the photo decomposition of both dyes in aqueous solution using ZnO catalyst at low dyes concentration and under light intensity and band gap of catalysts.

Mechanism

Photocatalyic degradation schemes for an azo-dye are characterized by nitrogen to nitrogen double bonds (N= N) that are usually attached to two radicals of which at least one but usually both are aromatic groups (benzene or naphthalene rings). The color of azo-dyes is determined by the azo bonds and their associated chromophores and auxochromes. Azo bonds are the most active bonds in azo-dye molecules and can be oxidized by positive hole or hydroxyl radical or reduced by electron

Experimental parameters		Experimental results		Pseudo-first order model			Experimental results		Pseudo-first order model		
				CR					BP4B		
		ADS / %	PCD /%	k_{app} /min⁻¹	$t_{1/2}$ /min	R^2/%	ADS/%	PCD/%	k_{app} min⁻¹	$t_{1/2}$/min	R^2/%
Light Intensity I/ J/cm²	50	12.20	77.11	0, 029	16.906	83.7	50.65	77.35	0.022	31.364	92.8
	60	11.19	83.78	0.027	12.603	64.7	32.41	80.24	0.022	31.364	96.9
	70	17.93	94.33	0.045	15.403	98.6	34.95	87.57	0.032	21.661	75.2
	80	24.07	98.69	0.055	12.603	83.1	26.14	86.52	1.032	0.672	75.2
	90	10.50	95.02	0.041	16.906	88.1	3.89	97.24	0.026	26.456	95.6

R^2: Regression coefficient, ADS: Adsorption and PCD: Photocatalytic degradation

Table 4: Kinetic parameters of photocatalytic degradation of dyes in aqueous solution as a function of light intensity ([ZnO]=0.5g/L (for CR) and 1g/L (for BP4B), pH =8, [CR]=[BP4B]=20-60mg/L, pH =8, T=298K and λ_{max}=365 nm).

Experimental parameters		Experimental results		Pseudo-first order model			Experimental results		Pseudo-first order model		
				CR					BP4B		
		ADS / %	PCD /%	k_{app} /min⁻¹	$t_{1/2}$ /min	R^2/%	ADS/%	PCD/%	k_{app} min⁻¹	$t_{1/2}$/min	R^2/%
Band gap of catalyst [Eg] (eV)	Zn O	10.50	95.02	0.041	16.906	88.1	3.89	97.24	0.026	26.456	95.6
	Ti O_2	13.47	93.34	0.045	15.40	98.5	28, 52	68.54	0.002	346.57	78.2
	Ce O_2	8.00	14.93	0.003	231.05	49.0	25, 59	47.19	0.033	20.94	70.7
	Al_2O_3	4.21	7.91	0.001	693.15	88.7	13, 73	50.21	0.011	63.01	46.4
	Fe_2O_3	5.15	4.72	0.001	693.15	34.0	1, 74	12.77	0.018	38.08	95.0

R^2: Regression coefficient, ADS: Adsorption and PCD: Photocatalytic degradation

Table 5: Kinetic parameters of photocatalytic degradation of dyes in aqueous solution as a function of band gap of semiconductors.([Catalyst]=0.5g/L (for CR) and 1g/L (for BP4B), [CR]=20mg/l, pH =8, T=298K, λ_{max}=365 nm and I=90j/cm²).

in the conduction band. The cleavage of N= N bonds leads to the decoloration of dyes [18].

When a semiconductor is irradiated with light having energy equal to or more than band gap energy (hν≥ Eg), a heterogeneous photocatalytic reaction occurs at the photocatalyst/adsorbate interface. The conduction band electrons (e⁻) and valence band holes (h⁺) are formed Equation (21). A part of the photogenerated charge carriers recombines in the bulk of the semiconductor, while the rest transfer in the photocatalyst surface, where the holes as well as the electrons act as powerful oxidants, respectively. The photogenerated electrons react with the adsorbed molecular O_2 on the ZnO photocatalyst particle sites, reducing it to a superoxide radical anion $O_2^{\cdot-}$ Equation (22), while the photogenerated holes can oxidize either the dye molecule directly or the OH⁻ ions and the water molecules adsorbed the ZnO surface to ·OH radicals Equation (23).

In the photocatalytic oxidation process, the generation of hydroxyl radicals occurs in two different pathways.

First pathway: In the first pathway, where U.V light is used in the photocatalytic reaction, electrons in the semiconductor are excited from the valence band to the conduction band leaving positive holes in the valance band. The photogenerated electrons react with the adsorbed oxygen molecules to form $O^{\cdot-}_{2ads}$ species, while the photogenerated holes that are able to migrate to the hydroxylated surface can create a highly reactive and short-lived hydroxyl radicals •OH. These processes could be represented in the following equations:

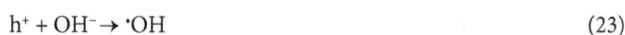

$$ZnO + h\nu \rightarrow ZnO\ (e^- + h^+) \tag{21}$$

$$O_{2ads} + e^- \rightarrow O^{\cdot-}_{2ads} \tag{22}$$

$$h^+ + OH^- \rightarrow {}^\cdot OH \tag{23}$$

Second pathway: In the second pathway where a solar radiation is used a photosensitization process takes place. In this process, the dye molecules act as a sensitizer by the absorption of UV light in the visible range to yield an excited state of the sensitizer Equation (24). The dye radicals inject electrons to the conduction band of the ZnO photocatalyst Equation (25-26) in where it is scavenged by O_2 to form active oxygen molecule as shown in Equation (27). The electron transfer from the excited dye molecule to the conduction band of ZnO usually is too fast (in the range of tens of femtoseconds). Further active oxygen molecule formed in Equation (27) subsequently reacts with H_2O to generate ·OH radicals Equation (28) and peroxide Equation (29).

The formed species oxidize the dye molecules, as follows:

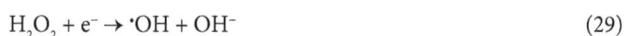

$$Dye + h\nu \rightarrow Dye^* \tag{24}$$

$$Dye^* \rightarrow Dye^+ + e^- \tag{25}$$

$$ZnO + e^- \rightarrow ZnO^{\cdot-} \tag{26}$$

$$ZnO^{\cdot-} + O_2 \rightarrow ZnO + O^{\cdot-} \tag{27}$$

$$O_2^{\cdot-} + 2H_2O + e^- \rightarrow 2H_2O_2 \tag{28}$$

$$H_2O_2 + e^- \rightarrow {}^\cdot OH + OH^- \tag{29}$$

These •OH radicals formed on the illuminated semiconductor surface via either a photoexcitement of semiconductor or photosensibilization of dye are highly effective oxidizing agent

which have been shown to be the primary cause of organic matter mineralization (Equation 30).

$$•OH + Dye \rightarrow degradation\ of\ the\ dye \tag{30}$$

These observations clearly demonstrate the involvement of holes (h⁺), hydroxyl (OH·) and superoxide anion ($O_2^{\cdot-}$) radicals as highly oxidizing agents in the ZnO-mediated photocatalytic oxidation of Congo red as well as Benzopurpurine 4B.

Conclusion

Congo red dye tends to aggregate in acidic aqueous solutions leading to dimer and higher order aggregates due to hydrophobic interactions between aromatic rings of dye molecules. Degradation photocatalytique of CR and BP4B was negligible when ZnO catalyst and UV light were used on their own, whereas, the observed high decomposition in the UV/ZnO process is exclusively attributed to the photocatalytic reaction of the semiconductor particles. The degradation kinetics of both dyes were fast with maximum efficiencies of 95.02% for CR and 97.24 % for BP4B have been achieved within 60 and 80 min using 0.5 and 1g/L of ZnO catalyst for CR and BP4B respectively at a pH of 8 and 298K. The degradation rate of dyes was obviously affected by the operating parameters (illumination time, initial dye concentration, and amount of photocatalyst, light intensity, pH and the band gap of semiconductors). L-H mechanism is the most suitable to describe the photocatalytic process of BP4B, so the oxidants involved in the photodegradation reactions were identified as holes (h⁺), hydroxyl radicals (·OH) and superoxide anion radicals ($O_2^{\cdot-}$). However the present kinetic data are not sufficient to conclude the validity of the L-H model in interpreting the results of heterogeneous photocatalysis of CR dye.

Acknowledgments

We greatly acknowledge the material support obtained from DR T.M. University.

References

1. Eslami A, Nasseri S, Yadollahi B, Mesdaghinia A, Vaezi F, et al. (2008) Photocatalytic degradation of methyl tert-butyl ether (MTBE) in contaminated water by ZnO nanoparticles. Journal Chemical and Technology Biotechnology 83: 1447-1453.

2. Lingzhi L, Bing Y (2009) CeO_2-Bi_2O_3 nanocomposite: Two step synthesis, microstructure and photocatalytic activity. Journal of Non-Crystalline Solids 355: 776- 779.

3. Djurišić A B, Chan Y, Li E H (2002) Progress in the room temperature optical functions of semiconductors. Material Science Engineering R 38: 237-293.

4. Sushil K, Kansal N K, Sukhmehar S (2009) Photocatalytic Degradation of Two Commercial Reactive Dyes in Aqueous Phase Using Nanophotocatalysts Nanoscale Research Letter 4: 709-716.

5. Kansal SK, Singh M, Sud D (2007) Studies on photodegradation of two commercial dyes in aqueous phase using different photocatalysts. Journal of Hazardous Materials 3: 581-590.

6. Poulios I, Kositzi M, Kouras A (1998) Photocatalytic decomposition of triclopyr over aqueous Semiconductor suspensions. Journal of Photochemestry and Photobiology A: Chem 115: 175-183.

7. Soboleva NM, Nosovich A A, Goncharuk V V (2007) The heterogenic photocatalysis in water treatment processes. Journal of Water Chemistry and Technology 29: 72-89.

8. Movahedi M, Mahjoub A R, Janitabar-Darzi S (2009) Photodegradation of Congo Red in Aqueous Solution on ZnO as an Alternative Catalyst to TiO_2. Journal Iranian of Chemical Society 6: 570-577.

9. Habibi M H, Hassanzadeh A, Zeini-Isfahani A (2006) Spectroscopic studies of Solophenyl red 3BL polyazo dye tautomerism in different solvents using UV–visible, ^{1}H NMR and steady-state fluorescence techniques. Dyes and Pigments 69: 93-101.

10. Akyol A, Yatmaz HC, Bayramoglu M (2004) Photocatalytic decolorization of Remazol Red RR in aqueous ZnO suspensions. Applied Catalysis B: Environment 54: 19-24.

11. Zhang F, Zhao J, Shan T, Hidaka H, Pelizzetti E, Serpone N (1998) TiO_2-assisted photodegradation of dye pollutants : II. Adsorption and degradation kinetics of eosin in TiO_2 dispersions under visible light irradiation Applied Catalysis B: Environment 15: 147-156.

12. Yates DE, Levine S, Healy TW (1974) Site-binding model of the electrical double layer at the oxide/water interface. Journal of Chemical Society Faraday Trans 70: 1807-1818.

13. Fernandez J, Kiwi J, Lizama, C, Freer J, Baeza J, Mansilla HD (2002) Factorial experimental design of Orange II photocatalytic decolouration. Journal of Photochemistry and Photobiology A: Chem. 151: 213-219.

14. Comparelli R, Fanizza E, Curri M L, Cozzoli P. D, Mascolo G, et al. (2005) UV-induced photocatalytic degradation of azo dyes by organic-capped ZnO nanocrystals immobilized onto substrates. Applied Catalysis B: Environment 60: 1-11.

15. Anandan S, Vinu A, Venkatachalam N, Arabindoo B, Murugesan V (2006)) Photocatalytic activity of ZnO impregnated H and mechanical mix of ZnO/H_ in the degradation of monocrotophos in aqueous solution. Journal of Molecular Catalysis A: Chem. 256: 312-320.

16. Vora J, Chauhan S K, Parmark K C, Vasava S B, Shaharma S, Bhutadiya L S (2009) Kinetic Study of Application of ZnO as a Photocatalyst in Heterogeneous Medium. E-Journal of Chemistry 6: 531-536.

17. Tomasevic A, Đaja J, Petrovic S, Kiss E E, Mijina D (2009) Study of the photocatalytic degradation of Methomyl by UV light. Chemical Industry and Chemical Engineering Quarterly 15: 17-19.

18. Pardeshi S K, Patil A B (2009) Solar photocatalytic degradation of resorcinol a model endocrine disrupter in water using zinc oxide. Journal of Hazardous Materials 163: 403-409.

21

Modeling of the Dynamics Adsorption of Phenol from an Aqueous Solution on Activated Carbon Produced from Olive Stones

Nouri Hanen* and Ouederni Abdelmottaleb

Research Laboratory: Engineering Process and Industrial System, National School of Engineers of Gabes, University of Gabes, Gabes, Tunisia

Abstract

A continuous fixed bed study was carried out by using granular activated carbon produced from olive stone for the removal of phenol from aqueous solution. The effects of initial phenol concentration (40-250 mg/l), feed flow rate (2.2-8.4 ml/min) and activated carbon bed depth (5-20 cm) on the breakthrough characteristics of the adsorption system were determined. The obtained results showed that the adsorption capacity increases with the bed depth and the initial concentration and it decreases at higher flow rate. Three models namely Clark, Thomas, and Yoon - Nelson were employed to predict the breakthrough curves and to determine the characteristic parameters of the column useful for column design. These models fitted well the adsorption data with coefficient of correlation R2>0.9 at different conditions. The activated carbon from olive stone was shown to be suitable adsorbent for adsorption of phenol using fixed bed adsorption.

Keywords: Phenol; Activated carbon; Olive stones; Adsorption; Breakthrough; Fixed bed

Introduction

Adsorption is known as an effective process for the removal of hazardous pollutants from wastewater. It can be carried out in batch systems with powdered adsorbents or in continuous flow in packed bed column. Fixed bed adsorption is simple to operate, and it can be relatively easily scaled up from a laboratory-scale study. Comparing with batch procedure, fixed bed is more effective for the cycle operation of adsorption/desorption, as it makes the best use of the concentration difference known to be a driving force for adsorption and allows more efficient utilization of the sorbent capacity and results in a better quality of the effluent. Also, the reuse of adsorbents is possible [1].

Fixed bed adsorption was tested for the elimination of several pollutants in effluents resulting from various industries.

Many studies of the adsorption capacities of activated carbon for heavy metals in fixed bed have been reported: arsenic As (V) [2,3], chromium Cr (VI) [4], fluoride [5,6], Pb (II) and Cu (II) [7]. Other researchers were interested in the elimination of the dyes which constitute a big discharge of the textiles industries. Hamdaoui [8], Banat et al. [9], Ferro [10], Han et al. [11], Song et al. [12] and Nasuha et al. [13] studied the elimination of the methylene blue from an aqueous solution on a fixed bed with various adsorbents and they found very satisfactory results.

The performance of fixed bed is usually described using the breakthrough curve. However, development of a model to accurately describe the dynamic behavior of adsorption in a fixed bed column is usually difficult. The use of simple models without numerical solutions appears to be more suitable and has practical benefits. Several models have been applied to describe fixed bed adsorption, the most useful are the Thomas model, Clark model and Yoon-Nelson model.

The common adsorbent which has good capacity in removing pollutants is the activated carbon which has been known and used for a long time, initially as adsorbent and later on as catalyst or support for catalyst. In its applications, the activated carbon can be presented in the form of fine powder, extruded or granulated particles, or fibers. Activated carbon can be prepared starting from various materials containing a height percentage of carbon and a small percentage of inorganic matter such as: peat, lignite, wood, coconut, etc.

The valorization of waste and industrial by-products currently knows a significant progress; some papers had reported several kinds of low cost adsorbents such as grape pomace [14], rice husk [15], residue of oils and lubricating oils [16], dates' stones [17]. Several studies were undertaken in the Mediterranean countries (Italy, Spain, Tunisia), considered as the main producer of the olive in the world, to use the olive residue as being lignocellulosic precursors being able to produce activated carbon of good quality and at a low cost considering the abundance of this vegetable matter. Thus, the production of activated carbon from olive stone constitute one of the most interesting research fields in our laboratory, several studies have been made to optimize these processes since 1996.

In this work, we were interested in the study of fixed bed adsorption of phenol on activated carbon from a simple aqueous solution. We used activated carbon manufactured from the olive stone in order to use a very abundant by-product in our region. The phenol is a model molecule of the phenolic pollutants which are strongly toxic products and known by their persistence and their aptitude for the bio-accumulation. In addition, the adsorption of phenol is a standard method used to evaluate the capacity of the activated carbon to eliminate organic molecules from water.

Our objective is to control, through an experimental study, the fixed-bed adsorption on activated carbon, and to develop on this basis a model making it possible to predict the performances and the characteristics of the bed (capacity of adsorption, breakthrough curve, depth of the adsorption zone) according to the operating conditions (initial concentration, flow rates, bed depth). In this purpose, we tested

*Corresponding author: Nouri Hanen, Research Laboratory: Engineering Process and Industrial System, National School of Engineers of Gabes, University of Gabes, Street Omar Ibn El Khattab-6029 Gabes, Tunisia

the most widely used models (Thomas, Clark, Yoon and Nelson) with a critical approach.

Experimental

Adsorbate: Phenol

The choice of phenol results from its frequency in waste water resulting from several industrial activities (Petrochemical, pharmaceutical, paper, plastic, agro-alimentary, etc). Because of its strong toxicity, the phenol appears in the category of high-risk product of water pollution.

Activated carbon

Olive stone used for preparation of granular activated carbon was obtained from local factory in Gabes, Tunisia. It was washed with hot distilled water to remove dust like pulp of olives and impurities then dried at ambient temperature.

The granular activated carbon (GAC) used in this study were produced in our laboratory by various processes.

Activated carbon produced by physical methods (GACA): In this process, there are two preparation stages: carbonization and activation. Carbonization or pyrolysis is carried out at a temperature of 600°C for 2 hours. Then the second stage, activation which is a controlled endothermic oxidation, takes place at a temperature of 700°C during 8 hours with water vapor as oxidant. Activation has as a role to free the pores and to eliminate the tarry residues blocking the fine structures. After this stage, coal acquires a significant and more accessible internal surface with a very developed porosity.

Activated carbon produced by chemical methods (GACB): In this process, the olive stone is impregnated by phosphoric acid (H_3PO_4) during 7 hours. After air drying, the semi-finished product undergoes a carbonization in an inert atmosphere at 410°C. The activated carbon obtained is then washed several times until reaching a pH of the filtrate close to 6. This condition of neutrality makes it possible to be ensured of the elimination of phosphates contained in the product. Then it is placed in a drying oven at 110°C for 24 hours.

Activated carbon produced by combined methods (GACC): This process combines the two ways already mentioned; it implies activation with the phosphoric acid like a chemical agent and the water vapor like a physical agent at the time of the thermal stage.

Table 1 contains the main characteristics of the used activated carbon.

Experimental set up

Batch adsorption: The batch adsorption was performed using water bath shakers at 40°C and constant agitation speed of 100 rpm. In this study 200 ml of phenol solution was agitated with 0.2 g of activated carbon (diameter lower than 100 μm) in 250 ml flask.

Properties	GACA	GACB	GACC
Surface BET m²/g	736.9	1285	1421
Pores Volume (cm³/g)	0.555	0.711	0.655
Micropores volume (cm³/g)	0.397	0.644	0.635
Bulk density (g/ml)	0.32	0.459	0.4
Acidic groups (meq/g)	0	4.79	1.562
Basic groups (meq/g)	1.46	0.51	0.625

Table 1: Characteristics of the activated carbon used.

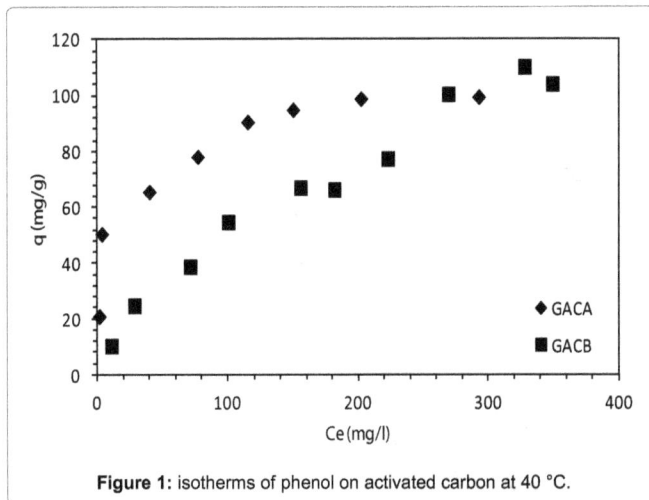

Figure 1: isotherms of phenol on activated carbon at 40 °C.

Equilibrium experiments were performed with different initial phenol concentration (20 mg/l-350 ml/g). The agitated time was 24 hours to reach equilibrium.

Phenol concentration in solution was analyzed using UV spectrophotometer (Shimadzu UV-1700) by monitoring the absorbance changes at the wavelength of maximum absorbance (λ=270 nm).

The amount of phenol onto the unit weight of the adsorbent was calculated using the following equations:

$$q = \frac{V(C_0 - C)}{m} \tag{1}$$

Where V is the solution volume in L, C_0 is the initial phenol concentration in mg/l, C is the phenol concentration at any time t in mg/l, and m is the dry weight of activated carbon in g.

Fixed bed adsorption: Fixed bed studies were conducted in a glass column (1 cm in internal diameter, length of 22 cm), packed with a known quantity of activated carbon. The phenol solution at a known concentration was continuously pumped into the column upwards at a constant temperature of 40°C. The desired flow rate was regulated with a variable peristaltic pump. Samples were collected at regular intervals and analyzed for the remaining concentration of phenol in effluent.

The effects of the following parameters were investigated: bed depth (5-20 cm), flow rate (2.2-8.4 ml/min) and initial phenol concentration (40-250 mg/l).

Results and Discussion

Equilibrium study

Equilibrium relationship between the concentration of phenol in the aqueous solution and the concentration of adsorbed phenol at a constant temperature is represented by an adsorption isotherm.

Figure 1 was the adsorption isotherm of phenol on the two activated carbons (GACA, GACB) produced from the olive stone.

Modeling of the equilibrium data has been done using the most widely used models: Langmuir and Freundlich. Figure 2 show that both of models described well the isotherms.

According to the isotherms, we found that in spite of its micro porosity and its great specific surface, the granular activated carbon produced chemically (GACB) does not adsorb as much phenol as the physical one which has a weaker specific surface (700 m²/g). This

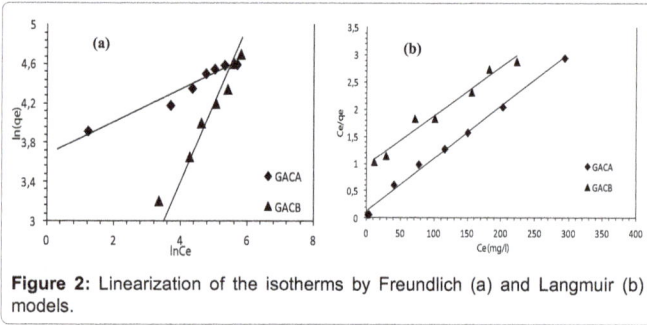

Figure 2: Linearization of the isotherms by Freundlich (a) and Langmuir (b) models.

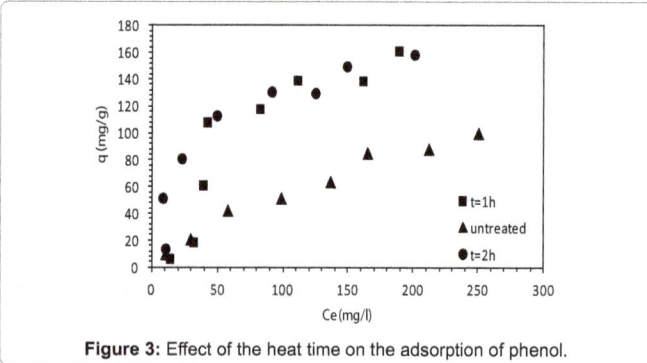

Figure 3: Effect of the heat time on the adsorption of phenol.

was explained by the basic character of the surface of the latter which ameliorates the adsorption of phenol since it is known to be a molecule of a slightly acid character. With this intention, we tried to modify the nature of the surface of the chemical activated carbon by carrying out an adequate treatment in order to improve its capacity of phenol adsorption with preserving a significant specific area.

Treatment of granular activated carbon type B

According to the bibliographical sources [18-21], we found that a simple heat treatment decreases the acid sites in favor of those basics which are favorable for the adsorption of the phenolic compound. Thus we chose this treatment in the objective to improve the alkalinity of the surface of activated carbon produced chemically (GACB). We used a cylindrical reactor placed horizontally in a muffle furnace maintained at a controlled temperature and with a continuous flow of nitrogen to preserve an inert atmosphere.

In order to determine the optimum conditions for the treatment we carried out several tests: first, we maintained the heat temperature at 700°C and we varied the duration of the treatment (1h, 2h). The isotherms obtained are illustrated in figure 3.

The significant effect of the heat treatment on the capacity of phenol's adsorption by GACB is very clear. For better understanding this phenomenon, we analyzed the functional groups of the treated one; the results are represented on figure 4.

The heat treatment reduced the quantity of the acid functions of coal's surface. The increase in the duration of the treatment decreases even more these functions. In fact, the acid groups passed from 90 %, for the raw coal; to 69 % for the treaty one along 2 hours. Where as one hour duration makes it possible to have a coal with 88% of the acid functions. Thus, we chose this processing time for the rest of the experiments. Second, we fixed the time of the treatment at two hours and we studied the influence of the temperature of treatment which

varied between 500°C and 900°C. The isotherms of phenol's adsorption are represented on figure 5.

An increase in the temperature from 500°C to 700°C induced an increase in the capacity of AC on all the interval of concentration studied. But a higher temperature (900°C) slightly degraded the performances obtained with 700°C. Consequently, we retained 700°C as a temperature of heat treatment for GACB in order to improve the relative quantity of the basic functions of surface.

The isotherms of phenol's adsorption by different AC and the treated one are represented on figure 6.

Thus, the beneficial effect of the heat treatment is quite clear since it makes it possible to increase the adsorbed capacity of GACB better than other types of coal.

Dynamic study

In order to study the dynamic behavior of the column of adsorption, we tried first to follow the evolution of the breakthrough according to

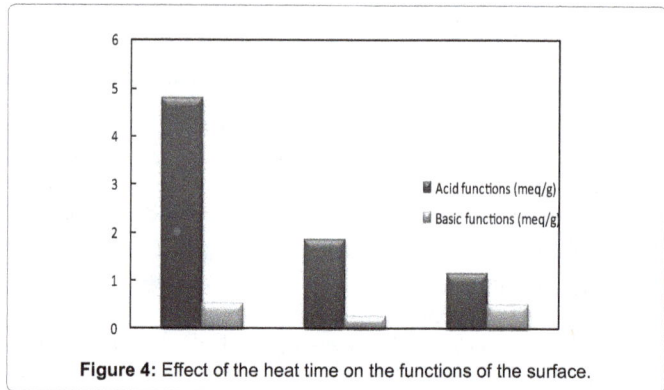

Figure 4: Effect of the heat time on the functions of the surface.

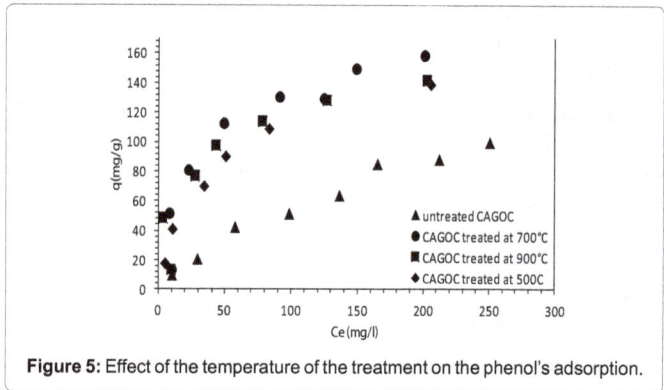

Figure 5: Effect of the temperature of the treatment on the phenol's adsorption.

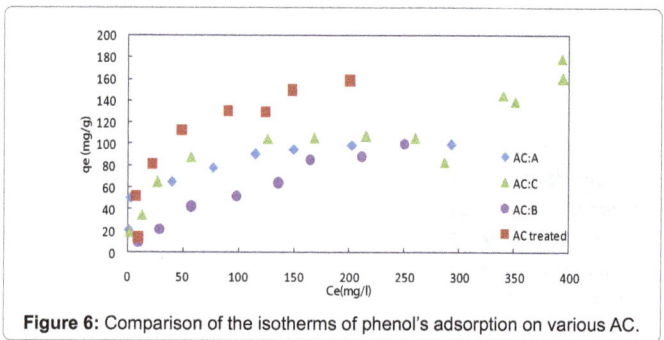

Figure 6: Comparison of the isotherms of phenol's adsorption on various AC.

the operational parameters (bed depth, initial concentration and flow rate) and to see their influence on the breakthrough time and the adsorbed quantity.

Comparison between the activated carbon: The breakthroughs of the different activated carbon were obtained for an initial concentration of 100 mg/l, a flow rate of 8.4 ml/min and a bed depth of 20 cm (figure 7).

These results come to confirm what we found in the equilibrium study: the beneficial effect of the presence of basic groups on the surface is confirmed by the breakthrough of treated activated carbon which presents the greatest value of the adsorbed quantity as well as the highest breakthrough and saturation time. It is seen clearly that the heat treatment appreciably improves adsorption of phenol on the activated carbon.

Effect of flow rate: The effect of the flow rate on the adsorption of phenol was investigated by varying the flow rate (from 2.2 to 8.4 ml/min) with a constant adsorbent bed depth of 5 cm and the initial phenol concentration of 100 mg/l, as shown by the breakthrough curve in figure 8.

As indicated in this figure, at the lowest flow rate of 2.2 ml/min, relatively higher uptake values were observed. At higher flow rates, the breakthrough occurred faster and the breakpoint time and total

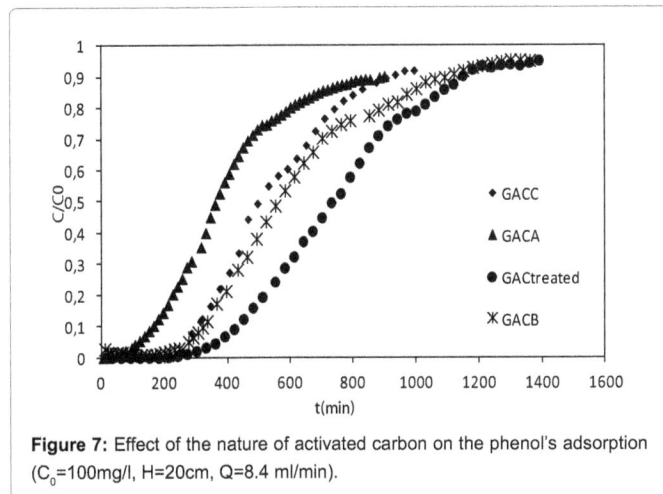

Figure 7: Effect of the nature of activated carbon on the phenol's adsorption (C_0=100mg/l, H=20cm, Q=8.4 ml/min).

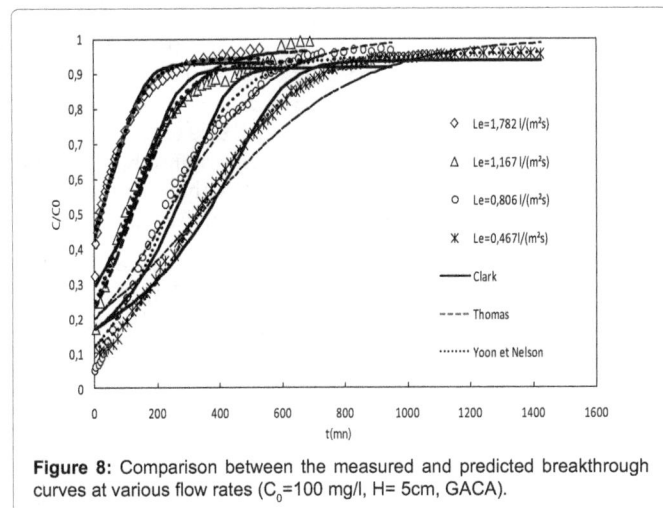

Figure 8: Comparison between the measured and predicted breakthrough curves at various flow rates (C_0=100 mg/l, H= 5cm, GACA).

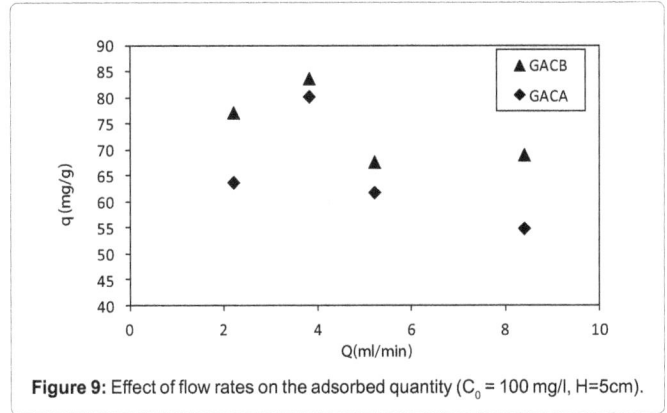

Figure 9: Effect of flow rates on the adsorbed quantity (C_0 = 100 mg/l, H=5cm).

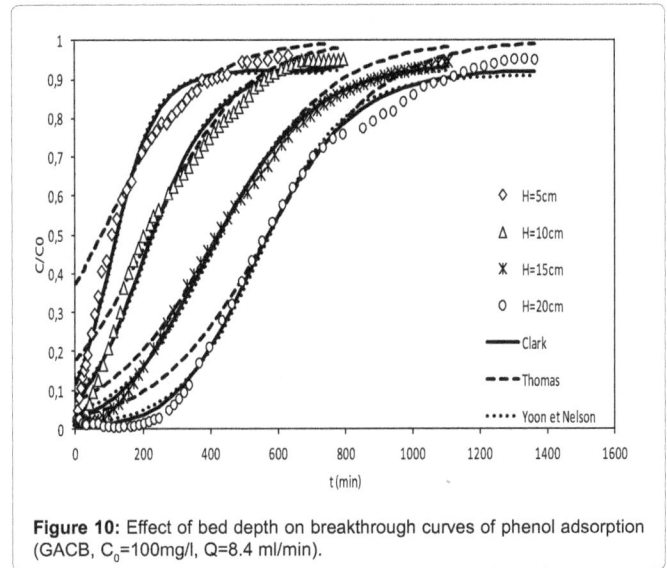

Figure 10: Effect of bed depth on breakthrough curves of phenol adsorption (GACB, C_0=100mg/l, Q=8.4 ml/min).

adsorbed quantity decreased. This behavior can be explained by insufficient residence time of the solute in the column: the residence time decrease with the increase of the flow, that leads to insufficient time for diffusion of the phenol into the pores of activated carbon and limits the number of the available active sites for adsorption, thus reducing the volume of the aqueous solution being treated and the solute left the column before equilibrium occurred.

Figure 9 shows the variation of the adsorbed quantity for different flow rates.

The increase in the flow causes a reduction of the resistance of film in the external transfer which causes an improvement in the kinetics of adsorption. But we should mention that this increase is favorable only until one value of the flow beyond which this external resistance is not any more limited. A rise in the flow beyond this value, localized in our conditions towards 3.8 ml/min, causes a fall in the adsorbed quantity since the residence time decreases and the phenol molecules do not penetrate deeply in the pores.

Effect of bed depth: The adsorption capacities of fixed bed column with bed depths of 5, 10, 15 and 20 cm were tested at a constant flow rate of 8.4 ml/min and influent concentration of 100 mg/l of phenol. The breakthrough curves were illustrated in figure 10. Increasing the height of the bed can be explained by more sites which were supplied for solute so the breakthrough time increased with bed depth.

Effect of influent concentration: The breakthrough curves obtained at different influent concentration (40-250 mg/l) and constant flow rate of 8.4 ml/min and a bed depth of 15 cm were shown in figure 11. The increase in the concentration made the breakthrough curves stiffer and decreased the breakthrough from 150 min to 48 min for the GACA. This was explained by the improvement of the diffusion of phenol in the pores of activated carbon.

The evolution of the quantity adsorbed according to the initial concentration is represented on figure 12.

The adsorbed quantity increased with the initial concentration. This is due to an increase in the gradient of concentration which is the driving force of the phenomenon of adsorption and consequently an improvement of the diffusion of the aqueous solution of the liquid phase towards the grains of activated carbon.

Breakthrough curves modeling

In order to describe the fixed bed column behavior and to scale it up for industrial applications, three models, Thomas, Clark, and Yoon-Nelson were used to fit the experimental data in the column. The predicted curves were represented by figures 8, 10 and 11.

Thomas model: Thomas model assumes plug flow behavior in the bed, and uses Langmuir isotherm for equilibrium, and second- order reversible reaction kinetics. This model is suitable for adsorption

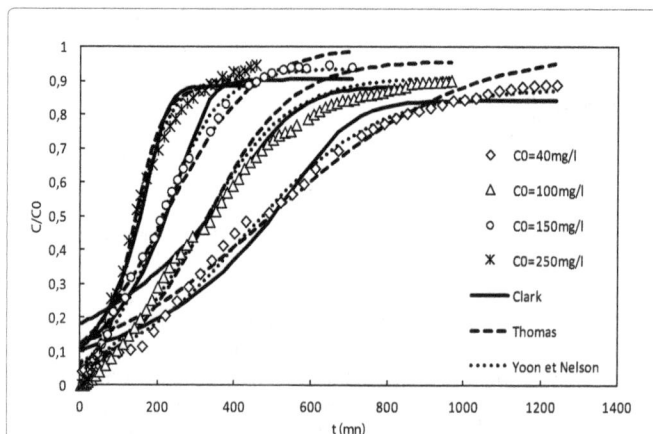

Figure 11: Comparison between the measured and predicted breakthrough curves with different models at various influent concentrations (GACA, Q=8.4 ml/min, H= 15cm).

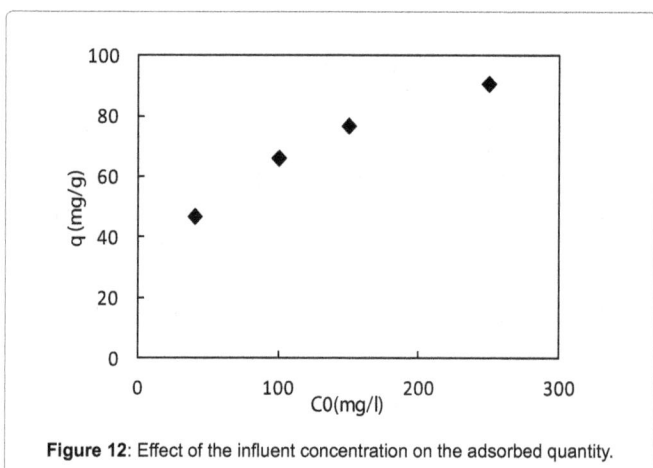

Figure 12: Effect of the influent concentration on the adsorbed quantity.

CA	Langmuir			Freundlich		
	KL (l/mg)	q (mg/g)	R²	KF	n	R²
GACA	0.071	111.11	0.994	39.57	6.06	0.948
GACB	0.004	200	0.914	1.906	1.379	0.989

Table 2: Parameters of the adsorption model for phenol adsorption onto Granular Activated Carbon.

process where the external and internal diffusion limitations are absent [22]. The Thomas model has the following form [1,15,23]:

$$\frac{C_t}{C_0} = \frac{1}{1 + \exp\left(K_{th}(q_0 x - C_0 V_{eff})/v\right)} \tag{2}$$

were K_{Th} (ml/min mg) is the Thomas rate constant, q_0 (mg/g) is the maximum solid-phase concentration of solute, x is the amount of adsorbent in the column (g), V_{eff} is the volume of effluent (ml), C_0 is the influent concentration and C_t is the effluent concentration at time t (mg/l). The values of K_{Th} and q_0 can be obtained from a plot of ln ((Ci/Ce)-1) against V_{eff} at a given flow rate and influent concentration. The results were summarized in table 2.

The values of K_{Th} and q_0 were influenced by both flow rate and influent concentration of phenol.

As the flow rate increased, the value of q_0 decreased and the Thomas rate constant K_{Th} increased. With the bed volume and influent concentration increasing, the capacity of the bed q_0 increased while K_{Th} slightly decreased with C_0 and remained constant with the bed depth.

The difference between the measured and predicted q_0 was negligible. The predicted breakthrough curves were in a good agreement with experimental data for all flow rates and influent concentration in the range of $(C/C_0)>0.3$ and it deviated for $(C/C_0)<0.3$. The result that the Thomas model did not predict the initial part of the initial part of the breakthrough curve well has also been reported by other researchers [1,18].

Clark model: This model was based on the use of a mass-transfer concept in combination with the Freundlich isotherm [24]. We used the following expression (3) where n is the Freundlich parameter and A and r are the Clark constants. Eq (3) was modified by introducing parameter "a" to take into account that the curve is not perfectly symmetrical at the point C = 0, 5 C_0.

$$\frac{C}{C_0} = \left(\frac{a}{1 + A_c.e^{-r_0 t}}\right)^{\frac{1}{n-1}} \tag{3}$$

According to figures 8, 10 and 11 it was clear that the model of Clark fit very well the totality of the breakthrough curve for the different operating conditions. However, there was a slight deviation for the low values of the flow and the concentration in the case of the GACA. The values of A and r in the Clark equation were determined by non-linear regression analysis and are shown in table 2. As both flow rates and influent concentration increased, the values of r increased.

Yoon-Nelson model: Yoon and Nelson developed a model based on the assumption that the rate of decrease in the probability of adsorption of adsorbate molecule is proportional to the probability of the adsorbate adsorption and the adsorbate breakthrough on the adsorbent [22].

The Yoon-Nelson a linearized model for a single component system is expressed as [1,25]:

$$\ln\frac{C}{C_0 - C} = K_{YN}t - \tau K_{YN} \tag{4}$$

Where K_{YN} is the rate constant (min^{-1}) and τ is the time required for 50 % adsorbate breakthrough (min).

The values of the rate constant K_{YN} increased with the flow rate and the initial concentration. Table 3 indicates that the values of τ obtained by the model are close to the experimental results and the predicted curves fitted well the experimental data. Thus, this model provided a good correlation of the effects of bed height and flow rate.

Characteristic of the adsorption zone

In order to properly design and operate fixed bed adsorption processes, we applied the concept of the Mass Transfer Zone (MTZ) developed by Mickaels for the exchanges of ions and applied by Luchkis for adsorption [26,3].

Fractional capacity (F): It determines the elimination efficiency of the granular activated carbon. It may be defined also by the ratio of the real quantity eliminated of solute on the potential capacity of elimination of carbon within the mass transfer zone. The fractional capacity is given by the ratio:

$$F = \frac{A_z}{A_{max}} = \frac{\int_{tb}^{ts}(C0 - Ci)dt}{C0(ts - tb)} \qquad (5)$$

The height of MTZ (H_z): It is the area where practically takes place all the phenomenon of adsorption. It determines the rate of elimination of the adsorbate by the adsorbent and gives indications about the diffusion resistances. The lower resistance to the transfer is, the faster the kinetics of adsorption is, the smaller the depth of the zone of transfer of matter is. The height of MTZ is given by:

$$H_Z = \frac{Hb(t_s - t_b)}{tb + F(t_s - t_b)} \qquad (6)$$

With : H_b: height of the adsorbent bed, t_s: saturation time, t_b: breakthrough time

The rate of the movement of the MTZ (U_z): The rate of the movement of the MTZ is a function of adsorption capacity of the adsorbent; it permits to calculate the rate of bed saturation. It is directly related to the height of mass transfer zone. The small is the bed depth of MTZ, the more quickly is the rate of transfer, and rapid is the saturation bed.

The rate of the movement of the mass transfer zone is given:

$$UZ = \frac{H_z}{t_z} = \frac{HZ}{t_s - t_b} = \frac{Hb}{t_s - t_f} \qquad (7)$$

With $t_z = t_s - t_b = \frac{V_s - V_b}{Q}$: the time required for the movement of the MTZ down its own length in the column.

t_f = the time required for the formation of the MTZ, it is given by: $t_f = (1-F)t_z$

The percentage of saturation of the column in the breakthrough point is:

$$\%saturation = \frac{H_b + (F - 1)H_Z}{H_b} 100 \qquad (8)$$

All these parameters were calculated: F, H_z and U_z from the experimental breakthrough curves for the GACA, for different initial concentrations, for a height of the bed H=15 cm and a flow rate Q = 8.4 ml/min.

We fixed a breakthrough concentration C_b = $0.2C_0$ and a concentration of saturation Cs = $0.8C_0$, Table 4.

Conclusion

The adsorption of phenol from aqueous solution by activated carbon produced from olive stone was investigated in a continuous-flow fixed-bed column.

The distribution of phenol between the liquid phase and solid phase was described by the Langmuir and the Freundlich models. It was seen that the sorption equilibrium data can be fitted by both models.

The breakthrough curves have been determined at various flow rates, bed depth and initial concentrations. Several models were applied to analyze experimental data and to determine the column characteristics. These models gave good approximations of experimental behavior; the Thomas model were in good agreement with the experimental data only as the relative concentration (C/C_0) higher than 0.3 while the whole breakthrough curve was well predicted by the Clark and the Yoon-Nelson models.

References

1. Aksu Z, Gönen F (2004) Biosorption of phenol by immobilized activated sludge in a continuous packed bed: prediction of breakthrough curves. Process Biochemistry 39: 599-613.

	t_z (h)	F(%)	H_z (cm)	U_z (cm/min)	%saturation
40	19.08	87	12.30	0.011	89
100	14.35	90	11.92	0.014	92
150	6.8	85	12.96	0.032	86
250	5.65	91	11.39	0.036	93

Table 3: Adsorption data for fixed bed of GACB for phenol sorption at different process conditions.

Process conditions			Models parameters						
			Thomas		Clark			Yoon -Nelson	
Q (l/min)	C_0 (mg/l)	H (cm)	K_{Th} (ml/(min mg))	q_0 (mg/g)	r (min^{-1})	A_c	a	K_{YN} (min^{-1})	τ the(min)	τ exp (min)
8.4	100	5	7 10^{-5}	31.92	0.014	1.82	0.96	0.016	112	103
5.2	100	5	5 10^{-5}	58.97	0.01	2.1	0.98	0.01	207	212
3.8	100	5	4 10^{-5}	93.54	0.005	2.57	0.98	0.006	382	352
2.2	100	5	3 10^{-5}	82.32	0.003	3.66	0.99	0.004	675	690
8.4	40	10	5 10^{-5}	41.33	0.003	2.01	0.95	0.004	524	576
8.4	100	10	7 10^{-5}	46.25	0.008	2.53	0.97	0.01	217	330
8.4	150	10	3.3 10^{-5}	114.74	0.006	2.99	0.98	0.008	307	304
8.4	250	10	2.8 10^{-5}	104.85	0.01	2.81	0.98	0.012	195	188
8.4	100	15	6 10^{-5}	59.78	0.006	4.54	0.97	0.007	410	412
8.4	100	20	6 10^{-5}	59.99	0.006	13.99	0.96	0.008	552	562

Table 4: Characteristics of the mass transfer zone.

2. Ayoob S, Gupta AK, Bhakat PB (2007b) Analysis of breakthrough developments and modeling of fixed bed adsorption system for As(V) removal from water by modified calcined bauxite (MCB). Separation and Purification Technology 52: 430-438.

3. Kundu S, Gupta AK (2005) Analysis and modeling of fixed bed column operations on As (V) removal by adsorption onto iron oxide-coated cement (IOCC). Journal of Colloid and Interface Science 290: 52-60.

4. Malkoc E, Nuhoglu Y (2006) Fixed bed studies for the sorption of chromium (VI) onto tea factory waste. Chem Eng Sci 61: 4363-4372.

5. Ayoob S, Gupta AK (2007) Sorptive response profile of an adsorbent in the defluoridation of drinking water. Chem Eng J 133: 273-281.

6. Ghorai S, Pant KK (2005) Equilibrium, kinetics and breakthrough studies for adsorption of fluoride on activated alumina. Sep Purif Technol 42: 265-271.

7. Quek SY, Al-Duri B (2007) Application of film-pore diffusion model for the adsorption of metal ions on coir in a fixed-bed column. Chemical Engineering and Processing 46: 477-485.

8. Hamdaoui O (2006) Dynamic sorption of methylene blue by cedar sawdust and crushed brick in fixed bed columns. Journal of Hazardous Materials 138: 293-303.

9. Banat F, Al-Asheh S, Al-Ahmad R, Bni-Khalid F (2006) Bench-scale and packed bed sorption of methylene blue using treated olive pomace and charcoal. Bioresource Technology 98: 3017-3025.

10. Ferrero F (2007) Dye removal by low cost adsorbents: Hazelnut shells in comparison with wood sawdust. J Hazard Mater 142: 144-152.

11. Han R, Wang Y, Zou W, Wang Y, Shi J (2007) Comparison of linear and nonlinear analysis in estimating the Thomas model parameters for methylene blue adsorption onto natural zeolite in fixed-bed column. J Hazard Mater 145: 331-335.

12. Song J, Zou W, Bian Y, Su F, Han R (2011) Adsorption characteristics of methylene blue by peanut husk in batch and column modes. Desalination 265: 119-125.

13. Nasuha N, Hameed BH (2011) Adsorption of methylene blue from aqueous solution onto NaOH-modified rejected tea. Chem Eng J 166: 783-786.

14. Soto ML, Moure A, Domı́nguez H, Parajo JC (2008) Charcoal adsorption of phenolic compounds present in distilled grape pomace. Journal of Food Engineering 84: 156-163.

15. Han R, Ding D, Xu Y, Zou W, Wang Y, et al. (2008) Use of rice husk for the adsorption of congo red from aqueous solution in column mode. Bioresource Technol 99: 2938-2946.

16. Ahmaruzzaman M, Sharma DK (2005) Adsorption of phenols from wastewater. Journal of Colloid and Interface Science 287: 14-24.

17. Alhamed YA (2009) Adsorption kinetics and performance of packed bed adsorber for phenol removal using activated carbon from dates' stones. Journal of Hazardous Materials 170: 763-770.

18. Salame II, Bandosz TJ (2003) Role of surface chemistry in adsorption of phenol on activated carbons. Journal of Colloid and Interface Science 264: 307-312.

19. Attia AA, Rashwan WE, Khedr SA (2006) Capacity of activated carbon in the removal of acid dyes subsequent to its thermal treatment. Dyes and Pigments 69: 128-136.

20. Martin-Gullon I, Marco-Lozar JP, Cazorla-Amoros D, Linares-Solano A (2004) Analysis of the microporosity shrinkage upon thermal post-treatment of H3PO4 activated carbons. Carbon 42: 1339-1343.

21. Merzougui Z, Addoun F (2008) Effect of oxidant treatment of date pit activated carbons application to the treatment of waters. Desalination 222: 394-403.

22. Ahmad AA, Hameed BH (2010) Fixed-bed adsorption of reactive azo dye onto granular activated carbon prepared from waste. Journal of Hazardous Materials 175: 298-303.

23. Thomas HC (1944) Heterogeneous Ion Exchange in a Flowing System. J Am Chem Soc 66: 1664-1666.

24. Clark RM (1987) Evaluating the cost and performance of field-scale granular activated carbon systems. Environ Sci Technol 21: 573-580.

25. Yoon YH, Nelson JH (1984) Application of gas adsorption kinetics. I.A Theoretical Model for respirator cartridge service life. Am Ind Hyg Assoc J 45: 509-516.

26. Namane A, Hellal A (2006) The dynamic adsorption characteristics of phenol by granular activated carbon. Journal of Hazardous Materials 137: 618-625.

22

Representative Sampling of Suspended Particulate Materials in Horizontal Ducted Flow: Evaluating the Prototype 'EF-sampler'

Claas Wagner[1]* and Kim H Esbensen[1,2]

[1]ACABS Research Group, Aalborg University, campus Esbjerg (AAUE), Denmark
[2]Geological Survey of Denmark and Greenland, Copenhagen, Denmark

Abstract

The 'EF-sampler' is a newly developed sampler for suspended particulate materials in horizontal pneumatic conveying systems, designed for maximum possible compliance with the Theory of Sampling (TOS). Hitherto no sampler for this deployment location exists on the market that ensures representative samples as defined by TOS. Because of confinement of the pressurised ducted flow and because of gravitative and flow segregation, unbiased sampling constitutes a serious challenge. In addition to the primary demand for representativeness, interference with the material flow needs to be minimized in order to prevent clogging effects and/or possible pressure surges. We here disclose all design principles of the 'EF-sampler' and validate a 1/3-scale prototype in a pneumatic test facility by presenting our first test campaign results. Testing focuses on assessing sampling representativeness (accuracy and precision) using wheat flour and pulverized alumina as the major test materials, both spiked with LDPE plastic pellets in the role as trace constituents, extraneous material or contaminants. Input pellet concentration levels served as nominal reference values for the accuracy evaluation. Test parameters include airflow rate, sample LDPE pellet concentration and different cross-cutting sampler velocities. Results show that the patented EF-sampler prototype enables to extract fit-for-purpose samples with a relative inaccuracy <5% for the stated test materials, which is highly acceptable for this most-difficult deployment context. Sampler velocity and especially the material flow dilution status impact the accuracy of sample extraction, while precision remains constantly good for all test conditions. The prototype EF-sampler is not a universal sampler, since it is designed to require situation-dependant adjustments based on specific material heterogeneity and flow regime characteristics. However, the first test campaign results on two widely different materials show conclusively that it accommodates a wide field of potential applicability for many similar types of materials.

Keywords: Horizontal sampler; Pressurised ducted flow; Representative sampling; Theory of sampling (TOS); Pneumatic conveying; Suspended particulate materials

Introduction

Pneumatic conveying systems are widely used in processing industries, i.e. systems that transport suspended aggregates/particulate matter forced by an air or gas stream through horizontal and vertical ducts. The advantages of pneumatic transportation compared to alternative mechanical conveying systems, e.g. conveyer belts, are potential economic benefits and higher flexibility in terms of rerouting or expansion, and especially complete enclosure of the material of interest. This is particularly important when dealing with pulverized materials in order to prevent material loss. Furthermore changes of the ducted material, e.g. moisture level, can be minimized by confinement.

In many cases it is important to have exact knowledge of the material properties of the material during pneumatic conveying. Reliable quality assurance of the transported material therefore makes it a requirement to be able to extract representative samples as defined by the Theory of Sampling (TOS). Pierre Gy's TOS Sampling Theory and Practice (STP) is the only comprehensive framework that allows a profound analysis of all sampling equipment, methods and procedures; this has therefore been used as the backbone for design, development, implementation and evaluation of the present sampler.

Pneumatic conveying systems are challenging, since the pressurized system cannot be arbitrarily intersected for sample extracting due to the risk of pressure loss or external discharge of material. Furthermore the sampling operation must not constrict the material flow in order to minimize the risk of clogging and/or pressure surges. Another important adverse factor is the fact that ducted horizontal flow causes significant vertical and sometimes also radial flow segregation. The prime objective in order to gain a representative sample is to overcome these adverse effects by an appropriate designing for a sampler for this specific process deployment.

In this study we describe a newly developed sampler for horizontal pneumatic conveying system, the "EF-sampler", which is in full compliance with the stringent principles laid down by TOS as possible.

A brief introduction of the most important aspects of TOS relevant for the design is given in the next paragraph, followed by a detailed description of the resulting design principles. This is followed by a discussion on the experimental test design used to validate the sampler while the main results are presented and discussed in last few paragraphs.

Representative Sampling–The Theory of Sampling (TOS)

Almost every measurement involves the process of taking samples. It is often questionable, indeed often undocumented if the extracted

*Corresponding author: Claas Wagner, ACABS Research Group, Aalborg University, Campus Esbjerg (AAUE), Denmark

samples are truly representative, or whether samples are in reality just 'specimens', which is a TOS' term for a non-representative sub-part of a lot; such specimens are uninteresting lot extractions.

All naturally occurring materials are heterogeneous, caused by compositional differences as well as grouping and segregation of the material in the lot. The heterogeneity phenomenon makes sampling far from trivial and requires solid knowledge about heterogeneity, and especially how to counteract its effects in the sampling process. Since more than 60 years, Pierre Gy's Theory of Sampling (TOS) has reigned as the only complete theoretical and practical framework for representative sampling. In particular, TOS shows how to sample in an unbiased fashion ("correct sampling") which is the prerequisite for representative sampling from all sorts of materials and lot types. TOS' principles demand an equal probability of increments of the lot

Terms of TOS	Definition
Sample	Correctly extracted material from the lot, which only originates from a qualified sampling process ("sampling correctness").
Composite sample	Aggregation of several increments – a composite sample constitutes "physical averaging".
Specimen	A 'sample' that cannot be documented to be representative.
Increment	Correctly delineated, materialised sampling units of the lot. Composite samples result from an increment aggregation process.
Fragment	Smallest separable unit of the lot, e.g. mineral grain, kernel, biological cell etc. that is not affected by the sampling process itself. By naming the smallest unit-of-interest a fragment, TOS allows to treat even the situation in which the sampling process results in fragmentation of (some) of the original units.
Lot	Sampling target, e.g. truck load, railroad car, ship's cargo, batch etc. Lot refers both to the physical, geometrical form as well as the physic-chemical characteristics of the material being subject to sampling. Lots can be either stationary or dynamic (moving).
Lot dimensionality	TOS distinguishes between 0-, 1-, 2- and 3-dimensional lots. A 0-dimensional lot can be manipulated (forcefully mixed, moved etc.) in its entirety without undue efforts.
Scale	Heterogeneity, and counter-acting sampling efficiency, is influential at all scales from increment to lot. Correct sampling is scale-invariant, i.e. the same principles apply to all relevant scales in the sampling pathway.
Heterogeneity	Heterogeneity is the prime characterisation of all naturally occurring materials, including industrial lots. Heterogeneity manifests itself at all scales related to sampling for nearly all lot and material types. The only exception is uniform materials[1], which however are such rare occurrences that no generalisation w.r.t. general sampling can be made here from.
Sampling correctness	Elimination of sampling bias, by correct design, performance and maintenance of the sampling process/equipment. In the event of sampling correctness, only sampling precision remains, which is a much easier issue to control for within specified limits.
Representativeness	Representativeness implies both correctness as well as a sufficiently small sampling reproducibility (sampling precision).

Table 1: Fundamental concepts and definitions in Theory of Sampling. See also literature cited immediately above for a more fully developed introduction to TOS and TSP.

[1]Uniform materials: Materials with a repeated (correct) sampling reproducibility lower than 2% (or lower still, various definitions pertain to different sciences and technology fields). Such materials do only very rarely occur naturally however (exception gasses and infinitely diluted solutions etc.).

to be sampled, which is the critical guarantee for non-biased sampling. TOS' derived Sampling Theory and Practice (STP) furthermore enables to analyse sampling methods as well as sampling procedures and equipment types with respect to the principle of representativeness. Non-representative samples are primarily caused by so-called 'incorrect sampling errors', which must be identified and eliminated, or at least be reduced significantly. Such sampling process errors will unavoidably lead to an inconstant sampling-bias, which cannot be corrected for, leading to uncontrolled inflation of the total sampling error. For a complete introduction to TOS the reader is referred to the following selected literature [1-7]. The main definitions of TOS, as used in the following, are provided in the following (Table 1).

In the development phase of any new sampler, preventing incorrect sampling errors by correct design principles must have the absolutely highest priority. Unfortunately not all OEMs of sampling equipment respect these principles, and thus do not necessarily manufacture bias-generating sampling equipment, a situation which is also partly caused by misleading and/or incomplete sampling standards [8]. For a better understanding of the design principles of the presented horizontal sampler, a brief of the main principles of TOS is given below.

The critical criterion for representativeness is expressed by the 'Fundamental Sampling Principle (FSP)', stating that all fragments (grains, particles) in the lot must have the same none-zero probability of ending up in the final sample [1]. This implies that elements not belonging to the material lot must have a zero probability of being extracted. FSP applies identically at higher scale dimensions, in particular governing extraction of increments, which are the lot volume elements of paramount interest for practical sampling. The final sample, termed a composite sample, should consist of an appropriate number of increments, which is related to the effective heterogeneity of the lot material. It is an often disregarded or fully unknown fact that the number of increments needed to counteract a specific material heterogeneity is not related to the size (mass) of the lot, but solely to the magnitude of the heterogeneity (sic). A recent comprehensive illustration of this issue can be found in [9-12].

In contrast to unitary sampling operations ('grab sampling'), which disobey the FSP and therefore never can achieve sampling correctness and therefore neither representativeness, composite sampling actively counteracts the inherent lot heterogeneity.

In contrast to many sampling standards and guides specifying measurement uncertainty, TOS defines the term 'representative' explicitly with full theoretical and practical rigour. According to TOS, sampling processes can be considered as representative if, and only if, samples are extracted by procedures, which are both 'accurate' and 'precise' [1]. Accuracy of a sampling process is achieved when the average sampling error equals zero or is effectively confined to be below a predetermined acceptable low value; otherwise the sampling process is biased. Likewise a sampling process can be rated as precise if the variance of the sampling error is below a predetermined acceptable value. Mathematical expressions of the defined terms can be found in the TOS literature.

The representativeness of a sampling process can be compromised by the effects of several types of sampling errors. In TOS these error sources are subdivided in 'incorrect sampling errors' and 'correct sampling errors' (and two, much more easily handled process sampling errors). The perhaps paradoxical term, 'correct sampling errors' expresses the situation that these errors occur even when the sampling process itself is 'correct'. The following figure gives a schematic

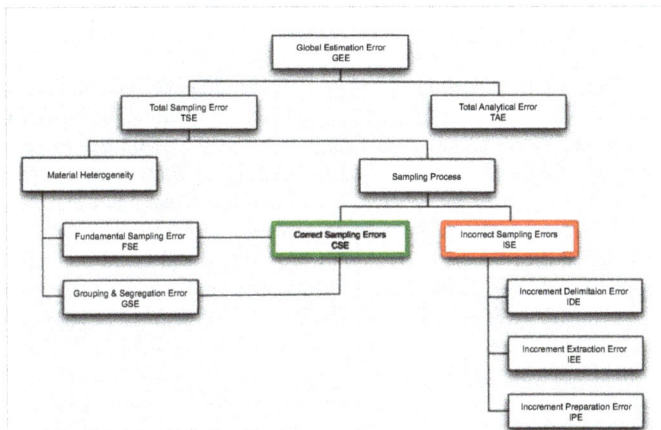

Figure 1: Relation of TOS' basic five sampling errors in stationary and process sampling situations, highlighting *correct* and *incorrect* sampling errors. Source: Wagner, Esbensen 2012.

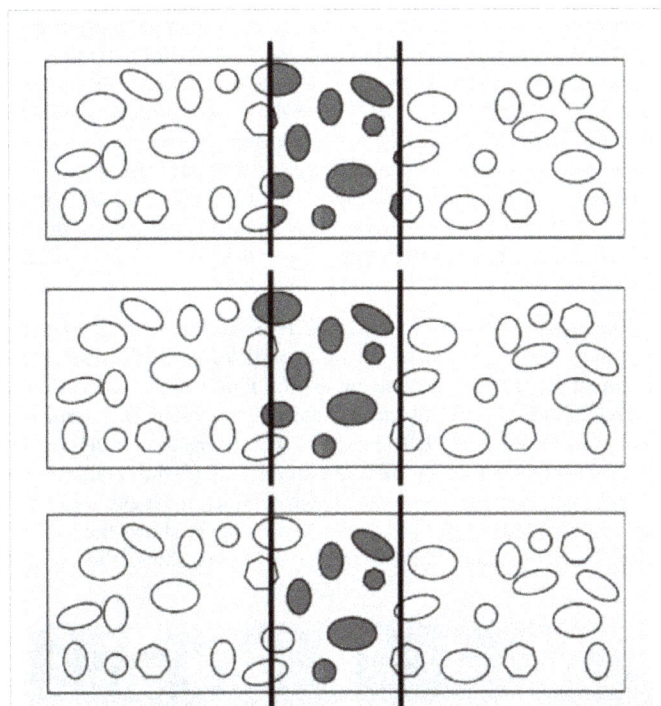

Figure 2: Illustration of the centre-of-gravity rule with grey shadings demonstrating material, which must end up in the final sample, lest IEE occur. Upper figure: correct theoretical extraction. Center: correct practical extraction. Bottom picture: *incorrect* extraction, disobeying the centre-of-gravity rule. Source: Gy, 1993, Pitard, 1993; Smith, 2001; Petersen et al., 2004.

overview of TOS' classification of the basic five sampling errors for stationary lot sampling. These are also the five major influencing errors occurring in process sampling.

The sum of the effects of all error sources is termed 'Global Estimation Error' (GEE), consisting of the 'Total Sampling Error' (TSE) and the 'Total Analytical Error' (TAE). Uncertainties of analytical results are expressed as the variance of the TAE, while all other error sources are summed up under the term 'Total Sampling Error'. It is essential to notice that for significantly heterogeneous materials TAE is nearly always much smaller than the sum of all sampling errors (up to 10-100 times smaller), while many guidelines and even some standards

dealing with "uncertainty estimation" still entirely focus on TAE as the only error source [7].

The schematic overview in (Figure 1) shows that both the material heterogeneity and the sampling process can cause these types of sampling errors to occur. The 'Correct Sampling Errors' (CSE) comprise the Fundamental Sampling Error (FSE) and the Grouping & Segregation Error (GSE), both caused by material heterogeneity. TOS differentiates between constitutional heterogeneity (CH) and distributional heterogeneity (DH). CH represents the component of heterogeneity, which depends on the physical and/or chemical differences between individual fragments in the lot material, causing the Fundamental Sampling Error. The constitutional heterogeneity increases when the compositional difference between fragments increases. FSE can only be reduced (but never completely eliminated)

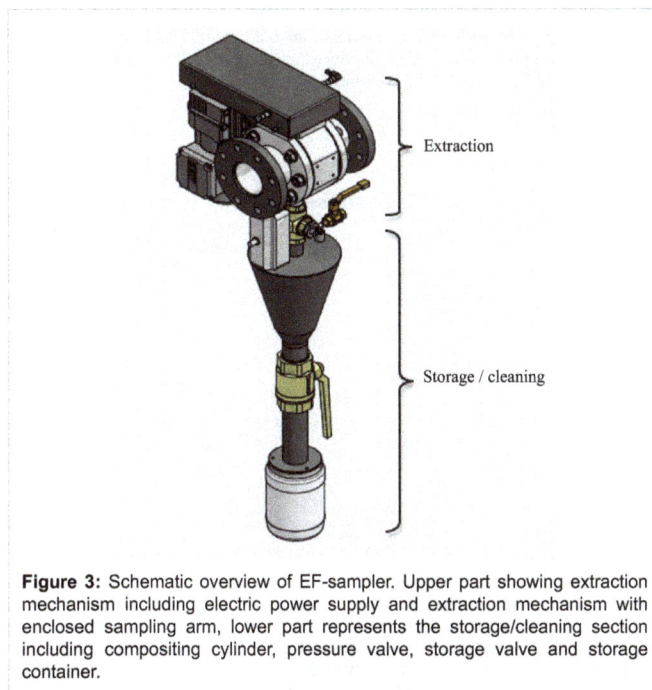

Figure 3: Schematic overview of EF-sampler. Upper part showing extraction mechanism including electric power supply and extraction mechanism with enclosed sampling arm, lower part represents the storage/cleaning section including compositing cylinder, pressure valve, storage valve and storage container.

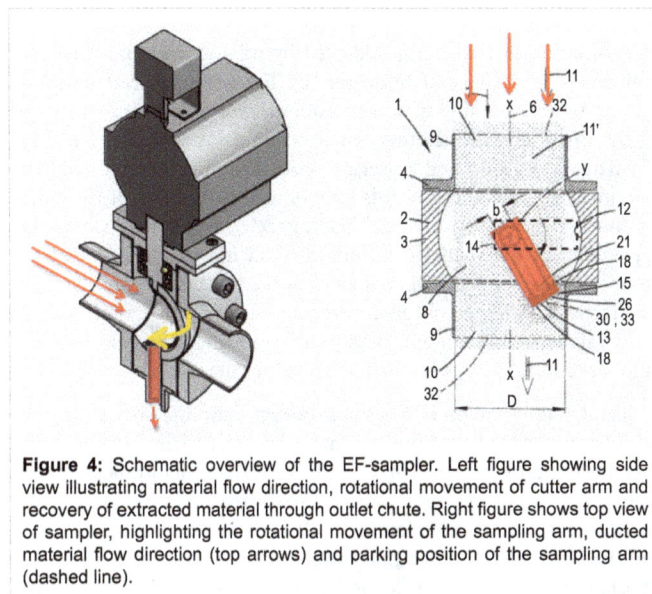

Figure 4: Schematic overview of the EF-sampler. Left figure showing side view illustrating material flow direction, rotational movement of cutter arm and recovery of extracted material through outlet chute. Right figure shows top view of sampler, highlighting the rotational movement of the sampling arm, ducted material flow direction (top arrows) and parking position of the sampling arm (dashed line).

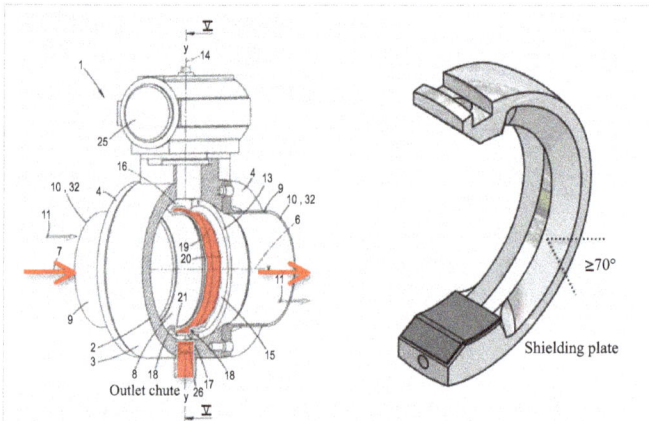

Figure 5: Side view of EF-sampler (left figure), highlighting cutter arm, material flow direction and outlet chute for material extraction. Right figure depicts details of the replacable sampling arm with inclination of cutter blade edges and shielding plate for the extraction opening.

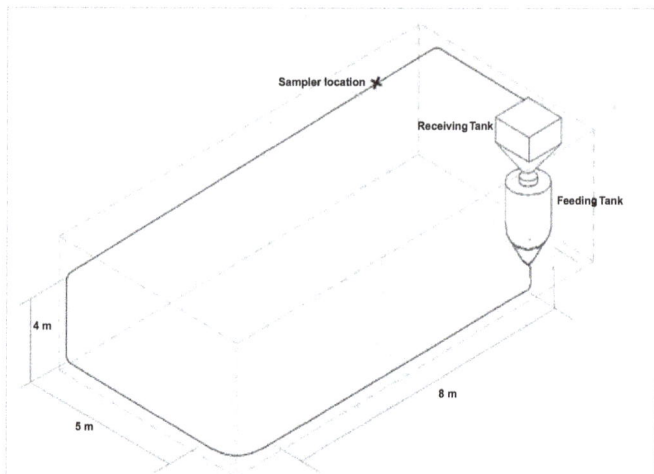

Figure 6: Schematic drawing of pneumatic transportation system (ca. 30m) including sampler location (marked with 'X'), receiving tank and feeding tank (volume ca. 1,5m3). Rotational feeder located below feeding tank is not shown. Source: POSTEC (modified by the current authors).

never constant and therefore cannot subject to the standard statistical "bias correction".

The 'Increment Delimitation Error' (IDE) can occur in connection with delineating the increments for physical extraction. In order to ensure that the Fundamental Sampling Principle is obeyed (equal likelihood for all fragments in the lot of ending up in the sample), IDE can only be avoided by ensuring that the geometrical delineation of the increment completely covers the relevant dimensions of the lot. In TOS lot dimensionality is not only related to the physical geometry of the lot only (e.g. material on conveyer belts or in pipes, stockpiles etc.) but also refers to the operative number of dimension that are 'covered' during the sampling process. Thus TOS considers 1-, 2-, and 3-dimensional sampling lots, whereas the special case of 0-dimensional refers to a lot that can be effectively manipulated: moved, mixed and sampled with complete correctness. Ideally every 2-D and 3-D lot should be transformed to a 1-dimensional sampling situation [2,1,7], i.e. into a sampling situation for which one dimension in space dominates (e.g. process streams, pipelines, conveyer belts etc.). This configuration, i.e. process sampling facilitates correct delineation of increments consisting of the material from the complete depth and width of the source stream, effectively reducing the lot heterogeneity to one dimension-the longitudinal dimension of the material flow direction. Horizontal pipe sections, as in the present context of the new EF-sampler, constitute typical one-dimensional lot examples. A correct delineation of increments can only be achieved by extracting a complete cross-sectional slice of the material stream with constant width, fully reproducible over time. A prerequisite is that the cutting planes define the increment sides, must be strictly parallel, perpendicular to the material flow.

The second incorrect sampling error is termed 'Increment Extraction Error' (IEE), occurs if/when particles inside the delineated increment in fact do not end up in the final sample. This is also referred to as the 'centre-of-gravity rule' in TOS stating that particles/fragments, which have their centre of gravity inside the delineated increment, must end up in the final sample [1,2]. (Figure 2) illustrates this rule; grey shading represents material, which must end up in the final samples, lest an IEE occurs. The upper figure shows the correct theoretical extraction, whereas the centre figure shows the correct

by comminution (crushing), followed by mixing, meaning that FSE is always present to a certain degree [1]. The second correct sampling error is termed 'Grouping & Segregation Error' (GSE) and is caused by the distributional heterogeneity, meaning the inherent tendency of particles to group and segregate at scales commensurate with the increment volume and upwards only limited by the size of the entire lot. Since it is the distributional heterogeneity, which basically is also dependent on the spatial distribution of all individual fragments or groups of fragments (i.e. increments) in the lot, that causes the GSE, an effective counteraction process for minimizing this error source is either mixing before sampling and/or using a higher number of smaller increment volume, together better "covering" the lot volume [13,6,7].

Contrasting FSE and GSE, the incorrect sampling errors can, and must by all means be minimized, indeed preferentially be eliminated from the sampling process - of course this is an issue of foremost interest in the design phase of a new sampler. These three bias-generating errors (IDE- Increment Delimitation Error, IEE-Increment Extraction Error and IPE- Increment Preparation Error) cannot be corrected for by any a posteori procedures, statistics or data analysis, since a sampling bias is

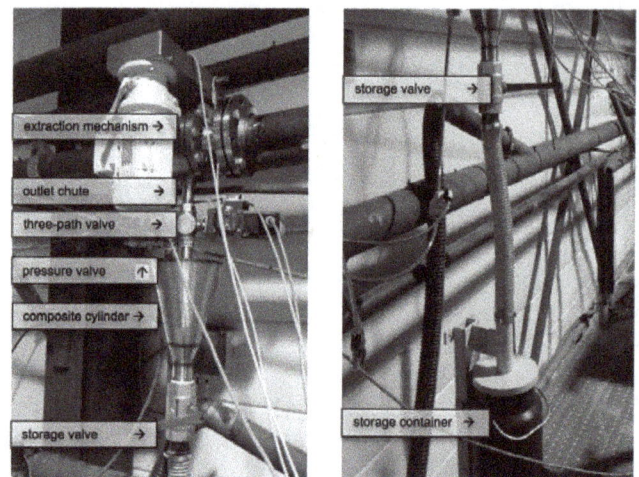

Figure 7: E-F-sampler installed at test facility. Extraction and upper storage section (left) including electric power supply, extraction mechanism and compositing cylinder. Right: close-up of lower storage section showing storage valve and storage container.

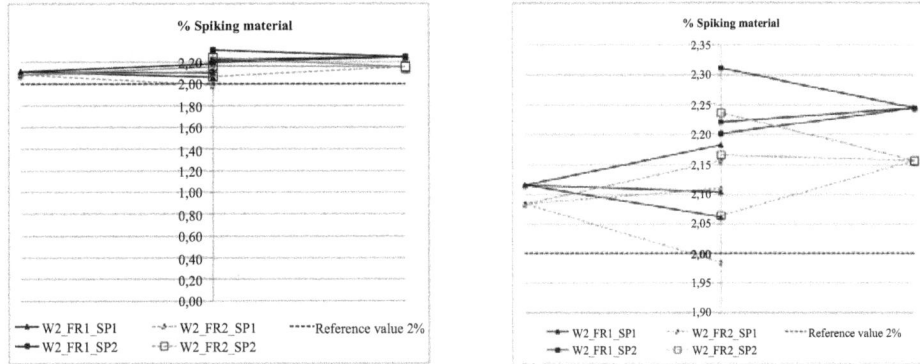

Figure 8: Sampling results of wheat flour with 2% spiking concentration, comparing SP1, SP2, FR1 and FR2 – 3 repetition rounds (left). Close-up of variations of sampling results (right). Horizontal stippled line represents the nominal reference value.

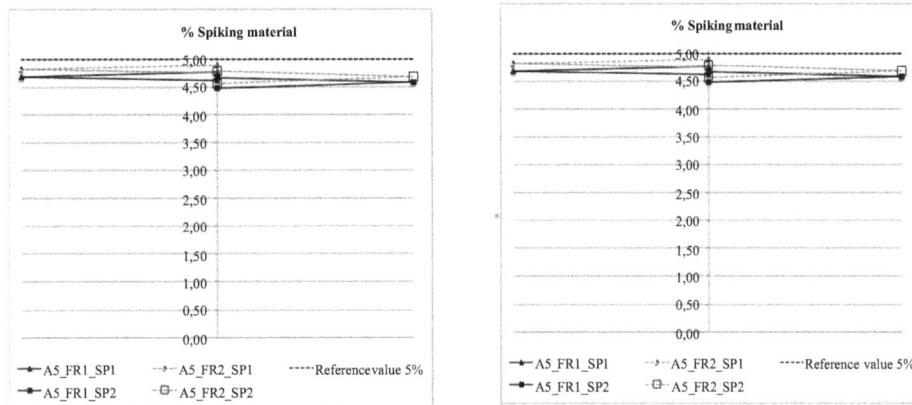

Figure 9: Sampling results of alumina with 5% spiking concentration, comparing SP1, SP2, FR1 and FR2 – 2 repetition rounds (left). Close-up of variations of sampling results (right). Horizontal stippled line represents the nominal reference value.

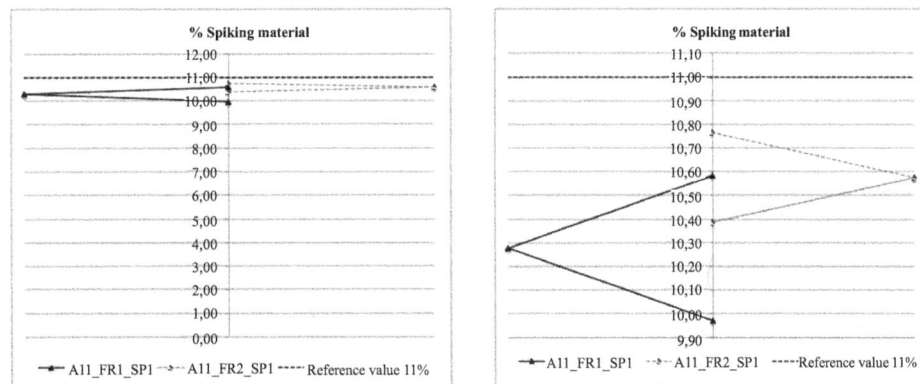

Figure 10: Sampling results of alumina with 11% spiking concentration, comparing SP1, FR1 and FR2 – 2 repetition rounds (left). Close-up of variations of sampling results (right). Horizontal stippled line represents the nominal reference value.

practical extraction. The bottom figure gives an example of incorrect extraction, disobeying the centre-of-gravity rule and therefore causing an IEE. Sufficient depth and volume of the sampler, correct inclination of cutter blade sides for preventing particles from climbing up the edges into the sample volume, as well as a limited, predefined cross-cutting velocity are factors that effectively can prevent IEE and which must therefore be carefully considered in the sampler design phase. Examples of the effects causing IEEs can be

1) Particles bouncing off the sampling tool edges

2) Particles ending up in the final sample, which do not belong to the delineated increment or

3) Fine particles that are blown away before extraction. The latter requires that the sampling extraction mechanism (e.g. cutter) is fully closed preventing loss of material. Since the material, which is interacting with the extraction cutter is of varying composition and

disposition (lot heterogeneity); IEE is an intermittent, irregularly varying error.

The third IEE is termed 'Increment Preparation Error' (IPE). IPE may occur when samples or increments are modified, in whatever fashion, after extraction, including alterations like evaporation, moisture absorption, loss of material as well as deliberate manipulation as fraud or sabotage. To limit the occurrence of a potential IPE appropriate (correct) sample handling and laboratory protocols are imperative, regulating i.e. sealing, storage and documentation of final samples. IPE is a type of sampling error, which can be brought under complete control, only depending on knowledge, willingness and diligence.

Pitard [2] contains a comprehensive, and perfunctory, survey and many in-depth examples of the issues involved regarding all ISE and the critical ways and means to counteract their effects for both stationary as well as dynamic lots.

In many cases elimination of incorrect sampling errors demands only minor but decisive changes in the sampling process, while other incorrect sampling errors are caused by a faulty design of the sampler or the sampling system itself. This study also serves the purpose to point out the importance of integrating TOS in the design phase of new samplers and shows an example of how this can be achieved for a very challenging deployment scenario.

Design Principles of EF-Sampler

The new horizontal sampler, the 'E-F-sampler', is a fully operational 1/3-scale version constructed to fit into horizontal pipe sections with a diameter of 76 mm (3-inches). The fundamental design principles of the sampling mechanism of a preliminary prototype were presented and preliminary discussed at the 5th World Conference on Sampling and Blending [14]. In the present work all design principles of the final scale-model sampler version are disclosed, which have now been verified in a dedicated test campaign.

Even though many industries use horizontal pneumatic transportation systems, no satisfactory general sampler exists on the market for the one-dimensional case of forcefully ducted horizontal material streams. As one example, the international power industry is currently converting coal-fired power plants to be able to use biomass as fuel, especially in CHP installations. In this endeavour transportation of various pulverized biomass-derived fuel types from macerating mills to the combustion chamber are carried out via pneumatic transportation. Reliable on-line information about particle size, moisture and sometimes also for a few other chemical/physical characteristics are often of critical importance for the combustion efficiency and therefore the economy of the power plant.

New industry standards currently being developed stipulate that sampling at this process position must be able to document a relative accuracy below 5%. This limiting demand has naturally served as the ultimate guideline for the development of the E-F-sampler.

While sampling in vertical flow streams, especially falling streams, is a relatively easy task when applying TOS correctly, horizontally ducted flows moved under a confining pressure present a much more difficult sampling challenge, indeed for many decades considered an (almost) impossible task. Firstly, all horizontally transported material streams must per force cause vertical segregation. Counteracting this constant, potentially severe segregation possibility must therefore be one of the prime design objectives for the EF-sampler. Furthermore

pneumatic transportation systems cannot be arbitrarily intersected for extracting increments, due to the risk of pressure loss and unwanted discharge of material. It is also important that the sampling process itself does not constrict the transportation flux in order minimize the risk of pressure surges or clogging of pipes.

In the near-future design phase, the current 1/-scale EF-sampler as presented here will be scaled up to fit standardized coal pipeline diameters and installed in a converted coal power plant now utilising biomass. A comprehensive long-term experiment will be set up to evaluate the behaviour of the sampler in this full-scale industrial environment.

The following (Figures 3, 4 and 5) show schematic overviews of the EF-sampler and its two main functioning sections: extraction and cleaning/storage.

The extraction mechanism consists of a scythe-shaped sampling arm, which rotates 180 degrees through the ducted flow stream around a vertical axis, as shown in the top view of (Figure 4). The direction of the sampling arm movement can be adjusted according to the material flow, so that both prograde and retrograde movements are possible (not necessary for many fixed industrial installations, but interesting for equally challenging other, propriety application scenarios under parallel development). An electric power supply with sufficient over-capacity has been installed to facilitate a constant rotational velocity of the sampling arm, no matter what flux encountered. By using this efficient power source, acceleration and deceleration time of the engine and sampling arm respectively have also been minimized. Two sensors on each parking side control the correct resting position of the sampling arm. In the unlikely case that the sampling arm gets stuck in the material flow, the position sensors detect this error immediately and stop the sampling operation so that no further material is extracted and remedial actions can be initiated.

To clarify the half-circular movement of the sampling arm, the material flow direction has been marked with arrows in (Figure 4 and 5). The grey box-shaped area on top of the left (Figure 4) represents the power supply in the first prototype, which has since been modified in the final prototype and is now facing downwards (compare Figure 3). The dashed rectangle in the left top view of (Figure 4) depicts the parking position of the sampling arm, i.e. the position of the sampling arm while being inactive. Due to a complete 180-degree movement through the ducted flow for each increment extraction step, the sampling arm can rest also on the opposite side until moving back through the material flow to its initial position. When in either of these symmetrical parking positions the sampling arm is not interfering with the transportation flux. It has been a crucial requirement that the sampling arm should not constrict the material flow, neither while sampling nor while resting in the parking position. By this requirement the risk of clogging of material as well as for pressure surges has been minimized.

These general issues will to some degree always be dependent upon the specific material being transported and its aggregate characteristics. As an example, it is obvious that a generally fibrous material will behave differently from a generally grain-aggregate material, the latter being very much easier to sample than the former. It has been clear from the outset of the design that a completely general sampler could not, indeed should not, be realised. However, many related types of material should be amendable for sampling with only standard changes and modification regarding the intrinsic dimensions of the scythe arm, e.g.

cutter width, depth etc. Also the cross-cutting velocity can be changed without any effort via a simple computer command.

The sampling arm has been designed with the primary objective of reliably sampling the complete vertical segregating slice of the material stream, be this segregation minor, significant, or severe. A detailed drawing of the sampling arm design features is shown in (Figure 5, right). The sampling arm consists of a u-shaped "box" with two parallel sidewalls, termed the 'cutter' hereafter. Depending on the size distribution characteristics of the material to be sampled, the opening width of the cutter must accommodate the requirement of being at least three times the nominal top diameter of the ducted material in order to prevent clogging [1,2]. Furthermore the cutter must have a sufficient volume (depth) depending on sampling speed and material flow rate, to completely eliminate any risk for over-spilling of material during the sampling operation. The outer angle of the two parallel cutter blade tips should have an angle of at least 70 degrees to prevent material not belonging to the delineated increment from 'climbing' up along the edges into the cutter [2,15]. The cutter blades of the sampling arm are designed to be analogous to parallel cutter blades on conventional cross-stream samplers, which have been proven to delineate increments correctly, ibid. Deviations from these design requirements would lead to both Increment Delimitation Errors (IDE) and Increment Extraction Errors (IEE), therefore contributing to an inconstant sampling bias.

The downward facing outlet chute of the sampling arm (rectangle marked '26' in (Figure 5, left)) is used for isokinetic extraction of the material captured in the cutter arm into the storage section of the sampler. In order to prevent material not belonging to the delineated increment of ending up in the sample during the sampling operation, the extraction opening is covered with a shielding plate when in the parking position (Figure 5, right).

Beneath the outlet chute a three-way valve is installed (Figure 7). During the active sampling operation the valve is opened, so that the delineated material can fall into the upper storage cylinder, termed the 'compositing cylinder' (Figure 3 and 7). While the sampler is inactive, in the parking position, the valve is closed, preventing material falling through this extraction tube into the composite cylinder. The third valve position is activated before the cutter arm can be cleared to start its active sampling movement. In this position, pressurized air is blown upwards through the outlet chute into the sampling arm, facilitating removal of unwanted material in the cutter before taking the next increment. After the required increments has been extracted and collected in the composite cylinder, the three-path valve is again set to the closed position and the sampling cycle is finished.

Before the storage valve, which is situated below the composite cylinder (Figure 7), can be opened in order for the material to fall from the compositing cylinder into the final storage container, the pressure in the composite cylinder has to be equalised to ambient pressure. For this purpose an additional valve with a filter is installed on the top of the storage cylinder, allowing to reduce the remaining pressure in the compositing cylinder without loosing fines. The final storage cylinder has been designed in such way that it can be easily exchanged and sealed directly after removal.

A PLS steering unit controls the entire sampling process. The number of increments to be used for composite sampling, rotation/cutting speed, pause interval, as well as cleaning duration must be defined before the sampling operation starts. Each of these parameters

can be separately set in a predefined range by a software user interface. After starting the sampling unit, the cleaning mechanism is always activated first in the predefined time period, followed by the opening of the three-path valve, which commences the actual sampling operation. Once the position sensor detects that the sampling arm has completed a 180-degree movement, the three-path valve below the outlet chute is closed and the predefined pause interval starts. This course of action (cleaning, sampling, and pausing) is repeated until the required number of increments has been extracted.

Estimating the required number of increments to be composited is a critical success factor for any process sampling operation, described many places in the relevant literature, see general process sampling references below. Most recently this issue was detailed in Esbensen et al., Minkkinen et al. and Esbensen et al. [10,11,12]. For significantly heterogeneous materials the authors arrive at appropriate number of increments between 42 and 100. In the prospective industrial setting the driver will always be optimal accuracy and precision, so the present testing phases resolved to use 75 increments, which is a reasonable projection from the full-scale industrial context, in which a representative sample will be required with a minimum resolution of 5 minutes (Vattenfall DK, DONG ENERGY pers. com.).

Experimental Test Design

All testing was carried out on a semi-industrial pneumatic transportation test system situated at the Department of Powder Science and Technology, Tel-Tek in Porsgrunn, Norway. A schematic overview of the full round pneumatic transportation system is depicted in (Figure 6), comprising a total distance of approximately 30 meters. The scale-version of the presented EF-sampler has been designed to fit the 3-inch piping system with pipe wall thickness of 3mm.

Before each transportation round, the material is filled from the receiving tank into the feeding tank. Below the feeding tank a rotational feeder is located (not depicted in (Figure 6)) allowing to adjust the feeding rate of the material into the piping system. A vibrating mechanism attached to the external wall of the feeding tank facilitates a constant material flux into the rotational feeder. By controlling the amount of compressed air ducted into the piping system, the dilution of the transported material, i.e. the flow regime, can be set to the required flow characteristics. The 'X' in (Figure 6) marks the installation position of the EF-sampler. For minimizing the effect of flow disturbances, mainly 'roping effects', the sampler position was chosen to be at least 3m away from any pipe bends (distance to previous bending was ca. 7m). The roping effect describes a particle segregation effect in pipe bends caused by the action of centripetal forces, whereby particles flowing to the outer wall of the bend form a relatively dense phase structure, termed rope [16].

Pneumatic conveying can be broadly categorized in dense phase vs. dilute/lean phase conveying. Despite the fact that the literature lacks a clear separation between these flow regimes, dilute phase can be best understood as a flow state where solid particles are fully suspended in the air or gas stream and behave as individual dynamic units, in contrast to dense phase where particles are severely influencing one-another in the aggregate stream flux, i.e. definitely not fully suspended [17]. Depending on the application, each flow regime carries its own advantages and disadvantages. The following experimental test campaign has been conducted with varying flow rates commensurate with the dilute phase regime only, since the EF-sampler was designed primarily for dilute phase biomass conveying (but with an eye towards more general applicability of course). One main advantage of dilute phase conveying in terms of representative sampling is the fact that vertical flow segregation is lower than in a dense phase flow. The results section clarifies this aspect more fully.

Figure 7 shows the EF-sampler with its main technical components as installed at the marked position on the pneumatic conveying system (compare Figure 6). The left photograph shows the extraction and upper storage section, while the right illustration depicts a close-up of the lower storage section.

For the all-important verification of accuracy and representativeness, sampling tests with two very different materials were performed: wheat flour and pulverized alumina. Wheat flour has been selected to represent a very fine, cohesive powder, similar to cement powder and fly ash for example, two further powders often conveyed in pneumatic systems. The second test material, alumina, was selected since presenting a quite dense material type but with good flow characteristics. The present campaign aims at a first general feasibility testing for which these two materials were deemed sufficient; a follow-up campaign with several other, market-relevant, materials is in progress.

Particle size distribution, material density as well as the total amount of conveyed materials is presented in (Table 2). In order to focus on validation of accuracy and precision, spiking material in predetermined precisely calibrated concentration levels was added to each of these materials. LDPE (Low Density Polyethylene) plastic pellets were selected as a well-near optimal spiking material, having a density in-between the two test materials and, crucially, a particle size which is significantly larger than both test materials, i.e. a significantly different minor to trace constituent. The concentration levels of LDPE pellets in wheat flour were set to 0%, 2% and 5% [w/w] respectively. For the subsequent alumina test campaign the concentration level range was expanded to 0%, 2%, 5%, 8% and 11% [w/w]. The physical characteristics of the plastic pellets are also stated in (Table 2). It was our deliberate intention to present the sampler with materials of a reasonably high degree of difficulty, for which reason wheat and alumina powders with added LDPE pellets actually constitute materials that are more difficult to sample that the nominated biomass target, when in routine transportation in the designated industrial setting. For feasibility and validation purpose however, this stringent test scenario will serve very well however.

Test material	Density	Particle size distribution			Amount of material
		D10	D50	D90	
Wheat flour	0.46g/cm³	13.3µm	66.5µm	161.3µm	240kg
Alumina	1.25g/cm³	35.7µm	85.7µm	134.5µm	150kg
Plastic pellets	0.58g/cm³	~3mm			Depending on concentration level

Table 2: Physical characteristics of test materials.

Test material	Naming of flow rate	Mass flow (kg/s)	Airflow (Nm3/h)	Mass-air-ratio
Wheat flour	FR 1	0.30	750	$\frac{1.2}{1}$
	FR 2	0.20	950	$\sim\frac{0.6}{1}$
Alumina	FR 1	0.50	750	$\frac{2}{1}$
	FR 2	0.35	950	$\sim\frac{1}{1}$

Table 3: Tested mass/air ratios of wheat flour and alumina.

Test material	Naming of sampling speed	Sampling speed (s)	Pausing time (s)	Cleaning time (s)
Wheat flour	SP 1	5	1	2
	SP 2	3	3	2
Alumina	SP 1	2	1	1
	SP 2	1	2	1

Table 4: Test parameters of EF-sampler.

Test material	Spiking concentration levels (%)	Flow rates	Sampling speed	Repetitions
Wheat flour	0, 2, 5	FR1, FR2	SP1, SP2	3
Alumina	0, 2, 5	FR1, FR2	SP1, SP2	2
	8, 11	FR1, FR2	SP1	2

Table 5: Overview of complete test campaign parameters.

RSV (%)	FR1, SP1	FR1, SP2	FR2, SP1	FR2, SP2
Wheat flour - 2% spiking	2,91%	2,63%	4,24%	4,04%
Alumina – 5% spiking	2,18%	2,98%	1,91%	3,43%
Alumina – 11% spiking	4,22%	-*	2,53%	-*

* These scenarios have been only tested with sampling speed 1.

Table 6: RSVs of the present test scenarios.

In order to validate whether variations of the mass-air ratio in the dilute phase regime affects the sampling results, the airflow was varied between 750 Nm³ (Flow rate 1) and 950 Nm³ (Flow rate 2) with a constant feeding rate for both test materials; a constant feeding rate ensures an equal volume flow of both test material into the piping system. Due to the higher density of alumina an identical constant feeding rate results in a higher mass-air-ratio for alumina when varying only airflow in the different test scenarios. (Table 3) depicts the realised mass- and airflow as well as the resulting mass-air ratios in the test scenarios for both materials.

The test campaign also comprised variations in sampling speed (rotational speed of cutter arm) as well as adjusting cleaning- and pausing times in order to allow that 75 increments could be extracted in the required time for each transportation round. Due to the different test material load realised, two sampling speeds were set for wheat flour and correspondingly for alumina. (Table 4) lists the varying sampling speeds and cleaning-/pausing times for each test material.

For all scenarios a sampling arm width of 14 mm was used with a depth of 15 mm ensuring a correct delineation of the increment and preventing any spill-over effects. For test materials with a larger nominal top diameter, the sampling arm needs to be adjusted accordingly. As a rule of thumb TOS states that the width of the cutter or sampling arm should be at least three times the nominal top diameter of the material of interest. The E-F sampler is designed so that replacement of the sampling arm can be carried out without undue efforts, i.e. without having to detach the sampler in its pipefitting a.o.

A complete overview of all test parameters including replication rounds is stated in (Table 5), resulting in a total of 36 test rounds for wheat flour and 32 test rounds for alumina. It is important to mention that in each transportation round the entire test material (lot) was transported from the feeding tank into the receiving tank, obeying TOS' principle of sampling correctness for the entire lot. The required amount of plastic pellets presenting different nominal spiking concentrations levels was inserted directly into the feeding tank, while discharging the test material from the receiving tank into the feeding tank. Since a vestige of a layering effect of the plastic pellets could not be completely avoided during insertion, the spiked material was transported passively for two full circulation rounds before starting the sampling test operations proper in order to achieve an effective 'in-line' mixing effect. By this type of successively sampling of the lot, the remaining segregation effects in the feeding tank could be minimized, and perhaps even fully counteracted. Any such lingering effects will of course show up in the test results, i.e. as inflated accuracy and precision measures, but as long as these stay within the pre-specified brackets this will be an acceptable price to pay for a sampler that can accommodate the highly taxing pneumatic horizontal flow scenario.

Results

In the following sections selected results of the test scenarios are presented and discussed. Besides estimation of the effective accuracy and precision range of the EF-sampler under the applied test conditions, the experimental results also show to which degree the predetermined test parameters influence upon these evaluation criteria. The range of each test parameter had to be pre-set before the actual test campaign. This implies that the globally optimal set up of test parameters may actually be found (with more experimentation) lie outside this initial test parameter range. Because of the feasibility test objective, it was considered acceptable to try to shoot for the critically important accuracy testing with a manageable set if inferred best test conditions first. Below are presented selected results, not all corresponding to what was eventually found, after all experiments were in, to be the optimal conditions.

(Figure 8) depicts the sampling test results for wheat flour, spiked with a plastic pellet concentration of 2%. The left figure shows sampling results using 0% pellet concentration as the origin on the y-axis origin, while the right figure shows a close-up of the variations in the occurring concentration level range. The y-axis displays the plastic pellet concentration of composite samples consisting of 75 increments. The x-axis illustrates the mean value concentration of the three replication rounds for each test scenario. On the left side of the y-axis sampling results for flow rate 1 (FR1) and flow rate 2 (FR2) using sampling speed 1 (SP1) are compared, while on the right side of the y-axis analysed concentration values of the samples gained from FR1 and FR2 with sampling speed 2 (SP2) are shown. Furthermore the nominal reference concentration level is depicted as a dashed line.

For these initial test scenarios all composite samples with respect to the analysed spiking concentration level have a relative inaccuracy smaller than 16% (highest rel. inaccuracy occurring in test scenario W2_FR1_SP2). For both sampling speeds (SP1 and SP2), a higher airflow rate (FR2) leads to an improved accuracy of the plastic pellet concentration in the wheat flour, while the longer sampling time 1 (SP1=5s) further improves the accuracy compared to the shorter cutting interval SP2 (sampling speed=3s).

Thus the highest accuracy for this scenario, rel. accuracy <4,5%, was achieved by using SP1 with FR2, i.e. the longer sampling time in the more dilute material stream. These results also point out that nearly all analysed concentration levels of the composite samples lead to a slightly higher concentration compared to the reference level of 2%. An explanation for this bias is given after presenting selected results for the alumina test scenarios. The precision for all test variations is good throughout.

(Figure 9) presents the results of composite samples extracted during the pneumatic transportation of alumina with a plastic pellet concentration of 5%. Each test scenario has again been repeated twice, varying sampling speed (SP1, SP2) and airflow rate (FR1, FR2).

The worst (highest) relative inaccuracy for the alumina tests is now below 11%, occurring in the composite sample gained with the faster sampling speed (SP2 for alumina=1s) and the lower airflow rate 1 (FR1). The close-up of (Figure 9) confirms that composite samples extracted during a higher airflow rate are much more accurate (rel. inaccuracy <4%), being further improved by the slower sampling speed SP1 (SP1 for alumina=2s). These observations are consistent with the results outlined by the initial wheat flour scenarios above. The positive effect of longer sampling times (SP1) can easily be explained by TOS, substantiating that an increase of the increment volume improves the accuracy of the composite sample.

In contrast to the plastic pellet concentration analysed in the wheat flour experiments, which are slightly overrepresented (also valid for 2% spiking concentration), the concentration levels in the alumina are slightly unrepresented (also valid for all other tested concentration levels). This effect can be also observed in (Figure 10), presenting the concentration levels for alumina spiked with 11% plastic pellets. For this scenario, which was repeated twice, only the slower sampling speed 1 (SP1) was used for increment extraction. The composite samples extracted during testing with the higher airflow rate 2 (FR2) lead again to a better accuracy, <4% rel.

Thus for both test materials, under the best sampling conditions obtainable (still sub-optimal with respect to the intended industrial conditions), a relative accuracy level below the pre-set 5% was indeed achieved under the test conditions available at the test site.

The corresponding lowest total sampling errors, including all sampling errors and total analytical errors, can be expressed as a convenient RSV (Relative Sampling Variance) measure[2], also termed the relative coefficient of variation (CV_{rel}). The CV_{rel}, meaning the standard deviation (STD) in relation to the average (X_{avr}), can be effectively expressed as a percentage:

$$CV_{rel} = \frac{STD}{X_{avr}} * 100$$

A CV_{rel} corresponding to 20-35% has been suggested for 'significantly heterogeneous materials', in particular for uncharacterized stationary systems [18], while for less heterogeneous systems, as also being the case for the presented test campaign, the CV_{rel} level should be set to a level specified as 15-20% [15].

Compared to these general guidelines, it is also the opinion that for process sampling these thresholds must be lower still due to the much more optimal sampling conditions that can be realised in this regimen. Due to the limited replication rounds of the present test scenarios, the results allow serve as first estimates, which are reported in (Table 6). These statistical results should optimally be based on a repetition of at least 10 repetition rounds, which was out of the possible range of test conditions available in this first trial in which the accuracy test had absolute priority. A full second test campaign on several additional material types and with further replications is under way.

All test scenarios have a relative sampling variance <4,5%, lying far below the suggested thresholds. Seen together with the <4.5% inaccuracy results, the RSV estimates show that the incorrect sampling

---

[2]Because only two, or three, replications could be achieved in the present test campaign, estimating a standard deviation on this basis is naturally a somewhat contrived endeavor. But if a first estimate of the relevant RSV is wanted, these have been calculated and reported in the text. They should be viewed in this particularly bracketed context of course.

errors in the EF-sampling system, the major source for bias and high sampling variance, have been very nearly eliminated, confirming the correctness of the EF-sampler's design principles as well as its performance.

Discussion

An explanation for the small bias detected for both test materials, over-representing the plastic pellet concentration in the wheat flour and under representing the concentration level in the alumina, can logically be attributed to the density differences between wheat flour, plastic pellets and alumina respectively, as was documented in (Table 2). Even though the sampling arm has been designed with the stated objective of being able to cope with the vertical segregation occurring in horizontal transportation sections, the test facility parameters do not allow to achieve a completely unbiased suspension of the compound mixed material; in the test facility used, the realisable air capacity and pipe dimension limit the dilution range possible. This means that it can be expected that the material with a higher density is (slightly) more likely to be transported at the bottom part of the piping system. Assuming an equal influx of material into the sampling arm, the more dense material most likely disturbs the less dense material flow inside the sampling arm, allowing a slightly higher proportion of dense material to fall directly into the outlet chute, located at the bottom of the sampling arm. This is of course a classical segregation effect in the framework of TOS.

Normally only one material type is transported in industrial pneumatic transportation rigs. The ultimate purpose of acquiring representative samples from CHP plant pneumatic transportation ducts is to assess the % of "fines" (the smallest biomass particle sizes). This scenario is thus not affected by the kind of density differences employed in the current test campaign, which were set up here for maximal test validity, i.e. the deliberate choice to test based on added minor contaminant concentrations of significantly different particle size as well as density with respect to the matrix material.

The obtained results show quite satisfactorily that the EF-sampler, even under these "more-than-necessary difficult" sampling conditions, extracts fit-for-purpose samples with performance results that can be considered as acceptably close to the pre-set acceptable accuracy threshold of 5%. The present results also reveal that higher airflow rate capacities would minimize the detected bias level. In the target case of converted coal power plants mass-air ratios up to 1 (mass) to 2 (air) are achieved, resulting in a very dilute flow regime favouring even better sampling accuracy. In a forthcoming comprehensive test campaign the EF-sampler will be tested under this very dilute flow regime in a fully up-scaled version for direct implementation in converted coal power plants, pneumatically transporting pulverized biomass from the mills to the combustion chamber. These results will serve to validate the performance of the EF-sampler under fully realistic industrial conditions, including larger process variations and varying compositions of the transported material.

For this up-scaled test campaign of the EF-sampler and also for any other sampler verification, the Theory of Sampling points out that the ultimate test regarding intrinsic parameters (e.g. chemical composition, physical grain-sizes a.o.) is to which degree a sampler is able to reproduce the lot grain size distribution quantitatively within the specified acceptance levels. Thus if the results with respect to the particle size distribution show an acceptable small bias (e.g. 5% rel. for all size bins), it follows that the EF-sampler is also similarly acceptably 'unbiased' with respect to most other chemical attributes.

Powder characterises like cohesiveness or moisture level variations are parameters, which need further investigation in respect to their influence on the sampling results however, as these are not intrinsic characteristics. Furthermore, the optimal sampling speed for the set flow regime of pulverized biomass in converted power plants will be determined applying the results of the planned further test campaigns.

Conclusions and Prospects

The EF-sampler is the first sampler for horizontal pneumatic transportation systems, with the objective to ensure acceptable TOS-representativeness of samples extracted from ducted pressurised material streams. The design principles prevent that neither the sampling operation itself nor the resting of the sampler in its inactive parking position cause any major disturbances of the material flux, which could otherwise lead to pressure surges or clogging effects. The automated extraction mechanism, including the possibility to vary sampling-, cleaning- and pausing intervals makes the EF-sampler flexible and adjustable to many different material types and flow characteristics. However the sampling arm width must be commensurate with the grain size distribution of the material of interest. For the test scenarios presented a sampling arm width of 14mm was sufficient.

The first experimental test campaign allowed full validation of the technical functionality, the accuracy (bias) and the precision of the EF-sampler under varying conditions for the most influential parameters. Sampler velocity as well as the material dilution ranges influenced significantly on the accuracy of composite samples. Particularly an increase of airflow capacity leading to a higher dilution of the transported spiked material improved the sampling accuracy. The higher dilution of the spiked test material in combination with a slower sampler velocity resulted in a relative inaccuracy for all tested scenarios below 5% with a generally very good reproducibility (precision), i.e. fully compliant with the pre-trial critical success criterion established. Furthermore, the relative sampling variance for all test scenarios is quantitatively also below 5% (<4,5%), signifying that perhaps all critical sampling errors have been successfully eliminated or minimized.

The prototype EF-sampler for horizontal ducted suspended material streams is not a universal sampler, but is designed to incorporate material-dependant adjustments based on the specific material heterogeneity and flow regime characteristics. The present feasibility validation of the EF-sampler installed in a pilot-scale transportation system showed however that variation of spiking concentration, dilute flow range and sampling speed are fully controllable parameters, which in all test scenarios lead to fit-for-purpose representative sampling. The EF-sampler has thus passed the first crucial test qualifications with respect to its intended primary biomass implementation scenario, and with a promising much wider potential application field as well. It will be highly relevant to test it on a wide range of other market-relevant material types. Such testing is in progress and will be reported elsewhere.

Acknowledgements

We would like to express our gratitude towards FLS midth Wuppertal, Germany (Jakob Hess, Lars Kristensen) for continued project involvement, and in particular Horst Faust for his essential technical advice and logistical support within the present work, as well as the Department of Powder Science and Technology, Tel-Tek, Porsgrunn (Chandana Ratnayake, Magne Saxegaard, Runar Holm), Norway for supporting the setup of the pneumatic test facility, without which the first test campaign would not have been possible.

References

1. Gy P (1998) Sampling for Analytical Purposes. John Wiley and Sons Ltd, Chichester, UK.

2. Pitard FF (1993) Pierre Gy's Sampling Theory and Sampling Practice (2 edition), CRC Press Ltd, Boca Raton, USA.

3. Esbensen KH, Minkkinen P (2004) Special issue: 50 years of Pierre Gy's theory of sampling: proceedings: first world conference on sampling and blending (WCSB1). Tutorials on sampling: theory and practice. Chemometrics and Intelligent Laboratory Systems 74: 236, 1242.

4. Petersen L, Dahl CK, Esbensen KH (2004) Representative mass reduction in sampling-a critical survey of techniques and hardware. Chemometrics and Intelligent Laboratory Systems 74: 95-114.

5. Petersen L, Minkkinen P, Esbensen KH (2005) Representative sampling for reliable data analysis. Theory of sampling. Chemometrics and Intelligent Laboratory Systems 77: 261-277.

6. Petersen L, Esbensen KH (2005) Representative process sampling for reliable data analysis-a tutorial. Journal of Chemometrics 19: 625-647.

7. Esbensen KH, Julius LP (2009) Representative sampling, data quality, validation-a necessary trinity in chemometrics. In Brown S, Tauler R, Walczak R (Eds.) Comprehensive Chemometrics, Wiley Major Reference Works 4: 1-20.

8. Wagner C, Esbensen KH (2012) A critical review of sampling standards for solid biofuels-Missing contributions from the Theory of Sampling (TOS). Renewable and Sustainable Energy Reviews 16: 504-517.

9. Minkkinen PO, Esbensen KH (2010) Grab vs. composite sampling of particulate materials with significant spatial heterogeneity-a simulation study of "correct sampling errors". Anal Chim Acta 653: 59-70.

10. Esbensen KH, Paoletti C, Minkkinen P (2012) Representative sampling of large kernel lots-I. Theory of Sampling and variographic analysis. Trends in Analytical Chemistry (TrAC) 32: 154-164.

11. Minkkinen P, Esbensen KH, Paoletti C (2012) Representative sampling of large kernel lots-II. Application to soybean sampling for GMO control. Trends in Analytical Chemistry (TrAC) 32: 165-177.

12. Esbensen KH, Paoletti C, Minkkinen P (2012) Representative sampling of large kernel lots-III. General Considerations on sampling heterogeneous foods. TrAC Trends in Analytical Chemistry 32: 178-184.

13. Smith PL (2001) A Primer for Sampling Solids, Liquids and Gases-Based on the Seven Sampling Errors of Pierre Gy. ASA SIAM, USA.

14. Wagner C, Faust H, Esbensen KH (2011) Proceedings of 5th World Conference on Sampling and Blending (WCSB 5). Experimental testing of 1/3 scale model sampler for horizontally ducted particulate material streams 445-451.

15. Pitard FF (2009) 'Pierre Gy's Theory of Sampling and C.O. Ingamells' Poisson Process Approach. Pathays to representative sampling and appropriate industrial standards. Dr. Tech Thesis Aalborg University, Campus Esbjerg, Denmark.

16. Akilli H, Levy EK, Sahin B (2001) Gas-solid flow behaviour in a horizontal pipe after a 90˚ vertical to horizontal elbow. Powder Technology 116: 43-52.

17. Fokeer S, Kingman S, Lowndes I, Reynolds A (2004) Characterisation of the cross sectional particle concentration distribution in horizontal dilute flow conveying-a review. Chemical Engineering and Processing 43: 677-691.

18. Esbensen KH, Geladi P (2010) Principles of proper validation: use and abuse of re-sampling for validation. Journal of Chemometrics 24: 168-187.

23

Comparative Study of Bivalent Cationic Metals Adsorption Pb(II), Cd(II), Ni(II) and Cu(II) on Olive Stones Chemically Activated Carbon

Thouraya Bohli[1]*, Isabel Villaescusa[2] and Abdelmottaleb Ouederni[1]

[1]*Laboratory of Research: Engineering Processes and Industrials Systems (LR11ES54), National School of Engineers of Gabes, University of Gabes, Gabes, Tunisia*
[2]*Departement d'Enginyeria Quimica, Agraria i Tecnologia Agroalimentaria, Universitat de Girona, Girona, Spain*

Abstract

In this work the ability of olive stone activated carbon (COSAC) to remove Pb(II), Cd(II), Ni(II) and Cu(II) metal ions from aqueous solutions was evaluated. The effect of initial pH, contact time and initial concentration on metal ions adsorption was investigated. The results indicated that pH 5 is the optimum value for metal removal. Adsorption kinetic rates were found to be fast; total equilibrium was achieved after 4 hours. Kinetic experimental data fitted very well the pseudo-second order equation and the values of adsorption rate constants were calculated. The equilibrium isotherms were evaluated in terms of maximum adsorption capacity and adsorption affinity by the application of Langmuir and Freundlich equations. Results indicate that the Langmuir model fits adsorption isotherm data better than the Freundlich model. The removal efficiency of heavy metal ions by COSAC decreases in the order Pb(II) > Cd(II) > Ni(II) ≥ Cu(II).

Keywords: Heavy metal ions; Adsorption; Activated carbon olive stones; Kinetic; Equilibrium

Introduction

Environmental pollution by heavy metals has become one of the most hazard problems due to their dangerous effect on aquatic flora and fauna even in relatively low concentrations. Several heavy metal ions such as copper, nickel, cadmium and lead have been included in the U.S. Environmental Protection Agency's list of priority pollutants [1]. Copper leads to toxic effects including liver damage and gastrointestinal disturbances. Nickel is a well-known human carcinogen, particularly in human lung cancer [2]. Cadmium causes renal dysfunctions, hypertension and diabetes [3], and Lead is highly hepatotoxic [4]. Many industrial processes, such as smelting, metal plating, mining pigments, cadmium-nickel-batteries, brass manufacture and discharge aqueous effluents containing high levels of these heavy metal ions. Various physical and chemical processes have been extensively used to remove heavy metal pollutants from aqueous solutions such as chemical precipitation, ion exchange, bioadsorption, adsorption, membrane filtration, etc. Adsorption on activated carbon is considered one of the most widely applied techniques for pollutants removal from contaminated media owing to its efficiency especially at low concentrations and process simplicity. Most research has been focused to the adsorption of heavy metals by activated carbon prepared from different precursors, these include apricot stone [5], bamboo [6], the olive pulp, cherry stones, rice husks etc. Many factors can influence the degree of adsorption of metal ions on activated carbon especially. In one side one must take account of: (i) solution pH, solutes nature and concentration, physicochemical properties of metal ions e.g. speciation, ionic radius, hydration energy and electro-negativity. In other side one must consider the sorbent characteristics as (i) the point of zero charge of the sorbent surface (pH_{pzc}); (ii) sorbent specific surface area and porosity; (iii) sorbent surface functional groups. Some of these adsorbent properties are related to the activation process and oxidation agent.

The aim of this work was to evaluate the adsorption capacity of olive stone carbon chemically activated as well as to investigate adsorption equilibrium and kinetics of copper, nickel, cadmium and lead divalent ions from aqueous solutions.

Experimental

Chemicals and equipments

All chemicals used in the present study were of analytical grade. Stock metal solutions of 1 gL^{-1} were prepared by dissolving in Milli-Q water quality appropriate amounts of $Cu(NO_3)_2.6H_2O$, $Ni(NO_3)_2.12H_2O$, $Cd(NO_3)_2.4H_2O$ and $Pb(NO_3)_2$. The stock solution of each nitrate salt was diluted to get the desired initial metal ion concentration (C_0). Different diluted metal concentrations prepared from Cu(II), Ni(II), Cd(II) and Pb(II) standard solutions (1 gL^{-1} for FAAS, Merck) were used to obtain the calibration curves for Flame Atomic Absorption Spectrometry (VARIAN Absorption Spectrometer (Model220FS)). Solutions of NaOH (0.1 M) and HNO$_3$ (0.1 M) were used for adjusting the pH.

Adsorbent

Olive stones activated carbon (COSAC) produced by chemical activation, according to Ghrib et al. [7], was used as an adsorbent. Before using, COSAC was washed with distilled water to remove soluble organic matter, dried at 60°C, then grinded and sieved, for a selected particle size range of 0.250-0.500 mm. Specific surface area and pores characteristics of olive stone activated carbon were determined by nitrogen adsorption and desorption isotherms at 77.7 K with an automatic Sorptiometer Autosorb-1C Quantachrome apparatus. The pH_{pzc} and surface functional groups were determined respectively by batch equilibrium technique [8], and Boehm titration [9].

*Corresponding author: Thouraya Bohli, Laboratory of Research: Engineering Processes and Industrials Systems (LR11ES54), National School of Engineers of Gabes, University of Gabes, St Omar Ibn Elkhattab, 6029 Gabes, Tunisia

Adsorption experiments

Equilibrium experiments were conducted on batch mode at temperature of 30 ± 2°C. To obtain the adsorption equilibrium isotherms, a sample of 0.3 g (with particle size range: 0.25-0.5 mm) of COSAC were directly placed into 250 ml glass-stoppered conical flasks containing 50 mL of metal ions solutions having different initial concentrations within the range of 0.2 to 5.0 mM for Pb(II), 0.2 to 8.0 mM for Cd(II) and 0.2 to 14.0 mM for both Cu(II) and Ni(II). The flasks were agitated for 10h with an agitation speed of 400 rpm. Then, samples were filtrated through a 0.45 μm cellulosic filter paper. During the experiments the temperature was maintained at 30 ± 2°C by using a thermostatic bath.

Evaluation of initial solution pH effect on the adsorption of Cu(II), Ni(II), Cd(II) and Pb(II) on COSAC was undergone in the same conditions as indicated above with an initial metal ions concentration of 1mM and an equilibrium time of 10h. Initial pH solutions were adjusted within the range of 2.0 to 5.5 by adding some volume of 0.1 M NaOH or 0.1M HNO_3.

Adsorption kinetics was investigated at three different initial metal concentrations (0.5, 1 and 1.5 mM) and a fixed adsorbent dose of 6 gL^{-1} COSAC in a batch mixed suspension with 500 mL of metal solution. The temperature of the suspension was maintained at 30 ± 2°C by using a thermostatic bath. Solution samples of 2 mL volume were withdrawn at a pre-set time intervals.

Initial concentrations and final concentrations at equilibrium were measured, after acidification with HNO_3 solution to prevent metal precipitation, by FAAS at the wavelengths 324.8, 232.0, 228.8 nm and 217.0 nm, for Cu(II), Ni(II), Cd(II) and Pb(II), respectively, by using air-acetylene flame.

The amount of metal ion uptake by COSAC was calculated using the following equation:

$$q_t = (C_0 - C_t)\frac{V}{m}$$ (1)

Where C_0 is the initial metal concentration in the solution (mmol), C_t is the metal concentration remaining in the solution at a given time (mmol), m is the weight of COSAC (g), q_t is the metal ion uptake (mmol.g^{-1}) and V is the volume of solution in the flask.

Linear equation of Langmuir and Freundlich models used to correlate adsorption equilibrium were calculated with Equation (2) and the Equation (3), respectively.

$$\frac{C_e}{q_e} = (\frac{1}{q_{max}K_L}) + (\frac{1}{q_{max}})C_e$$ (2)

$$\log q_e = \log K_F + \frac{1}{n}\log C_e$$ (3)

where K_L is the equilibrium adsorption constant related to the free energy of the adsorption (l mg^{-1}) and q_{max}: the maximum adsorption capacity (mmol g^{-1}), C_e the equilibrium concentration (mmol l^{-1}), and q_e amount adsorbed at equilibrium (mmol g^{-1}). K_F, (mmol l^{-1})(l g^{-1})$^{(1/n)}$ and n are the constants of Freundlich equation, respectively.

Each experiment was carried out in duplicate and the average results are presented herein.

Results and Discussion

Characterization of activated carbons

The adsorption behaviour of activated carbon is generally determined

Physical properties	
Total area (BET)	1086 m².g⁻¹
Micropore volume	0.51 cm³g⁻¹
Total pore volume	0.52 cm³g⁻¹
Average pore diameter	19.16 A°
pH$_{pzc}$	3
Surface functional groups (meq.g⁻¹)	
Carboxylic groups	0.100
Carbonyl groups	2.720
Lactonic groups	0.300
Phenolic groups	1.650

Table 1: Properties of COSAC.

by the surface area, the pH$_{pzc}$ and by the chemistry of the carbon surface [10]. The specific surface area of COSAC, determined by applying the BET method, was found to be 1086 m²g⁻¹. The micropores and mesopores volumes were calculated to be respectively 0.51 and 0.01 cm³g⁻¹ (Table 1). This indicates that COSAC is a micropores activated carbon.

The pH$_{pzc}$ gives information about the charge of the carbon surface [8], our carbon surface is positively charged at pH values below the pH$_{pzc}$. In contrast, at solution pH values higher than the pH$_{pzc}$, negative surface charge increases. COSAC has a pH$_{pzc}$=3, therefore the carbon surface is negatively charged for pH higher than 3.

Some functional groups occurred on the surface of COSAC (Table1). COSAC contained carboxylic, lactones, phenolic and carbonyl groups, resulting in 4.77 meq.g⁻¹ of total acidity. Total basicity of COSAC was 0.75 meqg⁻¹. Results of surface functional groups reported in table 1 indicate the predominance of acidic functional groups, especially carbonyl and phenolic functional groups on the COSAC surface. Physical COSAC properties show a higher specific surface area and developed micropores.

Effect of pH on the adsorption capacity of COSAC

The pH of metal solution is considered to be the most important parameter that can affect the adsorption capacity of activated carbon because of its influence on metal solubility for the dissociation degree of functional groups located on sorbent surface [11,12].

In general, a decrease in ions uptake at acidic pH is due to an increase of competition between hydrogen ions and metal ions for the same adsorption sites. However, an increase in alkalinity enhance metal adsorption rate, due to the predominant presence of hydrated species of heavy metals, changes in surface charge and the precipitation of the appropriate salt. Therefore, there is an optimum pH in which the competition of hydrogen ions is minimized and metal ions precipitation is avoided thus, enhancing metal adsorption.

The experiments were carried out at selected pH values below the pH where metal hydroxide chemical precipitation can occur: pH> 5 for Pb(II) and Cu(II), pH> 6 for Ni(OH)$_2$ and pH>7.8 for Cd(OH)$_2$. These pH values have been retrieved by the chemical species distribution diagrams calculated using the MEDUSA computer program [13].

Figure 1 shows the amount of metal adsorbed and the adsorption percentage of Pb(II), Cd(II), Ni(II) and Cu(II) versus initial solution pH. Results indicate that metal adsorption is strongly pH-dependent and increases with the pH increase until it reaches a maximum at around pH 5. Initially, as the pH of solution was increased from 2.2 to 5.0, the removal percentage of Cu(II), Ni(II), Cd(II) and Pb(II) increased respectively from 16.4 to 62%, from 21 to 78%, from 34 to 97% and from 50 to 100%. The low adsorption at low pH, mainly for

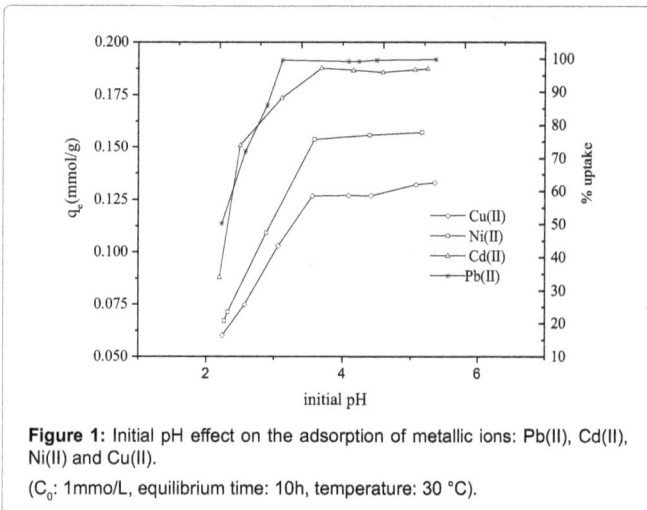

Figure 1: Initial pH effect on the adsorption of metallic ions: Pb(II), Cd(II), Ni(II) and Cu(II).

(C_0: 1mmo/L, equilibrium time: 10h, temperature: 30 °C).

copper and nickel, is due to the competition between protons and metal ions for the same adsorption sites. The effect of pH can also be explained by considering the sorbent surface charge. As said before, COSAC pH_{pzc} was pH 3, therefore, when solution pH is lower than 3, the surface of the COSAC is positively charged and metal adsorption is inhibited, due to electrostatic repulsion between metal ions and positively charged functional groups. Conversely, for pH > 3, the number of negatively charged sites on COSAC surface increases, and metal adsorption becomes more important.

In most of the reported studies concerning metal ions adsorption onto activated carbon adsorption uptake shows a maximum for initial pH around 5 [14-16].

The important uptake of cadmium and lead even at low pH may be related to higher affinity of the surface functional groups of COSAC for these two metal ions compared to the affinity towards copper and nickel.

Based on these results, the best initial pH for these metallic ions adsorption was considered to be 5 and both kinetics and equilibrium experiments were carried out at this pH.

Adsorption kinetics at different initial concentrations

Equilibrium time: The time necessary to reach equilibrium for copper, nickel and lead adsorption was investigated for different initial concentrations. For these experiments solid-liquid ratio 6 gL⁻¹ and pH 5 were kept constant. Results are shown in figure 2. As seen in the figure, the adsorption of Cu(II), Ni(II) and Pb(II) ions onto COSAC increases with time until about 200 min and, thereafter, becomes constant. The adsorption process reaches equilibrium in 270 min for all the studied initial concentrations. With changing the initial concentration of metal ion solution from 0.5 to 1.5 mM, the uptake amounts of Cu(II), Ni(II) and Pb(II) ion increase from 0.061 mmol.g⁻¹ to 0.101 mmol.g⁻¹, from 0.052 mmol.g⁻¹ to 0.090 mmol.g⁻¹ and from 0.0701 mmol.g⁻¹ to 0.109 mmol.g⁻¹, respectively.

From these results, the chosen contact time for further experiments was 10h in order to be sure that equilibrium was achieved.

Kinetics: The mechanism of adsorption depends on the physical and chemical characteristics of the adsorbent as well as on the fluid-solid mass transport process. Kinetic data were submitted to both pseudo-first and pseudo-second kinetic models [17], in order to ascertain which model fitted better to the experimental data.

The pseudo-first order used is that of Lagergren given by the following equation:

$$\frac{dq}{dt} = k_1(q_e - q_t) \tag{4}$$

The integration of equation (4) in the conditions (t=0, qt=0) and (t=t, q=qt) gives:

$$Ln(q_e - q_t) = Lnq_e - k_1 t \tag{5}$$

Where k_1 is the velocity constant of the pseudo-first order (min⁻¹).

The pseudo-second order or Ho equation is given by the flowing expression:

$$\frac{dq}{dt} = k_2(q_e - q)^2 \tag{6}$$

By integrating equation (6) into the boundary conditions (t=0, q=0) and (t=t, q=qt), we obtain:

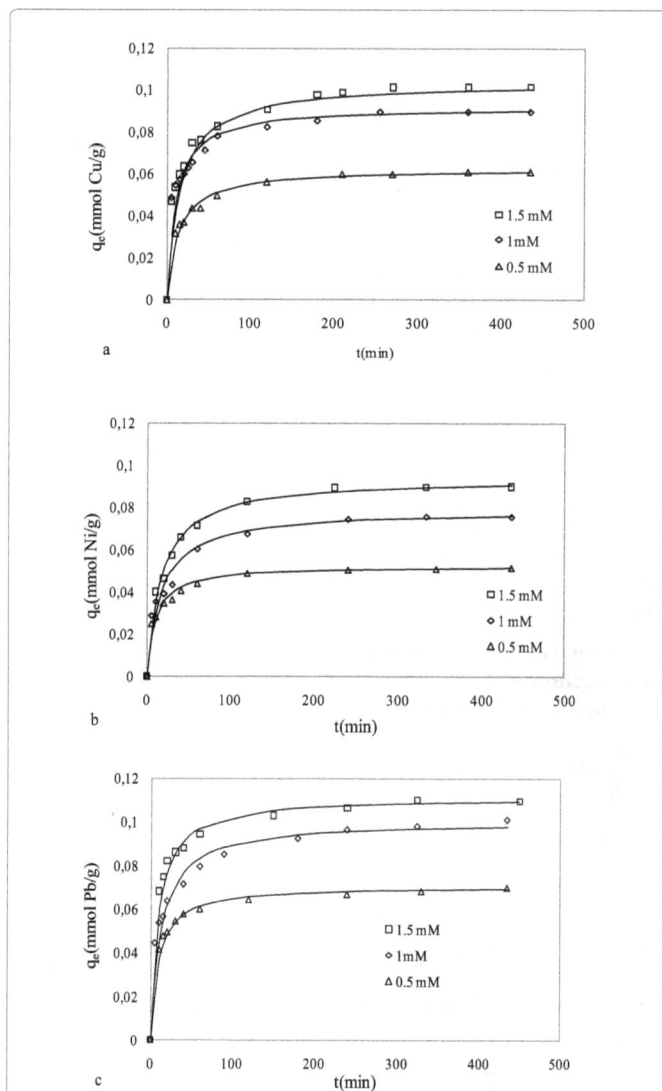

Figure 2: The effect of metal initial concentration on the adsorption of (a): Cu(II); (b): Ni(II); (c): Pb(II), symbols are experimental data and solid lines represent predicted data by pseudo-second- order model, (sorbent concentration:6 gL⁻¹, temperature: 30 °C, pH = 5).

$$\frac{1}{q_t} = \frac{1}{(k_2 q_e^2)} + \frac{t}{q_e} \qquad (7)$$

Where k_2 is the velocity constant of the pseudo-second order ($g\,mmol^{-1}min^{-1}$).

Pseudo-first-order and pseudo-second-order theoretical fitting parameters are listed in table 2. Results showed that the calculated amounts of metal adsorbed ($q_{e,cal}$) by the pseudo-first-order model differ substantially from those measured experimentally, whereas these obtained from the pseudo-second-order kinetic model are very close to experimental data and the correlation coefficient values quite good ($0.985 \leq R^2 \leq 0.999$) and much higher than those find by of pseudo first-order model, suggesting that the adsorbent systems can be well described

by the pseudo-second-order kinetic model. Figure 2 illustrates the fit by the pseudo-second-order model of the experimental results of the kinetics of Cu, Ni and Pb adsorption by COSAC.

Table 2 shows, in one hand, that for all the studied metals, k_2 values decrease with increasing initial concentrations. Thus may be due to the enhancement in the competition between metal cations themselves and between metal cations and H^+ protons for the same active sites. On the other hand, and for the same initial concentration k_2 values found for each metal ion decreased in the following order: Pb(II) >Ni(II) $\geq$ Cu(II). This indicated that lead was more easily and rapidly adsorbed by COSAC than nickel and copper.

Adsorption isotherms: Equilibrium adsorption studies were performed to determine the maximum metal adsorption capacities of COSAC. Experimental equilibrium isotherms of Pb(II), Cd(II), Ni(II) and Cu (II) ions determined at pH 5.0 and 30°C are showed in figure 3. Isotherms were evaluated using the usually used Langmuir and Freundlich models [18].

Figure 3 shows experimental isotherms data fitted by Langmuir and Freundlich models of the adsorption of Pb(II), Cd(II), Ni(II) andCu(II) metal ions from aqueous solutions on olive stones activated carbon at 30 °C and pH 5. The corresponding Freundlich and Langmuir parameters and correlation coefficients are regrouped in table 3.

Isotherms are of L-type of the Giles classification [19], describing high adsorbate-adsorbent interaction and showing initially a rapid adsorption tending to be asymptotic at higher concentration. Figure 3 also shows that the sites able to bind Pb(II) and Cd(II) are saturated (equilibrium reached) at high metal ions concentrations, as compared to nickel and copper. It is clear that Langmuir model fits to the experimental data very well with high correlation coefficients ($0.98 \leq R^2 \leq 0.999$) as compared to the ones given by Freundlich linear equation ($0.880 \leq R^2 \leq 0.947$). Table 3 also indicates that the theoretical maximum values of adsorption capacity of COSAC given by Langmuir equation (q_{max}) were equal to experimental ones and were found to decrease in the order Pb(II) $\geq$ Cd(II) > Ni(II) $\geq$ Cu(II). This trend matches with the order of ionic radius (r_i) of metals and hydration energy but does not match with the values of other characteristics listed in table 4. Pb(II) has the largest ionic radius (1.19A°) followed by those of Cd(II) (0.97A°), Ni(II) (0.72A°) and Cu(II) (0.69A°). For the large non hydrated ions, since the charge is more dispersed, hydration water is held less strongly. The bigger the ionic radius, the stronger the adsorption of the ion since the hydration capacity of that ion is smaller, resulting in weaker binding of the ion and water phase. The preference of adsorption exhibited for Pb(II) and Cd(II) over Ni(II) and Cu(II) may be also due to the difference hydration energy [20].

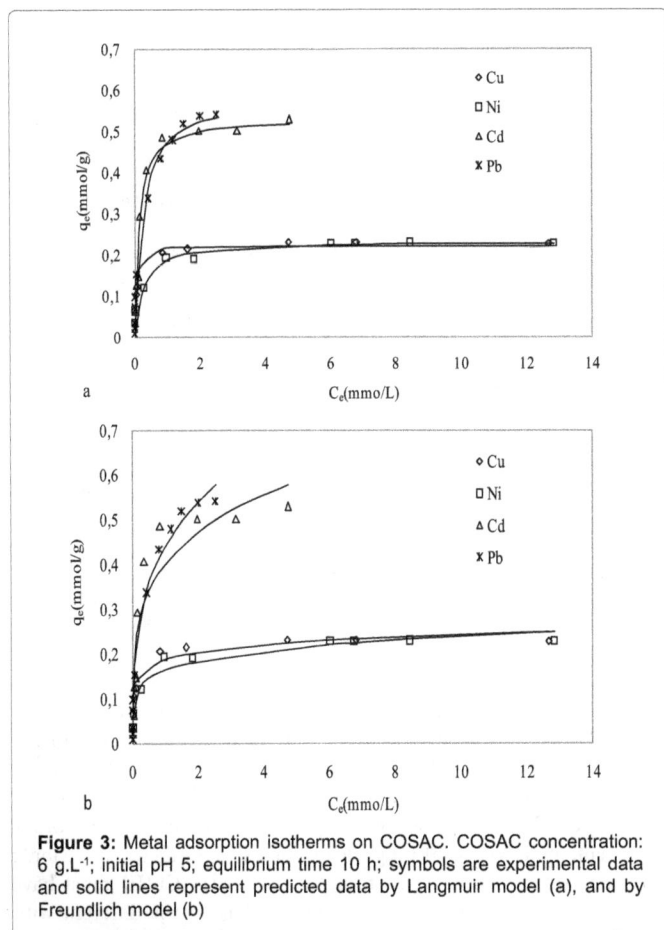

Figure 3: Metal adsorption isotherms on COSAC. COSAC concentration: 6 g.L⁻¹; initial pH 5; equilibrium time 10 h; symbols are experimental data and solid lines represent predicted data by Langmuir model (a), and by Freundlich model (b)

Metal	C_0 (mmol)	$q_{e,exp}$ (mmol g⁻¹)	Pseudo-first-order			Pseudo-second-order		
			k_1 (min⁻¹)	$q_{e,cal}$ (mmolg⁻¹)	R^2	k_2 (g.mmol⁻¹min⁻¹)	$q_{e,cal}$ (mmolg⁻¹)	R^2
	0.5	0.0609	0.013	0.0285	0.968	1.1690	0.0629	0.999
Copper	1	0.0894	0.013	0.0362	0.955	1.0870	0.0917	0.999
	1.5	0.1014	0.017	0.0576	0.946	0.6277	0.1040	0.985
	0.5	0.0516	0.016	0.0171	0.843	1.4703	0.0527	0.999
Nickel	1	0.0756	0.024	0.0414	0.982	0.7514	0.0788	0.999
	1.5	0.0900	0.016	0.0508	0.952	0.6208	0.0942	0.999
	0.5	0.0701	0.002	0.0312	0.947	1.5037	0.0708	0.999
Lead	1	0.1011	0.009	0.0457	0.971	0.7973	0.1009	0.998
	1.5	0.1098	0.01	0.0356	0.968	0.9841	0.1119	0.999

Table 2: Kinetic parameters for the adsorption of Pb(II), Ni(II) and Cu(II) onto COSAC.

Metal ion	q_{exp} (mmol g^{-1})	Langmuir constants			Freundlich constants		
		q_e (mmol g^{-1})	K_L (l mmol^{-1})	R^2	K_F (mmol l^{-1})(l g^{-1})$^{(1/n)}$	n	R^2
Pb(II)	0.5405	0.5417	14.091	0.980	0.4528	2.9069	0.880
Cd(II)	0.5283	0.5327	12.068	0.998	0.4013	3.1545	0.928
Ni(II)	0.2272	0.2306	6.609	0.999	0.1586	5.1282	0.947
Cu(II)	0.2268	0.2290	20.302	0.999	0.1837	7.6923	0.932

Table 3: Calculated Langmuir and Freundlich constants and correlation coefficients for the adsorption of Pb(II), Cd (II), Ni(II) and Cu(II) onto COSAC.

Metal	Ionic radius A°	Hydrated radius A°	Hydration energy (kJmol^{-1})	Solubility of hydroxides (pk$_s$)	Hydrolysis constants (pK$_h$)	Electro-negativity
Pb(II)	1.19	4.01	-1481	-	7.80	2.33
Cd(II)	0.97	4.26	-1807	-	11.70	1.69
Ni(II)	0.72	4.04	-2106	14.7	9.40	1.91
Cu(II)	0.69	4.19	-2100	19.6	7.53	2.00

Table 4: Chemical characteristics of the metal ions.

Adsorbent	conditions	Cu(II)	Ni(II)	Cd(II)	Pb(II)	References
Olive stones A.C (ZnCl2)		-	-	1.851	-	Kula et al. 2008
Granular A.C	pH 5	15.58		3.37	15.58	Faur-Brasquet et al., 2002
Powdered A.C		26.9		3.37	26.90	
Chestnut shell A.C		100.00				Özc et al. 2009
Grapeseed A.C		48.78	-	-	-	
apricot stone A.C	pH 6.5; 25 °C	24.08	26.9	33.57	22.84	Kobya et al. 2005
hazelnut shell A.C		51.52	-			Demirbas et al. 2009
hazelnut husks A.C	pH5.7; 18 °C	6.645	-	-	13.05	Imamoglu et al. 2008
Moso bombo A.C						
S1C1	7.54; 25 °C	0.401		0.174	0.407	
S2C1	7.4; 25 °C	0.400		0.257	0.396	
Ma bamboo A.C						Lo et al., 2011
S1M1	10.02; 25°C	0.634		0.398	0.404	
S2M1	7.95; 25°C	0.398		0.396	0.397	
immobilized tannic acid A.C		2.224		1.461		Ucer et al., 2006
Carbon Nanotube	pH 5	28.595		10.903	96.969	Li et al., 2003
Olive stone waste	pH 5	5.655	5.165	16.972	28.386	Fiol et al. 2006
Bamboo activated carbon	pH 5; 23 °C	7.371	6.457	5.170	-	Liu et al., 2005
COSAC	pH 5; 30°C	14.552	13.536	60.021	112.348	In this study

Table 5: Comparison of maximum adsorption capacities (mg.g^{-1}) with various adsorbents for different metal ions.

The affinity constant, K_L, which is related to the adsorbent energy, followed this order: Cu(II) (24.302 l.mmol^{-1}) > Pb(II) (14.091 l.mmol^{-1}) > Cd(II) (12.068 l.mmol^{-1}) > Ni(II) (6.609 l.mmol^{-1}). These results indicate that probably not all binding sites on COSAC surface may be available for Cu (II) binding due to its relatively higher hydration energy [20].

The ability of an adsorbent to adsorb heavy metals is reflected in K_F Freundlich constant. The values of this parameter follows the inverse order given by the maximum experimental capacity of COSAC to remove metal ions (Cu(II)>Ni(II)>Cd(II)>Pb(II)). Furthermore, the very high n values of the Freundlich model reflect a high affinity between adsorbate and adsorbent and also indicate the existence of chemisorptions process. We can also note either that the highest intensity was obtained for copper adsorption, which highlights the large affinity of COSAC for this metal.

In comparison to other activated carbons studied for the same metal ions adsorption, the adsorption order changes strongly from one adsorbent to another. Ucer et al. [14], found that Cu(II) metal ion was more adsorbed on tannic acid activated carbon than Cd(II) and related the difference to the electronegativity of metal ions. The same tendency was found, for copper and cadmium, by Lo et al. [6], when they

studied the adsorption of metal ions onto bamboo activated carbon. The selectivity of adsorption was found to decreases in the order: Pb(II)>Cu(II)>Cd(II). The adsorption capacities order, determined by Faur-Brasquet et al. [16], follow the trend Cu(II)>Ni(II)>Pb(II) on activated carbon cloths, since the activated carbon used is microporous adsorbent, metal ions go through adsorption sites easily when the ionic diameter becomes small. According to this order, Cu(II) must have the small ion diameter than Ni(II) and Pb(II). This is different with our experimental results. These authors explained the differences in adsorption based on the ionic radius, the large ionic radius of Pb(II) compared to that of Cu(II) induces a quick saturation of adsorption sites, because of steric overcrowding thus the surface available for Cu(II) ions is then larger than for Pb(II). The same interpretation was given by Gao et al. [20], when they studied the removal of heavy metal ions on oxidized carbon nanotubes; the uptake amount followed the sequence: Cd(II)>Cu(II)>Ni(II). Whereas the order related to the maximum adsorption of lead, nickel and cadmium onto peanut husks carbon determined by Ricordel et al. [18], was found to be in the order of ionic diameters of the heavy metals used: Pb(II)>Cd(II)>Ni(II). As seen in table 4, lead is the metal with large ionic radius and some authors reported that the adsorption of metals having a larger ionic radius was higher than for those with a smaller ionic radius (Table 3).

It seems that the difference in adsorption affinity order and adsorption amount may be related not only to metal ions properties but also to physical-chemical properties of the adsorbent (morphology, surface area, pores disruption, functional groups). The reason for this is that a substance which is a good adsorbent for one adsorbate may not be a good adsorbent for another (Table 5).

Table 5 lists a comparison of maximum adsorption capacities of different adsorbents towards the metals under study, with data have been collected at different conditions. It can be seen that COSAC has a relatively high adsorption capacities 13.54, 14.55, 60.02 and 112.35 mg.g^{-1} for Ni(II), Cu(II), Cd(II) and Pb(II) ions, respectively. Therefore, COSAC has a significant potential for the removal of these metal ions from aqueous solution.

It can be seen from this table 5 that adsorption amount of metal ions differ from one activated carbon to another and this is partly because of the difference in the functional groups, the specific surface area, pore volume, micropore volume, the average diameter, etc.

For the studied activated carbon, table 1 indicated the dominating of acidic functional groups and the most important compounds are carbonyl groups (2.720 meq.g^{-1}), whereas the carboxylic groups present only 0.100 meq.g^{-1}.

Many researchers have reported the importance effect of acid functional groups on the adsorption capacity for heavy metal ions [6,21,22]. Kadirvelu et al. [15], indicated that carboxylic groups on adsorbents can play an important role in the adsorption of Cu(II), Pb(II) and Ni(II). Results found by Faur-Brasquet et al. [16], when they studied the removal of Cu(II) and Ni(II) onto three activated carbons; indicated that activated carbon with high carboxylic functions and lower pH$_{pzc}$ is very efficient for copper and nickel metal ions removal.

Lo et al. [6], compared the removal efficiency of two groups of activated carbons; S1M1 and S2C1, based in the difference of the specific surface area, micropore volume and average pore diameter. Results show that for the same group of activated carbons, the adsorption capacity and removal efficiency of heavy metal ions were obtained by carbon that has the large specific surface area, pore volume and micropore area. Nevertheless, these interpretations were not always true when we compare two different groups of activated carbons. When comparing the results showed in table 5, we can conclude that adsorption of heavy metal ions depends on the adsorbent and its preparation by different oxidation agents that influence the surface area, pores distributions, and gives the acidic or basic characteristics of the functions groups formed on the surface of adsorbent. This can be improved by the high difference in the amount adsorption between olive stone waste [23], and the carbon studied derived either from olive stones and oxidised by ortho-phosphoric acid.

Conclusion

The removal of heavy ionic metals from aqueous solution by adsorption on olive stone activated carbon produced by thermo chemical process using phosphoric acid (COSAC) was investigated.

The sorbents adsorption performance was evaluated at different initial pH, contact time and initial ion concentration and the results were discussed on the basis of the sorbent physical and chemical properties. The obtained results show that metal adsorption is pH-dependent and maximum adsorption was found to occur at initial pH 5. Adsorption equilibrium was fast and it was achieved after 4hours for the three different initial concentrations studied. The kinetics of metal adsorption on COSAC follows pseudo-second order rate expression. The equilibrium adsorption data are better fitted by the Langmuir model as compared to Freundlich model and the adsorption capacity of COSAC decreased in the order: Pb(II)$\geq$ Cd(II)>Ni(II)$\approx$ Cu(II). The impregnation with phosphoric acid confers to COSAC characteristics of acidic functional groups (low pH$_{pzc}$), high surface area and developed micropores. All these features make COSAC a promising sorbent for the removal of heavy metal ion compared to other adsorbents.

References

1. Molinari R, Poerio T, Cassano R, Picci N, Argurio P (2004) Copper(II) removal from wastewaters by a new synthesized selective extractant and SLM viability. Ind Eng Chem Res 43: 623-628.

2. Sivulka DJ (2005) Assessment of respiratory carcinogenicity associated with exposure to metallic nickel: a review. Reg Toxicol Pharmacol 43: 117-133.

3. Swaddiwudhipong W, Limpatanachote P, Mahasakpan P, Krintratun S, Punta B, et al. (2012) Progress in cadmium-related health effects in persons with high environmental exposure in northwestern Thailand: A five-year follow-up. Environmental Research 112 194-198.

4. Mudipalli A (2007) Lead hepatotoxicity & potential health effects. Indian J Med Res 126: 518-527.

5. Kobya M, Demirbas E, Senturk E, Ince M (2005) Adsorption of heavy metal ions from aqueous solutions by activated carbon prepared from apricot stone. Bioresource Technology 96: 1518-1521.

6. Lo SF, Wang SY, Tsai MJ, Lin LD (2012) Adsorption capacity and removal efficiency of heavy metal ions by Moso and Ma bamboo activated carbons. Chemical Engineering and Design 90: 1397-1406.

7. Ghrib H, Ouederni A (2005) Transformation du grignon d'olive Tunisien en charbon actif par voie chimique à l'acide phosphorique. Récents Progrès en Génie des Procédés 92.

8. Milonjic SK, Cerovi LS, Cokeša DM, Zec S (2007) The influence of cationic impurities in silica on its crystallization and point of zero charge. Journal of Colloid and Interface Science 309: 155-159.

9. Oickle AM, Goertzen SL, Hopper KR, Abdalla YO, Andreas HA (2010) Standardization of the Boehm titration: Part II. Method of agitation, effect of filtering and dilute titrant. Carbon 48: 3313-3322.

10. Prahas D, Kartika Y, Indraswati N, Ismadji S (2008) Activated carbon from jackfruit peel waste by H3PO4 chemical activation: Pore structure and surface chemistry characterization. Chemical Engineering Journal 140: 32-42.

11. Tajar AF, Kaghazchi T, Soleimani M (2009) Adsorption of cadmium from aqueous solutions on sulfurized activated carbon prepared from nut shells. Journal of Hazardous Materials 165: 1159-1164.

12. Srivastava VC, Mall ID, Mishra IM (2008) Adsorption of toxic metal ions onto activated carbon, Study of adsorption behaviour through characterization and kinetics. Chemical Engineering and Processing 47: 1269-1280.

13. Puigdomenech I (2004) Hydra/Medusa chemical equilibrium diagrams and plotting software KTH.

14. Ücer A, Uyanik A, Aygün SF (2006) Adsorption of Cu(II), Cd(II), Zn(II), Mn(II) and Fe(III) ions by tannic acid immobilized activated carbon. Sep Purif Technol 47: 113-118.

15. Kadirvelu K, Faur-Brasqnet C, Le Cloirec P (2000) Removal of Cu(II), Pb(II) and Ni(II) by adsorption onto activated carbon cloths. Langmuir 16: 8404-8409.

16. Faur-Brasquet C, Kadirvelu K, Le Cloirec P (2002) Modeling the adsorption of metal ions (Cu2+, Ni2+, Pb2+) onto ACCs Using surface complexation models. Applied Surface Science 196: 356-365.

17. Azouaou N, Sadaoui Z, Djaafri A, Mokaddem H (2010) Adsorption of cadmium from aqueous solution onto untreated coffee grounds: Equilibrium, kinetics and thermodynamics. Journal of Hazardous Materials 184: 126-134.

18. Ricordel S, Taha S, Cisse I, Dorange G (2005) Heavy metals removal by adsorption onto peanut husks carbon: characterization, kinetic study and modelling. Separation and Purification Technology 24: 389-401.

19. Giles CH, Smith D, Huitson A (1974) A general treatment and classification of the solute adsorption isotherm. I. Theoretical. J Colloid Interf Sci 47: 755-765.

20. Gao Z, Bandosz TJ, Zhao Z, Han M, Qiu J (2009) Investigation of factors affecting adsorption of transition metals on oxidized carbon nanotubes. Journal of Hazardous Materials 167: 357-365.

21. Ahmendna M, Marshall WE, Rao RM (2000) Production of granular activated carbon from select agricultural by-products and evaluation of their physical chemical and adsorption properties. Bioresour Technol 71: 113-123.

22. Machida M, Yanazaki R, Aikawa M, Tatsumoto H (2005) Role of minerals in carbonaceous adsorbents for removal of Pb (II) ions from aqueous solution. Sep Purif Technol 46: 88-94.

23. Fiol N, Villaescusa I, Martınez M, Miralles N, Poch J, et al. (2006) Sorption of Pb(II), Ni(II), Cu(II) and Cd(II) from aqueous solution by olive stone waste. Separation and Purification Technology 50: 132-140.

24

Synthesis of Novel Series of Phthalazine Derivatives with Antibacterial and Antifungal Evaluation

Maher A El-Hashash[1], Dalal B Guirguis[1]*, Nayera AM Abd El-Wahed[2] and Mohamed A Kadhim[3]

[1]*Chemistry Department, Faculty of Science, Ain Shams University, Abassia 11566, Cairo, Egypt*
[2]*Department of Chemistry of Natural and Microbial Products, Division of Pharmaceutical Industries, National Research Center, El Behos street, Dokki, Egypt*
[3]*Chemistry Department, Faculty of Science, Anbar University, Iraq*

Abstract

The oxirane derivative (2) was allowed to react with hydrazine hydrate, 4-aminobenzoic acid and o-phenylene diamine to give β-hydrazine alcohol derivative (3) and β-amino derivatives (4) and (5). The hydrazide (8) reacted with glucose, phthalic anhydride and aromatic aldehydes to give the phthalazine derivatives (9), (15) and (16a-c). A new heterocyclic molecules were synthesized using ethyl acetoacetate, acetyl acetone and benzoyl chloride with the hydrazide (8) to give the pyrazole derivatives (12), (14) and the oxadiazole derivative (18). The new compounds were synthesized with the objective of studying their antifungal and antimicrobial activity. Some of them gave positive results. The newly synthesized compounds were characterized on the basis of their spectral (1H-NMR, Mass spectrum, IR and Elementary analysis).

Keywords: Pyrazole; Epichlorohydrin; Glucose; Oxadiazole

Introduction

Phthalazin-1(2H)-ones are of considerable interest due to their antidiabetic [1], antiallergic [2] , vasorelaxant [3], PDE4 inhibitors [4], VEGF (vascular endothelical group factor) receptor tyrosine kinases for the treatment of cancer [5,6], antiasthmatic agents with dual activities of thromboxane A2 (TXA2) synthetase inhibition and bronchodialation [7], herbicidal [8]. A number of established drug molecules like Hydralazine [9,10], Burdralazine [11,12], Azelastine [13,14], Ponalrest [15], and Zopolrest [16] are prepared from the corresponding phthlazinones. Several phthalazine derivatives have been reported to possess antitumor [17-20], antihypertensive [21,22], anticovulsant [23,24], antimicrobial [25], antitrypanosomal [26], and anti-inflammatory activities [27,28]. Most of the current nonsteroidal anti-inflammatory drugs (NSAIDs) show serious side effects including gastrointestinal disorders and kidney damage. These studies for developing safer NSAIDs lacking the gastrointestinal and renal side effects of current used ones have recently been of interest for many researches. Most of the classical NSAIDs exerts their side effects by inhibition of COX-1 enzyme since the COX-1 isoform is the constitutive one that is responsible for regulation of physiological processes, and the COX-2 isoform is discovered to be the enzyme induced by a inflammatory stimuli, selective inhibition of COX-2 provides a rationale for developing anti-inflammatory and analgesic agents. Although the diaryl heterocyclic compounds are mainly studied as new class of NSAIDs without gastric side effects, many studies have also focused on a different type of compounds to develop safer NSAIDs [27]. Also in terms of this aspect, many studies have been focused on pyridazin-(3H)-ones, which are characterized to possess good analgesic and anti-inflammatory activities. Beside pyridazinones, these studies have indicated that the heterocyclic ring substitutions at the six position, and the presence of acetamide side chain when linked to the lactam nitrogen of pyridazinone ring at the two position of the pyridazinone ring, improve the analgesic and antiflammatory activity along with nil or very low ulcerogenicity [29-34]. In view of the aforementioned facts, it seemed most interesting to synthesize some [4-(2-tetryl)-2-substituted phthalazin-1(2H)-one] derivatives with the aim to obtain more precise information about the course of reactions and biological activities.

Results and Discussion

Upon reacting (1) [35] with epichlorohydrin in the presence of anhydrous potassium carbonate in dry acetone on heating water bath afforded 2-(oxiran-2-yl methyl-4-tetryl phthalazin-1(2H)-one (2). The structure of (2) was inferred from correct microanalytical data as well as IR spectrum which revealed strong absorption bands at ν 1134, 1655, 2852, 2952 cm^{-1} attributable to O-C, C=O, and CH with devoid of any band for NH. ^{1}H NMR spectrum also showed δ: 1.26 (t, 4H, β-methylene protons of tetryl moeity), 1.26 (t, 4H, α- methylene protons of tetryl moeity), 2.86 (2H, octet), 3.5 (1H, octet stereogenic protons), 4.5 (2H, diastereotopic protons), and 7.2-7.8 (m, 7H, Ar-H). The EIMS showed m/z at 332 corresponding to M$^+$(molecular ion peak). The reaction possibly takes place via the following mechanism (Figure 1).

The mechanism involves opening of the more reactive oxirane nucleus followed by ring closure via SN$_2$ mechanism. The function of KCO$_3^-$ is to augment the removal of leaving group (chloride ion).

Also the structure of (2) was verified chemically with nitrogen nucleophiles namely hydrazine hydrate, p-aminobenzoic acid and o-phenylene diamine. Thus when (2) reacted with hydrazine hydrate in boiling ethanol yielded the β-hydrazino alcohol 2-(4-hydrazinyl-2-hydroxybutyl)-4-(5,6,7,8-tetrahydronaphthalen-2-yl)phthalazin-1(2H)-one (3) in good yield with high regioselectivity with preferential attack at the nonbenzilic type terminal carbon (less hindered). The IR spectrum which revealed strong absorption bands at ν 1158, 1663, 2857, 2927 and broad peak centered at 3374 cm^{-1} attributable to O-C, C=O, CH, NH, and OH. ^{1}H NMR spectrum also showed δ: 1.7 (s,

*Corresponding author: Dalal B Guirguis, Chemistry Department, Faculty of Science, Ain Shams University, Abassia 11566, Cairo, Egypt

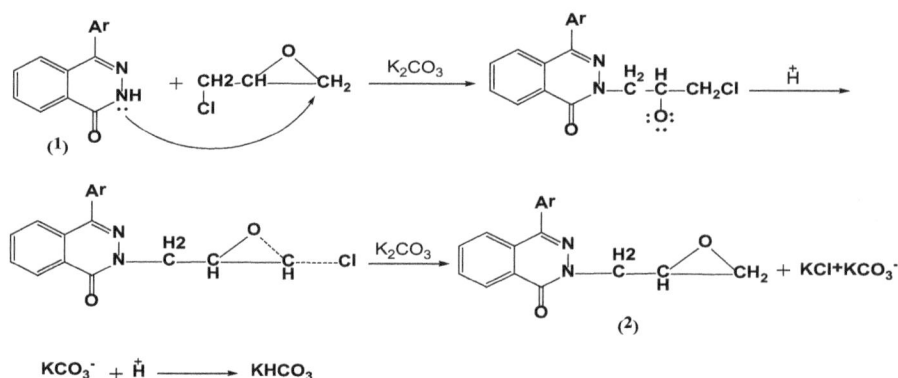

Figure 1: [4-(2-tetryl)-2-substituted phthalazin-1(2H)-one] derivatives.

4H, β- methylene protons of tetrylmoeity), 2.8 (s, 4H, α- methylene protons of tetryl moeity), 3.7 (m, 1H, stereogenic protons), 4.5 (m, 4H, diastereotopic protons), 7.2-7.8 (m, 7H, Ar-H), 8.5-(2, 2H, NH$_2$), and 10.4 (s, 1H, NH). Also upon interaction of oxirane derivative (2) the β-amino alcohol derivative 4-(3-hydroxy-4(1-oxo-4-(5,6,7,8-tetrahydronaphthalen-2-yl)phthalazin-2(1H)-yl)butyl amino)benzoic acid (4). The reaction exhibited regioselective hetero ring opening at carbon atom. Since the β-amino alcohol are versatile compounds with pharmaceutical and biological importance, this prompted us to extend the reaction of oxirane derivative (2) with o-phenylene diamine with the aim of obtaining the β-amino alcohol 2-(4-(2-aminophenylamino)-2-hydroxybutyl)-4-(5,6,7,8-tetrahydronaphthalen-2-yl)phthalazin-1(2H)-one (5). The reaction takes place via hetero ring opening of oxirane nucleus regioselectivity at the primary carbon atom leading to the desired product (5). EIMS showed m/z at 441 corresponding to M$^+$(molecular ion peak) at 332 corresponding to M$^+$(molecular ion peak). Also the ^{1}H NMR and IR specta were in good agreement with the desired structure (c.f. experimental). The structure of (5) was inferred chemically via its interaction with aromatic aldehydes namely naphthaldehyde and salicylaldehyde (6a,b) showing strong absorption bands at v 1650, 1687, 3052, cm^{-1} attributable to C=N, C=O, CH, NH, and OH.

The previously prepared hydrazide (8) [35] namely 2-(1-oxo-4(5,6,7,8-tetrahydronaphthalen-2-yl-phthalazin-2(1H)-yl) acetohydrazide was allowed to react with glucose which on turn reacted with acetic anhydride afforded the corresponding sugar hydrazones E-2-(1-oxo-4(5,6,7,8-tetrahydronaphthalen-2-yl)phthalazin-2(1H)-yl)-N'-(2,3,4,5,6-pentahydroxyhexylidene)acetohydrazide and 5-2-(2-(1-oxo-4-(5,6,7,8-tetrahydronaphthalen-2-yl)phthalzin-2(1H)-yl)acetyl)hydrazono)pentane-1,1,2,3,4-pentyl pentaacetate (9) and (10). The structure of (9) was inferred from correct microanalytical data as well as IR spectrum which revealed strong absorptipn bands at v 1625, 1643, 1687, 2917, 2938, 3313 and 3409 cm^{-1} attributable to C=N, C=O, CH, NH, and OH. On the other hand the IR spectrum of compound (10) revealed strong absorptipn bands at v 1658, 1752, 2860, 2936, 3300 cm^{-1} attributed to C=O, C=O, CH, NH and devoid any band for OH. Also the ^{1}H NMR was in good agreement with the desired structure (c.f. experimental).

Interaction of the hydrazide (8) with ethylacetoacetate in boiling ethanol for 3 hours afforded the ester derivative (E)-ethyl-3-(2-(2-(1-oxo-4-(5,6,7,8-tetrahydronaphthalen-2-yl)phthalazin-2(1H)-yl)acetyl) hydrazono)butanoate (11). The structure of (11) was inferred from correct microanalytical data as well as IR spectrum which revealed

strong absorption bands at v 1660, 1742, 2854 and 2924 and 3211 cm^{-1} due to two carbonyl group, CH and NH which agreed well with the proposed structure. On the other hand when the reaction was subjected for prolonged reflux (9 hours), 3-methyl-5-hydroxy pyrazole derivative 2-(2-(3-Methyl-5-oxo-4,5-dihydro-1H-pyrazol-1-yl)-2-oxoethyl)-4-(5,6,7,8-tetrahydronaphthalen-2-yl)phthalazin-1(2H)-one (12) was obtained. The structure of the pyrazole (12) was deduced from correct microanalytical data, IR spectrum revealed strong absorption bands at v 1629, 1666 and 3200 cm^{-1} attributed to two carbonyl group and OH . The EIMS exibits m/z at 414. Compound (12) present in enol form due to extended conjugation and intramolecular H-bond. Also the hydrazide (8) reacted with acetylacetone in boiling ethanol for 3 hours to give the acetone derivative Z-2-(1-oxo-4-(5,6,7,8-tetrahydronaphthalen-2-yl)phthalazin-2(1H)-yl-N'-(4-oxopentan-2-ylidene)acetohydrazide (13). The IR spectrum revealed strong absorption bands at v 1660, 1721, 2930 and 3210 cm^{-1} attributed to two carbonyl groups, CH and NH which agreed well with the proposed structure. The EIMS exhibit m/z 412 (M$^+$-H$_2$O). On the other hand when the reaction was carried out in boiling ethanol for 8 hours, the pyrazole derivative 2-(2-(3,5-dimethyl-4,5-dihydro-1H-pyrazol-1-yl)-2-oxoethyl)-4-(5,6,7,8-tetrahydronaphthalen-2-yl)phthalazin-1(2H)-one (14) was obtained. The IR spectrum revealed strong absorption bands at v 1654, 1680, 2925 cm^{-1} attributed to two carbonyl groups and CH. The EIMS exhibits m/z (412, M$^+$)

When the hydrazide (8) was allowed to react with phthalic anhydride in oil bath at 150°C, yielded the corresponding phthalimide derivative N-(1,3-dioxoisoindolin-2-yl)-2-(1-oxo-tetrahydronaphthalen-2-yl) phthalazin-2(1H)-yl)acetamide (15). The structure of (15) was inferred from the IR spectrum which revealed strong absorption bands at v 1650, 1733, 1791 cm^{-1} (mechanical coupling of imide moeity), 2860, 2929, 2999 and 3180 cm^{-1} attributable to C=O, CH, and NH. The EIMS showed m/z 474 (M$^+$-4H), 333 (M$^+$-phthalimide) as well as correct micro analytical data (c.f. experimental).

The hydrazide (8) reacted with some aromatic aldehydes namely benzaldehyde, anisaldehyde and piperonldehyde afforded 2-(2-arylmethyl diazinyl)-2-oxoethyl)-4-(5,6,7,8-tetrahydronaphthalen-2-yl)phthalazin-1(2H)-one namely N'-benzylidene-2-(4-(1,2,3,4-tetrahydronaphthalen-2-yl)-1-oxophthalazin-2(1H)-yl)acetohydrazide, N'-4-methoxybenzylidene-2-(4-(1,2,3,4tetrahydronaphthalen-2-yl)-1-oxophthalazin-2(1H)-yl) acetohydrazide, (Z)-N'-(benzo[d][1,3]dioxol-4-ylmethylene)2-(1-oxo-4-(5,6,7,8-tetrahydronaphthalen-2-yl)phthalazin-2(1H)-yl) acetohydrazide (16a-c). The structure of (16a-c) was inferred from

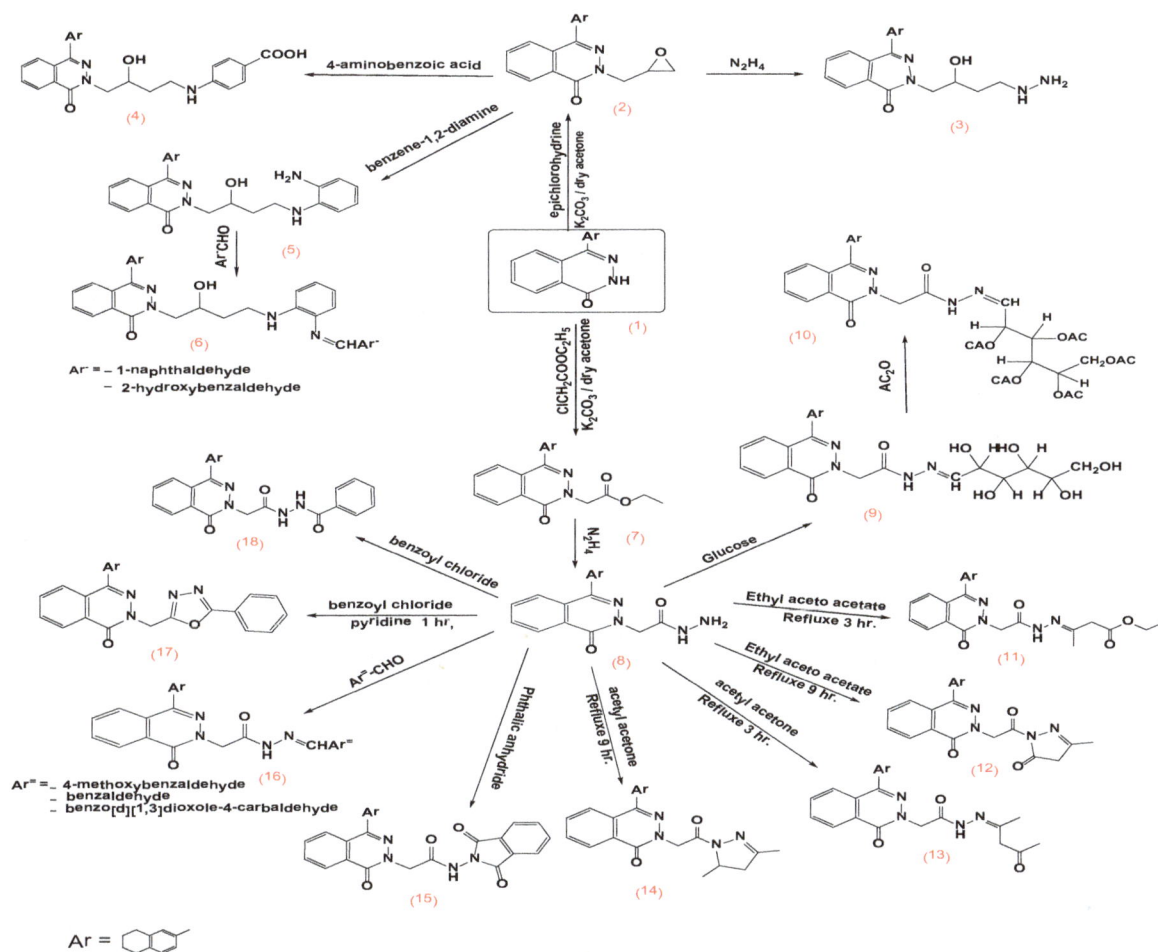

Scheme 1: The synthesis of several heterocyclic derivatives.

correct microanalytical and IR spectrum which revealed strong absorption bands in the region ν 1660-1647, 1669-1664, and 3196-3190 cm⁻¹ attributed C=O and NH.

The phthalazinone acetic acid hydrazide (**8**) used as versatile starting material for the synthesis of several heterocyclic derivatives through its reactions with a variety of activated reagents. Thus the hydrazide (**8**) reacted with benzoyl chloride to give N'(2-(1-oxo-4-(1,2,3,4-tetrahydronaphthalen-2-yl)phthalazin-2(1H)-yl)benzoyl acetohydrazide (**18**). The IR spectrum of (**18**) revealed strong absorption bands at ν 1615, 1687, 1715 cm⁻¹ attributed to three carbonyl groups (high values due to mutual inductions between the groups), 2925 and 3234 and cm⁻¹ attributable to CH and NH. The EIMS showed *m/z* 434 (M⁺-H₂O) in addition to elemental analysis as well as ¹H NMR (c.f. experimental). On the other hand when the hydrazide (**8**) reacted with benzoyl chloride in pyridine gave the cyclized oxaxdiazolophthalazine 2-(5-phenyl-2,3-dihydro-1,3,4-oxadiazol-2-yl)methyl)-4-(5,6,7,8-tetrahydronaphthalen-2-yl)phthalazin-1(2H)-one (**17**). The IR spectrum of (**17**) exhibited band at ν 1660 cm⁻¹ attributed to C=O, and devoid any band for NH (scheme 1).

Antimicrobial activity

The antibacterial activity of the synthesized compounds was tested against *Bacillus subtilis, Staphylococcus aureus* (Gram-positive bacteria),

Escherichia coli, Pseudomonas sp. (Gram-negative bacteria) using nutrient agar medium. The antifungal activity of the compounds was tested against *Candida albicans and Aspergillus niger* using Sabouraud dextrose agar medium.

Agar Diffusion Medium

All compounds were screened *in vitro* for their antimicrobial activity against, by agar diffusion method [36]. A suspension of the organisms were added to sterile nutrient agar media at 45°C and the mixture was transferred to sterile Petri dishes and allowed to solidify. Holes of 10 mm in diameter were made using a cork borer. An amount of 0.1 ml of the synthesized compounds was poured inside the holes. A hole filled with DMSO was also used as control. The plates were left for 1 h at room temperature as a period of pre-incubation diffusion to minimize the effects of variation in time between the applications of the different solutions. The plates were then incubated at 37°C for 24 h and observed for antimicrobial activity. The diameters of zone of inhibition were measured and compared with that of the standard. Ciprofloxacin (50 µg/ml) and Fusidic acid (50 µg/ml) were used as standard for antibacterial and antifungal activity respectively. The observe zone of inhibition is presented in Table 1.

Experimental

All melting points are uncorrected and were measured on an

Compounds	Microorganism					
	Gram +ve bacteria		Gram –ve bacteria		Fungi	
	Bacillus subtilis	*Staphylococcus aureus*	*Escherichia coli*	*Pseudomonas aeruginosa*	*Candida albicans*	*Aspergillus niger*
1	++++ ve	+++ ve	++++ ve	++++ ve	+++ ve	+ve
2	++++ ve	+++ ve	++++ ve	++++ ve	+++ ve	+ve
6b	+ ve	+ ve	+ ve	+ ve	+ ve	-ve
7	++ ve	++ ve	++ ve	++ ve	++ ve	-ve
8	++++ ve	+++ ve	++++ ve	++++ ve	+++ ve	+ve
9	++++ ve	+++ ve	++++ ve	++++ ve	+++ ve	+ve
12	++++ ve	+++ ve	++++ ve	++++ ve	+++ ve	+ve
13	+ ve	+ve	+ve	+ve	+ve	-ve
15	++++ ve	+++ ve	++++ ve	++++ ve	+++ ve	+ve
16c	++++ ve	+++ ve	++++ ve	++++ ve	+++ ve	+ve
17	++++ ve	+++ ve	++++ ve	++++ ve	+++ ve	+ve
18	+++ ve	++ ve	+++ ve	+++ ve	++ ve	-ve

Highly active (++++) =   (inhibition zone > 20 mm)
Moderately active (+++) = (inhibition zone 16- 20 mm)
Slightly active (++) =   (inhibition zone 13 - 15 mm)
Weakly active (+) = (inhibition zone 11 – 12 mm)
Inactive =   (inhibition zone < 11 mm)

Table 1: Antibacterial and antifungal activities of the newly synthesized compounds.

electrothermal melting point apparatus. Elemental analyses were performed using a Heraeus CHN Rapid analyzer at the Microanalytical unit, Cairo University. Thin-layer chromatography (TLC) was performed on Merck TLC aluminum sheets silica gel 60 F $_{254}$ with detection by UV quenching at 254 nm. IR spectra were measured on a Unicam SP-1200 spectrophotometer using KBr wafer technique. ^{1}H NMR spectra were measured in DMSO-d$_6$ on a Varian plus instrument (300 MHz). Mass spectra were recorded on a Shimadzu GC-MS QP 1000 EX instrument operating at 70 eV in EI mode.

2-(Oxiran-2yl methyl-4-tetryl phthalazin-1(2H)-one [2]

A mixture of (1) (0.01 mol) (2.7 g), epichlorohydrin (0.01 mol) (0.8 g) and potassium carbonate (0.04 mol) (6.7 g) in 30 mL dry acetone was refluxed for 24 hs. The resultant solid was filtered and crystallized from ethanol to give (2), 50% yields as colorless crystals mp 120-121°C. ^{1}H NMR (DMSO-d$_6$, 300 MH$_z$) δ: 1.26 (t, 4H, β- methylene protons of tetryl moeity), 1.26 (t, 4H, α- methylene protons of tetryl moeity), 2.86 (2H, octet), 3.5 (1H, octet stereogenic protons), 4.5 (2H, diastereotopic protons), 7.2-7.8 (m, 7H, Ar-H). IR (KBr) γ: 1655(C=O), 1579 (C=N), 1134 (C-O) cm^{-1}; MS, m|z (%); M$^+$, 332 (14), 290 (22), 276 (100). Anal cald for C$_{21}$H$_{20}$N$_2$O$_2$: C 75.87, H 6.07, N 8.43; found C 76.23, H 6.15, N 9.72

2-(4-Hydrazinyl-2-hydroxybutyl)-4-(5,6,7,8-tetrahydronaphthalen-2-yl)phthalazin-1(2H)-one [3]

A mixture of (2) (0.01 mol) (3.3 g) and 3 mL hydrazine hydrate in 30 mL ethanol was refluxed for 3 hs. The resultant solid was filtered and crystallized from ethanol to give (3), 30% yield as colorless crystals mp 85°C. ^{1}H NMR (DMSO-d$_6$, 300 MH$_z$) δ: 1.7 (s, 4H, β- methylene protons of tetryl moeity), 2.8 (s, 4H, α- methylene protons of tetryl moeity), 3.7 (m, 1H, octet stereogenic protons), 4.5 (m, 4H, diastereotopic protons), 7.2-7.8 (2m, 8H, Ar-H), 8.5 (s, 2H, NH$_2$, exchangeable with water), 10.4 (s, 1H, NH, exchangeable with water). IR (KBr) γ: 3376 (OH), 2927, 2857 (NH), 1669, 1647 (C=O), 1574 (C=N) cm^{-1}. Anal cald for C$_{22}$H$_{26}$N$_4$O$_2$: C 69.82, H 6.92, N 14.80; found C 70.43, H 6.35, N 14.40

4-(3-Hydroxy-4(1-oxo-4-(5,6,7,8-tetrahydronaphthalen-2-yl) phthalazin-2(1H)-yl)butyl amino)benzoic acid [4]

A mixture of (2) (0.001 mol) (0.3 g) and p-aminobenzoic acid (0.001 mol) (0.14 g) in 30 mL ethanol was refluxed for 3 hs. The resultant solid was filtered and crystallized from ethanol to give [4], 40% yield as colorless crystals mp 78°C. IR (KBr) γ: 3459 (OH), 3360 (NH), 1670, 1629 (C=O), 15798 (C=N), 1158 (C-O) cm^{-1} MS, m|z (%); M $^+$, 469 (1), 451 (2), 276 (5), 248 (2), 121 (100). Anal cald for C$_{29}$H$_{29}$N$_3$O$_4$: C 72.03, H 6.04, N 8.69; found C 71.75, H 6.35, N 8.31

2-(4-(2-Aminophenylamino)-2-hydroxybutyl)-4-(5,6,7,8-tetrahydronaphthalen-2-yl)phthalazin-1(2H)-one [5]

A mixture of (2) (0.01 mol) (3.3 g) and o-phenylenediamine (0.01 mol) (14 g) in 30 mL ethanol was refluxed for 3 hs. The resultant solid was filtered and crystallized from ethanol to give [5], 40% yields as colorless crystals mp 78°C. ^{1}H NMR (DMSO-d$_6$, 300 MH$_z$) δ: 1.2 (s, 4H), 2.48-2.8 (m, 8H, 4-methylene), 3.27 (s, 2H, NH$_2$ exchangeable with water), 4.31 (s, methine proton), 6.3-7.26 (m, 11H, Ar-H), 7.45 (s, 1H, NH), 7.9 (s, 1H, OH, exchangeable with water). IR (KBr) γ: 3384, 3362 (NH$_2$), 3287 (OH), 3280 (NH), 1633 (C=O), 1589 (C=N) cm^{-1} MS, m|z (%); M$^+$, 441 (13), 290 (16), 104 (21). Anal cald for C$_{28}$H$_{30}$N$_4$O$_2$: C 73.98, H 6.65, N 12.33; found C 73.43, H 6.25, N 11.72

2-(2-Hydroxy-3-(2-(naphtmetyleneamino)phenylamino) propyl)-4-(5,6,7,8-tetrahydronaphthalen-2-yl)phthalazin-1(2H)-one [6a]

A mixture of (5) (0.01mol, 4.5 g) and 1-naphthaldehyde (0.01 mol, 1.5 g), was refluxed in 20 mL of absolute ethanol for 3 hs. The solid obtained upon cooling was collected by filtration, dried, and crystallized from ethanol to give [6a], 60% yield, mp 195°C , IR ν : 1650, 1687, 3052 and 3400 cm^{-1} attributable to C=N, C=O, CH, NH, and OH. Anal cald for C$_{38}$H$_{34}$N$_4$O$_2$: C 78.87, H 5.92, N 9.68; found C 79.28, H 5.36, N 10.1

2-(2-Hydroxy-3-hydroxybenzylideneamino)phenylamino) propyl)-4-(5,6,7,8-tetrahydronaphthalen-2-yl)phthalazin-1(2H)-one [6b]

A mixture of (5) (0.01mol, 4.5 g) and 2-hydroxy benzaldehyde (salicylaldehyde) (0.01mol, 1.2g), was refluxed in 20 mL of absolute ethanol for 3 hs. The solid obtained upon cooling was collected by filtration, dried, and crystallized from the ethanol to give [6b], mp 235°C , yield 50% IR: ν 1650, 1687, 3052, cm^{-1} attributable to C=N,

C=O, CH, NH, and OH. Anal cald for $C_{34}H_{32}N_4O_3$: C 74.98, H 5.92, N 10.29; found C 75.32, H 6.15, N 9.71

E-2(1-oxo-4(5,6,7,8-tetrahydronaphthalen-2-yl)phthalazin-2(1H)-yl)-N'-(2,3,4,5,6-pentahydroxyhexylidene)acetohydrazide[9]

A mixture of (8) (0.001 mol) (0.3 g) and glucose (0.001 mol) (0.8 g) in 30 mL ethanol was refluxed for 3 hs and cooled at room temperature. The resultant solid was filtered and crystallized from ethanol to give [9], 70% yields as colorless crystals mp 200°C. IR (KBr) γ: 3409, 3313 (OH) (NH), 2917 (CH) 1687, 1643 (C=O), 1574 (C=N) cm⁻¹. Anal cald for $C_{26}H_{30}N_4O_7$: C 61.17, H 5.92, N 10.97; found C 60.88, H 5.46, N 10.50

5-2(2-(1-Oxo-4-(5,6,7,8-tetrahydronaphthalen-2-yl)phthalzin-2(1H)-yl)acetyl)hydrazono)pentane-1,1,2,3,4-pentyl pentaacetate [10]

A mixture of (9) (0.001 mol) (0.4 g) and 10 mL acetic anhydride was refluxed for 1 h, poured on water. The resultant solid was filtered and crystallized from benzene to give [10], 40% yield as colorless crystals mp 160-161°C. ¹H NMR (DMSO-d₆, 300 MH_z) δ: 1.72 (m, 4H, methylene), 2.2 (s, 1H, CH₃CO), 4.13-4.16 (m, methine and methylene CH₂O), 4.89 (s, 2H, NCH₂CO), 5.36 (s, 1H, azamethine), 7.27-7.92 (m, 7H, Ar-H), 8.36 (s, 1H, NH, exchangeable with water). IR (KBr) γ: 3300, 2936, (NH), 2860 (CH), 1752, 1658 (C=O), 1582 (C=N) cm⁻¹. Anal cald for $C_{35}H_{38}N_4O_{12}$: C 59.48, H 5.42, N 7.93; found C 58.9, H 5.82, N 7.45.

(E)-ethyl-3-(2-(2-(1-oxo-4-(5,6,7,8-tetrahydronaphthalen-2-yl)phthalazin-2(1H)-yl)acetyl)hydrazono)butanoate [11]

A mixture of (8) (0.001 mol) (0.34 g) and ethyl acetoacetate (0.001 mol) (0.13 g) in 20 mL ethanol was refluxed for 3 hs. The resultant solid was filtered and crystallized from ethanol to give (11), 40% yield as colorless crystals mp 160-162 °C. IR (KBr) γ: 3211 (NH), 1742, 1660, (C=O), 1581 (C=N) cm⁻¹. Anal cald for $C_{26}H_{28}N_4O_4$: C 67.81, H 6.13, N 12.17; found C 67.35, H 5.9, N 11.

2-(2-(3-Methyl-5-oxo-4,5-dihydro-1H-pyrazol-1-yl)-2-oxoethyl)-4-(5,6,7,8-tetrahydronaphthalen-2-yl)phthalazin-1(2H)-one [12]

A mixture of (8) (0.01 mol) (3.4 g) and ethyl acetoacetate (0.01 mol) (1.3 g) in 30 mL ethanol was refluxed for 9 hs. The resultant solid was filtered and crystallized from ethanol to give (12), 50% yields as colorless crystals mp 185-187°C. IR (KBr) γ: 1666, 1626 (C=O), 1583 (C=N) cm⁻¹. MS, m|z (%); M⁺, 414 (46), 399 (46), 371 (20), 289 (15), 125 (44), 97 (100). Anal cald for $C_{24}H_{22}N_4O_3$: C 69.55, H 5.35, N 13.52; found C 68.35, H 5.52, N 12.82

Z-2-(1-oxo-4-(5,6,7,8-tetrahydronaphthalen-2-yl)phthalazin-2(1H)-yl-N'-(4-oxopentan-2-ylidene)acetohydrazide [13]

A mixture of (8) (0.001 mol) (0.3g) and acetyl acetone (0.001 mol) (0.1g) in 20 mL ethanol was refluxed for 3 hs. The resultant solid was filtered and crystallized from ethanol to give (13), 50% yield as colorless crystals mp 87-88°C. IR (KBr) γ: 3210 (NH), 1742, 1660 (C=O), 1582 (C=N) cm⁻¹. MS, m|z (%); M⁺, 430 (1), 415 (5), 316 (100), 289 (50), 262 (15). Anal cald for $C_{25}H_{26}N_4O_3$: C 69.75, H 5.09, N 13.01; found C 70.35, H 5.52, N 12.82

2-(2-(3,5-Dimethyl-4,5-dihydro-1H-pyrazol-1-yl)-2-oxoethyl)-4-(5,6,7,8-tetrahydronaphthalen-2-yl)phthalazin-1(2H)-one [14]

A mixture of (8) (0.001 mol) (0.3g) and acetyl acetone (0.001 mol) (0.1 g) in 20 mL n-butanol was refluxed for 10 hs. The resultant solid was filtered and crystallized from ethanol to give (14), 50% yields as colorless crystals mp 130-135°C. ¹H NMR (DMSO-d₆, 300 MH_z) δ: 1.644 (s, 3H, methyl), 1.663 (s, 3H, methyl), 1.72 (m, 4H, methylene), 2.5 (t,4H), 5.8 (s, 2H), 4.89 (s, 2H, NCH₂CO), 7.27-7.92 (m, 7H, Ar-H). IR (KBr) γ: 1680, 1654 (C=O), 1583 (C=N) cm⁻¹. MS, m|z (%); M⁺, 412 (5), 316 (100), 290 (45), 261 (30), 247 (27). Anal cald for $C_{25}H_{26}N_4O_2$: C 72.43, H 6.33, N 13.52; found C 72.25, H 5.92, N 14.1,.

N-(1,3-dioxoisoindolin-2-yl)-2-(1-oxo-tetrahydronaphthalen-2-yl)phthalazin-2(1H)-yl)acetamide [15]

A mixture of (8) (0.001 mol) (0.3 g) and phthalic anhydride (0.001 mol) (0.14 g) in 15 mL acetic acid was refluxed for 8 hs and cooled at room temperature. The resultant solid was filtered and crystallized from ethanol to give [15] 60% yield as colorless crystals mp 165°C. IR (KBr) γ: 3180 (NH), 1793, 1733, 1650 (C=O), 1579 (C=N) cm⁻¹ MS, m|z (%); M⁺, 474 (9), 333 (21), 313 (21), 289 (100), 276 (22), 247 (42), 105 (25). Anal cald for $C_{28}H_{22}N_4O_4$: C 70.27, H 4.46, N 11.71; found C 70.35, H 5.12, N 12.43.

N'-benzylidene-2-(4-(1,2,3,4-tetrahydronaphthalen-2-yl)-1-oxophthalazin-2(1H)-yl)acetohydrazide [16a]

A mixture of (7) (0.001 mol) (0.3 g) and benzaldehyde (0.001mol) (0.1g) in 20 mL ethanol was refluxed for 3 hs and cooled at room temperature. The resultant solid was filtered and crystallized from ethanol to give (16a), 40% yield as colorless crystals mp 188-190°C. IR (KBr) γ: 3480 (NH), 1669, 1647 (C=O), 1574 (C=N) cm⁻¹. Anal cald for $C_{27}H_{24}N_4O_2$: C 74.29, H 5.54, N 12.84; found C 74.57, H 5.32, N 12.61

N'-4-methoxybenzylidene-2-(4-(1,2,3,4tetrahydronaphthalen-2-yl)-1-oxophthalazin-2(1H)-yl)acetohydrazide [16b]

A mixture of (8) (0.001 mol) (0.3 g) and p-methoxybenzaldehyde (0.001 mol) (0.13 g) in 20 mL ethanol was refluxed for 3 hs and cooled at room temperature. The resultant solid was filtered and crystallized from ethanol to give [16b], 60% yields as colorless crystals mp 140°C. IR (KBr) γ: 3195 (NH), 1663, 1603 (C=O), 1578 (C=N), cm⁻¹. Anal cald for $C_{28}H_{26}N_4O_3$: C 72.09, H 5.62, N 12.01; found C 72.57, H 5.32, N 12.81

(Z)-N'-(benzo[d][1,3]dioxol-4-ylmethylene)2-(1-oxo-4-(5,6,7,8-tetrahydronaphthalen-2-yl)phthalazin-2(1H)-yl)acetohydrazide [16c]

A mixture of (8) (0.001 mol) (0.3 g) and piperonaldehyde (0.001 mol) (0.18 g) in 20 mL ethanol was refluxed for 3 hs and cooled at room temperature. The resultant solid was filtered and crystallized from ethanol to give [16c], 40% yields as colorless crystals mp 220°C. ¹H NMR (DMSO-d₆, 300 MH_z) δ: 1.1 (t, 4H, β methylene of tetryl moeity), 2.5 (m, 4H, α-methylene of tetryl moeity), 5.2 (s, 2H, N-CH₂), 6.02 (s, 1H, N=CH-Ar) 6.92-8.75 (m, 11H, Ar-H), 9.8 (s, 2H, NH, exchangeable with water). IR (KBr) γ: 3076 (NH), 1628 (C=O), 1600 (C=N), 1257 (C-O) cm⁻¹. Anal cald for $C_{28}H_{24}N_4O_4$: C 69.99, H 5.03, N 11.66; C 69.53, H 5.50, N 11.33

2-(5-Phenyl-2,3-dihydro-1,3,4-oxadiazol-2-yl)methyl)-4-(5,6,7,8-tetrahydronaphthalen-2-yl)phthalazin-1(2H)-one [17]

A mixture of (8) (0.001 mol) (0.3 g) and benzoyl chloride (0.001 mol) (0.14 g) in 20 mL pyridine was refluxed for 1 h, then poured on ice and hydrochloric acid. The resultant solid was filtered and crystallized from ethanol to give [17], 30% yield as colorless crystals mp 243°C. IR (KBr) γ: 1660 (C=O), 1582 (C=N), 1205 (C-O) cm^{-1}. Anal cald for $C_{27}H_{24}N_4O_2$: C 74.29, H 5.54, N 12.84; found C 74.75, H 5.62, N 13.22.

N'(2-(1-oxo-4-(1,2,3,4-tetrahydronaphthalen-2-yl)phthalazin-2(1H)-yl)benzoyl acetohydrazide [18]

A mixture of (8) (0.001 mol) (0.3 g) and benzoyl chloride (0.001 mol) (0.14 g) in 20 mL ethanol was refluxed for 3 hs and cooled at room temperature. The resultant solid was filtered and crystallized from ethanol to give [18], 60% yields as colorless crystals mp 140°C. ^{1}H NMR (DMSO-d$_6$, 300 MH$_z$) δ: 1.1 (t, 4H, β methylene of tetryl moeity), 2.5 (m, 4H, α –methylene of tetryl moeity), 5.2 (s, 2H, N-CH$_2$), 7.13-8.25 (m, 13H, Ar-H), 10.3 (s, 2H, NH, exchangeable with water). IR (KBr) γ: 3234 (NH), 1715, 1687, 1651 (C=O), 1581 (C=N) cm^{-1} MS, $m|z$ (%); M $^{.+}$-H$_2$O) 434 (7), 276 (40). Anal cald for $C_{27}H_{23}N_4O_3$: C 71.67, H 5.35, N 12.38; found C 72.35, H 5.02, N 12.92.

References

1. Boland OM, Blackwell CC, Clarke BF, Ewing DJ (1993) Diabetes 42: 336-340.

2. Hamamto Y, Nagai K, Muto M, Asagami C (1993) Inhibitory effect of azelastine, a potent antiallergic agent, on release of tumor necrosis factor-alpha from activated human peripheral blood mononuclear cells and U937 cells. Exp. Dermatol 2: 231-235.

3. Del Olmo E, Barboza B, Ybarra MI, Lopez-Perez JL, Carron R, et al. (2006) Vasorelaxant activity of phthalazinones and related compounds. Med Chem Lett 16: 2786-2790.

4. Napoletano M, Norichini G, Pellacini F, Marchini F, Morrazoni G, et al. (2000) The synthesis and biological evaluation of a novel series of phthalazine PDE4 inhibitors I Bioorg. Med Chem Lett 10: 2235-2238.

5. Bold G, Altmann KH, Masso E, Roth R, Wood JM, et al. (2000) New anilinophthalazines as potent and orally well absorbed inhibitors of the VEGF receptor tyrosine kinases useful as antagonists of tumor-driven angiogenesis. Med Chem 43: 2310-2323.

6. Arif JM, Kunhi M, Bekhit AA, Subramanian MP, Al-Hussein K, et al.(2006) Evaluation of apoptosis-induction by newly synthesized phthalazine derivatives in breast cancer cell lines. Asian Pac J Cancer Prev 7: 249-252.

7. Yamaguchi M, Kamel K, Koga T, Alkima M, Maruyama A, et al. (1993) Novel antiasthmatic agents with dual activities of thromboxane A2 synthetase inhibition and bronchodilation. 1 2-[2-(1-Imidazolyl)alkyl]-1(2H)-phthalazinones Med 36: 4052-4060.

8. Li YX, Luo YP, Xi Z, Niu CW, He YZ, et al. (2006) Design and syntheses of novel phthalazin-1(2H)-one derivatives as acetohydroxyacid synthase inhibitors. Agric Food Chem 54: 9135-9139.

9. Leenen FHH, Smith DL, Faraks RM, Reeves RA, Marquez-Julio A A mJ, (1987) Vasodilators and regression of left ventricular hypertrophy. Hydralazine versus prazosin in hypertensive humans Med 82: 969-978.

10. Leiro JM, Alvarez E, Cano E, Orallo F, (2004) Antioxidant activity and inhibitory effects of hydralazine on inducible NOS/COX-2 gene and protein expression in rat peritoneal macrophages Int Immunopharmacol 4: 163-177.

11. Tanaka S, Tanaka A, Akashi A, (1989) Influence of antihypertensive treatment with budralazine on autoregulation of cerebral blood flow in spontaneously hypertensive rats. Stroke 20: 1724-1729.

12. Maroi R, Ono K, Saito T, Akimoto T, Sano M, (1977) Metabolism of budralazine, a new antihypertensive agent. II. Metabolic pathways of budralazine in rats. Chem. Pharm. Bull 25: 830-835.

13. Kemp JP, Meltzer EO, Orgel HA, Welch MJ, Bucholtz GA, et al. (1987) A dose-response study of the bronchodilator action of azelastine in asthma. Allergy Clin Immuonol. 79: 893-899.

14. Scheffler G, Engel J, Kustscher B, Sheldrick WS, Bell P, (1988) Synthese und Kristallstrukturanalyse von Azelastine Archiv. Der. Pharmazie 321: 205-208.

15. Kador PF, Kinoshita JH, Sharpless NE, (1985) Aldose reductase inhibitors: a potential new class of agents for the pharmacological control of certain diabetic complications J Med Chem 28: 841-849.

16. Mylari BL, Larson ER, Beyer TA, Zembrowski WJ, Aldinger CE, et al. (1991) Novel, potent aldose reductase inhibitors: 3,4-dihydro-4-oxo-3-[[5-(trifluoromethyl)-2-benzothiazolyl]methyl]-1-phthalazineacetic acid (zopolrestat) and congeners J Med Chem 34: 108-122.

17. Cockcroft X, Dillon KJ, Dixon L, Drzewiecki J, Eversley P, et al. (2005) Phthalazinones. Part 1: The design and synthesis of a novel series of potent inhibitors of poly (ADP-ribose)polymerase Bioorg Med Chem Lett 15: 2235-2239.

18. Cockcroft X, Dilton KJ, Dixon L, Drzwewiecki J, Kirrigan F, et al. (2005) Bioorg Med Chem Lett 15: 2235-2239.

19. Kim JS, Lee H, Suh M, Choo HY, Lee SK, et al. (2004) Synthesis and cytotoxicity of 1-substituted 2-methyl-1H-imidazo[4,5-g]phthalazine-4,9-dione derivatives Bioorg Med Chem 12: 3683-3686.

20. Haikal A, El-Ashery E, Banoub J, (2003) Synthesis and structural characterization of 1-(D-glycosyloxy)phthalazines Carbohydr Res 338: 2291-2299.

21. Demirayak S, Karaburun A, Beis R, (2004) Some pyrrole substituted aryl pyridazinone and phthalazinone derivatives and their antihypertensive activities. Eur Med Chem 39: 1089-1095.

22. Watanabe N, Kabasawa Y, Takase Y, Matsukura M, Miyazaki K (1997) J Med Chem 41: 105-107.

23. Kornet M, Shackelford J, (1999) Heterocycl Chem 36: 1095-1096.

24. Nassar OM, (1997) Indian J. Heterocycl. Chem 7: 105-108.

25. Cardia MC, Distinto E, Maccioni E, Delogu A, (2003) - Synthesis and biological activity evaluation of differently substituted 1,4-dioxo-3,4-dihydrophthalazine-2(1H)-carboxamides and –carbothioamides J Heterocyclic Chem 40: 1011.

26. Olio ED, Armas MG, Lopez-perez J, Ruiz G, Vargas F, et al. (2001) Anti-Trypanosoma Activity of Some Natural Stilbenoids and Synthetic Related Heterocyclic Compounds Bioorg Med Chem Lett 11: 2755-2757.

27. Dogruer DS, Kupeli E, Yesilada E, Sahin MF (2000) Arch Pharmacol 337: 303-310.

28. Napoletano M, Norcini G, Pellacini F, Marchini F, Moraazzoni G, et al. (2000) Bioorg. Med Chem Lett 10: 2235-2238.

29. Rubat C, Coudert P, Couquelet J, Bastide P (1988) Chem Pharm Bull 36: 1558.

30. Rubat C, Coudert P, Couquelet J, Albuisson E, Bastide J, et al. (1992) Synthesis of mannich bases of arylidenepyridazinones as analgesic agents J Pharm Sci 81: 1084.

31. Rohet F, Rubat C, Coudert P, Albuisson E, Couquelet J (1996) Synthesis and trazodone-like analgesic activity of 4-phenyl-6-aryl-2-[3-(4-arylpiperazin-1-yl) propyl]pyridazin -3-ones. Chem Pharm Bull 44: 980.

32. Courant P, Rubat C, Rohet F, Leal F, Fialip J, et al. (2000) J Pharm Pharmacol Commu 6: 387.

33. Dogruer SD, Sahin MF, Unlu S, Shigneru I, (2000) Studies on some 3(2H)-pyridazinone derivatives with antinociceptive activity. Arch Pharm 79: 333.

34. Glu MS, Ergun BC, Unlu S, Sahin MF, Keupel E, et al. (2005) Synthesis, analgesic, and anti-inflammatory activities of [6-(3,5-dimethyl-4-chloropyrazole-1-yl)-3(2H)-pyridazinon-2-yl]acetamides. Arch Pharm Res 28: 509.

35. El-Hashash MA, Guirguis DB, Kadhim MA (2013) J of American Science 9 (12).

36. Cruickshank R, Duguid JP, Marion BP, Swain RH (1975) A Medicinal Microbiology, twelfth ed., vol. II, Churchill Livingstone, London, 196-202.

25

Recovery of Cream of Tartar from Winemaking Solid Waste by Cooling Crystallization Process

Samira Kherici[1], **Djillali Benouali**[2*] **and Mohamed Benyetou**[3]

[1]Department of industrial organic chemistry USTO, POB 1505 El Mnaouar Oran, Algeria

[2]Department of physical chemistry USTO, POB 1505 El Mnaouar Oran, Algeria

[3]Laboratory of modeling and optimization of industrial systems (LAMOSI) USTO, Algeria

Abstract

Our work relates the recovery of potassium bitartrate (cream of tartar) from winemaking solid waste in Algeria. The process consists of crystallizing by cooling, separating and drying the cream of tartar. It uses water as solvent and wine tartar as raw material. This conventional method for the cream of tartar extraction has been carried out in our laboratory using batch process with a charge of 45 liters. For the various process unit operations, simple tools, not expensive and effective are used. Utilization of pipeline water filters with different cartridges (5μm, carbon/ diatomite) has facilitated the key process operation, the hot filtration. Also, we managed to control the dissolution step, so it was monitored by continuous pH and temperature measurements in the agitated vessel, using a pH meter model DDSJ308A. And the crystallization operation was controlled using an infrared turbidity sensor developed in our laboratory.

Keywords: Cream of tartar; Tartaric acid; Cooling crystallization; Infrared sensor

Introduction

Pollution from agricultural or industrial waste is considered as one of the major problems affecting the world and that may prove to be dangerous to fauna and flora. These wastes must be treated to preserve the natural balance and preserve our health. The winemaking solid waste is also known as "wine tartar"; It can be easily treated for valorization by recovering the cream of tartar and the free L (+) tartaric acid [1-7]. Thereby, reducing waste, protecting the environment and eliminating the disposal costs. In wine manufacturing, grapes are the raw material of choice. They contain tartaric acid in the potassium hydrogenotartrate form (cream of tartar). Because of their richness in this acid, they are the only source of obtaining it. This is the reason why it is recovered in the wine by-products [8-10]. The reasons which lead to its recovery are the existence of a market with reasonable prices for the calcium and potassium tartarates and the severity of the international environmental legislation for the wastes, on the other hand. Tartaric acid is the most important by-product and enjoys many foods, pharmaceutical and industrial applications. It is obtained via the decomposition of calcium tartarate by sulphuric acid. This acidification produces a sulphate precipitate of calcium (plaster) and an aqueous solution of tartaric acid. The concentration of this aqueous solution, by different technologies as electrodialysis, solvent extraction, and adsorption gives the required tartaric acid [11-15].

Cream of tartar is extracted from the wine tartar using a cooling crystallization, and water is the only solvent used. Small production does not seem suitable for continuous operation because of the problems with plugging, fouling, sedimentation of crystals in the pipes, etc.). It is better to implement in this case batch crystallization. Batch cooling crystallizers present the advantage of being simple, flexible, require less investment and, generally, involve less process development [16-20].

In our work, we have realized the optimisation of cream of tartar extraction at laboratory stage as follows:

- Volume solvent, heat temperature, mixing rate and time of extraction optimization.

- Study of the activated carbon addition or other mean of discolouration for the improvement of the quality of the final product.

Our work is based on the crystallizer behaviour study, and also to improve the performance of the method. So this laboratories work is exploited for the implementation of the process on a pilot of 45 litres. We describe different process operations for obtaining cream of tartar by batch cooling crystallization; some characteristics of equipments are discussed. Super saturation as the driving force of crystallization processes is an essential parameter, so the development of more accurate and sensitive sensors for real-time analysis of crystallization must allow significant advances in monitoring, control and optimization of crystallization processes. In a second part of our work, acquisition of temperature and pH parameters were used for control during the process. The optimization of the cooling time in the crystallizer was managed by turbidity measurements on-line was performed using an infrared sensor made in our laboratory.

Materials and Methods

Oenological waste

Oenological waste (Cream of tartar) was provided by an industrial plant (ONCV- hassi El- Ghala, Algeria). For the laboratory essay it has crushed, but for the pilot it was used in its raw form (Figure 1).

Description of the process

The agitated crystallizer vessel (C) (Figure 2), technologically

*Corresponding author: Djillali Benouali, Department of physical chemistry USTO, POB 1505 El Mnaouar Oran, Algeria

Figure 1: Cream of tartar.

simple, leads to different particle size distribution. In fact, it is the difficult stage of our process, because it requires control and monitoring.

The dissolver (D) is an agitated vessel mechanically stirred, stainless steel curved bottom, of capacity equal to 45 liters. This reactor consist of two concentric tanks, one is inside the other. The inner tank is mobile and coupled to a motor-IKA RW20 type which provides a speed range of around 300 rev/min (Figure 3).

The agitation is generated by the rotation of the inner tank which is nonsymmetrical with the outer vessel. The heating is made by two resistors of different capacities 2400 and 1000 Watts, installed between and inside the two tanks of the dissolver.

The reactor is covered by a sheet of glass, provided with holes and adaptations required to the liquid feed, the pH electrode, the temperature sensor and the condenser (F).

The dissolver is powered by the solvent (water) and tartar in its raw form, without crushing or grinding.

The crystallizer is a cylindrical steel reactor with a conical base in the form of a funnel (Figure 3), provided with a cooling coil. This later is connected to a refrigerating plant 1.6 KW. The total capacity of the crystallizer is 80 liters. It is stirred by a special stirrer, associating two types of agitators. A classic anchor and a Rushton turbine all coupled to a stirring motor with a speed reducer with 328 rev/min as range of agitation. This range of agitation was measured with a stroboscope COMPACT brand, model 461830.

The cooling crystallization process is based on a hot filtration. This operation is very sensitive to the presence of insoluble impurities causing the fooling of the filter and therefore, requiring frequent wash filter. To avoid this problem, we introduced a filter bag sized in the dissolver according to the internal walls of the tank. This bag is easy to wash and change. The dissolver is connected to the crystallizer consisting of agitated vessel with jacketed for heat exchange (cooling). The precipitate is collected and the solvent is recycled. After several crystallizations, cream of tartar is obtained in the pure form, and can be dried out and packaged.

Results and Discussions

Optimization of process parameters: Laboratory tests

Laboratory tests were achieved to optimize different parameters of the extraction operation: solvent volume, heating time, stirring time and stirring rate. The wine tartar, as a raw material, was provided by the different wine cellars in Western Algeria. Chemical composition

is given in Table 1. We introduced 10 grams of raw material crushed in different amounts of water (50-300 ml). After heating, boiling, hot filtrating and cooling the filtrate, the product crystallized. This product was filtered and dried at 105°C in an oven to constant weight for one hour. The results are presented in Figure 4.

According to preliminary tests we found that prolonged heating at boiling temperature caused a great loss of solvent. For this reason we carried out the extraction in the temperature range 70-90°C. As shown in Figure 5, the optimum volume is 200 ml. We studied the optimization of the heating temperature in Figure 5 in order to see the effect of low temperatures on the quality and quantity of the produced material.

The obtained results by setting the volume to 200 ml of water, confirm that the best performance may be obtained while heating to

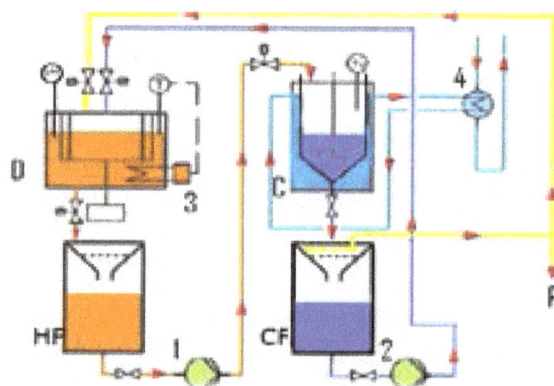

D: dissolver vessel; HF: hot filtration; C: Crystallizer; CF: cold filtration; P: product (cream of tartar); T: Temperature; 1, 2: circulating pump; 3: Heating device; 4: Condenser
Figure 2: Schematic of cream of tartar extraction process.

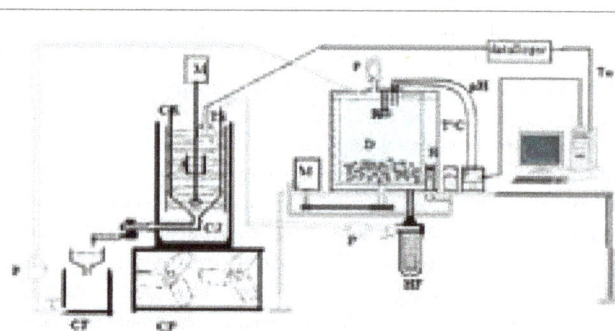

M: Engine type IKA-RW20; D: dissolver; R: Resistance electric capacity equal to (2400W + 1000W); F: steam condenser; HF: hot filtration; Cj: Cooling jacket; P: Reticulating pump; A: agitator; CR: crystallizer; CF: Cold filtration; TS: turbidity sensor, T °C: temperature captor; pH: pH electrode; datalloger: data acquisition
Figure 3: Semi-automated cooling crystallization pilot.

Parameters	Rate of optimization
Water (ml)	200/10g raw material
T(°C)	90-100
T (min)	10
V(rd/min)	900
Yield (%)	56 – 70

Table 1: Summary of experimental optimization results.

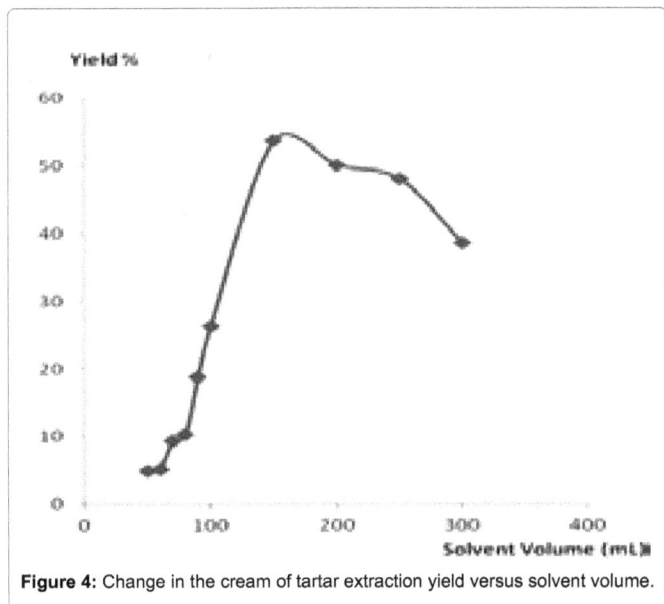

Figure 4: Change in the cream of tartar extraction yield versus solvent volume.

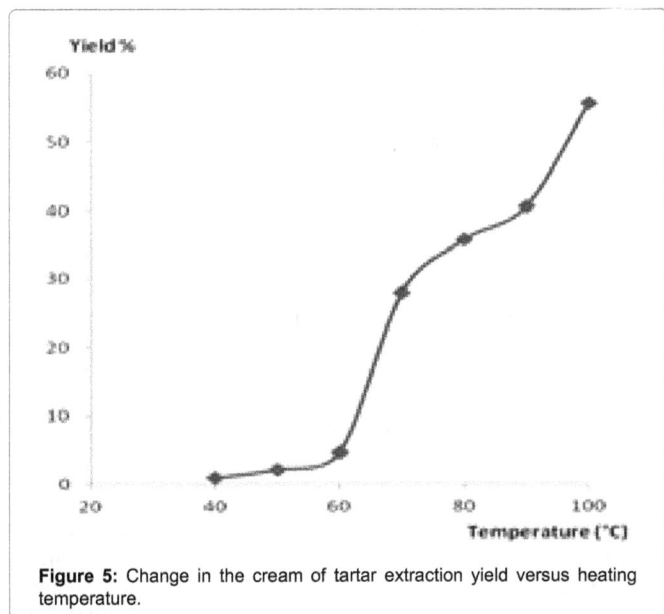

Figure 5: Change in the cream of tartar extraction yield versus heating temperature.

boiling temperature (100°C). Low temperatures extend the heating time and no improvement of the product was observed. However, we found that the yield at 90°C is approximately maximum. So, at such a temperature we avoid not only the evaporation of the solvent but energy consumption as well. Agitation is an important operation in our process. In fact it is also considered as a complex operation which is difficult to optimize and to translate the obtained results from the laboratory tests to the industrial equipment.

By maintaining the temperature at 90°C and the water volume 200 ml we performed the experiment by varying the stirring time (2 min, 5 min, 10 min) while varying the stirring rate. The rate stirring optimization results show that, a maximum yield may reached (approximately equal to 7 gr.). This can be achieved with a stirring time of 10 minutes along with agitations rates from 700 to 1000 rev/min (Figure 6).

In order to improve the quality of cream of tartar, we used the

bleaching clay (bentonite) during the dissolution stage. We used 10 to 50 mg of bentonite mixed to 10 grams of raw material. The results of the operation showed that the product quality (colour) has considerably improved from the first crystallization, but the filtration became more difficult. Besides, the insoluble residues caused the filter fooling to become too quickly.

However, the use of bentonite or other large-scale bleaching earth or powder active carbon (PAC) breaks the problem of disposal of sludge which should be avoided. The great disadvantage of bentonite and CAP was, sludge disposal and the impossibility of their regeneration. Currently, competitive to the bleaching clay is membrane filtration as reverse osmosis. In the microfiltration process diatomaceous earth is usually used as filter aid. So, our choice was ultimately focused on the use of diatomaceous earth, this choice needed to be justified and adapted to the pilot scale operation.

Dissolver reactor temperature and flow profiles

We calculated the cream of tartar dissolution heat; this amount of heat will be useful for determining the power needed to dissipate in our dissolver reactor (D). We have prepared saturated solutions at each temperature, and we have also determined the equilibrium constant at 100°C (K).

The value and the sign of ΔH (heat of solution) can be determined from the equation (3) below. After cream of tartar solubilisation the following equilibrium is established:

$$KHT(s) \rightarrow K^+ + HT^- \tag{1}$$

HT- : Rrepresents the hydrogen tartarate ion

$$K = [K^+][HT^-] \tag{2}$$

$$\ln\frac{k_2}{k_1} = \frac{\Delta Hs}{R}\left(\frac{1}{T_1} - \frac{1}{T_2}\right) \quad ; \quad \ln k = -\frac{\Delta Hs}{R}\left(\frac{1}{T}\right) + C \tag{3}$$

In equation (1), k_1 and k_2 are respectively the equilibrium constants at temperatures T_1 and T_2 in Kelvin, and R is the perfect gas constant. According to the graph of the equation (3), we have estimated that ΔH is 290.79 kJ/kg.

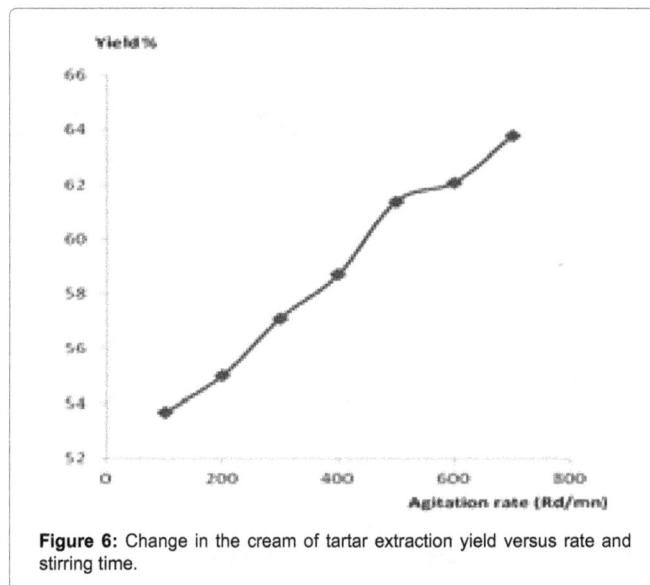

Figure 6: Change in the cream of tartar extraction yield versus rate and stirring time.

The generated flows profile by the rotation of the inside tank of the dissolver reactor (Figure 3) is presented in the Figure 7.

For mobile radial flow, the regime depends not only on the type of mixing device but on geometrical factors for the tank as baffles, eccentric shaft and size of the tank as well. In our case the flow regime is turbulent (based on the value of Reynolds number), which results in the absence of fluid motion in a direction different from that imposed by the inside tank rotation.

Control process and monitoring

We applied a slow agitation for the dissolver and we followed the evolution of the heating temperature versus time. This monitoring was performed using the software DRAW, with a recording time of a record each minute. At the end, the results were transferred to Excel software to plot the behaviour of heating temperature and pH in the dissolver. A time of 25 mn was necessary to reach 100°C as shown in Figure 8 and 9 respectively.

The pH curve show clearly the saturation of the solution in cream of tartar at a given temperature is determined. A saturated aqueous cream of tartar solution have a pH equal to 3.4.

The pH meter model DDSJ308A automatically performs the

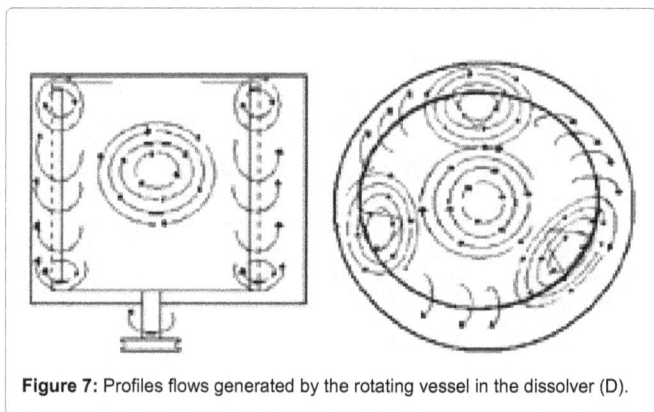

Figure 9: Change of pH versus heating time in the dissolver (D).

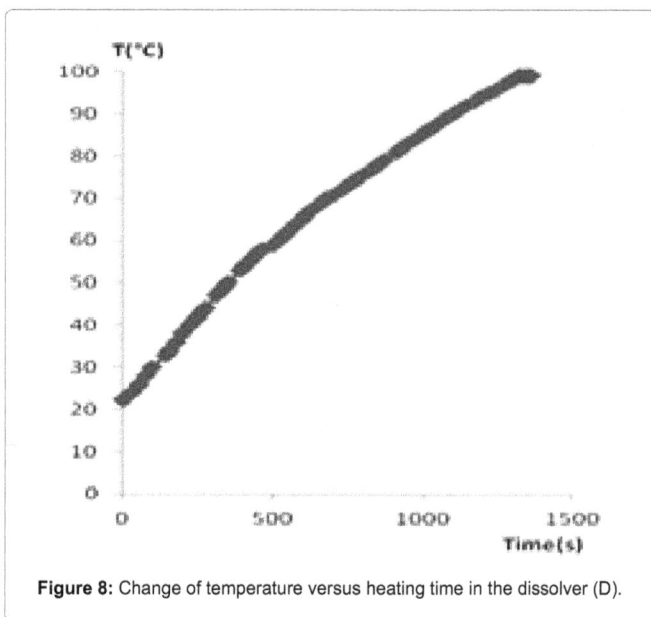

Figure 10: Change of turbidity versus cooling time in the crystallizer (C) at 10 °C.

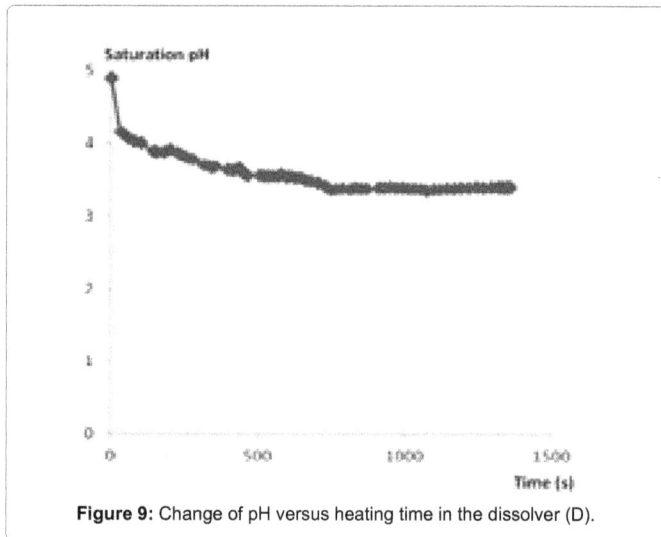

Figure 7: Profiles flows generated by the rotating vessel in the dissolver (D).

temperature correction. This method needs a special pH electrode, and represents an excellent way to optimize the dissolution time in the dissolver. According to our test, a heating time inferior at 30 mn is necessary to achieve the saturation temperature range 95-100°C and a maximum yield of 2 kg per charge.

In the crystallization stage, it should be noted that different element of the crystallizer complicates enormously this operation by depositing a very solid crystalline layer (tartar) leading to a number of adverse consequences in the process, in particular:

- The need to clean the crystallizer;

- The lost productivity by reducing the heat transfer performance of the device if the layer covers completely the surface of exchange;

- The Difficulties in filling and clogging as deposition occurs also in the filling elements and in the pipes transfer.

To avoid these problems, the washing is done with hot water or a solution of sodium bicarbonate to minimize the time stays in crystallizer. For this purpose, turbidity in crystallizer caused by the precipitation of cream of tartar is monitored on-line by an infrared

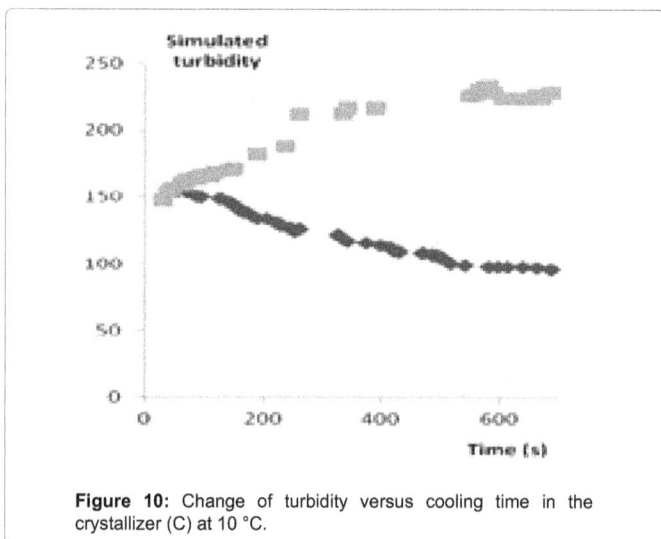

Figure 8: Change of temperature versus heating time in the dissolver (D).

turbidity sensor designed in our laboratory [19,20]. In Figure 10 it is shown that saturation is reached at 600 minutes at a temperature of cooling set at 10 °C (gray curve). The black curve in Figure 10 shows the effect of temperature on the measurement of turbidity using our infrared sensor; this curve represents the turbidity measurement of distilled water from 90 to 10°C. We noticed that the evolution of turbidity versus cooling time in the crystallizer can clearly indicates the end of the cooling operation.

Conclusion

In order to use any technique on an industrial scale it necessary to undertake laboratory and pilot trial. In our approach, we studied the behavior of a dissolver, a crystallizer and hot filtration in the laboratory. We began an audit of change of cooler temperature using an infrared camera (Ti FLUKE 10). We also realized crystallization of cream of tartar from wine waste in order to experience the batch crystallization process. Moreover, we also monitored the dissolver with the acquisition of temperature and pH on-line using a pH meter provided with RS232 output. The accuracy of off line methods for evaluating crystallization processes is strongly dependent on sampling. By using real-time analysis these types of errors can be greatly reduced. And the determination of the solubility of a solid in a specific solvent is a key step in the study of crystallization processes. So, we have used an infrared sensor of turbidity to determine the optimum cooling time made in our laboratory. Detailed study of these problems is one of our intended future works.

Acknowledgment

The authors are grateful to ONCV- (Hassi-El- Ghalla -Algeria) for their assistance, by making in our disposition 100 Kg of wine tartar.

References

1. Devesa-Rey R, Vecino X, Varela-Alende JL, Barral MT, Cruz JM, et al. (2011) Valorization of winery waste vs. the costs of not recycling. Waste Manag 31: 2327-2335.

2. Leber E (2004) Recovering value from winemaking by-products. ASEV 55th Annual Meeting San Diego, California.

3. Nerantzis ET, Tataridis P (2006) Integrated enology-utilization of winery by-products into high added 528 value products. eJournal of Science and Technology 3: 1-12.

4. Yalcin D, Ozcalik O, Altiok E, Bayraktar O (2008) Characterization and Recovery of Tartaric Acid 576 from Wastes of Wine and Grape Juice Industries. Journal of Thermal Analysis and Calorimetry 94: 767-771.

5. Braga FG, Fernando A, Lencarte S, Arminda A (2002) Recovery of Winery By-products in the Douro Demarcated Region: Production of Calcium Tartrate and Grape Pigments. Am J Enol Vitic 53: 41-45.

6. Rice AC (1976) Solid Waste Generation and By-Product Recovery Potential from Winery Residues. Am J Enol Vitic 27: 21-26.

7. Arvanitoyannis IS, Ladas D, Mavromatis A (2006) Potential uses and applications of treated wine waste: a review. International Journal of Food and Technology 41: 475-487.

8. DeBolt S, Cook DR, Ford CM (2006) L-tartaric acid synthesis from vitamin C in higher plants. Proc Natl Acad Sci U S A 103: 5608-5613.

9. Soyer Y, Koca N, Karadeniz F (2003) Organic acid profile of Turkish white grapes and grape juices. Journal of Food Composition and Analysis 16: 629-636.

10. Williams M, Loewus FA (1978) Biosynthesis of (+)-Tartaric Acid from l-[4-C] Ascorbic Acid in Grape and Geranium. Plant Physiol 61: 672-674.

11. Nurgel C, Canbas A (1998) Production of Tartaric Acid From Pomace of Some Anatolian Grape Cultivars. Am J Enol Vitic 49: 95-99.

12. Rivas B, Torrado A, Moldes AB, Domínguez JM (2006) Tartaric acid recovery from distilled lees and use of the residual solid as an economic nutrient for lactobacillus. J Agric Food Chem 54: 7904-7911.

13. Inci I, Asci YS, Tuyun AT (2011) Reactive Extraction of L (+) Tartaric Acid by Amberlite LA-2 in Different Solvents. E-Journal of Chemistry 8: 509-515.

14. Palma M, Barroso CG (2002) Ultrasound-assisted extraction and determination of tartaric and malic acids from grapes and winemaking by-products. Analytica Chimica Acta 458: 119-130.

15. MI Si, LI Hua, LIU Jing (2012) Optimization of Extraction Process for L(+)-Tartaric Acid from Vitis quinquangularis Lees via Response Surface Methodology. Food Science 33: 49-53.

16. Pollanen K, Hakkinen A, Reinikainen SP, Louhi-kultanen M, Nystrom L (2006) A study on batch cooling crystallization of sulphathiazole:Process monitoring using ATR-FTIR and product characterization by automated image analysis. Chemical Engineering Researsh & Desing 84: 47-59.

17. Fiordalis A, Georgakis C (2010) Optimizing Batch Crystallization Cooling Profiles: The Design of Dynamic, Experiments Approach. Proceedings of the 9th International Symposium on Dynamics and Control of Process Systems. Leuven, Belgium.

18. Bastos C, Costa B, Filho R M (2006) Cooling Crystallization: a Process-Product Perspective. 16th European Symposium on Computer Aided Process Engineering. Computer Aided Chemical Engineering 21: 967-972.

19. Benouali D, Kacha S, Kherici S, Benabadji N (2010) Study of the Flocculated Particles Sedimentation Assisted by Microcomputer. Open hydrology journal 4: 14-18.

20. Benouali D, Kherici S, Benabadji N (2012) Infrared Turbidimetric Titration Method for Sulfate Ions in Brackish Water. Pak J Anal Environ Chem 13: 118-122.

26

Prediction of Viscosities of Aqueous Two Phase Systems Containing Protein by Artificial Neural Network

Selvaraj Raja*, Varadavenkatesan Thivaharan, Vinayagam Ramesh and Vytla Ramachandra Murty

Department of Biotechnology, Manipal Institute of Technology, Manipal, Karnataka, India

Abstract

The viscosities of aqueous two phase system containing bovine serum albumin (BSA) were predicted by artificial neural network (ANN) as a function of concentration of poly-ethylene-glycol (PEG), concentration of BSA and temperature. A three layer feed forward neural network based on Levenberg-Marquardt (LM) algorithm which consisted of three input neurons, 10 hidden neurons and one output neuron (3:10:1) was developed. The performance parameters were calculated and compared with the conventional Grunberg-Nissan empirical model. The satisfactory values suggest that the proposed ANN model has the capability of predicting viscosity in a better way than the conventional empirical model.

Keywords: Aqueous two phase system; Artificial neural network; Levenberg-Marquardt (LM) algorithm; Grunberg-Nissan empirical model

Introduction

Aqueous two phase system (ATPS) is a separation method which is largely used to purify various biomolecules in a single step [1]. It is a liquid-liquid extraction method which can be prepared by mixing aqueous solution of water-soluble polymer and salt or two water-soluble polymers with water [2]. ATPS can also be formulated by mixing a water-soluble polymer and protein with water. One of such systems has been demonstrated by Johansson [3] for the partitioning of lactate dehydrogenase and some mitochondrial enzymes.

Viscosity is a transport property and the magnitude determines the mass transfer between phases. The various inter-related properties such as size, shape, structure, degree of polymerization and polymer-solvent interactions can be obtained from the viscosity data [4]. Studying and predicting the physical property, viscosity is very much essential for the design and optimization of an ATPS in large scale system. These data are helpful in designing the extraction process in large scale (since large volumes of phase components have to be handled and separated) and for the development of various mathematical models which can predict the partitioning behavior of biomolecules in aqueous two phase system [5].

To a large extent, viscosity of an ATPS depends on the concentration of phase component compositions and temperature. Substantial cost is involved in performing experiments in the entire range of phase components to determine the viscosity of solutions. In this context, mathematical equations should be available in predicting the physical properties. There are few polynomial equations and empirical equations are available in literature as a function of solute concentration for different temperatures [6]. However, empirical equations suffer because of their validity in a narrow range of dependent variables. Moreover the relative errors and deviations of these empirical models between experimental values and calculated values will be very high.

One of such empirical equations for viscosity is Grunberg-Nissan empirical model [7] which was successfully used by Gunduz [8] for an ATPS, consisting of poly-ethylene-glycol (PEG)- Bovine Serum Albumin (BSA)-Water system. The model (Eqn. 1) can be written as follows:

$$\ln(\eta_{sys}) = a_1 \ln(\eta_{PEG}) + a_2 \ln(\eta_{BSA}) + w_1 w_2 A \qquad (1)$$

Where, A is an adjustable parameter which is the characteristic of the intermolecular interactions between PEG and BSA. a_1 and a_2 are weight fractions of PEG and BSA respectively and η_{PEG} and η_{BSA} are viscosities of aqueous PEG and BSA solutions respectively.

In recent years, an artificial intelligence technique, ANN has been widely used to model and predict linear and non-linear systems. It is a simplified artificial model of the biological neuron system which can provide the relationship between the input and output variables from the given data set. The artificial neuron simulates the basic functions of biological neurons [9].

The major advantage of ANN is its high learning capacity and predictive ability from the limited information fed to the system. Moreover, in contrast to conventional empirical model, ANN can accommodate multiple input and output variables and it has been proven by researchers that a well-trained ANN can predict various processes in a better way [10-13].

Recently, ANN has been used to predict viscosity of PEG solutions [14], density of ionic liquids [15] and vapor liquid equilibrium data of ionic liquids [16].

To the best of our knowledge, there are no reports available for predicting viscosities of aqueous two phase systems containing protein using artificial neural network. Therefore the objective of the current study is to model the experimental viscosity of the aforementioned ATPS by using ANN and compare with the Grunberg-Nissan empirical model.

Methodology

Collection of viscosity data from literature

For the current study, the dynamic viscosity data set was chosen

---

*Corresponding author: Selvaraj Raja, Department of Biotechnology, Manipal Institute of Technology, Manipal, Karnataka-576104, India

from the literature [8] which included the viscosity for the given concentration of PEG (4-11 % w/w), concentration of BSA (1-8 % w/w) and the process temperature (15-35 °C). A total of 35 experimental data were taken for the present study.

Neural network topology

Artificial Neural Network modeling codes were developed by using Matlab R2013a software. Normally, multilayers are used in ANN models and the simplest form of them is the three-layer models which contain input layer (I), one hidden layer (H) and output layer (O). Each layer consists of many nodes which are connected to each other by two coefficients known as weights (w) and biases (b). The commonly used feed forward multilayer neural network (FFMNN) was chosen for the present study. Choosing the number of neurons in the hidden layer and the transfer functions for hidden and output layer is crucial in developing an ANN model. For the present system, the number of neurons in hidden layer was chosen as 10 which was optimum.

Tangent and linear transfer functions were chosen as hidden and output layers respectively. The output of neuron is calculated by

$$O_j = F(I_j) \tag{2}$$

Where I_j, O_j and F are input of ith neuron, output of ith neuron and transfer function respectively.

In the present study, hyperbolic tangent transfer function and linear transfer functions have been employed for neurons located in hidden layer and output layers respectively. The input of each neuron (I_j) is calculated by the following equation with respect to outputs of previous layers (O_j), weights connecting i^{th} neuron to the jth neuron (w_{ij}) and bias of the jth neuron (b_j)

$$I_j = \Sigma i \, w_{ij} y_i + b_j \tag{3}$$

In order to optimize the weights and biases, Levenberg-Marquardt (LM) back propagation algorithm has been used, which is one of the most popular tools [17].

The data set was divided into three sets, namely, training set (70%), validation set (15%) and testing set (15%). Training set is used to train the network and the network is adjusted according to its error. Validation set is used to measure the network generalization and to stop the training process when generalization stops improving. The testing set is used to evaluate the performance of trained model against new "unseen" data and hence gives an independent measure of network performance during and after training.

For the ANN model developed, the prediction accuracy and degree of fitness were evaluated by using the following factors namely, Mean Square Error (MSE), Root Mean Square Error (RMSE), Standard Error of Prediction (SEP), Average Absolute Relative Deviation (AARD) with Bias factor (B_f) and Accuracy factor (A_f) [18].

$$MSE = \frac{\sum_{i=1}^{n}\left[\mu_{i,exp} - \mu_{i,pred}\right]^2}{n} \tag{4}$$

$$RMSE = \sqrt{\frac{\sum_{i=1}^{n}\left[\mu_{i,exp} - \mu_{i,pred}\right]^2}{n}}$$

$$RMSE = \sqrt{\frac{\sum_{i=1}^{n}\left[\mu_{i,exp} - \mu_{i,pred}\right]^2}{n}} \tag{5}$$

$$SEP(\%) = \frac{RMSE}{\overline{\mu}_{i,exp}} x100 \tag{6}$$

$$AARD(\%) = 100 \; x \frac{\sum_{i=1}^{n}\left[\left|\frac{\mu_{i,exp} - \mu_{i,pred}}{\mu_{i,exp}}\right|\right]}{n} \tag{7}$$

$$B_f = 10^{\left[\frac{\sum_{i=1}^{n} log(\mu_{i,pred}/\mu_{i,exp})}{n}\right]} \tag{8}$$

$$A_f = 10^{\left[\frac{\sum_{i=1}^{n} log(\mu_{i,pred}/\mu_{i,exp})}{n}\right]} \tag{9}$$

where $\mu_{i,exp}$ and $\mu_{i,pred}$ are the experimental and predicted viscosities and 'n' is the number of experiments.

Results and Discussions

Neural network modeling of viscosity

Coding for the proposed ANN model was written by using Matlab R2013a software with the aim to minimize mean squared error (MSE) and maximize the correlation coefficient value (R^2). A low value of MSE and high value of R^2 is preferred for the best model.

Various numbers of hidden layers (5, 10 and 15) were checked for the performance of the neural network. The optimum number of hidden layers was found to be 10 which are depicted in the Table 1. Lowest value of MSE (0.0511) and highest value of R^2 (0.9979) was obtained for 10 number of hidden layers and therefore considered as optimum.

As discussed earlier, the experimental points from the literature [8] were taken and trained by ANN topology (3:10:1). The selected network consists of three input neurons namely PEG (% w/w), BSA (% w/w) and Temperature (°C), 10 hidden neurons and one output neuron (Viscosity). The connection between the various layers for the developed ANN is shown in Figure 1.

"trainlm" training function was used in this work which updates weight and bias values according to Levenberg-Marquardt (LM) optimization. It is considered to be one of the fastest back propagation algorithms available. The training set data was trained by adjusting the strength and connections between neurons with an objective to fit the output of the entire network to be closer to the desired target and to minimize the performance function. The performance function used was MSE which measures the performance of the network according to the mean squared errors. The training process was halted once the goal was met (i.e., reaching a smallest value of MSE). In the present study,

Data Set	No. of hidden layers=5		No. of hidden layers=10		No. of hidden layers=15	
	MSE	R²	MSE	R²	MSE	R²
Training	0.307	0.9857	0.013	0.9995	0.016	0.9802
Validation	0.149	0.9873	0.053	0.9965	0.243	0.6454
Testing	0.785	0.9251	0.24	0.997	0.202	0.8056
Overall	0.353	0.9839	0.051	0.9979	0.075	0.9481

Table 1: Optimum number of hidden layers based on MSE and R2 values.

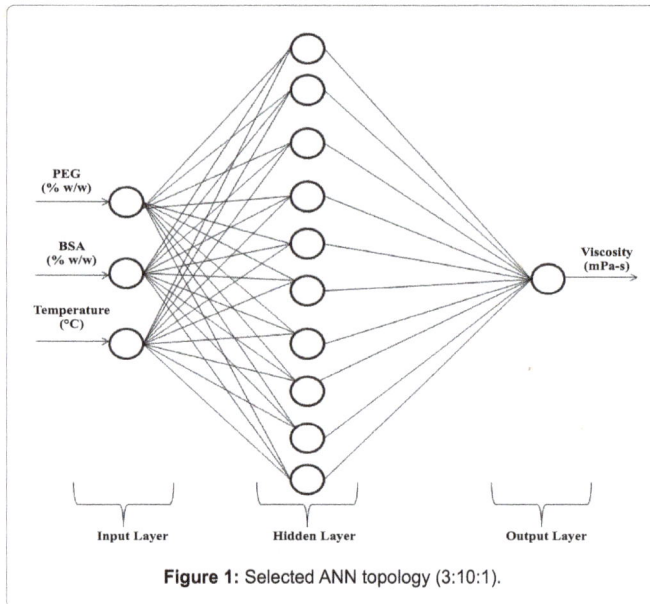

Figure 1: Selected ANN topology (3:10:1).

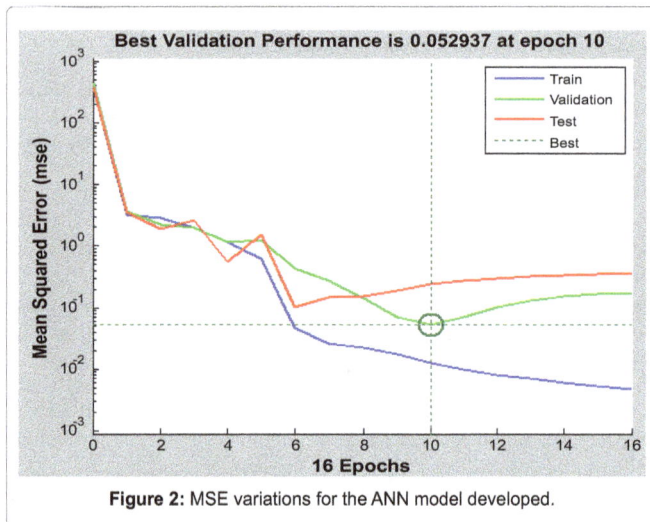

Figure 2: MSE variations for the ANN model developed.

the training was stopped after 16 iterations (epochs). Figure 2 shows the trend of MSE during the process of training, testing and validation. It can be noticed from the figure that MSE value reaches a minimum value of 0.0129 for training. Literature suggests that lower the value of MSE, higher the accuracy of the model. The best value of MSE (0.052937) was obtained after 10 epochs which was depicted (dotted lines) in the figure.

The optimal values of weights and biases of the ANN topology (3:10:1) based on LM algorithm is shown in Table 2.

In the Table 2, w_1 and w_2 are the input and hidden layer weight matrix, b_1 and b_2 are the biases of input and output layer respectively. The viscosity output values were predicted by using Eqn. 2 (data not shown) and the accuracy parameters were calculated.

The results obtained during training, validation and testing are shown in the Figure 3. The solid 45° line on the predicted versus experimental viscosities reveals that there is a perfect superpose of predicted and experimental data on each other. This confirms that the proposed model is valid. For all the data set, the value of R^2 was in acceptable range of greater than 0.99 (training: 0.9945, validation:

0.99647 Testing: 0.99697 and overall: 0.99790) which confirms a good agreement with the experimental data.

In addition to this, the prediction accuracy and degree of fitness were calculated (Equations 4 to 9) and shown in the Table 3.

The accuracy parameters of the ANN model are lesser than the Grunberg-Nissan model which suggests the best fitting. The B_f factor (1.01) and A_f factor (1.02) are closer to unity for the ANN model, which indicates that the model is "fail-safe" and shows a good concordance between the predicted and experimental values [19,20]. Because of these higher fitness and accuracy, the ANN models can be used to predict and model any linear or non-linear system in a better way than compared to polynomial or empirical equations.

Training Function	Weights and Biases of the neural networks					
	w_1			w_2	b_1	b_2
	-0.9447	-0.9069	1.9125	-2.2014	4.3511	1.996
	-2.7955	-1.191	0.3436	-1.8846	2.5251	
	-2.7081	2.9248	2.3544	-0.0795	0.1959	
	3.3501	-0.7734	-0.1839	0.1821	-0.2177	
Trainlm	-0.3902	0.3684	-2.4848	0.1775	-1.2122	
	2.0133	2.4695	-0.3276	0.1491	-0.3456	
	-1.6439	2.0489	-2.1602	-0.2154	-3.6291	
	1.1543	0.1692	-1.0892	0.4619	2.4491	
	1.992	1.6889	-0.0257	0.9454	3.4215	
	0.361	-2.2842	1.743	-0.2438	-3.343	

Table 2: Weights and biases of the proposed ANN model.

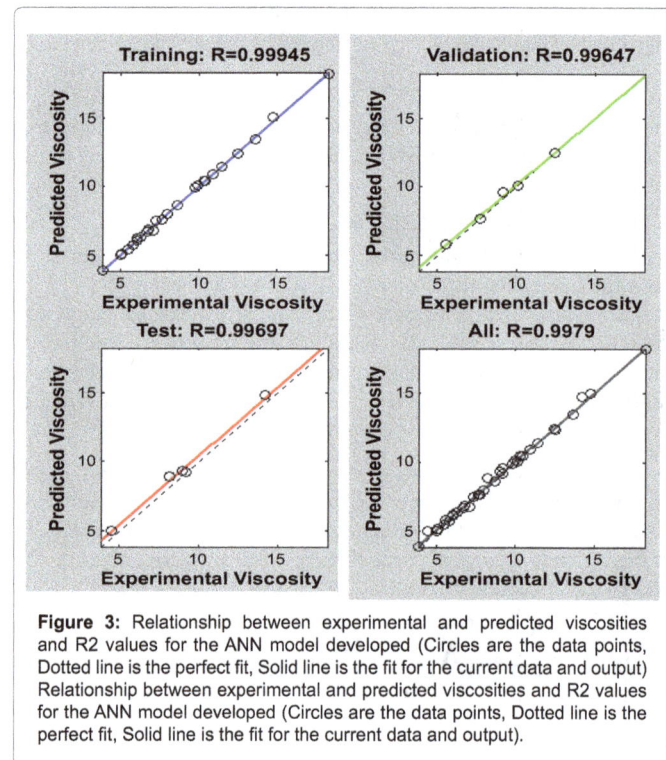

Figure 3: Relationship between experimental and predicted viscosities and R2 values for the ANN model developed (Circles are the data points, Dotted line is the perfect fit, Solid line is the fit for the current data and output) Relationship between experimental and predicted viscosities and R2 values for the ANN model developed (Circles are the data points, Dotted line is the perfect fit, Solid line is the fit for the current data and output).

Parameter	MSE	RMSE	SEP (%)	B_f	A_f	AARD (%)
Eqn. 1 [6]	0.32418	0.56937	6.46607	1.00596	1.06587	6.39
ANN	0.05106	0.22596	2.56611	1.01061	1.01689	1.71261

Table 3: Accuracy parameters for the viscosities from the literature [6] (Eqn.1) and present ANN model.

Temperature, °C	AARD (%)	
	Eqn. (1) [8]	ANN model
15	5.15	0.84
20	5.97	1.97
25	6.23	0.83
30	7.88	4.32
35	6.74	0.6
Overall AARD (%)	6.39	1.71

Table 4: AARD% values for the viscosities from the literature [8] (Eqn.1) and present ANN model.

PEG %	BSA %	Temperature °C	Value of parameter, A	Viscosity, mPa S			Absolute Relative Error, %	
				Experimental [8]	Eqn. 1 [8]	ANN model	Eqn. 1 [8]	ANN model
11	1	15	3.97	18.29	17.78	18.22	2.77	0.36
11	1	20	3.82	14.25	15.06	14.79	5.69	3.82
11	1	25	3.6	12.51	12.74	12.41	1.87	0.8
11	1	30	3.61	9.8	10.16	9.92	3.65	1.21
11	1	35	3.57	7.99	8.44	7.99	5.66	0.01

Table 5: Absolute relative error (%) values for the viscosities from the literature [8] (Eqn.1) and present ANN model.

Table 4 shows the comparison of AARD values of the literature [8] and the proposed ANN model. It is evident from the table that $AARD_{ANN} > AARD_{Eqn}$ for all the temperatures studied. Moreover, the overall AARD based on Grunberg-Nissan model and ANN model was 6.39% and 1.71% respectively which confirms that ANN model has lesser error for viscosity prediction.

In addition to these, the applicability of the ANN model for the generalized condition can be demonstrated by analyzing the absolute relative error (%) values for the proposed ANN model and Grunberg-Nissan model. A sample data values were shown in the Table 5. The interesting thing to be noted here is that the adjustable parameter A was not taken into consideration during the ANN topology. Even then, the absolute relative error (%) values are very low when compared to Grunberg-Nissan model. This significantly corroborates the validity of the ANN model.

Thence, the acceptable values of these parameters confirm that the developed ANN model defines the actual behavior of the system and the diversity of the model developed [18]. Therefore, the ANN model developed in the present study can be employed to predict the viscosity of PEG-BSA-Water ATPS with varying compositions and different temperatures.

Conclusions

An ANN model for the prediction of viscosity of ATPS which consisted of PEG-BSA-water was developed based on LM algorithm. The topology of ANN model developed was 3:10:1. Various accuracy parameters were calculated and compared with the existing empirical equation (Grunberg-Nissan model). The overall AARD (%) of the developed model was 1.71 which assures the validity of the model. A very high R^2 value (>.99) and low MSE justifies the selected ANN model for the prediction of viscosities. Therefore, it is concluded that ANN can be used as a predictive tool to evaluate the viscosities of aqueous two phase systems containing protein.

Acknowledgment

The authors gratefully acknowledge the Department of Biotechnology, MIT, Manipal University for providing the facilities to carry out the research work.

References

1. Albertsson PA (1986) Partitioning of Cell Particles and Macromolecules: seperation and purification of biomolecules, cell organelles, membranes, and cells in aqueous polymer two-phase systems and their use in biochemical analysis and biotechnology, 3rd edn. John Wiley and Sons, New York.

2. Hatti-Kaul R (2000) Aqueous Two Phase Systems: Methods and Protocols. Humana Press: Totowa, NJ.

3. Johansson G (1996) Aqueous two-phase systems with a liquid protein (bovine serum albumin) phase for partitioning of enzymes. J Chromatogr B Biomed Appl 680: 123-130.

4. Modarress H, Mohsen-Nia M, Mahdavi MA (2004) Experimental and theoretical studies of polymer/solvent viscosity mixtures. Journal of Applied polymer science 91: 1724-1729.

5. Gündüz U (2000) Viscosity prediction of polyethylene glycol-dextran-water solutions used in aqueous two-phase systems. J Chromatogr B Biomed Sci Appl 743: 181-185.

6. Kalaivani S, Srikanth CK, Regupathi I (2012) Densities and Viscosities of Binary and Ternary Mixtures and Aqueous Two-Phase System of Poly(ethylene glycol) 2000 + Diammonium Hydrogen Citrate + Water at Different Temperatures. Journal of Chemical engineering data 57: 2528-2534.

7. GRUNBERG L, NISSAN AH (1949) Mixture law for viscosity. Nature 164: 799.

8. Gündüz U (2004) Evaluation of viscosities of aqueous two-phase systems containing protein. J Chromatogr B Analyt Technol Biomed Life Sci 807: 157-161.

9. Haykin S (1998) Neural Networks: A Comprehensive Foundation. Prentice Hall, NJ, USA.

10. Hezave AZ, Lashkarbolooki M, Raeissi S (2012) Using artificial neural network to predict the ternary electrical conductivity ofionic liquid systems. Fluid Phase Equilibria 314: 128-133.

11. Khaye M, Cojocaru C (2012) Artificial neural network modeling and optimization of desalination by air gap membrane distillation. Separation and Purification Technology 86: 171-182.

12. Sharon Mano Pappu J1, Vijayakumar GK, Ramamurthy V (2013) Artificial neural network model for predicting production of Spirulina platensis in outdoor culture. Bioresour Technol 130: 224-230.

13. Kumar S, Pai PS, Rao BRS (2012) Radial Basis Function Network Based Prediction of Performance and Emission Characteristics in a Bio Diesel Engine Run on WCO Ester. Advances in Artificial intelligence.

14. Mohagheghian M, Ghaedi AM, Vafaei A (2011) Prediction of Kinematic Viscosity of Binary Mixture of Poly (Ethylene Glycol) in Water using Artificial Neural Networks. World Academy of Science, Engineering and Technology 49: 158-161.

15. Lashkarbolooki M, Hezave AZ, Babapoor A (2013) Correlation of density for binary mixtures of methanol + ionic liquids using back propagation artificial neural network. Korean Journal of chemical Engineering 30: 213-220.

16. Fazlali A, Koranian P, Beigzadeh R, Rahimi M (2013) Application of artificial neural network for vapor liquid equilibrium calculation of ternary system including ionic liquid: Water, ethanol and 1-butyl-3-methylimidazolium acetate. Korean Journal of chemical Engineering 30: 1681-1686.

17. Demuth H, Beale M (2002) Neural Network Toolbox for Use with Mathlab®, User's Guide, Version 4. Mathworks Inc.

18. Khaouane L, Si-Moussa C, Hanini S, Benkortbi O (2012) Optimization of Culture Conditions for the Production of Pleuromutilin from Pleurotus Mutilus Using a Hybrid Method Based on Central Composite Design, Neural Network, and Particle Swarm Optimization. Biotechnology and Bioprocess Engineering 17: 1048-1054.

19. Lu ZM, Lei JY, Xu HY, Shi JS, Xu ZH (2011) Optimization of fermentation medium for triterpenoid production from Antrodia camphorata ATCC 200183 using artificial intelligence-based techniques. Applied Microbiology and Biotechnology 92: 371-379.

20. Khoramnia A, Ebrahimpour A, Beh BK, Lai OM (2011) Production of a Solvent, Detergent, and Thermotolerant Lipase by a Newly Isolated Acinetobacter sp. in Submerged and Solid-State Fermentations. Journal of Biomedicine and Biotechnology 102.

27

Hydrothermal Degradation of Congo Red in Hot Compressed Water and its Kinetics

Asli Yuksel*

Izmir Institute of Technology, Faculty of Engineering, Department of Chemical Engineering, Gulbahce Campus, Urla, Izmir, Turkey

Abstract

A di-azo dye, Congo Red (CR) was used as a model compound to investigate the degradation mechanism in hot compressed water (HCW). The unique properties of HCW facilitated the degradation efficiency without addition of any organic solvent. The influences of reaction time, temperature, initial dye concentration and amount of hydrogen peroxide (H_2O_2) on the degradation of CR and the removal of total organic carbon (TOC) from the product solution were investigated. The presence of H_2O_2 was found to enhance the degradation of CR. The results showed that the degradation yield could reach 99.0% with a solution of 100 ppm CR and 50 mM H_2O_2 at 150°C at the end of 60 min. Maximum conversion of the total organic carbon was recorded as 62.2%. Moreover, the effect of the presence of several co-existing negative ions such as SO_4^{2-}, Cl^-, CO_3^{2-} were investigated. It was found that the presence of SO_4^{2-} accelerated evidently the degradation of CR. The other chosen anions (CO_3^{2-} and Cl^-) had an inhibitory effect on the decolorization of CR. Finally, kinetic study was carried out and the order of the reaction was calculated as 0.37.

Keywords: Congo red; Hot compressed water; Sub-critical water; Hydrothermal; Kinetics

Introduction

During the last decades, contamination of surface and ground water resources by various pollutant residues has become one of the major challenges for the preservation and sustainability of the environment. This contamination arises from surface runoff, leaching, wind erosion, deposition from aerial applications, industrial discharges and other sources [1]. Pesticides and heavy metal ions are one of the major threating groups of pollutants. Pesticides are among the most dangerous environmental pollutants because of their stability, mobility, capable of bioaccumulation and long-term effects on living organisms [2,3]. Removal of this kind of contaminants by adsorption [4-7], nanofiltration [8,9], using Fenton reagent [10], photo-catalytic degradation [11,12], biological treatment [13] and the combination of biological and photo-Fenton treatment [14] were widely studied. The other main contributors for environmental pollution are toxic heavy metals that are released from metallurgical, galvanizing, metal finishing, electroplating, mining industries [15]. Ion exchange resins [16-19], membrane filtration [20,21], adsorption [22-24], electrochemical [25] and biological treatments [26,27] are some of successful techniques for the removal of heavy metal ions (lead, arsenic, nickel, etc.) [28].

In addition to pesticides and heavy metal ions, textile dyes and other industrial dyestuffs constitute one of the largest groups of organic compounds that represent an increasing environmental danger [29,30]. About 10- 20% of the total world production of dyes is lost during the dyeing process and is released in the textile effluents [31]. This released colored wastewater creates serious environmental problems and can originate dangerous byproducts through oxidation, hydrolysis, or other chemical reactions taking place in the wastewater [32].

Organic dyes, particularly azo dyes, which contain one or more azo bonds, constitute one of the most important groups of pollutants in wastewater released from the industries such as textiles, paper, and leather [33,34]. These dyes are a major group of toxic, carcinogenic, colored and synthetic organic compounds [35,36]. Among azo dyes, in this study, it was given a special interest to a secondary di-azo dye Congo Red (CR) (sodium 3,3′-(1E,1′E)-biphenyl-4,4′-diylbis (diazene-2,1-diyl)bis(4-aminonaphthalene-1-sulfonate)). Because CR effluents are highly colored, have low biological oxygen demand (BOD) and high chemical oxygen demand (COD) while they contain high amounts of dissolved solids. Additionally, benzidine is a toxic metabolite of CR, which causes cancer of the bladder in humans [37]. The chemical structure of CR is shown in Figure 1.

The necessity of the removal of various types of dyestuffs has generated several studies in the last years including adsorption [38-40], ozonation [41,42], and photo-catalytic degradation [43-45] which are the most frequently used techniques for dye removal. However, these methods are inefficient and result in the production of secondary waste products that require further treatment.

Degradation of azo dyes by aerobic treatment is also not sufficient,

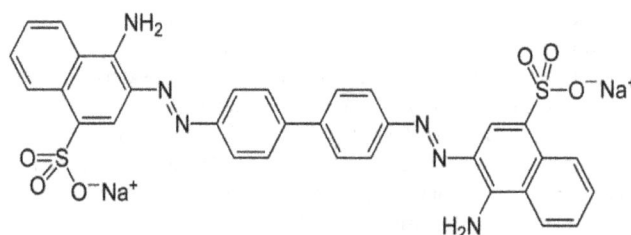

Figure 1: Chemical structure of the di-azo dye Congo Red (CR).

***Corresponding author:** Asli Yuksel, Izmir Institute of Technology, Faculty of Engineering, Department of Chemical Engineering, Gulbahce Campus, 35430, Urla, Izmir, Turkey

and may even be toxic to activated sludge [46]. However, their color can be efficiently removed by an anaerobic step [47,48]. Anaerobic degradation yields only azo reduction, so that mineralization does not occur [49]. This often results in toxic and colorless aromatic amines, with mutagenesis and carcinogenesis potential [50].

Combined biological treatments (anaerobic, then aerobic), with some physicochemical pre-treatments are the most economical ways to decolorize dyed effluents to date [51]. However, textile process wastewaters are generally not concentrated enough for a methanisation stage to be efficient [49].

Chemical-oxidative processes mainly referred as "Advanced Oxidation Processes (AOPs)" have gained more attention recently. The principle of this method is the generation of hydroxyl radicals, one of the strongest known oxidant, to oxidize and mineralize organic molecule completely into CO_2 and inorganic ions. AOPs such as H_2O_2/UV processes, Fenton and photo-Fenton catalytic reactions [52-54] have been widely used to destroy organic pollutants [55]. In most cases, however, these techniques are not sufficient for the conversion of organic carbon in the liquid product solution to inorganic carbon such as CO_2 and CO.

At this point, hot compressed water (or sub-critical water) attracts attention with interesting properties: below the critical point, water behaves as an acid-base catalyst precursor. Additionally, the high relative static dielectric constant of 78.5 at 25°C drops to a value of about 6 at the critical point thus enhancing ionic reactions for the degradation of organic compounds [56]. Moreover, high solubility of organic substances and low viscosity make sub-critical water an excellent medium for fast, homogeneous and efficient reactions [57]. Depending on temperature and pressure, HCW supports either free radical or polar and ionic reactions, which means that HCW is a "tuning solvent" [58].

The main purpose of this work is to study the degradation of CR and removal of total organic carbon (TOC) from the liquid product solution by using HCW, which is clean, non-toxic, cheap and abundant, as a reaction medium without addition of any organic solvent. In order to optimize reaction conditions, effects of reaction time, temperature, initial dye concentration and amount of H_2O_2 as an external oxidant were investigated. Moreover, the effect of the presence of several co-existing negative ions such as SO_4^{2-}, Cl^-, CO_3^{2-} on both the conversions of dye and TOC were examined. Finally, kinetic study was carried out to calculate the order of CR degradation reaction and the rate constants by initial rate and integration methods.

Experimental Procedure and Analysis

Experimental apparatus and procedure

In this study, model dye wastewater was prepared by dissolving 100 ppm of CR in 100 mL of de-ionized water, at first without addition of any external oxidizer and then by adding 20 mM of H_2O_2 to increase the degradation efficiency of organic contaminants. Because of the fact that, in real textile effluents, dying stuff coexists with various amounts of salts, separate experiments were carried out by adding 1 g/L of different sodium salts (NaCl, Na_2CO_3 or Na_2SO_4) with and without H_2O_2. In the structure of CR, sodium is included so in order to clarify the effects of salt addition on the removal of both dye and TOC from the product solution, only sodium salts were preferred to avoid an extra pollution with different ions. Studied reaction temperatures were 120, 150, 175, 200 and 250°C. The reaction pressure and temperature profile

for 250°C is given in Figure 2. To clarify the effect of oxidant addition, series experiment were done by the addition of 0, 10, 20 and 30 mM H_2O_2 and reactions were kept going for 30, 60, and 90 mins. For the kinetic study, calculations were done at three temperatures (120, 150 and 200°C) and initial CR concentrations (50, 100 and 125 ppm).

Degradation of CR in HCW was conducted by using a sealed 300-mL high temperature/high pressure batch reactor (Parr 5500, USA) made of SUS 316 stainless steel. In this system, the reactor vessel was made of titanium, which is quite well resistant to corrosion. Maximum temperature and pressure of the system were 350°C and 3000 psi, respectively. The batch reactor is illustrated in Figure 3. The model textile wastewater was prepared by adding 100 ppm (0.01 g) CR in 100 mL of de-ionized water in the titanium beaker (300 mL). For the experiments with external oxidizer, 20 mM H_2O_2 was added to the feeding solution. The experimental procedure described in Yuksel et al. [59], was applied for the hydrothermal degradation of CR.

Analytical methods

Acidity of the product solution was measured by a bench top pH meter (Thermo Scientific, Orion Star A111). In order to calculate dye

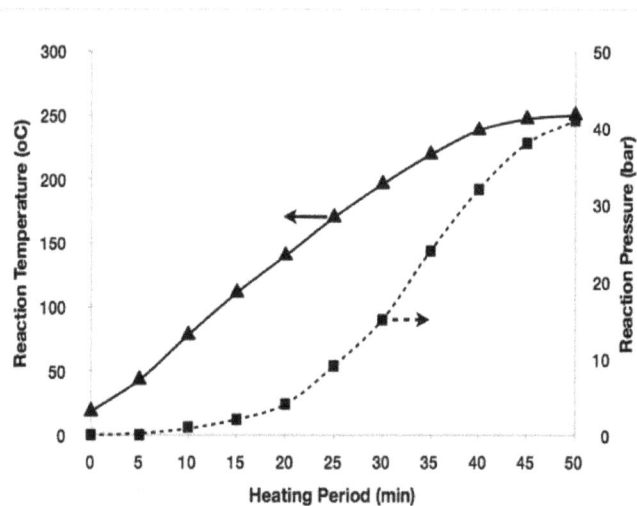

Figure 2: Typical temperature and pressure profiles for HCW at 250°C.

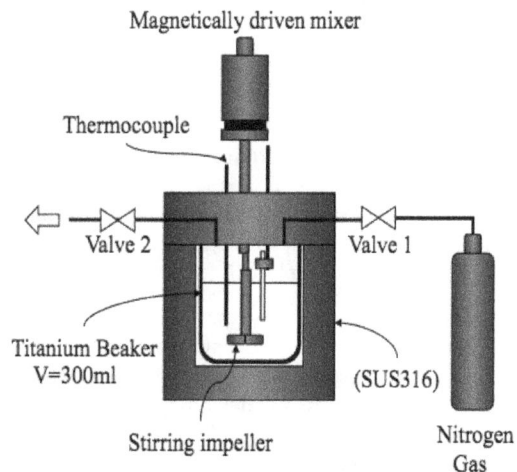

Figure 3: The experimental apparatus (batch reactor).

degradation, absorbance values were recorded between 200 and 800 nm using a scanning UV-visible spectrophotometer (Perkin Elmer, Lambda 45). Visible color removal from the product solution was measured at an optimum absorption wavelength of 499 nm.

The degradation percentage of CR was calculated based on the following Eq. 1:

$$\text{Degradation percentage (\%)} = [1 - (C_t / C_o)] \times 100 \qquad (1)$$

Where C_o (g/L) is the initial concentration of CR and C_t (g/L) is the concentration of CR at reaction time of t (min).

The organic and inorganic carbon content of the initial and final product solution were determined using a TOC analyzer (Shimadzu, TOC-VCPH). The instrument was operated using a 720°C furnace temperature by injecting 50 μL and 325 μL for total carbon (TC) and inorganic carbon (IC) measurements, respectively. It had a nitrogen-measuring unit (Shimadzu, TNM-1), which was used to determine the total nitrogen amount in the solution.

Results and Discussion

Effect of reaction temperature

The effect of reaction temperature (120, 150, 175, 200 and 250°C) on the removal of CR and TOC from the liquid product by using H_2O_2 as an external oxidant was investigated and experimental findings are represented in Figure 4. In these series of experiments, reactions were carried out with 100 ppm CR and 20 mM H_2O_2 for 60 min reaction time.

When reaction was carried out at 120°C, very small amount of dye (5.7%) was converted and most of the organic carbon (around 93%) was remained in the product solution. An increase in the dye degradation with rising reaction temperature was expected. On one hand, the collision frequency of molecules with the radicals increases when temperature rises. On the other hand, the fraction of molecules that possesses energy in excess of activation energy also increases; hence the dye degradation increases accordingly [60]. However, in this study, the conversion of CR at 150 and 250°C were almost same. It is deduced that the effect of reaction temperature was not significant to this reaction within the range studied. It may imply that the activation energy of the reactions could be very low. In this case, the main factor influencing the degradation mechanism was not temperature. It might be related to the concentration and life of radicals. In general, rise of temperature could help to vanish the active radicals. The negative effect of temperature on active species led to a decrease in the degradation percentage [61-64].

Because a lower reaction temperature is important for decreasing cost and increasing energy efficiency, the system temperature for the remainder of the experiments was set to 150°C.

In the case of TOC conversion, the highest efficiency was obtained at a temperature of 250°C. As shown in Figure 4, there was a significant difference in TOC conversions: at 120°C, it was measured as 7%, whereas at 250°C, 59% of TOC was removed from the product solution.

It must be pointed out that HCW at high temperatures has macroscopic properties like a non-polar solvent, but the single molecules are still polar. This change represents a new reaction medium that is suitable for organic species to be degraded [56]. Solubility of these non-polar substances in HCW is a consequence of temperature effect, which is directly related to entropy [61].

Reaction temperature also affects the formation of OH·. The life of this hydroxyl radical is very short and it was found that the decomposition rate of hydrogen peroxide increased with increasing reaction temperature. H_2O_2 decomposed very rapidly to the OH· that would rapidly attack to the dye molecule, leading to higher degradation rates.

Effects of oxidant (H_2O_2) concentration

To understand the role of an external oxidant on the conversion efficiency, separate experiments were carried out with different amounts of oxidant (0-30 mM) added to the model dying wastewater. For these experiments, H_2O_2 was used as an oxidant since it is strong and referred as an environmentally friendly oxidant with the oxidation byproducts of water and oxygen. Additionally, such strong oxidants can attack most of the organic structures found in the wastewater. In HCW, OH ions and radicals, which are responsible for the degradation of organic species (depending on temperature and pressure) have already been generated. With the addition of H_2O_2, the amount of these ions or radicals that would result in higher conversions is increased.

In these experiments, the temperature of the solution was adjusted to 250 and 150°C by heaters and the results are given in Figure 5. At first, reaction was carried out without any external oxidizer with a pH of 7.3. At 150°C, CR conversion was recorded as 6.8%, whereas at 250°C, it was 74.4%. When 10 mM of H_2O_2 was added to the feeding solution, a significant raise in the dye removal was noticed even at lower reaction temperature at a pH of 4.9. The dye conversion was increased to 89.6% from 6.8% at the end of 60 min. Similar trend was observed at 250°C, however, since the removal of dye from the product was already high without H_2O_2, the magnitude of the increase was not as high as it was at 150°C. When more H_2O_2 (20 mM) was used, CR conversion was found to be 99.0% at 150°C. With the increase in H_2O_2 amount to 30 mM (pH=4.4), it was observed that there was little bit decrease in CR conversion, which might follow a similar path for the addition of more H_2O_2 to the solution. In this study, the highest concentration of H_2O_2 was kept at 30 mM to avoid any risk of corrosion, since at HCW

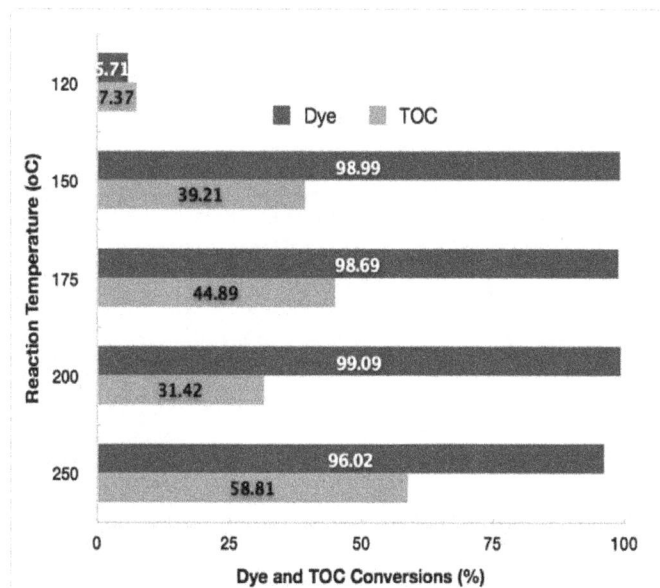

Figure 4: Effect of reaction temperature on the conversions of CR and TOC in the presence of 100 ppm dye and 50 mM H_2O_2.

because of high temperature and high pressure operating conditions; corrosion of the reactor is the common problem.

In the case of organic carbon removal from the product solution, TOC conversion was increased from 14.5% to 62.2% after the reaction with 30 mM H_2O_2 at 250°C. When the reaction was carried out at 150°C, resembling declination in TOC conversion was observed with a value of 34.3%. The reason of this decrease might be, with the addition of higher amounts of oxidant, more OH· radicals are produced to attack the organic species in the structure of dye. However, when the hydroxyl free radicals are finally saturated with hydrogen peroxide, hydrogen peroxide is used more than hydroxyl free radicals to produce hydro peroxide (HO_2·). As well known, HO_2· are less reactive than HO·, which explains well the decrease in the degradation yield with the addition of more H_2O_2. The hydroxyl radical propagation and termination involved are as follows [62-64]:

$$H_2O_2 + HO· \rightarrow HO_2· + H_2O$$

$$2HO· \rightarrow H_2O_2$$

$$HO_2· + HO· \rightarrow H_2O + O_2$$

$$HO· + dye \rightarrow Products$$

$$HO_2· + dye \rightarrow Products$$

$$HO_2· + H_2O_2 \rightarrow HO· + H_2O + O_2$$

As it can be clearly seen from Figure 6, the color removal was failed without H_2O_2 with the addition of 10 mM H_2O_2, decolorization was achieved but still it was insufficient. Finally, 20 mM H_2O_2 was found to be an optimum concentration for the best conversions of both dye and TOC.

Effects of reaction time and initial dye concentration

To investigate the effect of reaction time, experiments were done in 30, 60 and 90 minutes with the conditions of C_{CR}: 100 ppm; C_{H2O2}= 20 mM; reaction temperature: 150°C. The results are illustrated in Figure 7.

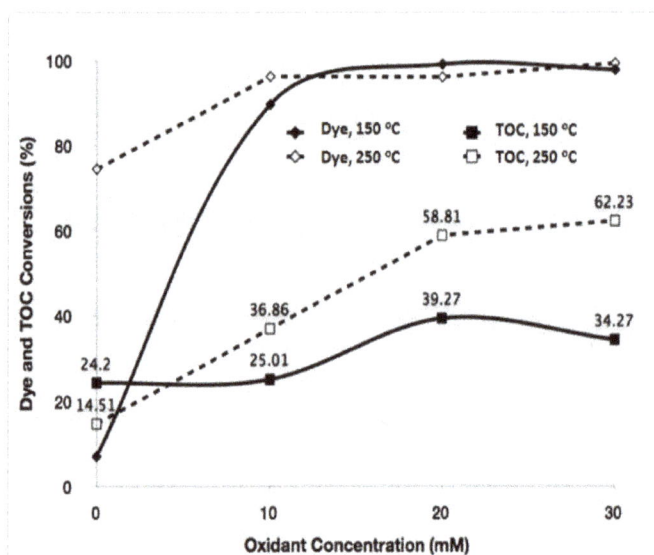

Figure 6: Product pictures after the treatment of 100 ppm CR with different amounts of H_2O_2 at 150°C.

According to Figure 7, after the treatment for 30 min, CR concentration reduced from 100 ppm to 25 ppm. When reaction was carried on for 60 and 90 min, the amount of CR were recorded as 0.9 and 1.1 ppm, respectively. As seen, almost all color (dye) disappeared at the end of 60 min. However, in the case of TOC conversion, longer reaction time (90 min) resulted in the removal of more organic carbon from the model wastewater with a value of 60.7%. This result indicates that the reaction trend was as follows: CR degraded initially resulting to intermediate products (containing aromatic rings) which in turn were further decomposed into simpler products till complete degradation to CO_2, water and inorganic salts.

In the decolorization of the aqueous solutions, TOC removal is as important parameter as the disappearance of the dye compound. After most of the treatment methods, although all color was almost removed from the wastewater, still significant amount of organic carbon remained in the productsolution. TOC conversion efficiencies are quite high in advanced oxidation processes based on the formation of HO·. In one study, the authors tried to degrade CR by galvanostic electrolysis and ozonation under similar conditions. They achieved around 85% of TOC removal by ozonation and by galvanostic electrolysis, almost complete mineralization occurred [65]. In another study, the mineralization of CR was done by decomposing ozone to the form of radical species with a very strong oxidation potential. Hence, both ozone and/or the free radicals broke the double bonds in the conjugated chains of the dye molecule. However, efficient mineralization had not been achieved with a TOC removal of 32% [41]. Seven azo dyes including CR (di-azo dye) were photo-catalytically degraded in TiO_2 suspension by Tanaka et al. [66]. 75% elimination of TOC was recorded after such a long reaction time of 240 min. The photo-catalytic degradation of azo dye proceeded predominantly through the oxidation by radicals and reduction by conduction band electron and electron injection from dye to conduction band. Same amount of conversion was achieved at 57 min for Acid yellow 17 (mono-azo dye) in the same study. As the number of azo bonds increased, the level of TOC conversion decreased because of the large dye molecules being broken down to small organic compounds for the same reaction conditions.

Figure 8 presents evolution with time of UV-vis spectra during hydrothermal degradation of dye wastewater contaminated with 100 ppm (0.1 g/L) CR for 30, 60 and 90 min. The reduction of CR characterized by the absorbance peak at 499 nm, clearly disappeared after 60 min. Although the absorbance at 499 nm decreased rapidly to near-zero level, the level of residual TOC was significant before 60 min. This behavior might be explained by considering that the active radicals and/or ions formed from water itself and H_2O_2 decomposition

Figure 5: Effect of oxidant concentration on the conversions of CR and TOC at 150 and 250°C.

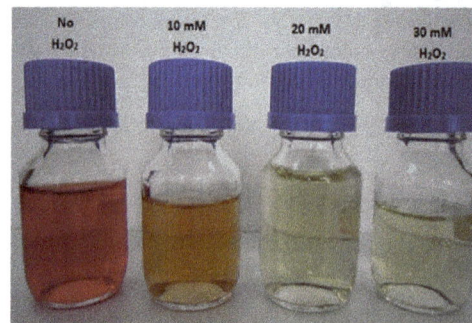

quickly attacked the azo dye, but the remainder of the molecule was not directly oxidized [67-70].

Bands situated in UV region of spectra can be attributed to π --π* transition in aromatic rings but the band of visible region can be due to azoic group transition. As it can be seen from Figure 8, continuous and simultaneous decrease of the band intensities was observed from the beginning of reaction in sub-critical water. These bands disappeared after 90 min. Total disappearance of the three bands (240, 340 and 499 nm) suggested that no more aromatic intermediates existed in the water. These results showed that total degradation of CR and its aromatic intermediates had been achieved by hydrothermal degradation [65]. At the first 30 min, CR started to decompose into aromatic intermediates. However, the degradation of these aromatic intermediates was not immediately started after they were formed, that could be proved with an increase in intensities of UV bands. After 60 min, almost complete destruction mechanism of azoic groups was involved and the degradation by-products absorbed at the same wavelengths with the initial compound [65].

The decolorization for CR was investigated at initial dye concentrations ranging from 50 ppm to 125 ppm at 150°C with 20 mM H_2O_2 concentration. It was clearly seen that with an increase in

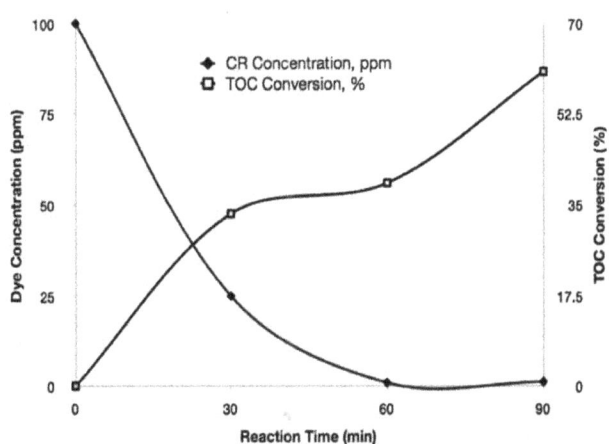

Figure 7: Effect of reaction time on the degradation of CR and TOC removal at 150°C.

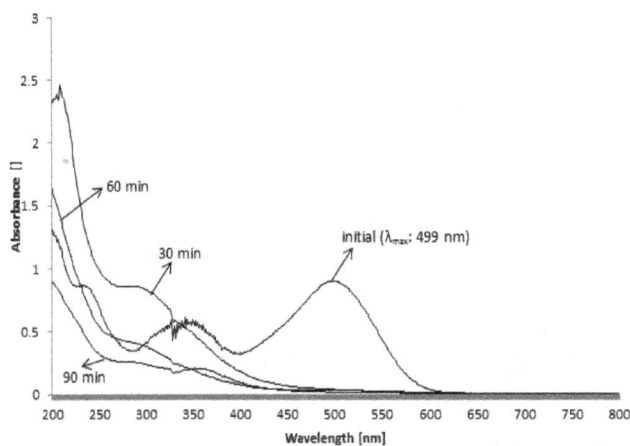

Figure 8: UV-visible spectra of the product solution at the end of 30, 60 and 90 mins. at 150°C.

the initial dye concentration, removal of dye and total mineralization efficiency decreased. With 50 ppm initial concentration of CR, the color was successfully removed with a TOC conversion of 47%. Under the same reaction conditions, addition of 125 mM of dye resulted in a decrease in the conversions of both dye and TOC to 85% and 32%, respectively.

This may be explained by the fact that, under hydrothermal condition, some radicals are generated; however, the active radical species have a nonselective property, which results in an attack on intermediate compounds produced rather than on the initial dye molecule while increasing the initial dye concentration [71]. Hence, the higher the concentration, the stronger the competition between the initial reactant and the intermediate products with active species. Such competition would be more pronounced in the presence of a high concentration level of reaction intermediates produced by the degradation of an increased initial dye concentration.

Effects of coexisting negative ions

In industrial wastewater, especially textile effluents, large amounts of inorganic salts are present with the dyes. Therefore, in this section, it was attempted to evaluate the influences of coexisting negative ions (SO_4^{2-}, Cl^-, CO_3^{2-}) on the degradation of CR and total mineralization efficiency at 150°C, at first without an external oxidizer, then with the addition of 20 mM H_2O_2. For this purpose, 1 g/L of NaCl, Na_2CO_3, and Na_2SO_4 salts were added to the model wastewater. Results of salt experiments are illustrated in Figure 9.

In the absence of H_2O_2, the addition of different sodium salts increased the decolorization rates of dye by enhancing the degradation of CR. When hydrothermal removal of CR was carried out without H_2O_2, degradation efficiency was very low around 7%. With the addition of 1 g/L of Na_2SO_4, it raised to 31%. In another experiments, the CR conversion values were recorded as 33% and 52% with the addition of NaCl and Na_2CO_3, respectively. In all cases, sodium salts were used.

At pH 8-10, most of the selected anions inhibited the degradation of CR in sub-critical water, when the model dye wastewater was prepared with 100 ppm CR and 20 mM H_2O_2. The inhibition by inorganic salts of the degradation of CR at basic pH was in the following decreasing order: chloride > carbonate > sulfate with inhibition percentages of 55%, 25% and 2%. Most of the color was removed when the reaction was carried out without addition of salt (98.99%) and with 1 g/L of Na_2SO_4 (97.16%).

In the literature, the inhibition of the degradation efficiency in the presence of ions is often explained by the scavenging of $HO^•$ radicals by ions [69]. For example, the reaction of CO_3^{2-} with $HO^•$ can be explained by the following equations [70]:

$$CO_3^{2-} + HO^• \rightarrow OH^- + CO_3^{•-}$$

$$HCO_3^- + HO^• \rightarrow H_2O + CO_3^{•-}$$

The addition of Na_2SO_4 together with H_2O_2 had almost no influence on the degradation of CR. This might be explained by the reaction between SO_4^{2-} and $HO^•$ under hydrothermal conditions.

$$SO_4^{2-} + HO^• \rightarrow OH^- + SO_4^{•-}$$

SO_4^{2-} formed from the equation above is less reactive than $HO^•$ radicals, therefore although in this study the amount of salts used (1 g/L) did not create extreme saline conditions, if excess Na_2SO_4 salt was

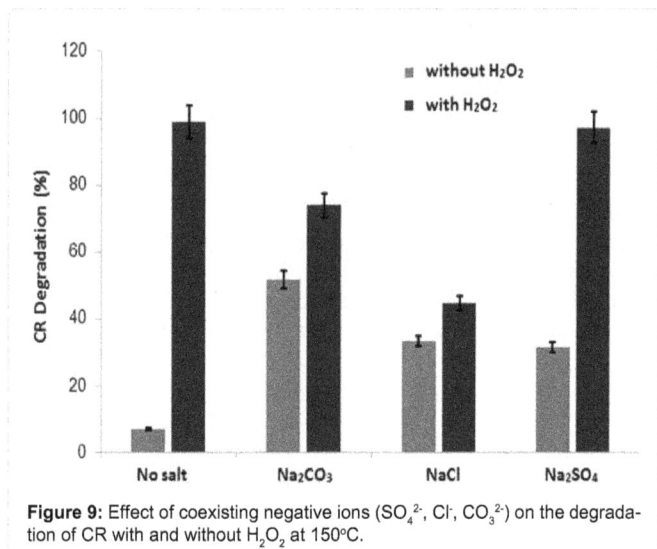

Figure 9: Effect of coexisting negative ions (SO_4^{2-}, Cl^-, CO_3^{2-}) on the degradation of CR with and without H_2O_2 at 150°C.

added, it would hinder the removal of color from the product solution [71,72].

The inhibitory effect of Cl^- was the highest (55%). As shown in Figure 10, the characteristic peak of CR at 499 nm corresponded to an absorbance value of 0.038 without any salt. This value increased to 0.52 with the addition of 1 g/L NaCl to the model wastewater, as an indication of the inhibitory effect of NaCl on the degradation of CR. This could be explained by the possible competitive reaction of Cl^- ions that could oxidize on the surface of reaction vessel made of titanium. In addition, at high temperatures, the mass transport controlled reduction of hypochlorite may be another reason for a decreased dye removal according to the following cathodic loss reaction [73,74]:

$$OCl^- + H_2O + 2e^- \rightarrow Cl^- + 2HO^-$$

The GC-MS analysis proved that no formation of secondary nitrogen containing organic pollutants was observed even in the presence of NaCl that showed the most inhibitory effect. The GC-MS spectrum of the model wastewater prepared by 100 ppm CR, 20 mM H_2O_2 and 1g/L NaCl and treated at 150°C for 60 min, is given in Figure 11.

In azo dyes, both nitrogen atoms are already formally at the zero oxidation degree. In the initial model wastewater, this oxidation degree favors the evolution of gaseous di-nitrogen with being close to zero combined with the existence of a –N=N– double bond [70]. N_2 evolution constitutes the ideal case for a decontamination reaction involving totally innocuous nitrogen-containing final product. Although in this study, GC-TCD analysis was not done, the results of GC-MS and TN analysis confirmed that, when the conversion of dye was high, almost complete mineralization was achieved.

Kinetic study

For the kinetic analysis, initial rate and integration methods were used. For the initial rate method, the reaction was carried out with different initial concentrations of CR (0.05, 0.1 and 0.125 g/L) at a reaction temperature of 150°C, whereas, for the integration method, initial concentration of CR was kept constant at 0.1 g/Land the final concentrations of CR were measured for different reaction times (30, 60 and 90 min).

For the initial rate method, following procedure was applied:

From the definition of rate: $-r_A = \dfrac{NAo}{V} \times \dfrac{dx}{dt}$ \hfill (2)

can be written. If conversion (x) vs time (t) was plotted, its slope at t=0 would give the $\left(\dfrac{dx}{dt}\right)$. Using Eq. 2, initial rate could be calculated. Figure 12 presents x vs t curves for different initial concentrations of CR (50, 100 and 125 ppm).

Calculated $(-r_A)_o$ values are given in Table 1.

For n^{th} order reaction rate, rate expression is given by Eq.3: $-rA0 = k$ (3)

Taking logarithm of both sides, the following Eq. 4 was obtained:

$\ln(-r_{A0}) = lnk + nlnC_{A0}$ \hfill (4)

From the slope of $\ln(-r_{A0})$ vs lnC_{A0} graph, the reaction order (n) was calculated as 0.38.

The order of reaction was also calculated from the integration method as follows:

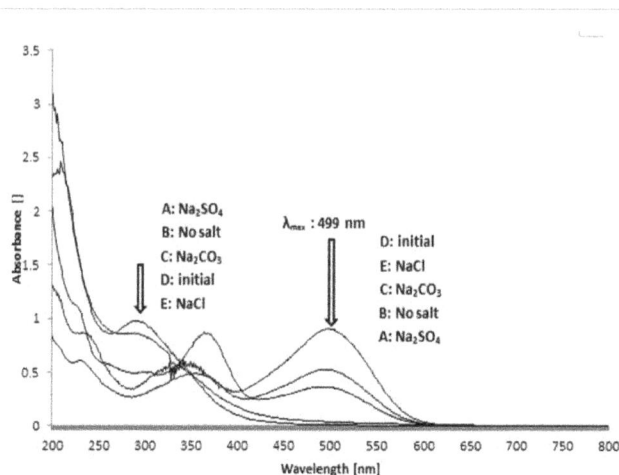

Figure 10: UV-visible spectra of the product solution with 1 g/L of NaCl, Na_2CO_3 and Na_2SO_4 salts at 150°C.

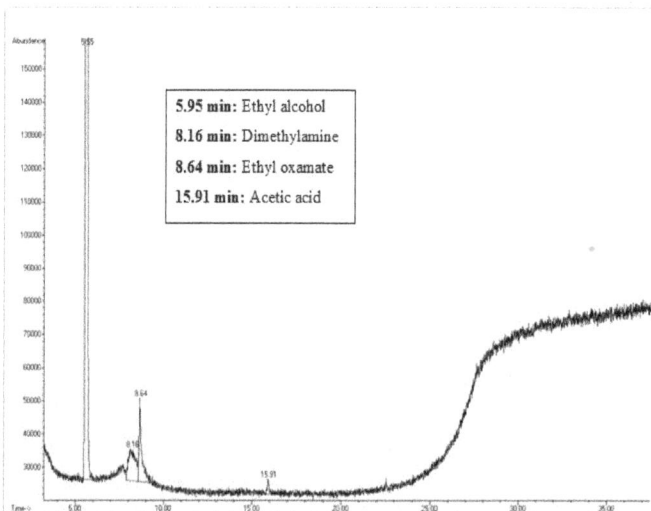

5.95 min:	Ethyl alcohol
8.16 min:	Dimethylamine
8.64 min:	Ethyl oxamate
15.91 min:	Acetic acid

Figure 11: GC-MS spectrum of the model dye wastewater (CR concentration: 100 ppm; salt addition: 1 g/L NaCl; oxidant concentration: 20 mM H_2O_2; reaction temperature: 150°C; reaction time: 60 min).

$$-\frac{dN_A}{dt} = r_A V \qquad (5)$$

For constant volume batch reactor, Eq. 5 becomes:

$$-\frac{dC_A}{dt} = r_A = kC_A^n \qquad (6)$$

where C_A is the final concentration of CR in the product solution, t is the reaction time, k is the rate constant and n is the order of reaction. When Eq. 6 is integrated, the final form of the equation for the calculation of n becomes:

$$lnC_A = \left[\frac{ln(1-n)}{n-1} + \frac{lnk}{n-1} + lnC_{A0}\right] + \left(\frac{1}{n-1}\right)lnt \qquad (7)$$

This expression $\left[\frac{ln(1-n)}{n-1} + \frac{lnk}{n-1} + lnC_{A0}\right]$ in Eq. 7 was a constant value. By the integration method, after drawing $ln(C_A)$ vs $ln(t)$, the order of reaction was calculated as 0.36. The calculated value was 0.38 from the initial rate method. Since two values were very close to each other, for the rest of the kinetic study, the mean value, which was **0.37**, was taken as the order of the degradation reaction of CR in sub-critical water.

For the calculation of reaction rate constants (k), reaction was carried out at three different temperatures of 120, 150 and 200°C. From the definitions of rate and conversion, Eq. 8 was obtained.

$$\frac{C_{A0}^{1-n}\left[(1-x)^{1-n}-1\right]}{n-1} = kt \qquad (8)$$

Since n was calculated as 0.37 and C_{Ao} was 100 ppm (1.435×10^{-4} mol/L), this expression can be written as:

$$-6.018 \times 10^{-3}\left[(1-x)^{0.63} - 1\right] = kt \qquad (9)$$

If the expression in the left hand side of Eq. 9 was plotted against time, k values at each temperature could be calculated from the slope. The conversion and k values at different reaction times can be seen in Table 2.

From Table 2 it was observed that, at 120°C, the color was decreasing relatively slowly with a rate constant of 0.0012 (mol/L)$^{0.63}$L^{-1}.

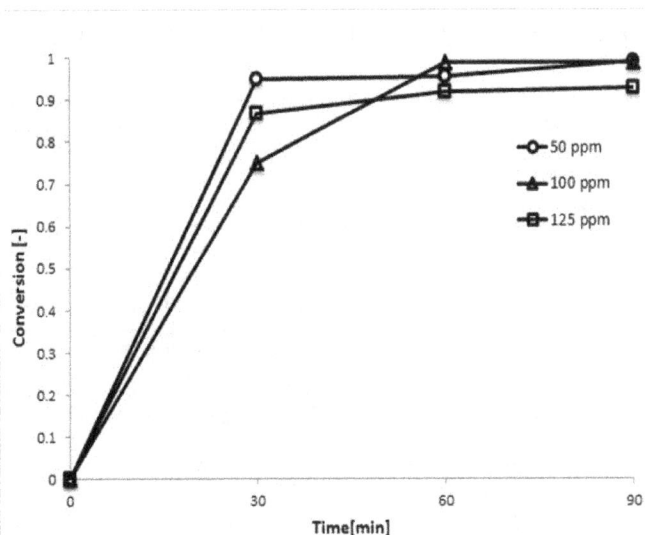

Figure 12: Conversion values of CR with the initial concentrations of 50, 100 and 125 ppm at 150°C.

C_{A0} (ppm)	$C_{A0}\left(\frac{mol}{L}\right)$	$\left(\frac{dx}{dt}\right)_{t=0}\left(\frac{1}{min}\right)$	$(-r_A)_0\left(\frac{mol}{Lmin}\right)$
50	7.177×10^{-5}	-2.8654	2.0565×10^{-4}
100	14.35×10^{-5}	-1.539	2.2162×10^{-4}
125	17.94×10^{-5}	-2.4561	4.4062×10^{-4}

Table 1: Calculated initial rates [$(-r_A)_0$] for 50, 100 and 125 ppm CR at 150°C.

	120 °C	150 °C	200 °C
t (min)	x	x	x
30	0.79	0.7516	0.8414
60	0.8157	0.9899	0.99091
90	0.8012	0.98889	0.9277
$k\left(\frac{mol}{L}\right)^{0.63}(min^{-1})$	**0.0012**	**0.0016**	**0.0015**

Table 2: Calculated conversions (x) and reaction rate constants (k) at 120, 150 and 200°C.

As temperature raised to 150°C, reaction rate constant increased to 0.0016 (mol/L)$^{0.63}$L^{-1}, as expected. However, high temperature does not always result in a higher reaction rate. Due to the mass transfer limitation of H_2O_2 at 200°C, rate constant dropped to 0.0015 (mol/L)$^{0.63}$L^{-1} at this temperature.

Conclusions

The degradation of Congo Red (CR) as a model compound of diazo dyes was studied in hot compressed water without any organic solvent under various operating conditions by using an autoclave. In the presence of 20 mM H_2O_2 as an external oxidant, at 150°C, 99% of CR was successfully removed from the model dye wastewater at the end of 60 min. The maximum amount of TOC conversion was recorded as 62%. From series of experiments, it was observed that the amount of external oxidant greatly affected the degradation of dye itself and TOC conversion. Without H_2O_2, only 7% of CR and 24% of TOC conversions were recorded after 60 min, respectively. The effect of the presence of several co-existing negative ions such as SO_4^{2-}, Cl$^-$, CO_3^{2-} on both the conversions of dye and TOC were also investigated and found that the presence of SO_4^{2-} accelerated evidently the degradation of CR, while the other CO_3^{2-} and Cl$^-$ anions had an inhibitory effect on the decolorization of CR with the inhibition percentages of 25% and 55%, respectively. Finally, kinetic study at temperatures of 120, 150 and 200°C was carried out and the order of the reaction was calculated as 0.37 as a mean value from the initial rate and integration methods. Under optimum conditions (reaction temperature: 150°C; oxidant amount: 20 mM; reaction time: 60 min), the rate constant (k) was calculated as 0.0016 (mol/L)$^{0.63}$L^{-1}. The results proved that without any organic solvent, as a green technology, water itself could be a good reaction medium for the degradation of organic contaminants by just tuning its temperature and pressure.

Acknowledgements

Izmir Institute of Technology (IZTECH) Scientific Research Committee supported this work with a project number of 2012IYTE01. I would like to thank to Filiz Kurucaovali, Fatma Ustun and Belgin Tuncel for their help in TOC and UV analysis of the product solution.

References

1. Salman JM, Njoku VO, Hameed BH (2011) Adsorption of Pesticides from Aqueous Solution onto Banana Stalk Activated Carbon. Chem Eng J 174: 41-48.

2. Naushad Mu, Alothman ZA, Khan MR (2013) Removal of Malathion from Aqueous Solution using De-Acidite FF-IP Resin and Determination by UPLC–MS/MS: Equilibrium, Kinetics and Thermodynamics Studies. Talanta 115: 15-23.

3. Marianne KS, Marta V, Miren LA, Raquel CS, Francesc V, et al. (2013) Occurrence and Behavior of Pesticides in Wastewater Treatment Plants and Their Environmental Impact. Sci Total Environ 458-460: 466-476.

4. Sergio S, Pedro JG, Antonio R, Nuno R, Arminda A (2011) Organochlorine Pesticides Removal from Wastewater by Pine Bark Adsorption After Activated Sludge Treatment. Environ Technol 32: 673-683.

5. Jusoh A, Hartini WJH, Ali N, Endut A (2011) Study on the Removal of Pesticide in Agricultural Run Off by Granular Activated Carbon. Bioresour Technol 102: 5312-5318.

6. Saha A, Shabeer A, Gajbhiye VT, Gupta S, Kumar R (2013) Removal of Mixed Pesticides from Aqueous Solutions using Organoclays: Evaluation of Equilibrium and Kinetic Model. Bull Environ Contam Toxicol 91: 111-116.

7. Tavakkoli H, Yazdanbakhsh M (2013) Fabrication of Two Perovskite-type Oxide Nanoparticles as the New Adsorbents in Efficient Removal of a Pesticide from Aqueous Solutions: Kinetic, Thermodynamic, and Adsorption Studies. Microporous and Mesoporous Materials 176: 86-94.

8. Bruggen BV, Everaert K, Wilma D, Vandecasteele C (2001) Application of Nanofiltration for Removal of Pesticides, Nitrate and Hardness from Ground Water: Rejection Properties and Economic Evaluation. J Membrane Sci 193: 239-248.

9. Sanches S, Penetra A, Rodrigues A, Cardoso VV, Ferreira E, et al. (2013) Removal of Pesticides from Water Combining Low Pressure UV Photolysis with Nanofiltration. Sep Purif Technol 115: 73-82.

10. Zapata A, Velegraki T, Sanchez-Perez JA, Mantzavinos D, Maldonado MI, et al. (2009) Solar Photo-Fenton Treatment of Pesticides in Water: Effect of Iron Concentration on Degradation and Assessment of Ecotoxicity and Biodegradability. Appl Catal B: Environ 88: 448-454.

11. Malato S, Blanco J, Caceres J, Fernandez-Alba AR, Agüera A, et al. (2002) Photocatalytic Treatment of Water-Soluble Pesticides by Photo-Fenton and TiO2 Using Solar Energy. Catal Today 76: 209-220.

12. Oller I, Gernjak W, Maldonado MI, Perez-Estrada LA, Sanchez-Perez JA, et al. (2006) Solar photocatalytic degradation of some hazardous water-soluble pesticides at pilot-plant scale. J Hazard Mater B 138: 507-517.

13. Fuentes MS, Briceno GE, Saez JM, Benimeli CS, Diez MC, et al. (2013) Enhanced Removal of a Pesticides Mixture by Single Cultures and Consortia of Free and Immobilized Streptomyces Strains. Biomed Res Int: 1-9.

14. Martin MMB, Perez JAS, Lopez JLC, Oller I, Rodriguez SM (2009) Degradation of a Four-pesticide Mixture by Combined Photo-Fenton and Biological Oxidation. Water Res 43: 653-660.

15. Naushad M (2014) Surfactant Assisted Nano-composite Cation Exchanger: Development, Characterization and Applications for the Removal of Toxic Pb2+ from Aqueous Medium. Chem Eng J 235: 100-108.

16. Nabi SA, Naushad M, Ganai SA (2007) Use of Naphtol Blue-black-modified Amberlite IRA-400 Anion-exchange Resin for Separation of Heavy Metal Ions. Acta Chromatograph 18: 180-189.

17. Nabi SA, Naushad M, Khan AM (2006) Sorption Studies of Metal Ions on Napthol Blue–black Modified Amberlite IR-400 Anion Exchange Resin: Separation and Determination of Metal Ion Contents of Pharmaceutical Preparation. Colloids Surfaces A: PhysicochemEng Aspects 280: 66-70.

18. Naushad M, Mitra R, Raghuvanshi J (2009) Use of Neutral Red Modified Strong Acid Cation Exchange Resin for Separation of Heavy Metal Ions. Ion Exch Lett 2: 31-34.

19. Naushad M, Alothman ZA, Khan MR, Wabaidur SM (2013) Removal of Bromate from Water Using De-acidite FF-IP Resin and Determination by Ultra-Performance Liquid Chromatography-Tandem Mass Spectrometry. Clean-Soil, Air, Water 41: 528-533.

20. Oehmen A, Vergel D, Fradinho J, Reis AM, Crespo JG, et al. (2013) Mercury Removal from Water Streams through the Ion Exchange Membrane Bioreactor Concept. J Hazard Mater 264: 65-70.

21. Qiu YR, Mao LJ (2013) Removal of Heavy Metal Ions from Aqueous Solution by Ultrafiltration Assisted with Copolymer of Maleic Acid and Acrylic Acid. Desalination 329: 78-85.

22. Rosales E, Pazos M, Sanroman MA, Tavares T (2012) Application of Zeolite-Arthrobacter Viscosus System for the Removal of Heavy Metal and Dye: Chromium and Azure B. Desalination 284: 150-156.

23. Mohan D, Pittman CU (2007) Arsenic Removal from Water/wastewater Using Adsorbents-A Critical Review. J Hazard Mater 142: 1-53.

24. Hua M, Zhang S, Pan B, Zhang W, Lv L, et al. (2012) Heavy Metal Removal from Water/wastewater by Nanosized Metal Oxides: A Review. J Hazard Mater 211-212: 317-331.

25. Karami H (2013) Heavy Metal Removal from Water by Magnetite Nanorods. Chem Eng J 219: 209-216.

26. Piccirillo C, Pereira SIA, Marques APGC, Pullar RC, Tobaldi DM, et al. (2013) Bacteria Immobilisation on Hydroxyapatite Surface for Heavy Metals Removal. J Environ Manag 121: 87-95.

27. Li M, Cheng X, Guo H (2013) Heavy Metal Removal by Bio-mineralization of Urease Producing Bacteria Isolated from Soil. Int Biodeterior Biodegrad 76: 81-85.

28. Fu F, Wang Q (2011) Removal of Heavy Metal Ions from Wastewaters: A Review. J Environ Manag 92: 407-418.

29. Neppolian B, Choi HC, Sakthivel S, Arabindoo B, Murugesan V (2002) Solar Light Induced and TiO2 Assisted Degradation of Textile Dye Reactive Blue 4. Chemosphere 46: 1173-1181.

30. An TC, Zhuand XH, Xiong Y (2002) Feasibility study of Photoelectrochemical Degradation of Methylene Blue with Three-dimensional Electrode-photocatalytic Reactor. Chemosphere 46: 879-903.

31. Houas A, Lachheb H, Ksibi M, Elaloui E, Guillard C, et al. (2001) Photocatalytic Degradation Pathway of Methylene Blue in Water. Appl Catal B: Environ 31: 145-157.

32. Saquib M, Muneer M (2003) Titanium Dioxide Mediated Photocatalyzed Degradation of a Textile Dye Derivative, Acid Orange 8, in Aqueous Suspensions. Desalination 155: 255-263.

33. Pereira WS, Freire RS (2005) Ferro zero: Uma nova abordagem para O Tratamento De Aguas contaminadas com compostos organicos Poluentes. Quim Nova 28: 130-136.

34. Carneiro PA, Osugi ME, Fugivara CS, Boralle N, Furlan M, et al. (2005) Evaluation of Different Electrochemical Methods on the Oxidation and Degradation of Reactive Blue 4 in Aqueous Solution. Chemosphere 59: 431-439.

35. Sun J, Qiao L, Sun S, Wang G (2008) Photocatalytic Degradation of Orange G on Nitrogen-doped TiO2 Catalysts Under Visible Light and Sunlight Irradiation. J Hazard Mater 155: 312-319.

36. Dakiky M, Nemcova I (2000) Aggregation of o,o'-dihydroxyAzo Dyes III. Effect of Cationic, Anionic and Non-ionic Surfactants on the Electronic Spectra of 2-hydroxy-5-nitrophenylazo-4-3-methyl-1-(4"-sulfophenyl)-5- pyrazolone. Dyes Pigments 44: 181-193.

37. 37. Maiti S, Purakayastha S, Ghosh B (2008) Production of Low-cost Carbon Adsorbents from Agricultural Wastes and Their Impact on Dye Adsorption. Chem Eng Commun195: 386-403.

38. Bilal A (2004) Adsorption of Congo red from Aqueous Solution onto Calcium-rich Fly Ash. J Colloid Interface Sci 274: 371-379.

39. Senthilkumaar S, Kalaamani P, Porkodi K, Varadarajan PR, Subburaam CV (2006) Adsorption of Dissolved Reactive Red Dye from Aqueous Phase onto Activated Carbon Prepared from Agricultural Waste. Bioresour Technol 97: 1618-1625.

40. Chatterjee S, Chatterjee S, Chatterjee BP, Guha AK (2007) Adsorptive Removal of Congo Red, A Carcinogenic Textile Dye by Chitosan Hydrobeads: Binding Mechanism, Equilibrium and Kinetics. Colloids and Surfaces A: Physicochem Eng Aspects 299: 146-152.

41. Khadhraoui M, Trabelsi H, Ksibi M, Bouguerra S, Elleuch B (2009) Disoloration and Detoxicification of a Congo Red Dye Solution by Means of Ozone Treatment for a Possible Water Reuse. J Hazard Mater 161: 974-981.

42. Muthukumar M, Sargunamani D, Selvakumar N, Nedumaran D (2004) Effect of Salt Additives on Decolouration of Acid Black 1 Dye Effluent by Ozonation. Indian J ChemTechnol 11: 612-616.

43. Kamel D, Sihem A, Halima C, Tahar S (2009) Decolourization of An Azoique Dye (Congo Red) by Photochemical Methods in Homogenous Medium. Desalination 247: 412-422.

44. Sun J, Wang X, Sun J, Sun R, Sun S, et al. (2006) Photocatalytic Degradation and Kinetics of Orange G Using Nano-sized Sn(IV)/TiO2/AC Photocatalyst. J Mol Catal A 260: 241-246.

45. Hachem C, Bocquillon F, Zahraa O, Bouchy M (2001) Decolourization of Textile Industry Wastewater by the Photocatalytic Degradation Process. Dyes Pigments 49: 117-125.

46. Forgacs E, Cserhati T, Oros G (2004) Removal of Synthetic Dyes from Wastewaters: A Review. Environ Int 30: 953-971.

47. Manu B, Chaudhari S (2003) Decolorization of Indigo and Azo Dyes in Semi-continuous Reactors with Long Hydraulic Retention Time. Process Biochem 38: 1213-1221.

48. Chakraborty S, Basak B, Dutta S, Bhunia B, Dey A (2013) Decolorization and Biodegradation of Congo Red Dye by a Novel White Rot Fungus Alternaria Alternata CMERI F6. Bioresour Technol 147: 662-666.

49. Huber P, Carre B (2012) Decolorization of Process Waters in Deinking Mills and Similar Applications: A Review. Bioresour 7: 1366-1382.

50. Maas R, Chaudhari S (2005) Adsorption and Biological Decolourization of Azo Dye Reactive Red 2 in Semi-continuous Anaerobic Reactors. Process Biochem 40: 699-705.

51. Robinson T, McMullan G, Marchant R, Nigam P (2001) Remediation of Dyes in Textile Effluent: A Critical Review on Current Treatment Technologies with a Proposed Alternative. Bioresour Technol 77: 247-255.

52. Paterlini WC, PupoNogueira RF (2005) Multivariate Analysis of Photo-Fenton Degradation of the Herbicides Tebuthiuron, Diuron and 2,4-D. Chemosphere 58: 1107-1116.

53. Farre MJ, Domenech X, Pera J (2006) Assessment of Photo-Fenton and Biological Treatment Coupling for Diuron and Linuron Removal from Water. Water Res 40: 2533-2540.

54. Ramirez JH, Costa CA, Madeira LM (2005) Experimental Design to Optimize the Degradation of the Synthetic Dye Orange II Using Fenton's Reagent. Catal Today 107–108:68-76.

55. Sakkas VA, Islam MA, Stalikas C, Albanis TA (2010) Photocatalytic Degradation Using Design of Experiments: A Review and Example of the Congo Red Degradation. J Hazard Mater 175: 33-44.

56. Kruse A, Dinjus E (2007) Hot Compressed Water as Reaction Medium and Reactant Properties and Synthesis Reactions. J Supercrit Fluids 39: 362-380.

57. Krammer P, Vogel H (2000) Hydrolysis of Esters in Subcritical and Supercritical Water. J Supercrit Fluids 16: 189-206.

58. Toor SS, Rosendahl L, Rudolf A (2011) Hydrothermal Liquefaction of Biomass: A Review of Subcritical Water Technologies. Energy 36: 2328-2342.

59. Yuksel A, Sasaki M, Goto M (2011) Complete Degradation of Orange G by Electrolysis in Sub-critical Water. J Hazard Mater 190: 1058-1062.

60. Sum OSN, Feng J, Hu X, Yue PL (2005) Photo-assisted Fenton Mineralization of an Azo-dye Acid Black 1 Using a Modified Laponite Clay-based Fe Nanocomposite.

61. Matubayasi N, Nakahara M (2000) Super- and Sub-critical Hydration of Non-polar Solutes: I. Thermodynamics of Hydration. J ChemPhys112: 8089-8109.

62. Kasiri MB, Khataee AR (2011) Photooxidative Decolorization of Two Organic Dyes with Different Chemical Structures by UV/H2O2 Process: Experimental Design. Desalination 270: 151-159.

63. Daneshvar N, Khataee AR (2006) Removal of Azo Dye C.I. Acid Red 14 from Contaminated Water Using Fenton, UV/H2O2, UV/H2O2/Fe(II), UV/H2O2/Fe(III) and UV/H2O2/Fe(III)/Oxalate Processes: A Comparative Study. J Environ Sci Health A 41: 315-328.

64. Gao J, Wang X, Hu Z, Deng H, Hou J, et al. (2003) Plasma Degradation of Dyes in Water with Contact Glow Discharge Electrolysis. Water Res 37: 267-272.

65. Elahmadi MF, Bensalah N, Gadri A (2009) Treatment of Aqueous Wastes Contaminated with Congo Red Dye by Electrochemical Oxidation and Ozonation Processes. J Hazard Mater 168: 1163-1169.

66. Tanaka K, Padermpole K, Hisanaga T (2000) Photocatalytic Degradation of Commercial Azo Dyes. Water Res 34: 327-333.

67. Peralta-Hernandez JM, Meas-Ving Y, Rodriguez FJ, Chapman TW, Maldonado MI, et al. (2008) Comparison of Hydrogen Peroxide-based Processes for Treating Dye-containing Wastewater: Decolorization and Destruction of Orange II Azo Dye in Dilute Solution. Dyes Pigments 76: 656-662.

68. Malik PK, Saha SK (2003) Oxidation of Direct Dyes with Hydrogen Peroxide Using Ferrous Ions as Catalyst. Sep Purif Technol 31: 241-250.

69. Bahnemann DW, Cunningham J, Fox MA, Pelizzetti E, Pichat P, et al. (1994) In: Zeep RG, Helz GR, Crosby DG (ed.) Aquatic Surface Photochemistry, Lewis Publishers, Boca Raton 261.

70. Guillard C, Lachheb H, Houas A, Ksibi M, Elaloui E, et al. (2003) Influence of Chemical Structure of Dyes, of pH and of Inorganic Salts on Their Photocatalytic Degradation by TiO2 Comparison of the Efficiency of Powder and Supported TiO2. J Photochem Photobiol A 158: 27-36.

71. Hu C, Yu JC, Hao Z, Wong PK (2003) Effects of Acidity and Inorganic Ions on the Photocatalytic Degradation of Different Azo Dyes. Appl Catal B: Environ 46: 35-47.

72. Gopinath KP, Kathiravan MN, Srinivasav R, Sankaranarayanan S (2011) Valuation and Elimination of Inhibitory Effects of Salts and Heavy Metal Ions on Biodegradation of Congo Red by Pseudomonas sp. Mutant. BioresourTechnol102: 3687-3693.

73. Rajkumar D, Song BJ, Kim JG (2007) Electrochemical Degradation of Reactive Blue 19 in Chloride Medium for the Treatment of Textile Dyeing Wastewater with Identification of Intermediate Compounds. Dyes Pigments 72: 1-7.

74. Rajkumar D, Kim JG (2006) Oxidation of Various Reactive Dyes with in Situ Electro-generated Active Chlorine for Textile Dyeing Industry Wastewater Treatment. J Hazard Mater B 136: 203-212.

28

Effect of Two Waves of Ultrasonic on Waste Water Treatment

Kumar R[1]*, Yadav N[1], Rawat L[1] and Goyal MK[2]

[1]*Forest Ecology and Environment Division, Forest Research Institute, Dehradun, India*
[2]*National Institute of Pharmaceutical Education and Research, Mohali, India*

Abstract

In the area of water purification, ultra sonication offers the possibility of an efficient removal of pollutants and germs. Ultrasound treatment is one of several technologies that promote hydrolysis – the rate-limiting stage during wastewater treatment. The basic principal of ultrasound is based on the destruction of both bacterial cells and difficult-to-degrade organics. In wastewater, various substances and agents collect in the form of aggregates and flakes, including bacteria, viruses, cellulose and starch. Wastewater is composed largely of the substances responsible for the offensive, pathogenic and toxic materials. In the present research work we found that the ultrasonic treatment is very effective for the waste water purification. The bacterial populations in sludge were decreased according to the frequency (35 KHz and 130 KHz) and time period (5, 10, 20 and 30 min). As the frequency and time period increase the bacterial population was decreased. It was also observed that 130 KHz frequency was more effective than 35 KHz.

This technique would play a major role in sustainable development in a large scale of water purification.

Keywords: Waste water; Ultrasound; Bacteria; Frequency; Cellulose; Sludge

Introduction

Biological treatment processes are widely used in the wastewater treatment industries. However, aerobic and anaerobic treatment processes can result in wastewater that is difficult to treat and handle. Ultrasound treatment is one of several technologies that promote hydrolysis -the rate-limiting stage during wastewater treatment. The basic principal of ultrasound is based on the destruction of both bacterial cells but difficult-to-degrade organics. In wastewater, various substances and agents collect in the form of aggregates and flakes, including bacteria, viruses, cellulose and starch. The energy produced during ultrasound treatment causes these aggregates to be mechanically broken down, altering the constituent structure of the wastewater and allowing the water to be separated more easily, because ultrasound attacks the bacterial cell walls, the bacterial cells release iso-enzymes that biocatalyst hydrolytic reactions [1]. This results in acceleration in the breakdown of organic material into smaller readily biodegradable fractions. The subsequent increase in biodegradable material improves bacterial kinetics resulting in lower wastewater quantities and, in the case of anaerobic digestion, increased biogas production. Therefore, its use is most suited to streams containing large quantities of refractory material and/or cellular matter.

Wastewater are conditioned with polymers to enhance the efficiency of dewatering with presses and centrifuges but the relationship between fluid dynamics and the polymer/ wastewater interaction has been shown to be critical in mixing and conditioning performance[2]. Processing and disposal of wastewater is one of the most complex environmental problems faced by the engineers as well as scientists in this field. Wastewater is composed largely of the substances responsible for the offensive, pathogenic and toxic materials.

Ultrasonic wave in context with bacteria

During the past 20 years the effects of underwater shock waves on living cell have been the subject of money in investigations. Destruction effect of ultrasonic wave on microorganisms and the observed effect in renal infections after extracorporeal shock waves [1]. A possible application could be in a new non-thermal preservation. The bactericidal effect of ultrasonic wave has been evaluated on *E. coli* ATCC 10536, *Salmonella typhimurium* ATCC 14028 and *Listeria monocytogenes*. Our result indicates ultrasonic produce pressure variation, cavitation and the radiation resulting from the underwater shock wave reduce the viability of these microorganisms.

Bacterial spores are so resistant to sonic and ultrasonic waves that such treatment used in past to eliminate vegetative cells from suspensions. Another useful application of ultrasonic wave treatment might to break up aggregates. Ultrasonic treatment however induces changes in certain characteristics of spores: swelling occurs, the surface is eroded and growth is stimulated. Some spore may be killed if the treatment is severe enough.

There is no doubt that microbial world constituted of diverse forms of microbes that dwell ubiquitous habitats i.e. water, air, soil, living/non-living system etc. Attempts were made to probe the stratosphere in the years immediately prior to the space age. Although it was claimed that bacteria and fungi could be found over the altitude range 18-39 km, such results were generally dismissed on the basis of contamination. Several early investigations were undertaken to attempt to determine the relationship between the number of viable bacteria found in the air and various meteorological parameters such as temperature, humidity and wind speed. Since that time more sophisticated techniques for measuring air pollutants and viable airborne microorganisms have been developed. These developments have led to more accurate studies, which show that correlations do exist between viable microorganisms and air pollutants. It is reported that the correlations exist between bacterial density and carbon monoxide, hydrocarbons, nitric oxide, nitrogen dioxide, and sulfur dioxide. Light hart and co-workers investigated the effects of various concentrations of carbon monoxide

*Corresponding author: Kumar R, Forest Ecology & Environment Division, Forest Research Institute, Dehradun-248006, India

and sulfur dioxide on various microorganisms in the laboratory and showed that these agents reduce bacterial density extensively in log-phase cultures and only partially in stationary-phase cultures [3]. Most frequently isolated organisms and their percent of occurrence were *Micrococcus* (41%), *Staphylococcus* (11%), and *Aerococcus* (8%). The bacteria isolated were correlated with various weather and air pollution parameters using the Pearson product-moment correlation coefficient method. Statistically significant correlations were found between the number of viable bacteria isolated and the concentrations of nitric oxide (-0.45), nitrogen dioxide (0.43), and suspended particulate pollutants (0.56). Calculated individually, the total number of *Micrococcus*, *Aerococcus*, and *Staphylococcus*, number of rods, and number of *cocci* isolated showed negative correlations with nitric oxide and positive correlations with nitrogen dioxide and particulates. Statistically significant positive correlations were found between the total number of rods isolated and the concentration of nitrogen dioxide (0.54) and the percent relative humidity (0.43). The other parameters tested, sulfur dioxide, hydrocarbons, and temperature showed no significant correlations [4].

Phosphate solubilising activity

Improvement in soil fertility leads to increase in agricultural and forest production or primary production, which support the heterotrophs. In many cases the total nutrients of each type of soil remain the same even then some nutrients become limiting to the primary producers. These nutrients (elements) get locked in unavailable forms due to biological immobilization and chemical precipitation.

Many soil microorganisms are able to solubilize unavailable forms of calcium bound phosphorus by their metabolic activities by excreting organic acids, which either directly dissolves phosphorus locked in rocks or chelates Ca^{++} to bring phosphorus into solution. The production of microbial metabolites results in decrease in soil pH, which probably plays a major role in solubilization. Besides change in pH, chelation by organic acids, which bind phosphate anions, also bring about phosphate in soil solution [5,6].

Soil inoculation with phosphate solubilising bacteria has been shown to improve solubilisation of fixed soil Phosphate and applied phosphates resulting in higher crop yield. High pH, high salt concentration and high temperature lead to poor growth and survival of phosphate solubilising bacteria. The decrease in pH clearly indicates the production of acids, which is considered to be responsible for Phosphate-solubilisation. It has been found that microorganisms, which decrease the medium pH during growth, are efficient Phosphate-solubilizers [7] suggested that calcium activity is an important factor controlling the rate and extent of dissolution of rock phosphate. In lab condition, maximum solubilization of phosphate occurs after 3 days of incubation of phosphate solubilizing bacteria. Further incubation of up to 5 days does not improve the extent of solubilization.

Mineral phosphate solubilizing (MPS) genes have been cloned from *Erwinia herbicola*. They have been identified as the genes that code for PQQ biosynthesis [8]. The PQQ is required for the functioning of glucose dehydrogenase. This system is involved in energy generating incomplete oxidation pathway that produces gluconic acid, which at the periplasmic space of some bacteria generates protons that functions to dissolve the insoluble phosphate. It is felt that direct oxidation pathway can be exploited to mine the mineral phosphates to release the inorganic phosphate for plant uptake. Two strains of *Rhizobium leguminosarum* bv. *phaseoli* colonizing maize and lettuce root have been shown to have in vitro phosphate solubilizing ability [9].

Materials

During my research work all the chemicals were used of A.R. grade and were supplied by E. Merck (India), Himedia (India), S.D. Fine chemicals (India), Qualigens (India) or Sigma (U.S.A). Waste water sample of sewage was collected from Paper and Pulp industry, Meerut (U.P., India).

Methods

Ultrasound is simply mechanical waves at a frequency above the threshold of human hearing. It can be generated at a broad range of frequencies (35 and 130 KHz) and acoustic intensities.

Ultrasonication conditions

Ultrasonic (US) pre-treatment specifications were taken as under [10].

Treatment time (min.): $5(t_1)$, $10(t_2)$, $20(t_3)$ and $30(t_4)$, Sample volume 100 ml.

US Frequency: 35 kHz and 130 kHz.

US Power: 250 W.

US Intensity: power supplied per transducer area (50.95 watt per cm sq.)

US Density: power supplied per sample volume (2500 watt per lit)

US Dose: Energy supplied per sample volume (j/l)

In the present study, the following conditions were useful for the isolation, characterization and treatment of isolated bacteria which were obtained from industry waste water.

Media preparation and bacterial isolation

LB Medium (Luria-Bertani)

Ingredients	Amount g/L
Yeast extract	5.0
Sucrose	0.5
NaCl	5.0
Tryptone	10.0
PH of solution	7.8

Waste water samples were treated at 35 KHz and 130 KHz frequency for time- 5, 10, 20 and 30 min respectively; including control (without ultrasonic treatment). 5μl of sample was spread on LB plates aseptically under laminar air flow and incubated at 32 °C for overnight.

Estimation of Biochemical Oxygen Demand (BOD)

Treated sample dilution was prepared in a 300 ml BOD flask.1 ml each of phosphate buffer, magnesium sulphate, calcium chloride and ferric chloride solution was added in 1 liter of sample and pH was adjusted to 7.0. Alkaline potassium iodide, manganous sulphate, starch (indicator) and sulphuric acid (to digest the precipitate) was added in a sample to remove the DO content from sample. Two set of samples were prepared. First set of sample was titrated by sodium thiosulphate solution and second set was kept in BOD incubator at 20 °C for 5 days. After the completion of 5 days, second set was also titrated with sodium thiosulphate.

Coliform Test of Domestic Sewage Sample

This technique is mainly use for the detection of presence or absence of coliform bacteria in water from treatment plants. 100 ml

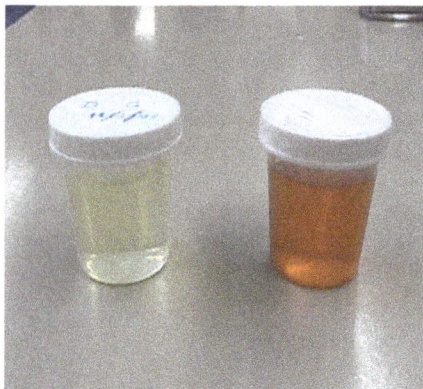

Figure 1: Control and Treated Sample of *E. coli* Bacterial Population.

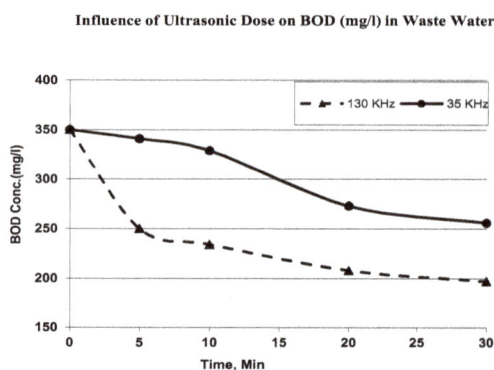

Figure 2: Influence of Ultrasonic Dose on BOD (mg/l) in Waste Water.

sample was collected in flask and treated with ultrasonic waves. Sample was transferred into ziplag bag and then in sterile disposable bottle (100 ml capacity). Entire quantity of dehydrated medium (Lauryl Tryptose Broth) was added slowly to water sample by swirling to dissolve the powder completely. After dissolution the sample was incubated at 30-35 °C for 24-48 h.

Escherichia Coli form (*E. coli*)

Culture medium for *E. coli*: 10 g peptone, 10 g lactose, 2 g KH_2PO_4, 15 g Agar, 0.4 g Eosin and 0.065 g Methylene blue were added in 1litre of distilled water and maintain pH up to 7.1. Sample was kept for autoclaving and poured in sterilized petriplates under laminar air flow. Plates were kept an incubator for overnight. One drop of diluted sample was spread on the plates and kept it for overnight an incubator.

Ultrasound Treatment of Waste water: Waste water sample of sewage was collected from Paper and Pulp industry, Meerut (U.P.). This is followed by ultrasonic treatment at four different duration of time (5, 10, 20 and 30 min) using ELMA, multi frequency ultrasonic bath (according to manufacture Instruction), The Sonication of wastewater was performed in borosilicate glass vessel (250 ml). During sonication, waste water temperature was increasing gradually with time.

pH and electric conductivity of sample were measure by electrode based probe (Water and Soil analysis kit, Electronics India, Model 161E), the rise in temperature of sample on ultrasonic treatment was measure by mercury filled thermometer. In order to characterize this

sample pH, electric conductivity, total solid content (gm/l), COD (mg/l), Total Nitrogen, Total Phosphorus, BOD ,Gram staining and Phosphatase activity test were measured.

Screening of Isolates for Phosphatase Activity: The basic principle behind the determination of phosphatase activity is to supply insoluble phosphorus source in agar based medium for the growth of the bacteria. Use of yeast extract is avoided in the medium. Phosphatase activity of all the isolates was tested after growth in Goldstein solid agar medium [9] which is specifically used for screening phosphate solubilizers.

Coli form Test of Sewage Sample: This technique is mainly use for the detection of presence or absence of coliform bacteria in waste water by ultrasonic treatment. After 24-48 h. incubation, if colour changes of the medium from reddish purple to yellow, indicating the presence of coliform bacteria which is shown in (Figure 1). But due to the ultrasonic treatment no any population of *E. coli* was present in the sample. It could be because no any colour changes in the medium. If colour becomes light yellow than it shows positive test.

Medium Composition:

Solution 1

General purpose Agar	20.0 g/l
Glucose	10.0 g/l
NaCl	1.0 g/l
NH_4Cl	5.0 g/l
$MgSO_4.7H_2O$	1.0 g/l
pH	7.0

Solution 2	K_2HPO_4	5.0 g/50 ml
Solution 3	$CaCl_2$	10.0 g/100 ml

All the above solutions were made separately and autoclaved. Solutions were cooled down to about 50°C. Solution- 2 was added to solution -1 and then solution -3 was added to this mixture. The resulting solution was poured into petriplates and was allowed to solidify. After a day plates were inoculated with cultures by streaking. Plates were incubated at 30°C for one week. Plates were observed for the zone of solubilisation of insoluble phosphate (Halozone).

Procedure for *E. coli* test by Lauryl media:

1. 100 ml sample was collected in flask and treated with ultrasonic waves.

2. Sample was transferred into ziplag bag and then in sterile disposable bottle (100 ml capacity).

3. Entire quantity of dehydrated medium (Lauryl tryptose Broth) was added slowly to water sample by swirling to dissolve the powder completely.

4. After dissolution the sample was incubated at 30-35°C for 24-48 h.

Results

Biochemical Oxygen Demand (BOD)

When DO was determined on the very first day, the concentration of treated waste water sample increased and then decreased according to their treatment time. It is observed that of ultrasound frequency at 130 KHz is more effective than at 35 KHz. After 5 days of incubation the concentration of DO decreased, as shown in (Figure 2), after determining the DO content in wastewater.

Bottle No.	Sample vol. in 1 liter	Dilution in %	Initial DO m/l	Incubated	Final DO mg/l	DO drop	BOD
A (35KHz)	10	1	9.25	5 days	5.89	3.36	336
B (35KHz)	10	1	8.85	5 days	5.62	3.23	323
C (35KHz)	10	1	8.45	5 days	5.8	2.6	260
D (35KHz)	10	1	8.3	5 days	5.8	2.5	250
A¹(130KHz)	10	1	9.25	5 days	6.75	2.5	250
B¹ (130KHz)	10	1	7.42	5 days	6.55	2.3	230
C¹ (130KHz)	10	1	7.3	5 days	6.05	2.0	200
D¹(130KHz)	10	1	7.29	5 days	5.74	1.9	190
Control	10	1	9.5	Untreated	6.0	3.5	350

Table 1: Dissolve Oxygen of sonicated wastewater after 5 days incubation.

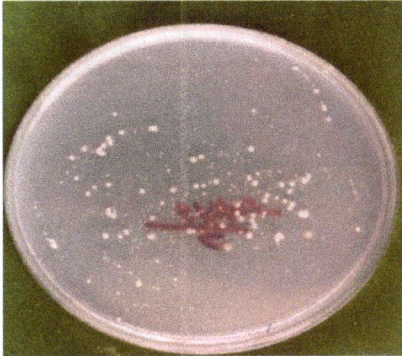

Figure 3a: Bacterial colonies in waste water sample after ultrasonic treatment at 35 KHz for 5 min.

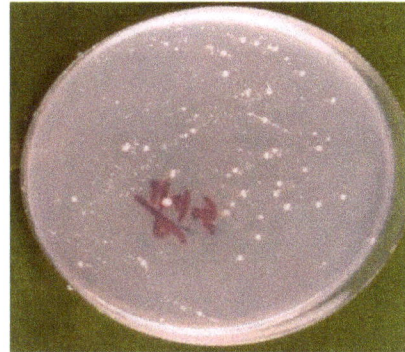

Figure 3b: Bacterial colonies in waste water sample after ultrasonic treatment at 35 KHz for 10 min.

Isolation and growth of bacteria

Sludge samples were treated at 35 KHz and 130 KHz frequency for time- 5, 10, 20 and 30 min respectively including control (without ultrasonic treatment) 5 µl of sample was spread on LB plates aseptically and incubated at 32°C for overnight. Bacterial colonies obtained were counted followed by Gram staining, 35 KHz = A, B, C, D and 130 KHz= A¹, B¹, C¹, D¹. It was observed that the number of bacterial colonies decreased with increase of treatment time and frequency. The number of bacterial colonies are described in (Table 1) and (Figures 3A-3D) A,B,C,D,A¹,B¹,C¹,D¹andcontrol (Figures 4A¹-4D¹).

Phosphatase activity test

With a view to screen Phosphate-solubilization activity, simple

plate test based, on the formation of halo-zone around the colonies were conducted. Out of six isolates, three isolates showed active Phosphate -solubilizing character (Table 2 and Figure 7).

Discussion

All the sample of wastewater was treated with ultrasonic waves of 35 KHz and 130 KHz (Figures 5 and 6). Ultrasound creates large cavitation bubbles which collapsed upon and initiate powerful jet streams exerting strong shear forces in the liquid. The decreasing wastewater disintegration efficiency observed at higher frequency was attributed to smaller cavitation bubbles, which allow the initiation of strong shear forces of comparatively smaller magnitudes. This is the decomposition

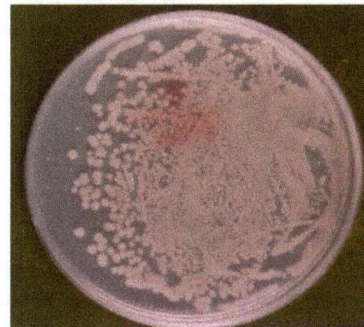

Figure 3c: Bacterial colonies in waste water sample after ultrasonic treatment at 35 KHz for 20 min.

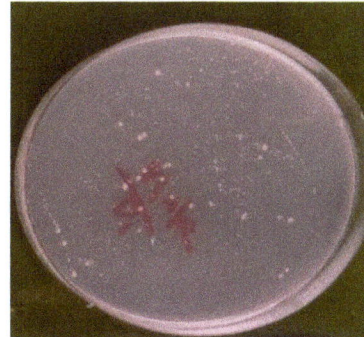

Figure 3d: Bacterial colonies in waste water sample after ultrasonic treatment at 35 KHz for 30 min.

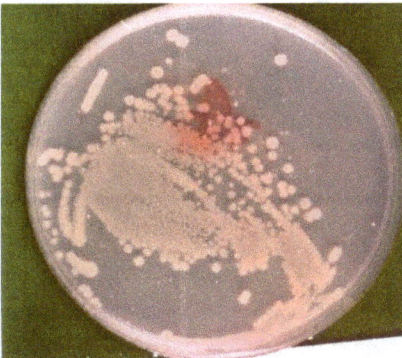

Figure 4a: Bacterial colonies in waste water sample after ultrasonic treatment at 130 KHz for 5 min.

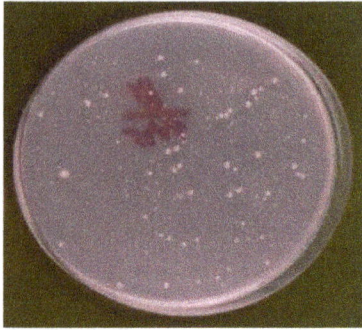

Figure 4b: Bacterial colonies in waste water sample after ultrasonic treatment at 130 KHz for 10 min.

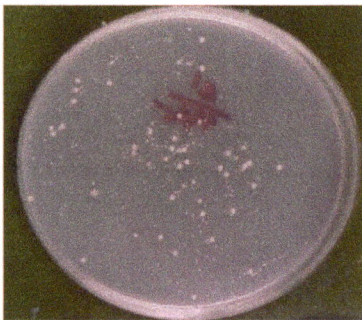

Figure 4c: Bacterial colonies in waste water sample after ultrasonic treatment at 130 KHz for 20 min.

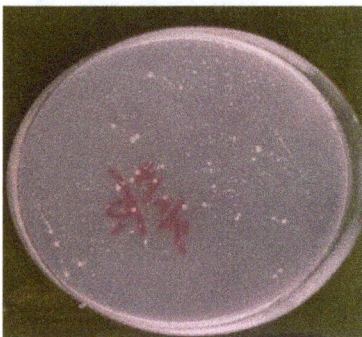

Figure 4d: Bacterial colonies in waste water sample after ultrasonic treatment at 130 KHz for 30 min.

S.No	No. of colonies at 35 KHz	Temp °C	No. of colonies at 130KHz	Temp °C	Time, Min
1.	1086	31	1298	32	5
2.	827	32	756	35	10
3.	742	34	492	37	20
4.	556	37	373	40	30
5.	Control- 1525	30		30	

Table 2: Average colonies obtained on LB plates.

of the flocculent structures within the sludge and or breaking down of the microorganisms embodied in the sludge (Figure 8). Shorter sonication times resulted in wastewater floc deagglomeration without the destruction of bacteria cells. Longer sonication may cause the break-up of cell walls with the probability less for sonication of shorter

duration. It is anticipated that ultrasonic shock waves hits the microbial cell walls. The larger bubbles upon implosion give high mechanical effect, which leads to disintegration of microbial cells. Temperature rise in each sample with the increase in treatment time. Particulate wastewater material was broken down into smaller pieces. We also measured the highest degree of disintegration at 35 KHz. The significant increase of the BOD was attributed to the back up of microbial cells leading to the release of intracellular material [11]. The efficiency of wastewater disintegration decreased with increasing frequency.

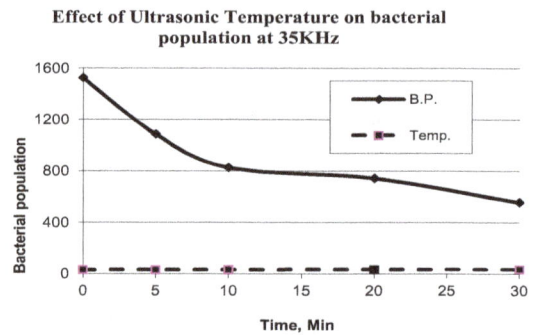

Figure 5: Effect of Ultrasonic Temperature on bacterial population at 35 KHz.

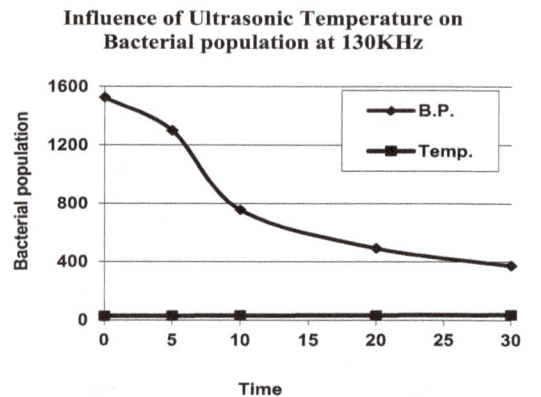

Figure 6: Influence of Ultrasonic Temperature on Bacterial population at 130 KHz.

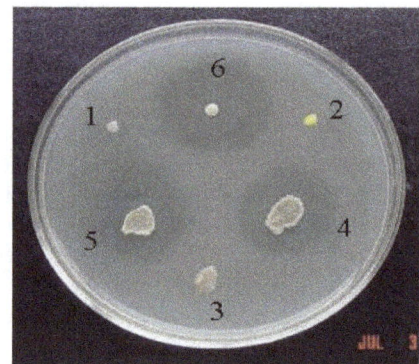

Figure 7: Plates showing halo zones by Phosphate-solubilizers. 1. I, 2. II, 3. III, 4. IV, 5. V, 6. VII.

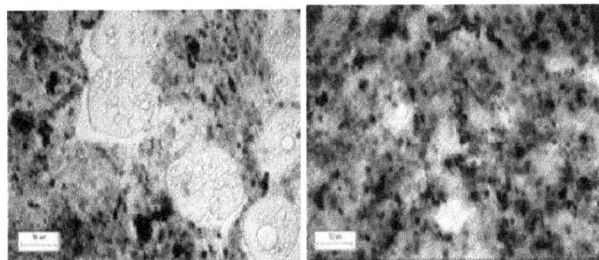

Figure 8: TWAS- Trickened Waste Activated Sludge.

The extent of bacterial cell rupture or cell disintegration were measured on the basis of rate of utilization of dissolve oxygen in wastewater samples. If microbial activities are more the rate of oxygen utilization will be more and vice-versa.

In this experiment, variation of dissolve oxygen in both control and treated samples of wastewater was measured. The dissolve oxygen concentration rates (DOCR) of control sample were more than the treated samples. This shows after treatment bacterial population underwent cell disintegration hence microbial activity decreases so DOCR fell down. The reason behind this is at low treatment time less ultrasonic energy is supplied to the medium (wastewater); due to this deagglomeration of bacterial wastewater flock occurs without the disintegration of bacterial cells. The microorganisms from inside of the wastewater flocs are expressed to the surface and have a better access to the oxygen and hence the oxygen consumption increases.

Conclusion

The ultra-sonication is one of the very useful techniques for the treatment of waste water. Using ultrasonic waves, we can decrease the bacterial population in waste water (paper and Pulp industry, Meerut, Uttar Pradesh, India). Experiment was oriented towards waste degradation through an alternative advanced oxidation technology.

The waste water was treated at two different frequencies (35 KHz, 130 KHz) for different time periods (5, 10, 20 and 30 min). The treated sludge was tested for different parameters (COD, BOD, Total phosphorus and Total nitrogen). Treated sample of waste water was spread on the LB plates, and bacterial isolation followed by morphological identification of Gram staining.

As a result, ultrasonic treatment was found to be treating very effective for the waste water. The bacterial populations in sludge were decreased according to the frequency (35 KHz and 130 KHz) and time period (5, 10, 20 and 30 min). As the frequency and time period increase the bacterial population were decreased. So, it was also observed that 130 KHz frequency was more effective than 35 KHz.

At this premature stage of technique and research beside advantages there are some disadvantages too that must be looked into. After disintegration and anaerobic digestion, waste water has more chemical oxygen demand (COD). So, waste water must be treated.

Acknowledgment

Authors are thankful to the Director FRI for his kind permission to publish this research articles. We also acknowledged the help of Mr. Rajendra kumar and Arun kumar kandwal for providing some of the chemicals, reagents etc.

References

1. Tiwari DK, Behari, Sen P (2008) World Applied Sciences Journal 3: 417-433.

2. Gronroos J, Mokrini (2005) Sonochemical destruction of waste water, Ultrasound in Environmental Engineering; 28: 250-259.

3. Lighthart B, Hiatt VE, Rossano AT Jr (1971) The survival of airborne Serratia marcescens in urban concentrations of sulfur dioxide. J Air Pollut Control Assoc 21: 639-642.

4. Mancinelli RL, Shulls WA (1978) Airborne bacteria in an urban environment. Appl Environ Microbiol 35: 1095-1101.

5. Abd-alla MH (1994) Phosphates and the utilization of organic phosphate by Rhizobium leguminosorum bivoar viceae. Lett Appl Microbial 18: 294-296.

6. Yadav KS, Dadarwal KR (1997) Phosphate solubilization and mobilization through soil microorganisms. In: Biotechnological approaches in soil microorganism for sustainable crop production; Scientific Publisher, Jodhpur India 293-308.

7. Farooq R, Rehman F, Baig S, Sadique M, Khan S, et.al (2009) The Effect of Ultrasonic Irradiation on the Anaerobic Digestion of Activated Sludge. World Applied Sciences Journal 6: 234-237.

8. Goldstein AH (1986) Bacterial solubilization of mineral phosphates; historical perspective and future prospects. American J Agriculture 1: 51-57.

9. Chabot R1, Antoun H, Kloepper JW, Beauchamp CJ (1996) Root colonization of maize and lettuce by bioluminescent Rhizobium leguminosarum biovar phaseoli. Appl Environ Microbiol 62: 2767-2772.

10. Hua I, Hoffmann M R (1997) Optimization of Ultrasonic radiation as an advanced oxidation Technology. Environmental Science Tech 31: 2237-2243.

11. Tiehm A1, Nickel K, Zellhorn M, Neis U (2001) Ultrasonic waste activated sludge disintegration for improving anaerobic stabilization. Water Res 35: 2003-2009.

29

Catalytic Performance of Carbon Nanotubes Supported 12-Tungstosilicic Acid in the Electrooxidation of Cyclohexane to Cyclohexanone and Cyclohexanol

Al-Mayouf AM, Saleh MSA, Aouissi A* and Al-Suhybani AA

Chemistry Department, College of Science, King Saud University, Riyadh, Saudi Arabia

Abstract

Carbon nanotubes-supported 12-tungsto-silicic materials (referred as CNT- SiW_{12}-x, where x is the loading of $H_4SiW_{12}O_{40}$, abbreviated as SiW_{12} were prepared and characterized by means of FTIR, XRD, and polarography. The prepared catalysts and attached onto glassy carbon (GC) electrodes by using polyvinylidene difluoride (PVDF) as binder. The resulting carbon supported SiW_{12} modified electrodes were investigated by cyclic voltammetry (CV) and tested for the electrooxidation of cyclohexane to cyclohexanone (K) and cyclohexanol (A). It has been found that the cyclohexanone, cyclohexanol and cyclohexyl hydroperoxide (CyOOH) are formed as major products of the reaction. Exchanged electronic charge and reaction time had an obvious influence on the catalytic performance of the catalyst. Short reaction time and low exchanged electronic charge favor the formation of cyclohexanol. High exchanged electronic charge and long reaction time favor the electrooxidation of cyclohexanol to cyclohexanone. The optimum condition for the formation of the cyclohexanone, which is needed for the production of the ε-caprolactam, is high exchanged electronic charges.

Keywords: Heteropolyanions; Carbon nanotubes; Cyclohexane electrooxidation; Modified carbon electrode

Introduction

The partial oxidation of low-cost raw hydrocarbons to more valuable oxygenated products is an economically interesting process [1]. However the chemical inertness of the hydrocarbons makes the activation of its C-H bonds very difficult, usually requiring drastic reaction conditions, such as high temperature and pressure [2]. Most hydrocarbon oxidations are unselective, whether conducted in the gas or liquid phase [3,4]. Among hydrocarbon oxidation reactions, the oxidation of cyclohexane to cyclohexanone and cyclohexanol is a very attractive reaction. In fact, Cyclohexanone (K) and cyclohexanol (A) (known as K/A oil mixture) are important chemicals used in the manufacture of nylon-6 and nylon-66, respectively [5,6]. The industrial production of the K/A oil mixture is achieved in a homogeneous catalysis process using soluble transition metal salts (such as cobalt naphthenate) at high temperature and pressure. However due to the fact that the cyclohexanol and cyclohexanone products are more reactive than the cyclohexane reactant, high selectivities (>80%) of the K/A oil mixture only could be observed at low cyclohexane conversion (<5%) [7]. Moreover, this homogeneous industrial process produces more amount of cyclohexanol, and additional steps are needed to improve the K/A mole ratio in the final products [8]. The major challenge in this field is to find an alternative heterogenous process that can improve both, the selectivity of K/A oil and the conversion of cyclohexane. An interesting approach is to use electrochemical process. Recently great attention has been paid to the research in the field of organic electrosynthesis owing to the advent of nuclear power, which will make electricity cheaper compared to chemical oxidants and reductants. Furthermore, in electrochemical processes less hazardous chemicals are used, and high product purity and selectivity of products can be obtained in mild reaction conditions. This research work deal with the electrosynthesis of K/A oil mixture by cyclohexane electrooxidation. To do this, a series of modified carbon nanotubes supported heteropolyanions electrodes have been fabricated and tested.

Materials and Experimental Methods

Preparation of the catalysts

The 12-tungstosilicic acid ($H_4SiW_{12}O_{40}$) was prepared according to a now well-known method [9]. In order to bind SiW_{12} on the carbon nanotubes (CNTs), oxygenated groups must be created on the carbon support (functionalization). The process of carbon functionalization was performed by using concentrated nitric acid according to the following steps: 0.1 g sample of Carbon was suspended in 100 ml nitric acid (65%), and heated for 5 hours at 80°C, then cooled at room temperature. The treated carbon was then washed with deionised water to pH 7, and dried at 100°C overnight. The resulting functionalized CNTs were then added to the desired amount of the prepared SiW_{12} already dissolved in acetone under stirring for 30 min. After removing the excess acetone by heating at 60°C, the prepared catalyst was dried in an oven at 80°C. A series of SiW_{12} catalysts supported on CNTs having various compositions have been prepared. They are denoted CNT-SiW_{12}-x where x is the weight in mg of SiW_{12} per 100 mg of CNTs:

CNT-SiW_{12}-50; CNT-SiW_{12}-100; CNT-SiW_{12}-150; CNT-SiW_{12}-200; CNT-SiW_{12}-300.

Preparation of working electrode

Prior to modification, Glassy Carbon electrode was cleaned by polishing with (0.5-0.05 μm) Al_2O_3 powder. Then after a suspension of

*Corresponding author: Ahmed Aouissi, Prof in Physical Chemistry, Chemistry Department, College of Science, King Saud University, Riyadh11451, Saudi Arabia

the desired amount of the catalyst dissolved in acetone was added to a suspension of the desired amount of polyvinylidene difluoride (PVDF) as a binder dissolved in N-methylpyrrolidone (NMP). The resulting mixture was stirred until colorless. Then after it 10 μl was pipetted onto the surface of the glassy carbon electrode and the solvent was allowed to evaporate at 80°C for 12 hours in the oven. After preparation the modified electrodes were examined by cyclic voltammetry and tested for the cyclohexane electrooxidation reaction.

Characterization of the catalysts

The Characterization of the CNTs supported SiW_{12} catalysts have been performed by means of infrared (IR) spectroscopy, XRD, Polarography, SEM, and TEM. IR spectra were recorded with an infrared spectrometer SHIMADZU FT-IR NICOLET-6700 (4000-400 cm^{-1}) as KBr pellets. The XRD powder patterns were recorded on an Ultima IV, X-ray diffractometer: Rigaku) using Cu-Kα radiation. Polarography measurements were performed by means of METROHEM 797 VA COPMUTRACE (Version 1.2) three-electrode apparatus using a mercury dropping electrode as the working electrode and a saturated calomel electrode (SCE) as the reference electrode. The sample was dissolved in aqueous 1 M HCl/dioxan mixture (50/50 v/v) (30 mg of

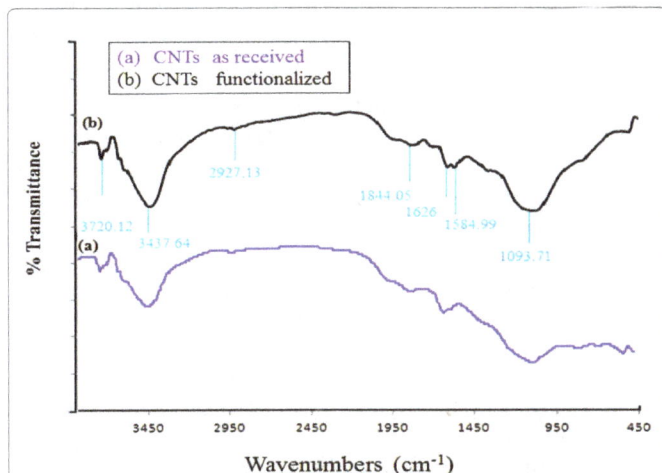

Figure 1: FT-IR spectrum of CNTs: (a) before oxidation; (b) after oxidation. Range temperature 25-800°C; heating rate 20 min/°C.

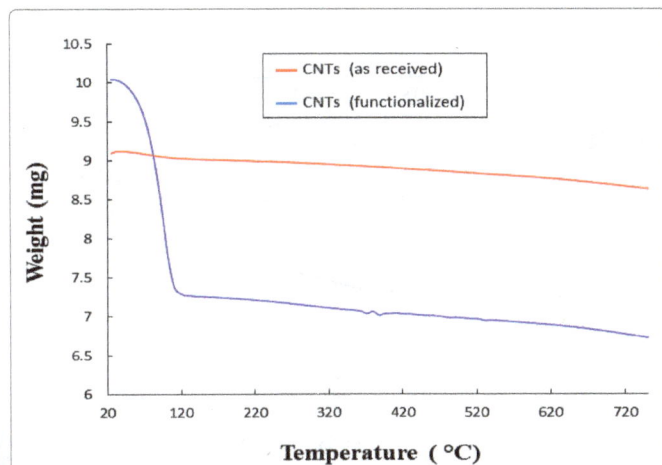

Figure 2: TGA of CNT (a) before oxidation; (b) after oxidation. Range temperature 25-800°C; heating rate 20 min/°C.

sample into 50 ml solution, i.e., a concentration of around 0.05 M. Under these conditions $SiW_{12}O_{40}^{4-}$ anion exhibits reversible waves in the range [-0.15 to -0.800 V].

Electrochemical experiments

Cyclic voltammetry: The cyclic voltammetry was performed in a conventional three-electrode single-compartment Pyrex glass cell using a computerized potentiostat/galvanostat (Autolab, PGSTAT30) with NOVA 1.8 software. The reference and the auxiliary electrodes were SC and pure Pt-foil, respectively. The cell was filled with 0.5 M H_2SO_4 until the lower ends of the electrodes were immersed.

Electrooxidation of cyclohexane: The prepared electrodes were tested for the electrooxidation of cyclohexane in an electrochemical jacketed cell fitted with a reflux condenser. The standard procedure is as follow: 5 ml of cyclohexane, 10 ml of hydrogen peroxide (30% in aqueous solution) and 5 ml of *tert*-butanol were charged in the cell and were heated at 50°C under stirring. After 2 hours of reaction time, the mixture was cooled and analyzed by means of a Gas Phase Chromatograph (Thermo Scientific Trace GC Ultra) equipped with a TCD and FID detectors. The products were separated with a capillary column (TR 5, ID 0.53 mm Film 1 μM).

Results and Discussion

Characterization of the carbon support

FTIR: FTIR spectra of CNTs before and after oxidation are shown in Figure 1. The peaks which are identified at 1386, 1720 and 3448 cm^{-1} characterize C-O, C=O and O-H bonds of the oxidized carbon [10]. Peaks at 1720 and 3448 cm^{-1} can be attributed to acidic groups like carboxyl and phenol. Peak at 1580 cm^{-1} assigns C=C bond in CNTs which appeared after disappearing of bond symmetry because of connection of oxygenated functional groups [11-13]. FTIR results showed that the oxidation treatments produce oxygenated groups such as carboxylic and hydroxyl, in the carbon surface. These groups are responsible for changing both the acid-basic character of the carbon black [14].

Thermogravimetric analysis (TGA): The thermograms of the CNTs before and after oxidation are shown in Figure 2. It can be seen from the figures that, for both CNTs the loss of weight for the oxidized carbon is more important than that for no oxidized carbon. This is due to elimination of oxygen groups in the form of water and oxygenated compounds.

Characterization of the series of catalysts

FTIR: The FT-IR spectra of the samples are shown in Figure 3. Figure 3(a) showed the characteristic features of functional groups created in the functionalized CNTs support. The peaks at 1700 cm^{-1}, 2848 cm^{-1}, and 2920 cm^{-1} indicate the existence of the carboxylic acid groups, aldehyde group, and methylene group respectively. Peaks around 1450-1320 cm^{-1} are an indication of the presence of aromatic groups [10,14]. Aldehyde and derivatives of benzene are detected by peaks at 875 and 761 cm^{-1}. The infrared spectrum of $H_4SiW_{12}O_{40} \cdot 13H_2O$ is shown in Figure 3(b). The main characteristic features of the Keggin structure are observed at 917 cm^{-1} (gas Si-Oa), at is 970 cm^{-1} (gas Mo-Od), at 850 cm^{-1} (gas Mo-Ob-Mo) and at 767 cm^{-1} (gas Mo-Oc-Mo). This result is in agreement with those reported in the literature [9,15] for this heteropoly acid.

The typical pattern of $H_4SiW_{12}O_{40}$ is partly obscured by the carbon bands. In particular the band at 1018 cm^{-1}, assigned to γ_{as} Si-O, is

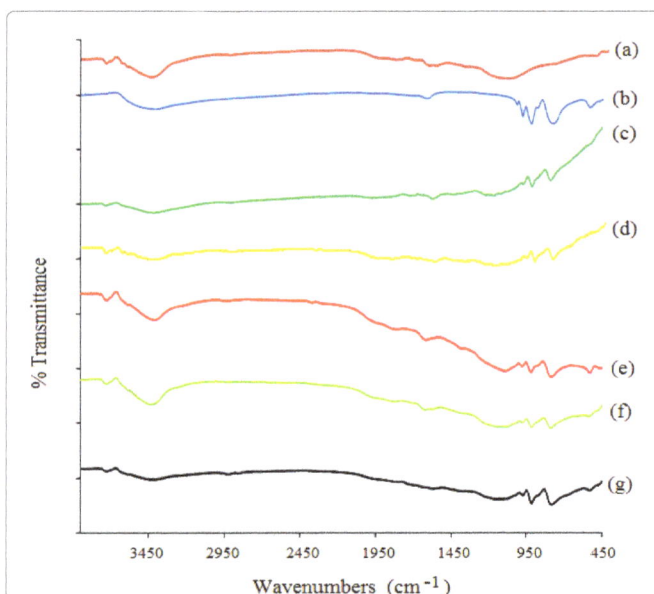

Figure 3: IR spectra of (a) Functionalized CNTs; (b) $H_4SiW_{12}O_{40}$; (c) CNT-SiW_{12}-50; (d) CNT-SiW_{12}-100; (e) CNT-SiW_{12}-150; (f) CNT-SiW_{12}-200; (g) CNT-SiW_{12}-300.

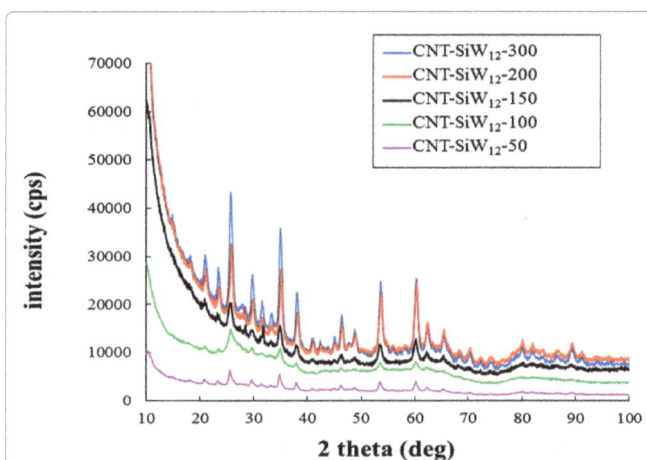

Figure 4: XRD spectra of the prepared CNT-SiW_{12}-x series of catalysts.

Catalyst	Nominal mass (mg)	Experimental mass (mg)
CNT-SiW_{12}-50	50	25.0
CNT-SiW_{12}-100	100	90.3
CNT-SiW_{12}-150	150	110.1
CNT-SiW_{12}-200	200	141.5
CNT-SiW_{12}-300	300	251.5

Table 1: Determination of the amount of SiW_{12} loaded on the CNTs support by polarography.

completely masked into the strong 1100 cm^{-1} band of the carbon: In the 1000-300 cm^{-1} range, subtraction of the carbon absorption is possible for the samples, showing that the Keggin structure is preserved on the support.

X-Ray diffraction

The XRD patterns of the CNT- SiW_{12}-x- series are shown in Figure

4. In each one of the ranges of 2θ, 16-23°, 25-30°, and 31-38°, the compounds showed the characteristic peak of the Keggin structure [16-19]. This result which indicated that the Keggin structure of the heteropolyanions in the synthesized catalysts was not altered is in agreement with that obtained by FTIR.

Polarography

The loading of SiW_{12} on CNTs support was measured by polarography. The results reported in Table 1, showed that the nominal loadings are slightly different of the experimental ones.

Cyclic voltammetry of the prepared materials

Before testing the prepared series of electrocatalysts for the electrocatalytic oxidation of cyclohexane, cyclic voltammetry measurements were performed to test the electrochemical properties of the CNTs support and the series of the CNTs supported catalysts. Cyclic voltammetry was carried out in 0.5 M H_2SO_4 at 25°C at a sweep rate of 50 mV s^{-1} under high-purity nitrogen.

Cyclic voltammetry of the carbon support: The obtained cyclic voltammograms of the functionalized and fresh CNTs support are shown in Figures 5 and 6. It can be seen that the capacitance of the functionalized CNTs was higher than that of the as received CNTs. This fact is due to the created functional groups on the surface of the CNTs [20].

Cyclic voltammetry of the CNT-SiW_{12}-x series: Figure 7 shows cyclic voltammograms of the series of CNT-SiW_{12}-x. It can be seen from this figure that the series exhibit three redox couples with the formal potentials (Ef) of -615.6 mV, -229.9 mV, and, -452.3 mV respectively, similar to those of SiW_{12} dissolved in aqueous solution. The three redox couples correspond to two one-electron processes and one two-electron process, respectively [21-23]. The first one-electron process has no proton participation, the second one-electron process is accompanied by one proton participation, and the third two-electron process is accompanied by two protons participation. The uptake of proton during the SiW_{12} reduction is to avoid charge concentration of SiW, which is commonly found for heteropolyanion compounds [24,25]. According to the above results, the three-redox processes of the series of CNT-SiW_{12}-x can be described as follows:

$$SiW_{12}O_{40}^{\,4-}+e^- \leftrightarrow SiW_{12}O_{40}^{\,5-}$$

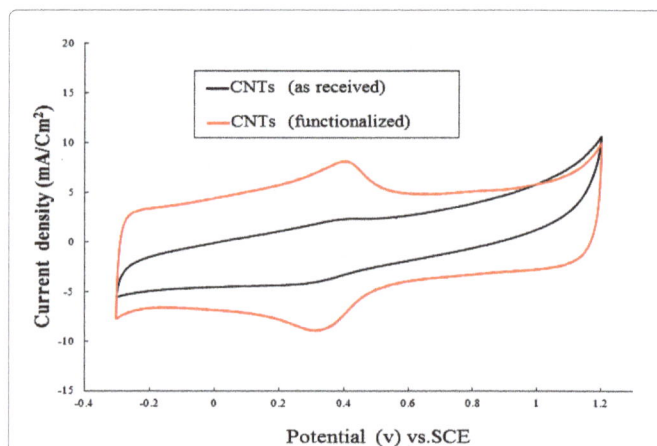

Figure 5: Cyclic voltammograms of CNTs before and after functionalization. Cyclic voltammetry was performed in a 0.5 M H_2SO_4 solution a scan rate of 50 mV s^{-1}.

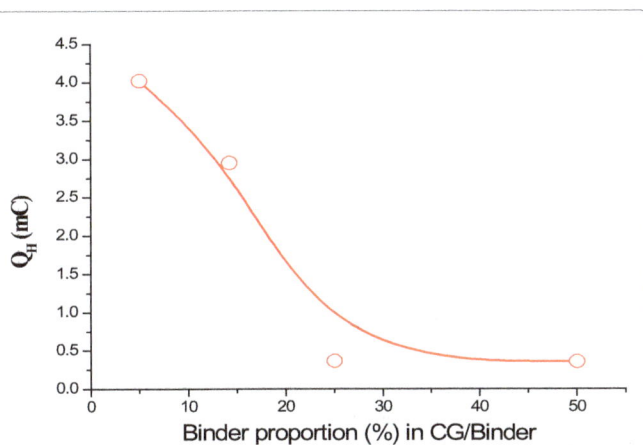

Figure 6: The electronic charges exchanged in the CNT-SiW$_{12}$-300 catalysts with various PVDF proportions.

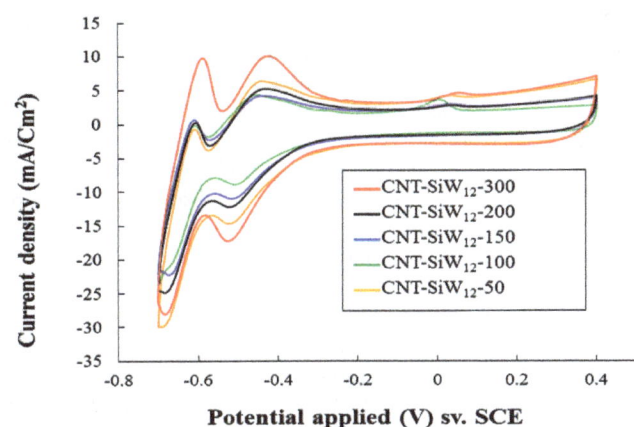

Figure 7: Cyclic voltammograms of CNT-SiW$_{12}$-x series of catalysts in 0.5 M H$_2$SO$_4$ solution with an average scan of 50 mVs^{-1}.

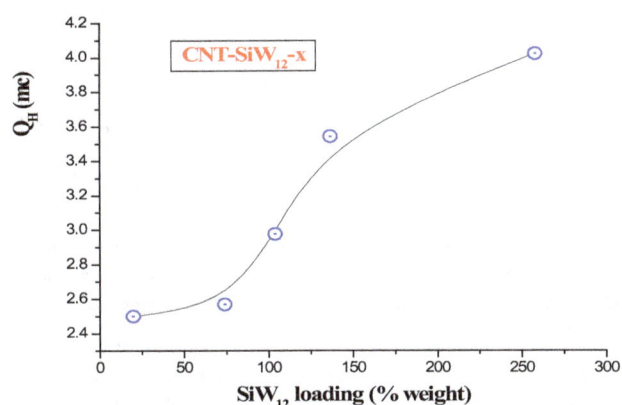

Figure 8: Effect of SiW$_{12}$ loading on the CNTs supported CNT-SiW$_{12}$-x catalysts on the exchanged electronic charges.

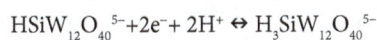

$$SiW_{12}O_{40}{}^{5-}+e^-+H^+ \leftrightarrow H\,SiW_{12}O_{40}{}^{5-}$$

$$HSiW_{12}O_{40}{}^{5-}+2e^-+2H^+ \leftrightarrow H_3SiW_{12}O_{40}{}^{5-}$$

The overall-redox processes of the CNT-SiW$_{12}$-x can be described as follows:

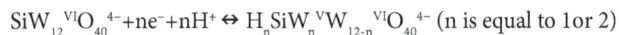

$$SiW_{12}{}^{VI}O_{40}{}^{4-}+ne^-+nH^+ \leftrightarrow H_nSiW_n{}^VW_{12-n}{}^{VI}O_{40}{}^{4-} \text{ (n is equal to 1or 2)}$$

Figure 8 shows the dependence of SiW$_{12}$ loading on the CNTs support of the series CNT-SiW$_{12}$-x on the exchanged electronic charges. It can be seen that at lower loadings the exchanged electronic charge increased rapidly whereas at higher ones the charge increased slightly.

Electro catalytic oxidation of cyclohexane

The electro catalytic activity of the CNT-SiW$_{12}$-x series of catalysts was investigated in the electrooxidation of cyclohexane. The reactions were carried out using H$_2$O$_2$ as oxidant at 50°C and the product cyclohexanone (K), cyclohexanol (A) and cyclohydroperoxide (CyOOH) were formed (Scheme 1).

The results of the effect of exchanged electronic charge of the CNT-SiW$_{12}$-x series of catalysts on the conversion of cyclohexane are shown in the Figure 9. It can be seen that the selectivity of the products as well as the conversion of cyclohexane depends strongly on the exchanged electronic charge. The increase of the exchanged electronic charge increased the conversion and the selectivity of cyclohexanone. As for the cyclohexyl hydroperoxide and the cyclohexanol, the results showed that they increased when the exchanged electronic charges increased from 2.5% to 3.0% mC then after they decreased when the charge was increased up to 4.0 mC. These results suggest that both cyclohexanol and cyclohexyl hydroperoxide were further oxidized to cyclohexanone. So, optimum conditions to obtain high selectivity of cyclohexanone, suitable for the production of the ε-caprolactam is operating at high exchanged electronic charges.

Figure 10 depicts the effect of the SiW$_{12}$ loading on the conversion and selectivity. It can be seen that the conversion and the selectivity of cyclohexanone increased significantly with the SiW$_{12}$ loading. The conversion and the selectivity of cyclohexanone were 33.4% and 17.5% respectively when CNT-SiW$_{12}$-25 was used, and they reached a value

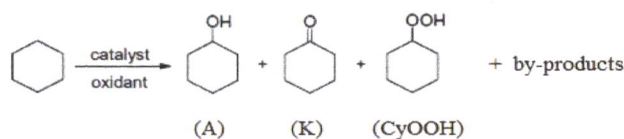

Scheme 1: Cyclohexane oxidation products.

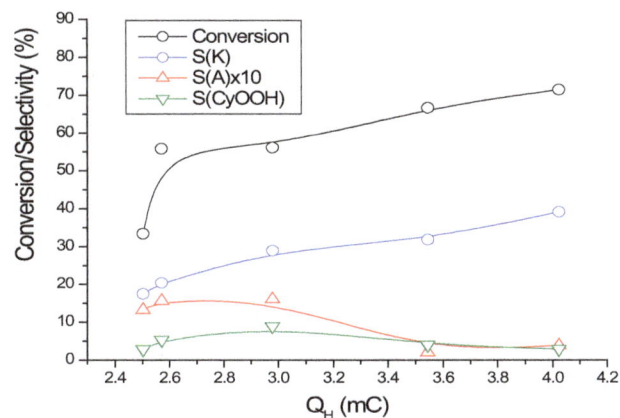

Figure 9: Effect of the exchanged electronic charge of the series of catalysts on the conversion and the selectivity of the products.

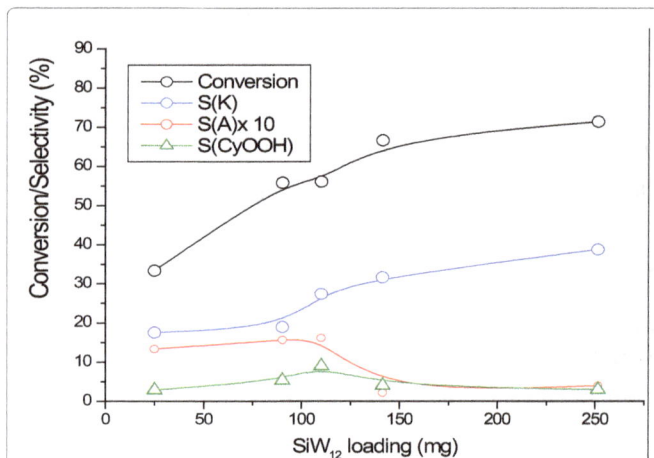

Figure 10: Effect of SiW$_{12}$ loading on the CNTs support of the series of catalysts on the conversion and the selectivity of the products.

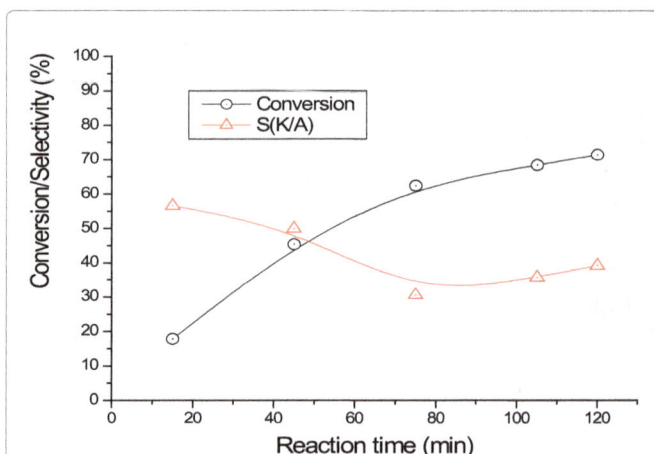

Figure 11: Effect of reaction time on the conversion and K/A selectivities for the electrooxidation reaction over CNT-SiW$_{12}$-300 catalyst at 40°C.

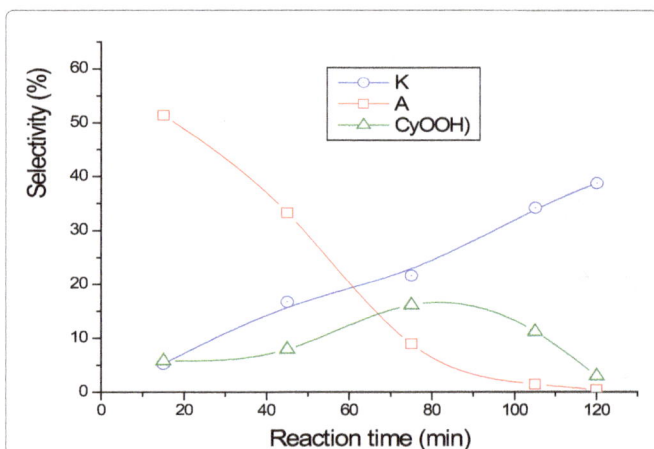

Figure 12: Effect of reaction time on the selectivities of the products for the electrooxidation reaction over CNT-SiW$_{12}$-300 catalyst at 40°C.

of 71.4% and 38.75% with CNT-SiW12-300. As for the cyclohexyl hydroperoxide and cyclohexanol, it has been found that their selectivity decreased at high loadings. This result seems to indicate that both

cyclohexanol and cyclohexyl hydroperoxide have undergone further oxidation leading to cyclohexanone.

The results of the effect of reaction time on the cyclohexane electrooxidation over CNT-SiW$_{12}$-300 catalyst at 313 K are shown in Figures 11 and 12. Figure 11 illustrates the changes of the conversion and selectivity of the K/A oil mixture with reaction time. It can be seen that the conversion increased with reaction time, it increased from 17.8% to 71.4% when the reaction time was extended from 15 min to 120 min. Contrary to the conversion, the K/A oil mixture decreased with the reaction time. It decreased from 56.6% to 39.1% within the same range of time. The observed decrease of the K/A oil mixture while the increase of the conversion indicated that the K/A oil are further oxidized (to adipic acid and hexanoic acid...). So, the suitable conditions for the production of K/A oil mixture is short reaction time. Compared to the conversion (3.9%) and the K/A oil selectivity (78%) obtained at the industrial process [7], the conversion (17.8%) and the K/A oil selectivity (56.6%) obtained over CNT-SiW$_{12}$-300 catalyst can be considered as a positive completion.

The effect of the reaction time on the product distribution is shown in the Figure 12. The results showed that the selectivity of cyclohexanol decreased with time in favor of that of cyclohexanone. This result clearly indicates that the cyclohexanol was further oxidized to cyclohexanone. As for CyOOH, The selectivity of CyOOH increased to reach a culmination and then declined gradually. This may indicates that CyOOH was further oxidized to cyclohexanone or to both cyclohexanol and cyclohexanone. This result confirm that CyOOH is an intermediate product in the electrooxidation of cyclohexane; the cyclohexane electrooxidation process is composed of the process of converting cyclohexane to cyclohexanol to cyclohexanone and that of cyclohexyl hydro peroxide to cyclohexanol to cyclohexanone. These results are in agreement with those reported in the literature [26,27]. In fact, it has been reported that CyOOH is the intermediate of the cyclohexane oxidation and two mechanisms were proposed for CyOOH decomposition: heterolytic (CyOOH → cyclohexanone) and hemolytic (CyOOH → cyclohexanol). Taking into account that cyclohexanol is more active than cyclohexane and can be converted to cyclohexanone easily, two possible pathways were proposed for the products formation [27]:

(1) CHHP → cyclohexanol → cyclohexanone

(2) Cyclohexanone ← CHHP → cyclohexanol (then cyclohexanol → cyclohexanone)

Conclusion

A series of CG-SiW$_{12}$-x catalysts were synthesized, characterized by means of FT-IR, XRD, and polarography. The results showed that the Keggin structure of the incorporated SiW in the carbon support was not altered

Stable SiW$_{12}$-modified carbon electrodes in the reacting media were successfully fabricated by using polyvinylidene difluoride (PVDF) as binder.

Characterization by cyclic voltammetry (CV) and tests in the cyclohexane oxidation showed that all the series of catalysts were electroactive. Electrocatalytic tests showed that the electrooxidation of cyclohexane led to cyclohexanone, cyclohexanol, and cyclohexyl hydroperoxide as major products for all the series of catalysts. The obtained conversion is relatively high whereas the selectivity is relatively low compared to those obtained in the industrial process.

Exchanged electronic charge and reaction time had an obvious influence on the catalytic performance of the catalyst. Short reaction time and low exchanged electronic charge favor the formation of cyclohexanol. High exchanged electronic charge and long reaction time favor the electrooxidation of cyclohexanol to cyclohexanone. Thus, the optimum condition for the formation of the cyclohexanone, which is needed for the production of the ε-caprolactam, is high, exchanged electronic charges.

Acknowledgement

The Investigators extend their appreciation to the National Plan for Science and Technology at King Saud University for funding the work through the research group project No 09-NAN863-02.

References

1. Feig AL, Lippard SJ (1994) Reactions of Non-Heme Iron(II) Centers with Dioxygen in Biology and Chemistry. Chem Rev 94: 759-805.

2. Fox BG, Froland WA, Dege JE, Lipscomb JD (1989) Methane monooxygenase from Methylosinus trichosporium OB3b. Purification and properties of a three-component system with high specific activity from a type II methanotroph. J Biol Chem 264: 10023-10033.

3. Lyons JE, Parshall GW (1994) Catalysis for industrial chemicals. Catal Today 22: 313-334.

4. Dartt CB, Davis ME (1994) Catalysis for Environmentally Benign Processing. Ind Eng Chem Res 33: 2887-2899.

5. Ebadi A, Safari N, Peyrovi MH (2007) Aerobic oxidation of cyclohexane with γ-alumina supported metallophthalocyanines in the gas phase. Appl Catal A Gen 321: 135-139.

6. Murrell JC, McDonald IR, Gilbert B (2000) Regulation of expression of methane monooxygenases by copper ions. Trends Microbiol 8: 221-225.

7. Zhao R, Ji D, Lv G, Qian G, Yan L, et al. (2004) A highly efficient oxidation of cyclohexane over Au/ZSM-5 molecular sieve catalyst with oxygen as oxidant. Chem Commun (Camb): 904-905.

8. Shylesh S, Samuel PP, Singh AP (2007) Chromium-containing small pore mesoporous silicas: Synthesis, characterization and catalytic behavior in the liquid phase oxidation of cyclohexane. Appl Catal A Gen 318: 128-136.

9. Rocchiccioli-Deltcheff C, Fournier M, Franck R, Thouvenot R (1983) Vibrational investigations of polyoxometalates. 2. Evidence for anion-anion interactions in molybdenum(VI) and tungsten(VI) compounds related to the Keggin structure. Inorg Chem 22: 207-216.

10. Pagona G, Sandanayaka ASD, Araki Y, Fan J, Tagmatarchis N, et al. (2007) Covalent Functionalization of Carbon Nanohorns with Porphyrins: Nanohybrid Formation and Photoinduced Electron and Energy Transfer. Adv Funct Mater 17: 1705-1711.

11. Zhao C, Ji L, Liu H, Hu G, Zhang S, et al. (2004) Functionalized carbon nanotubes containing isocyanate groups. Journal of Solid State Chemistry 177: 4394-4398.

12. Mawhinney DB, Naumenko V, Uznetsova A, Yates JT (2000) Infrared Spectral Evidence for the Etching of Carbon Nanotubes: Ozone Oxidation at 298 K. J Am Chem Soc 122: 2383-2384.

13. Chingombe P, Saha B, Wakeman RJ (2005) Surface modification and characterisation of a coal-based activated carbon. Carbon 43: 3132-3143.

14. Carmo M, Linardi M, Poco JGR (2009) Characterization of nitric acid functionalized carbon black and its evaluation as electrocatalyst support for direct methanol fuel cell applications. Appl Catal A Gen 355: 132-138.

15. Rocchiccioli-Deltcheff C, Fournier M (1991) Catalysis by polyoxometalates. Part 3-Influence of vanadium(V) on the thermal stability of 12-metallophosphoric acids from in situ infrared studies. J Chem Soc Faraday Trans 39: 13-3920.

16. Fournier M, Jantou CF, Rabia C, Herve G, Launay S (1992) Polyoxometalates catalyst materials: X-ray thermal stability study of phosphorus-containing heteropolyacids H3+xPM12–xVxO40•13-14H2O (M = Mo,W; x= 0-1). J Mater Chem 2: 971-978.

17. Atia H, Armbruster U, Martin A (2008) Dehydration of glycerol in gas phase using heteropolyacid catalysts as active compounds. J Catal 258: 71-82.

18. Zhao X, Xiong HM, Xu W, Chen JS (2003) 12-Tungstosilicic acid doped polyethylene oxide as a proton conducting polymer electrolyte. Mater Chem Phys 80: 537-540.

19. Wu QY, Meng GY (2000) Preparation and conductivity of vanadotungstogermanic heteropoly acid. Solid State Ionics 136: 273-277.

20. Yang J, Wang SC, Zhou XY, Xie J (2012) Electrochemical Behaviors of Functionalized Carbon Nanotubes in LiPF6/EC+DMC Electrolyte . Int J Electrochem Sci 7: 6118-6126.

21. Cheng L, Liu J, Dong S (2000) Layer-by-layer assembly of multilayer films consisting of silicotungstate and a cationic redox polymer on 4-aminobenzoic acid modified glassy carbon electrode and their electrocatalytic effects. Analytica Chimica Acta 417: 133-142.

22. Keita B, Nadjo L (1985) Activation of electrode surfaces: Application to the lectrocatalysis of the hydrogen evolution reaction. J Electroanal Chem 191: 441-448.

23. Dong S, Xi X, Tian M (1995) Study of the electrocatalytic reduction of nitrite with silicotungstic heteropolyanion. J Electroanal Chem 385: 227-233.

24. Yoo E, Habe T, Nakamura J (2005) Possibilities of atomic hydrogen storage by carbon nanotubes or graphite materials. Science and Technology of Advanced Materials 6: 615-619.

25. Serp P, Corrias M, Kalck P (2003) Carbon nanotubes and nanofibers in catalysis. Appl Catal A 253: 337-358.

26. Pohorecki R, Baldyga J, Moniuk W, Podgorska W, Zdrojkowski A, et al. (2001) Kinetic model of cyclohexane oxidation. Chem Eng Sci 56: 1285-1291.

27. Tian P, Liu Z, Wu Z, Xu L, He Y (2004) Characterization of metal-containing molecular sieves and their catalytic properties in the selective oxidation of cyclohexane. Catalysis Today 93: 735-742.

30

Optimization of Biodiesel Production from Sunflower Oil Using Response Surface Methodology

Mojtaba Mansourpoor[1]* and Dr. Ahmad Shariati[2]

[1]*Pars Oil and Gas Company, Process Engineering Unit, Asalouye, Iran*
[2]*Gas Engineering Department, Petroleum University of Technology, Ahwaz, Iran*

Abstract

Biodiesel produced by transesterification of triglycerides with alcohol, is the newest form of energy that has attracted the attention of many researchers due to various advantages associated with its usages. Response surface methodology, based on a five level, three variables central composite design is used to analyze the interaction effect of the transesterification reaction variables such as temperature, catalyst concentration and molar ratio of methanol to oil on biodiesel yield. The linear terms of temperature and catalyst concentration followed by the linear term of oil to methanol ratio, the quadratic terms of catalyst concentration and oil to methanol ratio and the interaction between temperature and catalyst concentration and also the interaction between temperature and molar ratio of methanol to oil had significant effects on the biodiesel production ($p<0.05$). Maximum yield for the production of methyl esters from sunflower oil was predicted to be 98.181% under the condition of temperature of 48°C, the molar ratio of methanol to oil of 6.825:1, catalyst concentration of 0.679 wt%, stirring speed of 290 rpm and a reaction time of 2h.

Keywords: Biodiesel; Energy; Petro-Diesel; Transesterification; Vegetable Oil; Response Surface Methodology; Central Composite Design

Abbreviations: FFA: Free Fatty Acid; ASTM: American Society for Testing and Materials; cSt: CentiStoke; RSM: Response Surface Methodology; CCD: Central Composite Design; ANOVA: Analysis of Variance

Introduction

The exponential growth of world population would ultimately lead to increase the energy demand in the world. Petroleum is a non-renewable energy source, which means that the resources of this kind of fossil fuel are finite and would be run out upon continuous use. Both of the shortage of resources and increase of petrol price have led to the findings of new alternative and renewable energy sources [1]. Biodiesel is defined as a fuel comprised of mono-alkyl esters of long chain fatty acids derived from vegetable oils or animal fats [2]. It is not toxic, biodegradable and available, has a high heat value, high oxygen content (10 to 11%) and does not contain sulfurs and aromatic compounds [3]. Biodiesel is a plant derived product, and it contains oxygen in its molecule, making it a cleaner burning fuel than petrol and Diesel [4]. Several studies have showed that biodiesel is a better fuel than fossil-based diesel in terms of engine performance, emissions reduction, lubricity, and environmental benefits [5,6]. The current feed stocks of production of biodiesel or mono-alkyl ester are vegetable oil, animal fats and micro algal oil. In the midst of them, vegetable oil is currently being used as a sustainable commercial feedstock. Among more than 350 identified oil-bearing crops, only sunflower, safflower, soybean, cottonseed, rapeseed, and peanut oils are considered as potential alternative fuels for diesel engines [7].

Vegetable oil is one of the renewable fuels which have become more attractive recently because of its environmental benefits and the fact that it is made from renewable resources [8].Vegetable oil has too high a viscosity for use in most existing Diesel engines as a straight replacement fuel oil. One of the most common methods used to reduce oil viscosity in the biodiesel industry is called transesterification [9]. Many of researchers have studied the transesterification for production of biodiesel. These studies [10-12] show that transesterification consists of a number of consecutive, reversible reactions. Triglycerides are first reduced to diglycerides. The diglycerides are subsequently reduced to mono-glycerides. Optimum conditions for the transesterification of vegetable oils to produce methyl ester were determined by the previous researchers which yielded a maximum conversion of various oils to the methyl esters.

The conventional catalysts used for transesterification are acids and alkali, both liquid and heterogeneous, depending on the oil used for biodiesel production. The use of acid catalysts has been found to be useful for pretreating high free fatty acid feedstocks but the reaction rates for converting triglycerides to methyl esters are very slow. Fatty acid contents are the major indicators of the properties of biodiesel since the amount and type of fatty acid content in the biodiesel largely determine its viscosity. Biodiesel from the waste cooking oil contained the highest amount of FFA content, an average 4.4%. The pure vegetable oils contained only about 0.15%, which are within permitted levels for being used directly for reaction with an alkaline catalyst to produce biodiesel [13].

Hossain et al. obtained the highest approximately 99.5% biodiesel yield required under optimum conditions of 1:6 volumetric oil to methanol ratio and 1% KOH catalyst at 40°C reaction temperature. The research demonstrated that biodiesel obtained under optimum conditions from pure sunflower cooking oil and waste sunflower cooking oil was of good quality and could be used as a diesel fuel which considered as renewable energy and environmental recycling process from waste oil after frying [14]. Therefore, the objectives of our work

*Corresponding author: Mojtaba Mansourpoor, Pars Oil and Gas Company, Process Engineering Unit, Asalouye, Iran

Figure 1: LR 2000 P Lab Reactor.

were to evaluate the effects of the reaction parameters of temperature, catalyst concentration and molar ratio of methanol to oil on the biodiesel yield and to optimize the reaction conditions using RSM. The properties of produced methyl ester were analyzed and the quality of biodiesel was compared with petro-diesel.

Materials

Sunflower oil was purchased from local shop. Methanol with a purity of 99.5% and Potassium Hydroxide (KOH) were purchased from Merck Company.

Equipments

Device that used in this work include reactor, EUROSTAR power control-visc P7 overhead stirrer. The reactor employed was a LR 2000P modularly expandable laboratory reactor. The IKA laboratory reactor was double-walled jacketed 2 liter vessels available made of stainless steel, with bottom discharge valve. A mixer with 8 to 290 rpm model EUROSTAR Power control-Visc P7 Overhead stirrer for mixing the medium of reaction was used. Temperature inside the reactor was controlled by a hot water bath. Figure 1 shows the LR 2000 P reactor.

Experiments and Methods

Several types of oils can be used for production of biodiesels. The most common types of oils are sunflower oil. The batch reaction kinetic experiments were employed to optimize various parameters in the production of the methyl esters. The transesterification reactions are performed in various conditions to determine the optimum conditions of transesterification. Two liter of sunflower oil poured in the reactor and allowed to equilibrate to the temperature of reaction at 290 rpm. Hot water circulated in the jacket of the reactor provided the necessary heat for the reaction. Variable quantities of catalyst were dissolved in various amount of methanol as described in each test. After attaining a required temperature, the potassium methoxide was added to the reactant and was maintained for 2 hours for completion of the reaction.

After 2 hours the transesterification reaction was completed and mixture was withdrawn from the reactor and poured in the funnel separator to separates biodiesel from glycerol. Separation of two phases which is performed by gravity requires at least 4 hours. Glycerol and biodiesel have a deep red and bright yellow color. After separation of

biodiesel, it should be washed out from impurities and unreacted agents. The biodiesel was washed out 10 times. At the first time, washing of the biodiesel should be done slowly and carefully to avoid soap formation. One liter of warm distilled water was used per 1 liter of biodiesel. At the next times, the washing procedure can be done more quickly until the color of water shifts to white. Finally, biodiesel was dried completely by silica jell.

Physical Properties

Physical properties of the biodiesel which were produced in the optimum conditions were measured in the Abadan's oil refinery lab and shown in Table 1. These results are tabulated and compared with ASTM standard and petro-diesel. It is found which results are drawn on the ASTM standard and compared well with petro-diesel. It was seen that the flash point of biodiesel is 170 when that of petro-diesel is about 60 which helps the transportation of biodiesel but it should be blend with petro-diesel for better combustion in engine. Kinematic viscosity of biodiesel at 40, were higher than that of petro-diesel and is 3.6 cSt. This biodiesel has a copper strip corrosion of 1a which indicate that this fuel is not corrosive.

Heating value of biodiesel was 39.58 MJ/Kg which is less than that of petro-diesel but only 9 percent. The reason for the lower heating value of biodiesel is the presence of chemically bound oxygen in vegetable oils which lowers their heating values.

In this work, biodiesel has a cetane index 49.2 which is higher than that of ASTM. Cloud point of biodiesel is -5 when maximum cloud point of petro-diesel is 2. It was found that biodiesel had a cloud point higher than petro-diesel because biodiesel produced from vegetable oils which have FFA that these free fatty acids cause a higher cloud point of biodiesel than petro-diesel. Whatever saturated fatty acids be higher in the oil, the produced biodiesel has a poorer cloud point. Carbon residue of biodiesel was lower than that of petro-diesel which is about 0.002 wt%. Biodiesel has a negligible amount of sulphur relative to petro-diesel. Methanol was produced from corn and corn has sulphur; hence, biodiesel has a negligible amount of sulphur.

Response Surface Method

A central composite design of the RSM is the most commonly used in optimization experiments. The method includes a full or fractional factorial design with center points that are augmented with a group of star points. As the distance from the center of the design space to a

Test	Method	Unit	Biodiesel	ASTM	Petro-Diesel
Flash point	D93	°C	170	130 min	54 min
Kinematics Viscosity @40°C	D445	cSt.	3.6	1.9-6.0	2.0-5.5
Total Sulfur	UOP 357	ppm	3.0	15	500
Copper Strip Corrosion	D130	-	1a	No.3 max	1a
Cetane Index	D976	-	49.2	*CN=47 min	*CN=50 min
Cloud Point	D2500	°C	-5	**N/A	2 max
Pour point	D97	°C	-7	**N/A	-3 max
Carbon Residue	D189	Wt%	0.002	0.050 max	0.010 max
Acid Number	D974	mg KO H/g	0.270	0.800	0.002
Heating value	D240	MJ/Kg	39.58	**N/A	43.73
Specific gravity	D1298	-	0.884	**N/A	0.835

*CN-Cetane Number
**N/A-Not Available

Table 1: Physical properties of the Biodiesel.

Symbols	Independent Variables	Coded levels				
		-1.68	-1	0	1	1.68
X1	T(°C)	33.2	40	50	60	66.8
X2	M	3.64	5	7	9	10.36
X3	C(%wt)	0.464	0.6	0.8	1	1.136

Table 2: Codes, ranges and levels of independent variables of temperature (T), molar ratio of methanol to oil (M) and catalyst concentration (C) in RSM design.

Run no.	X1	X2	X3	Yield	
				Experimental	Predicted
1	0	0	0	95.359	95.359
2	0	0	-1.68	91.819	93.642
3	0	0	0	95.359	95.359
4	0	1.68	0	89.939	89.194
5	1.68	0	0	96.720	97.038
6	1	1	-1	95.226	94.591
7	-1	1	1	87.315	88.478
8	1	-1	1	91.848	92.118
9	0	0	0	95.359	95.359
10	0	0	1.68	91.371	89.548
11	-1	-1	1	87.098	87.733
12	0	0	0	95.359	95.359
13	-1.68	0	0	90.555	90.237
14	1	1	1	90.284	91.278
15	-1	-1	-1	90.289	89.295
16	0	0	0	95.359	95.359
17	0	-1.68	0	87.824	88.569
18	0	0	0	95.359	95.359
19	-1	1	-1	91.150	90.880
20	1	-1	-1	95.754	94.591

Table 3: Experimental and predicted data for the yield of biodiesel obtained from the central composite experimental design.

factorial point is defined as ±1 unit for each factor, the distance from the center of the design space to a star point is ±α with $|α|>1$. In this study, the central composite design was used to optimize operating variables (temperature, catalyst concentration and oil to methanol ratio) to achieve high value of biodiesel yield. The coded values of the variables were determined by the following equation.

$$x_i = \frac{x_i - x_0}{\Delta x} \tag{1}$$

Where x_i is the coded value of the ith variable, X_i is the encoded value of the ith test variable and X_0 is the encoded value of the ith test variable at center point. The range and levels of individual variables were given in Table 2. The experiment design was given in Table 3. The value of biodiesel yield is the response.

The regression analysis was performed to estimate the response function as a second order polynomial.

$$Y = \beta_0 + \sum_{i=1}^{n} \beta_i x_i + \sum_{i=1}^{n}\sum_{j=1}^{i-1} \beta_{ij} x_i x_j \tag{2}$$

Where Y is the predicted response, β_i and β_{ij} are coefficients estimated from regression, they represent the linear, quadratic and cubical effect of $x_1, x_2, x_3...$ on response.

All results are expressed as mean ± SD for six mice in each group. To determine the effect of treatment, data were analyzed using one way analysis of variance (ANOVA) repeated measures. P-Values of less than

0.05 were regarded as significant. Significant values were assessed with Duncan's multiple range tests. Data were analyzed using the statistical package "SPSS 16.0 for Windows".

Results and Discussion

Fitting the model

As mentioned earlier, RSM was used to optimize Transesterification reaction and the experimental results were presented in Table 3. Experimental yields were analyzed to get a regression model. The predicted values of biodiesel yield were calculated using the regression model and compared with the experimental values. The estimated coefficients of the regression model are given in Table 4. The large value of the coefficient of multiple determination ($R^2=0.927$) reveals that the model adequately represents the experimental results.

The effect of the variables as linear, quadratic, or interaction coefficients on the response was tested for significance by ANOVA. As shown in Table 4, it can be found that the variable with the most significant effect on the oil yield was the linear term of Temperature ($p<0.001$), methanol to oil ratio ($p<0.05$) and Catalyst concentration ($p<0.01$), followed by the quadratic terms of methanol to oil ratio ($p<0.001$) and catalyst concentration ($p<0.01$) and the interactions between temperature and methanol to oil ratio ($p<0.05$) and temperature and catalyst concentration ($p<0.05$) had significant effects on the oil yield.

Response surface analysis

Response surface has been applied successfully for optimization of biodiesel production in fat and oil feedstocks, including mahua oil [15], Jatropha oil [16], waste rapeseed oil [17] and animal fat [18]. RSM can be illustrated with three-dimensional plots by presenting the response in function of two factors and keeping the other constant. It is visualized by the yield of biodiesel in relation to the temperature, methanol to oil ratio and Catalyst concentration in Figure 2 to 4. Figure 2 denotes the surface plot of the pomegranate Transesterification reaction yield as a function of temperature and methanol to oil ratio at Catalyst concentration of 0.679%wt. This figure show that temperature and molar ratio of alcohol to oil have a direct effect on the yield of methyl ester but until the near boiling temperature of alcohol and 6:1 molar ratio, then yield of biodiesel decreased with increasing the temperature and molar ratio of alcohol to oil. A few works reported the reaction at room temperature;

Regression coefficient	Value	Standard error	P-Value
β1	95.359	0.475	
Linear			
β1	2.024	0.316	0.000
β2	0.186	0.316	0.028
β3	-1.219	0.316	0.003
Quadratic			
β11	-0.610	0.308	0.075
β22	-2.295	0.308	0.000
β33	-1.334	0.308	0.001
Interaction			
β12	-0.396	0.412	0.039
β13	-0.228	0.412	0.020
β23	-0.210	0.412	0.212
Rsq	0.927		
Rsq(adj)	0.862		

Table 4: Regression coefficients of the fitted quadratic equation and standard errors.

most of the researches have focused on the transesterification at near boiling point of alcohol. Temperature has an important influence on speed of reaction and led to higher conversion of ester. With increasing temperature of reaction, yield of biodiesel increased quickly to near the boiling point of alcohol. At low temperatures, relatively low conversion to methyl ester evident due to the subcritical state of methanol. At higher temperature than boiling point of methanol, alcohol evaporates and the yield was decreased. Also methanol to oil ratio had a significant effect which yield decrease dramatically in high value of those. With increasing molar ratio of methanol to oil, OH group present in the alcohol reacts with triglycerides and lead to hydrolysis reaction which in turn leads to soap formation. The interaction of these variables is interest which that in the optimum value of temperature variation of methanol to oil ratio changes the reaction yield greatly. With higher molar ratio of alcohol to oil, triglycerides convert to fatty acid methyl ester. Hence reverse reaction performed which leads to formation of soap which is difficult to separate and yield of ester decreased.

Yield of biodiesel reduced when concentration of KOH increased than 0.679 wt%, as shown in Figure 3, because with increasing concentration of KOH, soap was formed exponentially with catalyst concentration and lower amount of biodiesel can separate from

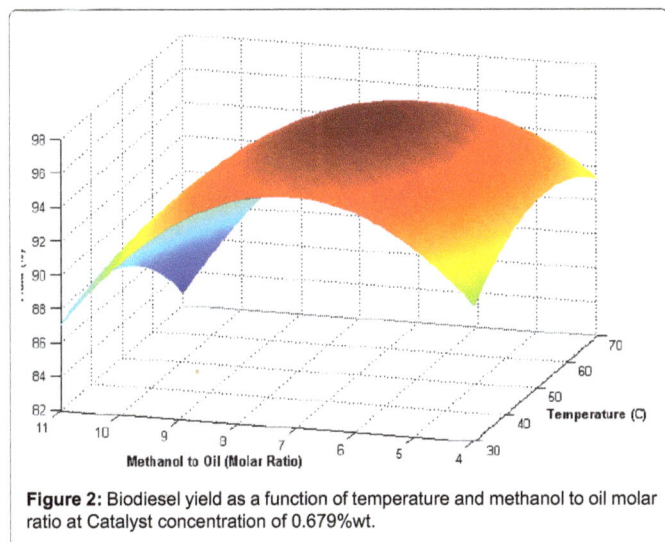

Figure 4: Biodiesel yield as a function of Catalyst concentration and methanol to oil ratio at temperature of 48°C.

glycerol. The resulting soaps do not only lower the conversion of ester, but also cause other problems associated with phase separation. The interaction effect of methanol to oil ratio and catalyst concentration had shown in Figure 4. It seems that the effect of catalyst concentration on the methanol to oil ratio is rare and the value of interaction coefficient ($p>0.05$) demonstrate this fact.

Optimization of extraction condition

In order to optimize reaction condition, the first partial derivatives of the regression model were equated to zero according to X_1, X_2, and X_3 respectively. The result was calculated as follows: $X_1=48$, $X_2=6.825$ and $X_3=0.679$. Under such condition, the yield of biodiesel was predicted to be 97.54%. The experimental work at this condition was performed due to maximum experimental yield. In this work, highest yield of methyl ester at temperature of 48°C, catalyst concentration of 0.679%wt, 290 rpm of stirrer, 2h and methanol to oil ratio of 6.825:1 is obtained 98.181%.

Conclusion

Response surface methodology was successfully applied for transesterification of methanol. The high regression coefficients of the second-order polynomial showed that the model was well fitted to the experimental data. The ANOVA implied that molar ratio of alcohol to oil; reaction temperature and concentration of catalyst have the great significant factor affecting the yield of biodiesel. The biodiesel production has a negative quadratic behavior by temperature, molar ratio of alcohol to oil and concentration of catalyst. It was predicted that the optimum reaction condition within the experimental range would be the molar ratio of 6.825:1 and temperature of 48°C and concentration of KOH equal to 0.679wt%. At the optimum condition we can reach to yield of 98.181%. The methyl ester which produced at optimum conditions has acceptable properties and compared well with petro-diesel. It has lower sulfur, carbon residue and acid number than petro-diesel, but kinematic viscosity, cetane number and heating value of petro-diesel is some better relative to biodiesel. Finally, we can conclude which biodiesel will be a suitable alternative for replacement of petro-diesel without any modification in engine.

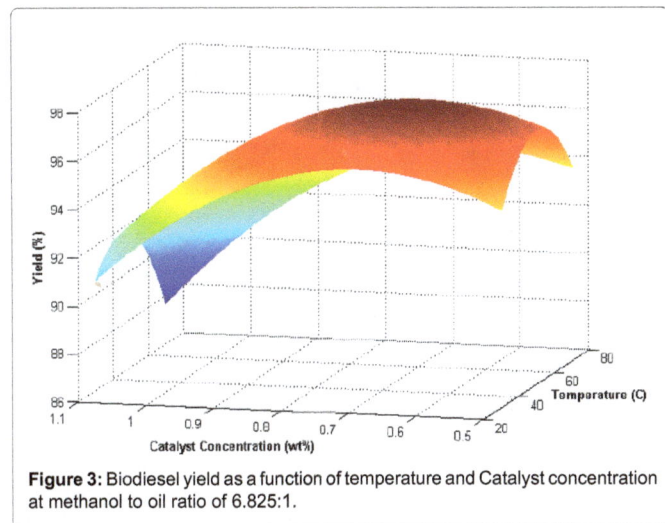

Figure 2: Biodiesel yield as a function of temperature and methanol to oil molar ratio at Catalyst concentration of 0.679%wt.

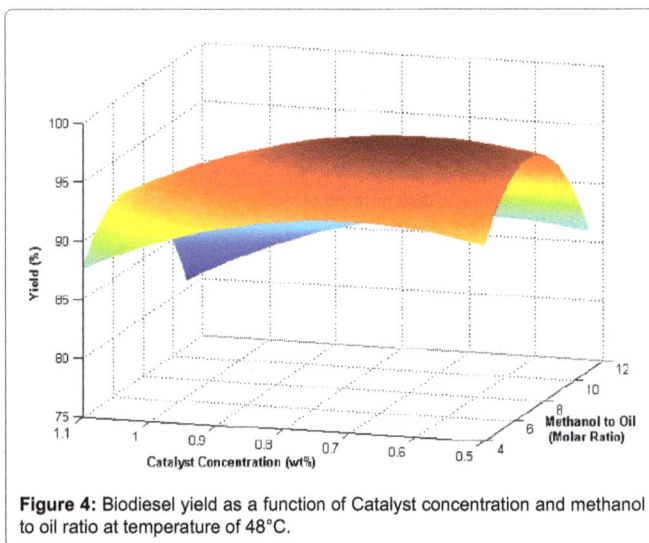

Figure 3: Biodiesel yield as a function of temperature and Catalyst concentration at methanol to oil ratio of 6.825:1.

References

1. Hossain ABMS, Mekhled MA (2010) Biodiesel fuel production from waste canola cooking oil as sustainable energy and environmental recycling process. AJCS 4: 543-549.

2. Vicente G, Martinez M, Aracil J (2007) Optimization of integrated biodiesel production. Part 1. A study of the biodiesel purity and yield. Bioresour Technol 98: 1742-1733.

3. Demirbas A (2003) Biodiesel fuels from vegetable oils via catalytic and non-catalytic supercritical alcohol transesterifications and other methods: a survey. Energy Conversion and Management 44: 2093-2109.

4. Sastry GSR, Krishna Murthy ASR, Ravi Prasad P, Bhuvaneswari K, Ravi PV (2006) Identification and determination of biodiesel in Diesel. Energy Sources, Part A 28: 1337-1342.

5. Canakci M, Gerpen JV (2000) The effect of yellow grease methyl ester on engine performance and emissions. Final Report: Recycling and Reuse Technology Transfer Center. Publication 2000-134.

6. Peterson CL, Reece DL, Hammond BL, Thompson J, Beck SM (1997) Processing, Characterization, and Performance of Eight Fuels from Lipids. Appl Eng Agric 13: 71- 79.

7. Demirbas A, Kara H (2006) New Options for conversion of Vegetable Oils to Alternative Fuels. Energy Sources, Part A: Recovery, Utilization and Environmental Effects 28: 619-626.

8. Demirbas A (2005) Biodiesel production from vegetable oils via catalytic and non-catalytic supercritical methanol transesterification methods. Progress in Energy and Combustion Science 31: 466-487.

9. Demirbas A (2008) Comparison of transesterification methods for production of biodiesel from vegetable oils and fats. Energ Convers Manage 49: 125-130.

10. Noureddini H, Zhn D (1997) Kinetics of transesterification of Soybean Oil. Applied Eng in Agric 74: 1457-1463.

11. Freedman B, Butterfield RO, Pryde EH (1986) Transesterification kinetics of soybean oil. Journal of the American Oil Chemists' Society 63: 1375-1380.

12. Darnoko D, Cheryan M (2000) Kinetics of Palm Oil transesterification in a batch reactor. Journal of the American Oil Chemists' Society 77: 1263-1267.

13. Gerpen JV (2005) Biodiesel processing and production. Fuel Processing Technology 86: 1097-1107.

14. Hossain ABMS, Boyce AN (2009) Biodiesel Production from Waste Sunflower Cooking Oil as an Environmental Recycling Process and Renewable Energy. Bulgarian Journal of Agricultural Science 15: 312-317.

15. GhadgeSV, Raheman H (2006) Process optimization for biodiesel production from mahua (Madhuca indica) oil using response surface methodology. Bioresource Technol 97: 379-384.

16. Tiwari AK, Kumar A, Raheman H (2007) Biodiesel production from Jatropha oil with high free fatty acids: An optimized process. Biomass Bioenergy 31: 569-575.

17. Yuan X, Lui J, Zeng G, Shi J, Tong J, et al. (2008) Optimization of conversion of waste rapeseed oil with high FFA to biodiesel using response surface methodology. Renew Energy 33: 1678-1684.

18. Jeong GT, Yan HS, Park DH (2009) Optimization of transesterification of animal fat ester using response surface methodology. Bioresource Technol 100: 25-30.

31

Isosteric Heats of Water Vapor Sorption in Two Castor Varieties

J. Ojediran[1]*, A.O Raji[2] and H.I Owamah[3]

[1]Agricultural Engineering Department, Landmark University, Omu-Aran, Kwara State, Nigeria
[2]Agricultural and Environmental Engineering Department, University of Ibadan, Ibadan, Nigeria
[3]Civil Engineering Department, Landmark University, Kwara State, Nigeria

Abstract

Equilibrium moisture content-water activity data for two varieties of castor (GSS and WBS) at temperatures 30, 40, 50, 60 and 70°C and water activity range of 0.07-0.98 were determined using the static gravimetric method. A non-linear regression programme was used to fit four sorption isotherm models (modified Halsey, modified Henderson, modified Oswin and modified Guggenheim-Anderson-de Boer (GAB)) to the experimental data. These models were then compared using standard error of estimate and randomness of residuals. The sorption isosteric heats were determined by the application of Clausius-Clapeyron equation. The modified Oswin and modified Halsey proved the best for predicting equilibrium moisture content of castor with standard error of estimate of 0.238 and 0.242 for GSS and WBS respectively. The isosteric heats decreased exponentially with increase in moisture content and approached the latent heat of pure water at a free water point of between 13 and 16% moisture content (d.b). The difference in isosteric heats of both varieties was not significant.

Keywords: Castor; Isosteric heat; Water Activity; Equilibrium Moisture Content

Introduction

Castor (*Ricinus communis L*) is an under-exploited tropical crop. The products derivable from it range from synthetic resins and fibres, lubricants, embalming fluids, soap manufacture, opal wax in polishes, cosmetics and brake fluids [1]. The leaves have insecticidal properties, while the stalks are a source of pulp and cellulose [2]. The seeds are ovoid, compressed dorsally, tick-like shining, pale grey (hereafter referred to as Grey Small Size, GSS) predominant in the north-eastern part of Nigeria, or yellowish white and big size (hereafter known as White Big Size, WBS) common in the eastern part of Nigeria.

Drying operations are involved after harvest in the processing of this crop as reported in earlier studies [3-5]. The design of more efficient drying systems and storage systems for crops require knowledge of energy requirements, state and mode of moisture sorption within them [6]. This could be achieved by stating the correct mathematical models to estimate the heat and mass transfer mechanisms. It is also essential to determine the energy required to remove moisture from agricultural crops (isosteric sorption heat) to complete the drying simulation model [7].

Isosteric heat of sorption (KJ.mol⁻¹) defined as the heat of sorption at constant specific volume hence a measure of moisture-solid site binding strength [8]. It is the total energy required to transfer water molecules from vapor state to a solid surface and vice-versa, as well as a measure of work done by a system to accomplish adsorption or desorption process. It is therefore used as an indicator of the state of adsorbed water by solid particles [6]. Isosteric heat is useful in predicting drying models, calculating energy consumption during the drying or wetting of agricultural materials, the design of drying equipment and describing any heat and mass transfer related processes [6,9-11].

The most widely used method for evaluation of isosteric heat is the application of the Clausius-Clapeyron equation which relates water activity and temperature at constant moisture content [12,13]. The other methods of estimation of sorption isosteric heat are calorimetric technique and Riedel equation [7,14,15]. It is noted that there is a good agreement between the isosteric sorption heats obtained by the calorimetric techniques and the Clausius-Clapeyron equation

[15,16]. Several authors have reported and studied the isosteric heats of agricultural products [6,7,15-23].

The purpose of the present work is to examine the application of Clausius-Clapeyron equation to sorption isotherms of two varieties of castor and to compare the calculated isosteric sorption heats using the adsorption isotherms. This is with a view to providing information useful in the drying process of the less explored castor varieties. Whole castor seeds are extremely poisonous they however become fit for human consumption when properly processed.

Materials and Methods

Sample preparation

Two castor varieties namely GSS and WBS obtained from castor farms in Borno and Anambra States, Nigeria, were used in this study. Triplicate samples, each weighing 10 g were used to determine the moisture content of whole seeds using the oven drying method as recommended in the ASABE standard. This involved drying in an oven at 103 ± 2 °C until constant weight was obtained [24]. The initial moisture contents were found to be 4.80 and 5.24 ± 0.2% (db) respectively for the two varieties. The bulk quantity of seeds was cleaned and divided into two portions. One portion was prepared for use in determining the desorption equilibrium moisture content by rewetting it to a higher moisture content. A calculated amount of water was added to this portion and the grains were then sealed in polyethylene bags and stored for 24 hours. This enabled the moisture content to be raised to stable and uniform levels of 17.00 and 18.00% (db) for GSS and WBS respectively. The polyethylene bags were marked and transferred into

*Corresponding author: J. Ojediran, Agricultural Engineering Department, Landmark University, Omu-Aran ,Kwara State, Nigeria

a refrigerator at 4°C and when needed for experiments, the grains were allowed to equilibrate in the ambient condition for six hours. The other portion was prepared for use for experiments on adsorption. The sample was dried at 80°C for two days to obtain a lower moisture content of about 3.23 and 3.95% (db) GSS and WBS respectively. The seeds from this portion were also sealed in marked polyethylene bags and kept in a refrigerator. They were equally allowed to equilibrate in the ambient conditions for six hours, when needed for experiments.

Equilibrium moisture content-equilibrium relative humidity determination

The desorption equilibrium moisture contents (EMC) of the two castor varieties were determined at temperatures of 40, 50, 60 and 70°C over a water activity (a_w) range of 0.07-0.98 using the static gravimetric method. This method involves the use of saturated salt solutions to maintain constant relative humidity (r.h.) in enclosed still moist air at a certain temperature to obtain the complete sorption isotherms. The triplicate samples of GSS and WBS weighing about 15 g were put into specimen baskets and placed inside glass desiccators. Saturated solutions of salts as used by Brooker et al. [25] and Rizvi [12] were used to maintain constant relative humidity levels in the desiccators [26,27]. Excess salt was maintained in each solution.

The samples were placed in an open flat container and arranged in each desiccator bottle so that they will not be in contact with the salt solution. The desiccators containing the salt solutions and castor seed samples were marked and placed inside temperature-controlled Gallenkamp DV 400 ovens (Weiss Gallenkamp, UK) which were set at 30, 40, 50, 60 and 70°C. The oven temperatures were monitored to within ± 1.0°C to maintain constant temperature for each setting. The samples were weighed daily using an analytical balance (Mettler PC2200 DeltaRange, Mettler-Toledo Inc., USA) with an accuracy of 0.001 g. Equilibrium was considered to have been attained when three identical consecutive measurements were obtained. This took between 7 and 12 days.

The dry matter content was then determined by oven drying the sample at 103 ± 2 °C until constant weight was attained [24]. The EMC was calculated on a dry basis from the weight change and dry matter weight, and the average values at each temperature and water activity were determined. At higher water activity above 0.85, microbial growth may occur in sample and may ruin it. Two methods are often used to control this: (1) Using antimicrobial agents but these could change the a_w and moisture profile; (2) Vacuum method, which was used in this study to create an anaerobic environment in the desiccator. Where the occurrence of mould was noticed, during the experiment such specimen pan was discarded.

Analysis of data

The experimental EMC-a_w data of GSS and WBS castor seeds were fitted to four moisture sorption isotherm models namely the Modified Halsey [19,28], Modified Henderson [29,30], Modified Oswin model [31,32] and the GAB model [33,34] as modified by Jayas and Mazza [35]. These models are:

The Modified Halsey a_w model
$$a_w = \exp\left[-\exp\left(A + BT\right)M^{-C}\right] \tag{1}$$

The Modified Henderson a_w model
$$a_w = 1 - \exp\left[-A(T + B)M^C\right] \tag{2}$$

The Modified Oswin a_w model

$$a_w = \frac{1}{\left[\dfrac{(A+BT)}{M}\right]^C + 1} \tag{3}$$

The Modified GAB a_w model
$$a_w = \frac{2 + \dfrac{C}{T}\left(\dfrac{A}{M} - 1\right) - \left[\left(2 + \dfrac{C}{T}\left(\dfrac{A}{M} - 1\right)\right)^2 - 4\left(1 - \dfrac{C}{T}\right)\right]^{1/2}}{2B\left(1 - \dfrac{C}{T}\right)} \tag{4}$$

Symbols are defined in Notation section.

Model fitting

The desorption and adsorption equilibrium moisture content data of castor seeds were fitted to moisture sorption models using the non-linear regression procedure in SPSS 16.0 for Windows, which minimizes the sum of squares of deviations between experiment and theory in a series of iterative steps. All the models are three parameter equations, which can be solved explicitly for a_w as a function of temperature and moisture content. The non-linear regression procedure required that initial parameter estimates be chosen close to the true values. The initial parameter estimates were obtained by linearization of the models through logarithmic transformation and application of linear regression analysis, or solving a quadratic form of the equation in the case of modified GAB model. The least-squares estimates or coefficients of the terms were used as the coefficients of the terms of the sorption models tested for the initial parameter estimates in the non-linear regression procedure. Model parameters were estimated by taking the water activity (a_w) to be the dependent variable.

Heat of vaporization

Igathinathane et al. [8] have reported the procedure for calculating isosteric sorption heat. The Clausius-Clapeyron equation as applied to vapor from free water is given by [11,12,36]:

$$\frac{1}{h_{fg}}\frac{\partial P_s}{P_s} = \frac{\partial T}{RT^2} \tag{5}$$

If in Equation (5), the P_s is replaced by the actual vapor pressure (P_o) in the material and the right hand side of the equation remains the same for both vapor from free water and vapor from moisture in material [6], then the following equation can be obtained:

$$\frac{1}{h_{fg}}\frac{\partial P_s}{P_s} = \frac{1}{L}\frac{\partial P_o}{P_o} \tag{6}$$

Integrating Equation (6) at constant moisture content yields

$$Ln\left(P_o\right) = \frac{L}{h_{fg}}Ln(P_s) + c_1 \tag{7}$$

Where c_1 is a constant. The saturation vapor pressure (P_s) at different temperatures can be obtained from Rogers and Mayhew (1981) and subsequently, vapor pressure in the material determined as:

$$P_o = a_w P_s \tag{8}$$

The values of a_w were obtained from the moisture sorption isotherm models which best describes the equilibrium moisture relations of GSS and WBS castor varieties respectively.

The slope of the logarithmic plot of P_o against P_s gives the ratio of heat of vaporization of moisture in a material to the latent heat of saturated vapor. The non-linear regression procedure in SPSS 16.0 for Windows was used to fit the Gallaher (1951) model (Equation 9) to the data obtained and to relate the latent heat ratio to the material moisture

| Temperature (°C) | Water activity | Equilibrium Moisture Content (%db) | | | |
		Desorption		Adsorption	
		WBS	GSS	WBS	GSS
	0.07	7.25	6.20	7.25	6.20
	0.11	8.45	7.50	8.45	7.50
	0.22	9.32	8.98	9.17	8.50
	0.32	9.89	9.30	9.68	8.93
30	0.51	10.71	9.51	10.35	9.15
	0.55	10.96	9.66	10.50	9.23
	0.70	11.49	9.97	11.05	9.50
	0.80	12.00	10.54	11.75	10.05
	0.92	13.99	12.53	13.99	12.53
	0.97	16.12	15.20	16.12	15.20
	0.07	6.25	5.80	6.25	5.80
	0.11	7.10	6.75	7.10	6.75
	0.22	8.50	7.75	8.20	7.50
	0.32	8.80	8.20	8.43	7.78
40	0.51	9.20	8.80	8.60	8.15
	0.61	9.45	9.10	8.80	8.39
	0.71	9.75	9.32	9.15	8.65
	0.82	10.05	10.17	9.72	9.50
	0.89	10.65	11.50	10.65	11.50
	0.98	13.50	13.70	13.50	13.70
	0.07	5.84	5.65	5.84	5.65
	0.11	6.60	6.17	6.60	6.17
	0.21	7.49	6.49	7.20	6.22
	0.31	7.80	6.71	7.40	6.28
50	0.50	8.25	6.99	7.70	6.44
	0.60	8.40	7.33	7.90	6.70
	0.70	8.50	7.86	8.05	7.18
	0.81	9.00	8.50	8.50	8.31
	0.89	9.60	9.98	9.60	9.98
	0.96	10.50	10.97	10.50	10.97
	0.07	5.15	4.30	5.15	4.30
	0.11	6.00	5.11	5.80	4.68
	0.21	6.60	5.50	6.25	4.96
	0.31	6.87	5.73	6.40	5.20
60	0.50	7.18	6.08	6.59	5.40
	0.60	7.28	6.34	6.75	5.60
	0.70	7.53	6.87	700	6.11
	0.80	7.80	7.74	7.40	6.99
	0.88	8.20	8.93	8.20	8.93
	0.95	9.50	9.97	9.50	9.97
	0.07	4.10	3.47	4.10	3.47
	0.11	4.79	4.21	4.70	3.80
	0.20	5.32	4.61	4.90	4.00
	0.30	5.68	4.88	5.00	4.20
70	0.50	6.00	5.10	5.25	4.49
	0.60	6.10	5.25	5.50	4.63
	0.70	6.25	5.49	5.70	4.96
	0.80	6.48	5.89	6.05	5.73
	0.88	7.16	7.13	7.16	7.13
	0.94	8.30	8.09	8.30	8.09

Table 1: Desorption and adsorption equilibrium moisture contents of castor seed.

content. Other researchers too have used this equation to relate the free energy or latent heat of vaporization of agricultural crops to moisture content [6,11,37]. The model can be stated thus:

$$\frac{L}{h_{fg}} = 1 + a * \exp(-bM) \tag{9}$$

The values of coefficients 'a' and 'b' were obtained through the process of logarithmic transformation and non-linear regression and the values of L/h_{fg} were computed and plotted against moisture content.

Results and Discussion

Equilibrium moisture content and water activity relationships

The experimental values of equilibrium moisture content-water activity at different temperatures are as presented in table 1. It was observed that the moisture content of castor seeds measured after each emc-a_w determination was lower than the moisture content before the determination. Ajibola [27] observed same and opined that this difference could be due to loss of moisture that accompanied the evacuation process.

EMC increased with increase in a_w and were lower at higher temperatures in both castor seed varieties. Similar trends for many seeds have been reported in the literature [28,38-42]. At low water activity values, the equilibrium moisture contents for both crop varieties were higher at low temperatures. These results are in agreement with the general characteristics of food isotherms, and the theory of physical adsorption which predicts that the quantity of sorbed water at a given water activity increases as the temperature decreases [43,44].

The experimental values of EMC-a_w for castor seeds at the different temperatures considered were taken as the average of two readings for each. These were used to estimate the best a_w model. The EMC models were obtained by logarithmic transformations of the a_w models and in the non-linear regression analysis, the errors were assumed to be normally and independently distributed. The Modified Oswin and Modified Halsey were found to be the best a_w models for GSS and WBS castor respectively with Standard Error of Estimate (SEE) of 0.238 and 0.242 (Table 2) and the residual plots showed high degree of randomness (The random plot for GSS adsorption and WBS desorption are presented in Figures 1a and 1b. The modified Henderson model which has been published for many agricultural seeds did not fit well. These two models (Modified Oswin and Modified Halsey), therefore, became the models with which the a_w values as presented in Equations (1) and (3) were determined for use in the analysis of isosteric heats and are represented as:

GSS castor

$$a_w = \frac{1}{\left[\frac{(4.809 - 0.118T)}{M}\right]^{0.457} + 1} \tag{10}$$

| Parameter and Criteria | GSS | WBS |
	Modified Oswin	Modified Halsey
A	44.809	58.776
B	0.118	0.131
C	10.457	8.339
R²	0.943	0.941
SEE	0.238	0.424
Residual plot	Random	Random

Table 2: Estimated parameters for water activity models of castor seeds (adsorption).

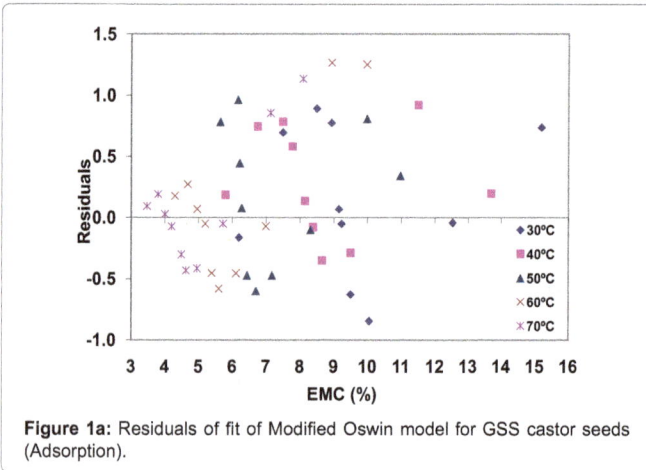

Figure 1a: Residuals of fit of Modified Oswin model for GSS castor seeds (Adsorption).

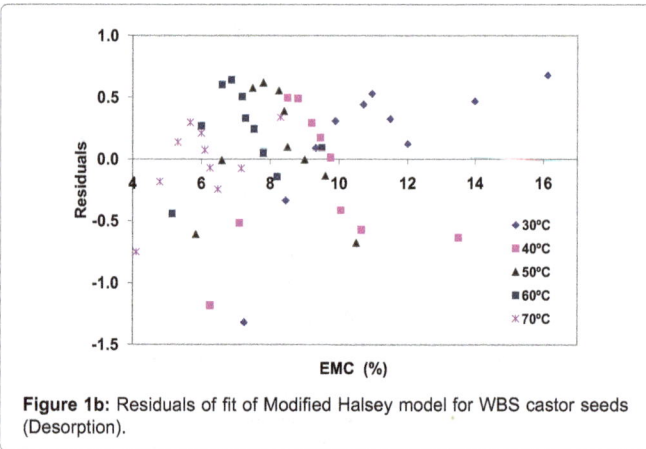

Figure 1b: Residuals of fit of Modified Halsey model for WBS castor seeds (Desorption).

WBScastor $a_w = \exp\left[-\exp\left(58.776 - 0.131T\right) M^{-8.339}\right]$ (11)

The log plots of the vapour pressures (actual, $\ln(P_o)$ and saturated, $\ln(P_s)$) for both castor varieties at various moisture levels are presented in figures 2 and 3 as discussed in Equation (7). The natural log of actual vapor pressure $\ln(P_o)$ increased linearly with increase in the natural log of saturated water vapor pressure $\ln(P_s)$ and moisture content in both cases. Other authors reported similar findings for winged bean seed and gari [6], unripe and ripe plantain [23], melon seed and cassava [45], cowpea and plantain. Therefore the relationship existing between $\ln(P_o)$ and $\ln(P_s)$ at the four moisture contents 4, 8, 12 and 16% d.b. respectively can be expressed by the regression equations (Equations 12-19 with their corresponding R^2 as indicated):

GSS

$\ln(P_o) = 4.6706 \ln(P_s) - 39.57$ $(R^2 = 0.99)$ (12)

$\ln(P_o) = 1.7499 \ln(P_s) - 7.50$ $(R^2 = 0.96)$ (13)

$\ln(P_o) = 1.0235 \ln(P_s) - 0.23$ $(R^2 = 0.99)$ (14)

$\ln(P_o) = 1.0010 \ln(P_s) - 0.01$ $(R^2 = 1.00)$ (15)

WBS

$\ln(P_o) = 1.0004 \ln(P_s) - 1.01$ $(R^2 = 1.00)$ (16)

$\ln(P_o) = 0.7749 \ln(P_s) - 1.36$ $(R^2 = 0.97)$ (17)

$\ln(P_o) = 1.0187 \ln(P_s) - 0.19$ $(R^2 = 0.99)$ (18)

$\ln(P_o) = 1.0010 \ln(P_s) - 0.01$ $(R^2 = 1.00)$ (19)

The ratios of heat of vaporization of saturated water at each of the moisture content were obtained from the slopes of the plots in figure 2 and 3. Using non-linear regression procedure (SPSS 16.0 for windows), Equation (9) was used to relate L to material moisture content. The values of the coefficients 'a' and 'b' in Equation (9) presented in Equations (20 and 21) were 19.173 and 41.298 for GSS and 70.916 and 102.994 for WBS castor seeds. The standard errors of estimate for L/h_{fg} were 0.650 and 0.702 for GSS and WBS castor respectively. The ratios of the heat of vaporization of moisture in the castor seed varieties to the latent heat of free water in Equation (9) are then obtained as:

GSS

$$\frac{L}{h_{fg}} = 1 + 19.173 \exp\left(-41.29M\right)$$ (20)

WBS

$$\frac{L}{h_{fg}} = 1 + 70.916 \exp\left(-102.94M\right)$$ (21)

The effect of moisture content on L/h_{fg} is shown in figure 4. The figure shows that L decreased with increasing moisture content in both castor varieties. This confirms the fact that at higher moisture levels, the strength of water binding decreases. The heat of sorption of both GSS and WBS castor varieties approached that of pure water at moisture

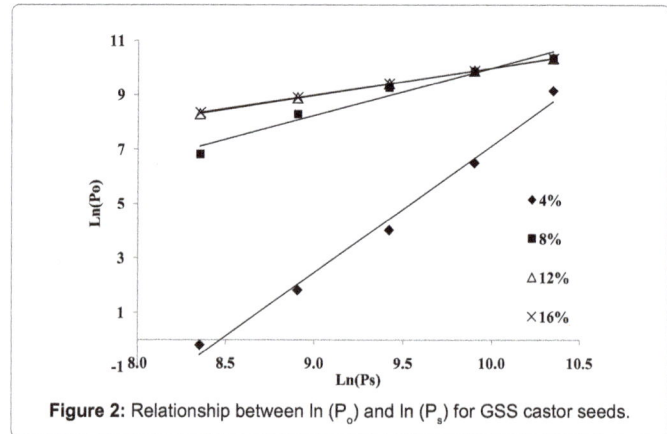

Figure 2: Relationship between $\ln(P_o)$ and $\ln(P_s)$ for GSS castor seeds.

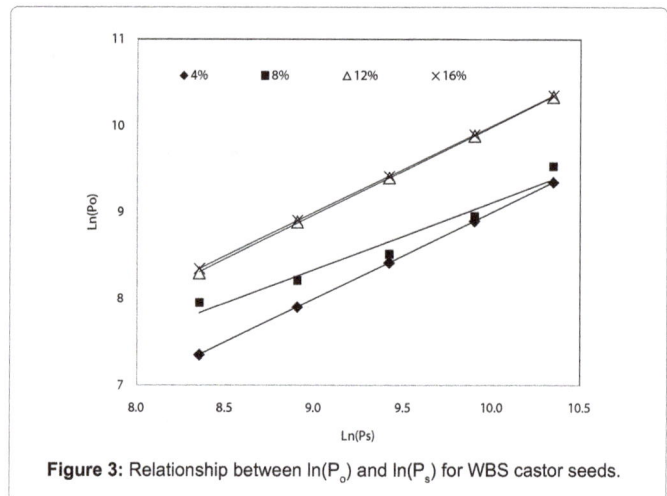

Figure 3: Relationship between $\ln(P_o)$ and $\ln(P_s)$ for WBS castor seeds.

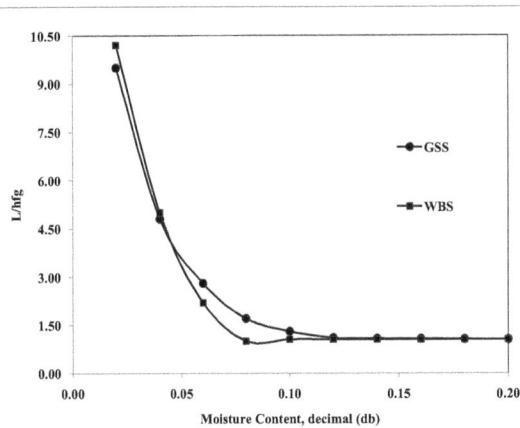

Figure 4: Effect of moisture content on the heat of vaporization castor seeds.

contents of between 13 and 16 % dry basis. Fasina and Sokhansanj [46], explained that beyond this moisture range, water existing in the seeds may not be in a bound form but existing as mono or multi-layer. There was no significant difference between the isosteric sorption heat of both castor varieties. Iglesias and Chirife [19], explained that the level of moisture content at which the heat of sorption approaches the heat of vaporization of water can be a sign of water existing in free form in a food product and Aviara et al. [37], denoted this point as the free water point, explaining further that the presence of dissolved soluble components can cause the actual point to be masked.

Conclusions

1. The equilibrium moisture content of GSS and WBS castor varieties determined using the static gravimetric method increased with increase in a_w but decreased with increasing temperature.

2. The Modified Oswin and the Modified Halsey which gave the least standard error of estimate and randomness of residuals proved the best a_w models for predicting adsorption a_w of GSS and WBS castor seeds respectively

3. For both varieties of castor seeds isosteric heats decreased exponentially with increase in moisture content and approached the latent heat of vaporization of pure water at a moisture content between 13 and 16 % (dry basis), an energy equivalent to that for free water may be adequate to dislodge moisture from castor seeds. There was no significant difference between the isosteric heats of the castor varieties.

Nomenclature

a_w water activity (mass fraction)

A, B, C constants of sorption models

T absolute temperature (K)

emc equilibrium moisture content (% dry basis)

P_o actual water vapor pressure (Pa)

P_s saturated water vapor pressure (Pa)

R gas constant for water vapor (0.462kJ/kg K)

h_{fg} latent heat of vaporization of free water in material (J/kg water)

L heat of vaporization of moisture in the product (J/kg water)

a,b,c_1 constants

M moisture content in decimal, dry basis

References

1. Ogunniyi DS (2006) Castor oil: A vital industrial raw material. Bioresource Technology 97: 1086-1091.

2. Ogunniyi DS, Njikang (2000) Preparation and Evaluation of alkyd resins from castor oil. Park J Sci Ind Res 43: 378-380.

3. Akpan UG, Jimoh A, Mohammed AD (2006) Extraction, characterization and modification of castor seed oil. Leonardo J Sci 5: 43-52.

4. Ojediran JO, Raji AO (2010) Thin layer drying of millet and effect of temperature on drying characteristics. International Food Research Journal 17: 1095-1106.

5. Ojediran JO, Raji AO (2011) Thin-layer drying characteristics of castor (Ricinus communis) seeds. Journal of Food Processing and Preservation 35: 647-655.

6. Fasina OO, Ajibola OO, Tyler RT (1999) Thermodynamics of moisture sorption in winged seed and gari. J Food Process Eng 22: 405-418.

7. Oztekin S, Soysal Y (2000) Comparison of Adsorption and Desorption Isosteric Heats for Some Grains.

8. Iganathane C, Womac AR, Sokhansanj S, Pordesimo LO (2007) Moisture sorption thermodynamic properties of corn stover fractions. Transactions of the ASABE 50: 2151-2160.

9. Bloom PD, Shove GC (1971) Near equilibrium simulation of shelled corn drying. Transactions of the ASAE 14: 709-713.

10. Sokhansanj S, Lang WG, Lischynski DE (1991) Low temperature drying of wheat with supplemental heat. Canadian Agricultural Engineering 33: 265-271.

11. Yang WH, Cenkowski S (1993) Latent heat of vaporization for canola as affected by cultivars and multiple drying – rewetting cycles. Canadian Agricultural Engineering 35: 195-198.

12. Rizvi SSH (1986) Thermodynamic properties of Food in Dehydration. In: Engineering properties of Foods (MA Rao and SSH Rizvi eds.). Marcel Dekker Inc; New York, USA.

13. 13. Bell LN, Labuza TP (2000) Moisture sorption: Practical Aspects of Isotherm Measurement and Use. (2nd edn).

14. Riedel L (1977) Calometric measurements of heats of hydration of foods. Chemie Microbiologie und Technologie der Lebensmittel 5: 97-101.

15. Mullet A, Garcia-Reverter J, Sanjuan R, Bon J (1999) Sorption isosteric heat determination by thermal analysis and sorption isotherms. J Food Sci 64: 64-68.

16. Sanchez ES, Juan NS, Simal S, Rossello C (1997) Calorimetric techniques applied to determination of isosteric heat of desorption for potato. J Sci Food Agric 74: 57-63.

17. Gallaher GL (1951) A method of determining the latent heat of agricultural crops. Agric Eng 32: 34-38.

18. Labuza TP (1968) Sorption phenomena in foods. Food Technol 22: 15-24.

19. Iglesias HA, Chirife J (1976) Isosteric heats of water vapor sorption in dehydrated foods. Part 1. Analysis of the differential heat curve. Lebensmittel Wissenchaft Technologie 9: 68-72.

20. Falabella MC, Aguerre RJ, Suarez C (1989) Determination of the heat of water vapor sorption by means of electronic hygrometers. Lebensmittel Wissenchaft Technologie 22: 11-14.

21. Wang N, Brennan JG (1991) Moisture sorption isotherm characteristics of potatoes at four temperatures. J Food Eng 14: 269-287.

22. Viollaz PE, Rovedo CO (1999) Equilibrium sorption isotherms and thermodynamic properties of starch and gluten. J Food Eng 40: 287-292.

23. Aviara NA, Ajibola OO (2000) Thermodynamics of moisture sorption in plantain. Journal of Agricultural Engineering and Technology 8: 65-75.

24. ASABE (2003) Moisture Measurement - Unground Grain and Seeds. American Society of Agricultural and Biological Engineers (ASABE) Standard S352.2, St. Joseph, MI).

25. Brooker DB, Bakker-Arkema FW, Hall CW (1974) Drying Cereal Grains. AVI Pub Co.

26. Young JF (1967) Humidity control in the laboratory using salt solutions, a review. J Appl Chem 17: 241-245.

27. Ajibola OO (1989) Thin layer drying of melon seed. Journal of Food Engineering 9: 305-320.

28. Menkov ND, Durakova AG, Krasteva A (2004) Moisture sorption isotherms of walnut flour at several temperatures. Bioprocess engineering and modelling 18: 201-206.

29. Thompson HL (1972) Temporary storage of high moisture shelled corn using ontinuous aeration. Transactions of the ASAE 15: 333-337.

30. Gely MC, Santalla EM (2008) Hygroscopic properties of castor seeds (Ricinus communis L.). Brazilian Journal of Chemical Engineering 26: 181-188.

31. Chen CC, Morey RV (1989) Comparison of four EMC/ERH equations. Transactions of the ASAE 32: 983-990.

32. Mehta S, Singh A (2006) Adsorption isotherms for red chill (Capsicum annum L.). Eur Food Res Technol 223: 849-852.

33. Van den Berg C (1984) Description of water activity of foods for engineering purposes by means of the GAB model of sorption, in Engineering and Foods, McKenna, B.M. (ed) (Elsevier Science Publishers, New York), pp. 119 - 131. (Elsevier Science Publishers, New York).

34. Timoumi S, Zagrouba F, Mihoubi D, Tlili MM (2004) Experimental study and modelling of water sorption/desorption isotherms on two agricultural products: apple and carrot. J Phys IV France 122: 235-240.

35. Jayas DS, Mazza G (1993) Comparison of five three-parameter equations for the description of adsorption data of oats. Transactions of the ASAE 36: 119-125.

36. Kapsalis JG (1987) Influences of hysteresis and temperature on moisture sorption isotherms, in Water Activity: Theory and Applications to Food,

Rockland, L.B. and Beuchat, L.B., Beuchat, L.B. (eds) (Marcel Dekker Inc, New York), pp. 173 - 213. (Marcel Dekker Inc, New York).

37. Aviara NA, Ajibola OO, Dairo UO (2002) PH-Postharvest technology: Thermodynamics of moisture sorption in sesame seed. Biosystems Eng 83: 423-431.

38. Vertucci CW, Leopold AC (1987) Water binding in legume seeds. Plant Physiology 85: 224-231.

39. Mazza G, Jayas DS (1991) Evaluation of four three-parameter equations for the description of the moisture sorption data of Lathyrus pea seeds. Lebensmittel-Wissenschaft and Technologie 24: 562-565.

40. Suthar SH, Das SK (1997) Moisture sorption isotherms for karingda (Citrullus lanatus Thumb) Mansf) seed, kernel and hull. Journal of Food Process Engineering 20: 349-366.

41. Walters C, Hill LM (1998) Water sorption isotherms of seeds from ultra dry experiments. Seed Science Research 8: 69-73.

42. Raji AO, Ojediran JO (2011) Moisture sorption isotherms of two varieties of millet. Food and Bioproducts Processing 89: 178-184.

43. Yu L, Mazza G, Jayas DS (1999) Moisture sorption of characteristics of freeze-dried, osmofreeze-dried and osmo-air dried cherries and blueberries. Transactions of the ASAE 42: 141-147.

44. Aviara NA, Ajibola OO, Aregbesola OA, Adedeji MA (2006) Moisture sorption isotherms of sorghum malt at 40 and 50°C. Journal of Stored Products Research 42: 290-301.

45. Aviara NA, Ajibola OO (2002) Themodynamics of moisture sorption in melon seed and cassava. Journal of Food Engineering 55: 107-113.

46. Fasina OO, Sokhansanj S (1993) Equilibrium Moisture Relations and Heat of Sorption of Alfafa Pellets. J Agric Engg Res 56: 51-63.

32

Ultrasound-Assisted Desulfurization of Commercial Kerosene by Adsorption

Mahesh Shantaram Patil*

Vishwakarma Institute of Technology, Department of Chemical Engineering, Pune, India

Abstract

Sorption of hexyl mercaptan sulfur onto carbon-based adsorbents (activated carbon and carbon nanotubes) by ultrasonic irradiation was investigated. The adsorptive capacity was examined. Carbon nanotubes show higher adsorptive capacity. The experimental data were fitted to the Langmuir and Freundlich adsorption isotherm model.

Keywords: Ultrasound; Hexyl mercaptan sulfur; Adsorption; Activated carbon; Carbon nanotubes

Introduction

The topic of research deals with removal of sulphur from commercial kerosene by adsorption "concerned with the desulphurisation" from a hydrocarbon strea. In 2006 the EPA reduced the allowable sulphur levels in liquid fuels. Gasoline sulphur limit was reduced from 300 ppmw to 30 ppm and diesel fuel sulphur limit was reduced from 500 ppmw to 15 ppmw [1]. Deep desulfurization of liquid hydrocarbon fuels is becoming an important subject worldwide. The desulfurization performance of sold super acid type adsorbent (sulphated alumina) for commercial kerosene was evaluated on batch system and on continuous flow system [2]. The removal of sulphur components from gasoline by carbon nanotubes for use as support in catalysis was conducted in batch conditions [3]. A novel approach to ultra-deep desulfurization of transportation fuels by sulphur-selective adsorption for pollution prevention at the source studied [4]. The mercaptan in kerosene is partially oxidized and the remaining was removed by a carbon impregnated with an oxidation catalyst [5]. Hexyl mercaptan was selected as a solute (adsorbate) since this mercaptan was present in substantial amount in typical naphtha [6].

Experimental

Materials

Hexyl mercaptan, activated carbon and carbon nanotubes were procured from Sigma Aldrich.

Preparation of carbon adsorbents

The activated carbon and carbon nanotubes were then washed with acidified distilled water to remove the greasy material and dust and then washed with distilled water till the washing give a clear transparent liquid free from turbidity. Usually 8-10 times of washings are required for this cleaning operation. The washed carbon adsorbents were dried in an oven for 24 hrs at 100-110°C.

Properties of commercial kerosenes

Table 1 shows sulfur contents of Thiophene type and benzothiophene type of kerosene A and kerosene B were 2.8 mass ppm (mg-sulfur/kg-kerosene) and 4.7 mass ppm respectively [2].

Apparatus and procedure

Sono-sorption batch experiments were performed using ultrasonic bath with frequency of 22.5 kHz and a nominal power of 120 Watt (ULTRASONICS LABLINE CL 500).

Properties	Commercial kerosene A	Commercial kerosene B
Total sulfur content [mass ppm]	5.6	6.4
Sulfur content of TP type and BT type [mass ppm]	2.8	4.7
Boiling point range [°C]	146.5-278.0	158.0-271.5
Density at 15°C [g/ml]	0.794	0.7940
Aromatics content [vol%]	17.8	16.9

Table 1: Properties of Commercial Kerosenes.

The stock solution of hexyl mercaptan in kerosene was prepared (10 g/l) and further diluted to desired concentrations. The hexyl mercaptan solution (100 ml) in Kerosene was taken in the ultrasonic bath. The initial hexyl mercaptan concentration was taken in the range of 100-4000 mg/l. The solution containing hexyl mercaptan in kerosene and carbon adsorbents was irradiated for 10 min and 1 h to reach equilibrium.

FTIR TEM UV spectroscopic and X-ray diffraction spectroscopy: Powder XRD grams of activated carbon and carbon nanotube were recorded by means of X-ray diffractometer (Brucker D8). Fourier transform infrared ray (FTIR) spectroscopic measurement was performed on a spectrometer (FTIR-8400; Shimadzu) with a resolution of 4.00 cm^{-1}. The adsorbent materials were characterized by Fourier transform infrared (FTIR) spectroscopy in KBr phase. Infrared spectra were recorded in the range 3800-600 cm^{-1}. Transmission electron microscopy (TEM) image was taken on a 120 kv JEOL1210 equipped with EDS analyzer Link QX-2000.

UV–VIS spectrophotometer (SHIMADZU 160A model) was used for determination of Hexyl mercaptan concentration .The wavelength of maximum absorbance of hexyl mercaptan was 230 nm.

Results and Discussion

FTIR spectra of activated carbon and carbon nanotubes in the re

---

*Corresponding author: Mahesh Shantaram Patil, Vishwakarma Institute of Technology, Department of Chemical Engineering, 666, Upper Indira Nagar, Pune-411037, India

region of 600-3800 cm^{-1} are shown in figure 1. Sharp peaks are observed in activated carbon and carbon nanotubes. To characterize the functional elements absorbed by activated carbon and carbon nanotube, FTIR is used. FTIR of a) shows sharp peak at 900, 1000, 1200, 1500, 1790 and 2800 cm^{-1} which corresponds to hydroxyl, carbonyl, aliphatic, ethers, aromatic C=C stretching and carboxylic groups respectively. FTIR of b) shows dominant peaks at 800, 900, 1250, 1500, 1600, 1800, 2800 and 3300 cm^{-1} which corresponds to Si-O, C-N, N-CH$_3$, CNT, C-O, and C-Hx respectively. The nanometric length of the carbon nanotubes was found in the range of 100 nm (Figure 2) [7].

X-ray diffraction patterns activated carbon and carbon nanotubes are reported in figure 3. It was observed as a broad intense peak in the case of activated carbon and a small peak in the case of carbon nanotubes. X-ray studies have shown that many so-called amorphous substances have crystalline characteristics even though they may not show certain features such as crystal angles and faces usually associated with crystalline state. Although interpretation of the X-ray diffraction patterns is not free from ambiguities there is general agreement that

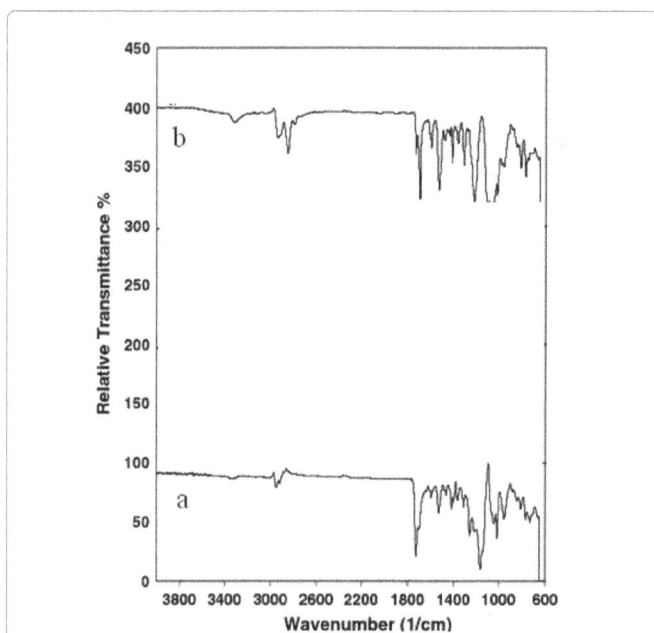

Figure 1: FTIR spectra of activated carbon and carbon nanotubes. a – activated carbon, b- carbon nanotubes.

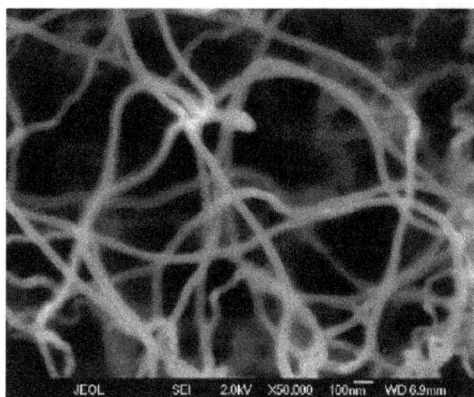

Figure 2: TEM image of carbon nanotube.

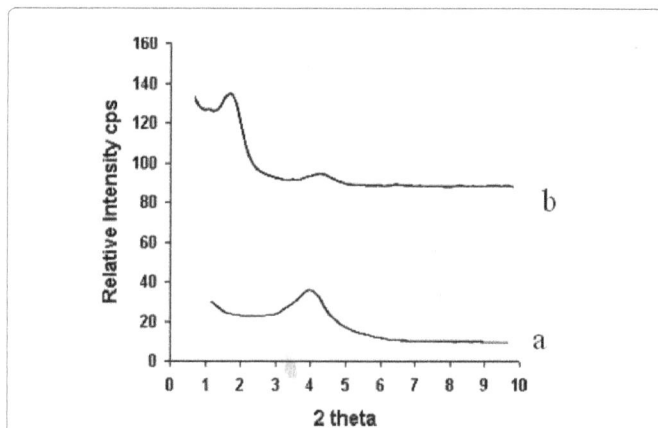

Figure 3: XRD gram of activated carbon and carbon nanotubes. a – activated carbon, b- carbon nanotube.

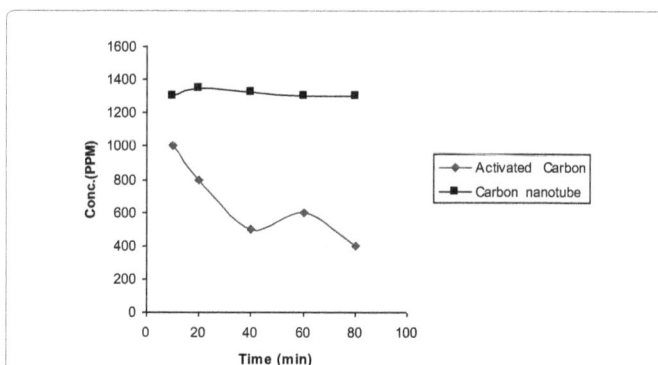

Figure 4: Time Vs carbon phase concentration curves for carbon adsorbents (Co = 2000ppm, 0.5 g adsorbents).

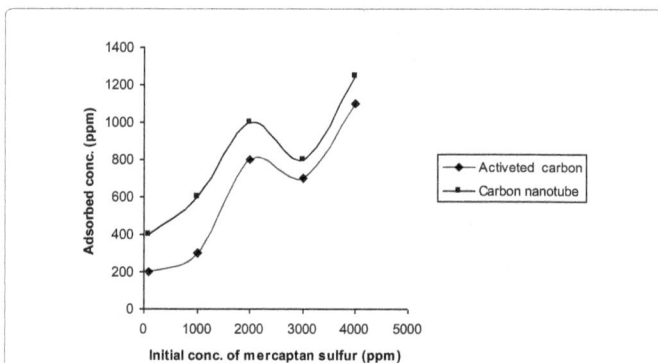

Figure 5: Effect of initial concentration on different adsorbents (adsorbents quantity -0.5g).

amorphous carbon consists of plates in which the carbon atom are arranged in a hexagonal lattice each atom except those at the edge being held by covalent linkages to three other carbon atom.

The time versus concentration data has shown in figure 4. The equilibrium reaches within 20 min for both adsorbents.

Figure 5 shows the effect of different initial concentrations of hexyl mercaptan in kerosene ranging from 100 to 4000 ppm. The mixture of hexyl mercaptan in kerosene and 0.5 g adsorbent was irradiated to ultrasound. It is clearly shows that activated carbon affect the adsorption for higher range of hexyl mercaptan concentration.

Figure 6 shows the effect of adsorbents onto the adsorption capacity of activated carbon and carbon nanotubes at 2000 ppm concentration of hexyl mercaptan. At 0.1 and 0.3 mg of adsorbent activated carbon shows less adsorption in comparison to carbon nanotubes at 2000 ppm concentration of hexyl mercaptan.

Isotherm models

The Langmuir isotherm model for carbon-based adsorbents sono-sorption (Figure 7) is shown by the linear plot of Ce/qe versus Ce. The parameters Qo and b were determined from the slope and intercept of the plot and were presented in table 2.The sono-sorption follows Freundlich isotherm model for activated carbon and carbon nanotubes is shown by the linear plot of log qe versus log Ce. From the intercept and slope k and 1/n were calculated for carbon nanotubes (Figure 8) the constants are 1.012 and 0.0053 (slope and intercept) [8]. It is generally stated that values of n (1/n=0.0053; n=188) less than 1 poor adsorption [9]. From R^2 values it is found that these two adsorbents favour both Langmuir and Freundlich adsorption isotherm.

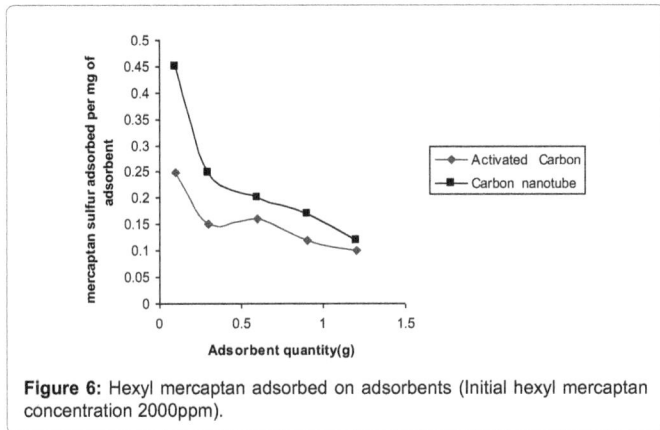

Figure 6: Hexyl mercaptan adsorbed on adsorbents (Initial hexyl mercaptan concentration 2000ppm).

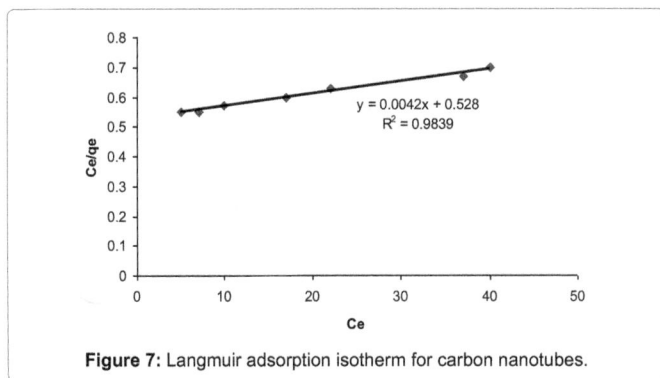

Figure 7: Langmuir adsorption isotherm for carbon nanotubes.

Langmuir parameters	Langmuir isotherm activated carbon	carbon nanotubes
Q_0(mg/mg)	0.0078	0.528
B(l/mg)	5.6363	0.0042
Correlation coefficient (R^2)	0.9795	0.9839
Freundlich parameters	**Freundlich isotherm activated carbon**	**carbon nanotubes**
Slope(1/n)	5.3023	0.0053
Intercept log k	0.0198	0.1913
Correlation coefficient (R^2)	0.9755	0.9813

Table 2: The Freundlich and Langmuir parameters of adsorption isotherm models.

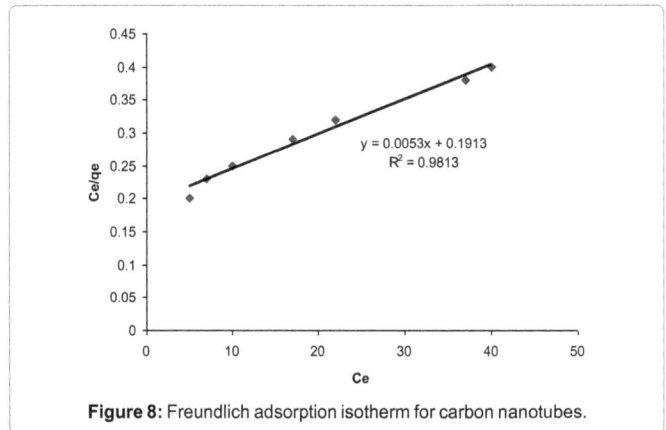

Figure 8: Freundlich adsorption isotherm for carbon nanotubes.

Intercalating agent	ΔH (kJ/mol)	ΔS (J/mol K)	-ΔG (kJ/mol) at temperature			
			308K	319K	329K	338K
activated carbon	7.48	16.41	20.00	13.39	15.00	15.32
carbonnanotubes	22.00	22.72	30.18	26.08	26.17	28.03

Table 3: Thermodynamic parameters for adsorption of hexl mercaptan (2000ppm) onto carbon adsorbent (5g/L).

Evaluation of thermodynamic parameters

Considering the equilibrium constant Thermodynamic data such as adsorption free energy change Ko can be obtained from the following equation:

$$\Delta G^0 = RT\ln K^0 \quad (2)$$

Where ΔG^0 is the free energy change (kJ/mol), R the universal gas constant (8.314 J/mol K), K^0 the thermodynamic equilibrium constant and T the absolute temperature (K). The corresponding values of ΔG^0 are presented in table 3. The values of ΔG^0 obtained were ranged from -13 to -30 kJ/mol which indicates feasibility of adsorption. The Gibbs free energy shows the favourable adsorption in the following order activated carbon < carbon nanotube.

Conclusion

The hexyl mercaptan adsorbed per mg of carbon nanotubes is higher (0.40 mg) at higher hexyl mercaptan concentration (4000 ppm). The sono-sorption data of hexyl mercaptan on carbon-based adsorbents studied in this work fits well in Langmuir and and Freundlich model as can be seen from the regression coefficient values R^2 which is in the range of 0.97 to 0.98.

References

1. Ambal J, Gokhan OA, Margarita D, Mathew S, John M, et al. (2008) Desulfurization of liquid fuels by adsorption. TDA research Inc, USA.

2. Toida Y, Nankamura K, Matsumoto K (2010) Adsorptive Desulfurization of Commercial Kerosene with Sulfated Alumina Producing Heavier Organic Sulfur Compounds. J Japan petroleum Institute 53.

3. Manteghian M, Nouri Asl P, Pahlavanzadeh H, Soroush Samaneh (2008) Sulfur components removal from gasoline by carbon nanotubes. Iran.

4. Song, Chunshan, et al. (2006) EPA project. Pennsylvania State University.

5. Guy Francoise Tom varadi, Merichem Company (1993) Hydrocarbon Technology International.

6. Nelson WL (1958) Petroleum Refinery Engineering.

7. Misra A, Tyagi PK, Singh MK, Misra DS (2006) FTIR studies of nitrogen doped carbon nanotubes. Proceedings of the Applied Diamond Conference/NanoCarbon 2005-ADC/NanoCarbon 2005.

8. Sonawane S, Chaudhari P, Ghodke S, Ambade S, Guliga S, et al. (2008) Combined effect of ultrasound and nanoclay on adsorption of phenol. Ultrasonics Sonochemistry 15: 1033-1037.

9. Robert E. Treybal (1981) Mass-Transfer Operations. (3rd edn), Mac- Graw Hill Publication, Singapore.

Permissions

All chapters in this book were first published in JCEPT, by Omics International; hereby published with permission under the Creative Commons Attribution License or equivalent. Every chapter published in this book has been scrutinized by our experts. Their significance has been extensively debated. The topics covered herein carry significant findings which will fuel the growth of the discipline. They may even be implemented as practical applications or may be referred to as a beginning point for another development.

The contributors of this book come from diverse backgrounds, making this book a truly international effort. This book will bring forth new frontiers with its revolutionizing research information and detailed analysis of the nascent developments around the world.

We would like to thank all the contributing authors for lending their expertise to make the book truly unique. They have played a crucial role in the development of this book. Without their invaluable contributions this book wouldn't have been possible. They have made vital efforts to compile up to date information on the varied aspects of this subject to make this book a valuable addition to the collection of many professionals and students.

This book was conceptualized with the vision of imparting up-to-date information and advanced data in this field. To ensure the same, a matchless editorial board was set up. Every individual on the board went through rigorous rounds of assessment to prove their worth. After which they invested a large part of their time researching and compiling the most relevant data for our readers.

The editorial board has been involved in producing this book since its inception. They have spent rigorous hours researching and exploring the diverse topics which have resulted in the successful publishing of this book. They have passed on their knowledge of decades through this book. To expedite this challenging task, the publisher supported the team at every step. A small team of assistant editors was also appointed to further simplify the editing procedure and attain best results for the readers.

Apart from the editorial board, the designing team has also invested a significant amount of their time in understanding the subject and creating the most relevant covers. They scrutinized every image to scout for the most suitable representation of the subject and create an appropriate cover for the book.

The publishing team has been an ardent support to the editorial, designing and production team. Their endless efforts to recruit the best for this project, has resulted in the accomplishment of this book. They are a veteran in the field of academics and their pool of knowledge is as vast as their experience in printing. Their expertise and guidance has proved useful at every step. Their uncompromising quality standards have made this book an exceptional effort. Their encouragement from time to time has been an inspiration for everyone.

The publisher and the editorial board hope that this book will prove to be a valuable piece of knowledge for researchers, students, practitioners and scholars across the globe.

List of Contributors

Elham Khaghanikavkani
Professor, Univeristy of Auckland, New Plymouth, New Zealand

Mohammed M. Farid
Professor, Univeristy of Auckland, New Plymouth, New Zealand

John Holdem
Professor, Univeristy of Auckland, New Plymouth, New Zealand

Allan Williamson
Professor, Univeristy of Auckland, New Plymouth, New Zealand

Nguyen Van Suc Ho Thi Yeu Ly
Ho Chi Minh City University of Technical Education, Vietnam

Mohamed M El-Toony
National center for radiation research and technology, Atomic energy authority, Nasr City, Cairo, Egypt

A.S. Al-Bayoumy
National center for radiation research and technology, Atomic energy authority, Nasr City, Cairo, Egypt

Archana Dixit
Department of Chemistry, Maulana Azad National Institute of Technology, Bhopal, India

Savita Dixit
Department of Chemistry, Maulana Azad National Institute of Technology, Bhopal, India

CS Goswami
Department of Chemistry, Kamal Radha Girls College, Gwalior, India

Maher A El-Hashash
Ain Shams University, Science Faculty, Chemistry Department, Cairo, Egypt

Sameh A Rizk
Ain Shams University, Science Faculty, Chemistry Department, Cairo, Egypt

Maher I Nessim
Egyptian Petroleum Research Institute, Evaluation and Analysis Department, Cairo, Egypt

Basavaraj K Nanjwade
Department of Pharmaceutics, KLE University's College of Pharmacy, Belgaum-590010, Karnataka, INDIA

Ritesh Udhani
Department of Pharmaceutics, KLE University's College of Pharmacy, Belgaum-590010, Karnataka, INDIA

Jatin Popat
Department of Pharmaceutics, KLE University's College of Pharmacy, Belgaum-590010, Karnataka, INDIA

Veerendra K Nanjwade
Department of Pharmaceutics, KLE University's College of Pharmacy, Belgaum-590010, Karnataka, INDIA

Sachin A Thakare
Department of Pharmaceutics, KLE University's College of Pharmacy, Belgaum-590010, Karnataka, INDIA

Elaziouti Abdelkader
LCPCE Laboratory, Faculty of sciences, Department of industrial Chemistry, University of the Science and Technology of Oran (USTO M.B), BP 1505 El M'naouar 31000 Oran, Algeria

Laouedj Nadjia
Dr. Moulay Tahar University, Saida, Algeria

Bekka Ahmed
LCPCE Laboratory, Faculty of sciences, Department of industrial Chemistry, University of the Science and Technology of Oran (USTO M.B), BP 1505 El M'naouar 31000 Oran, Algeria

Wahid Djeridi
Laboratory of Research: Engineering Processes and Industrials Systems (LR11ES54), National School of Engineers of Gabes, University of Gabes, Gabes, Tunisia

Abdelmottaleb Ouederni
Laboratory of Research: Engineering Processes and Industrials Systems (LR11ES54), National School of Engineers of Gabes, University of Gabes, Gabes, Tunisia

AK Minocha
Environmental Science and Technology Division, Central Building Research Institute, and NIPER, Punjab, India

Manish Kumar Goyal
Environmental Science and Technology Division, Central Building Research Institute, and NIPER, Punjab, India

Benamar Dahmani
Spectrochemistry and Structural Pharmacology Laboratory, Department of Chemistry, Science Faculty, University of Tlemcen, Algeria

Mustapha Chabane
Spectrochemistry and Structural Pharmacology Laboratory, Department of Chemistry, Science Faculty, University of Tlemcen, Algeria

Sokainah Rawashdeh
Applied science department, Faculty of Engineering Technology, Al-Balqaa Applied University, Amman, Jordan

Ismail Al-Raheil
Department of Physics, University College in Lieth, Umm Al-Qura University, Makkah , KSA

Shiv Sankar Bhattacharya
School of Pharmaceutical Sciences, IFTM University, Moradabad, Uttar Pradesh, India

Naveen Bharti
School of Pharmaceutical Sciences, IFTM University, Moradabad, Uttar Pradesh, India

Subham Banerjee
Division of Pharmaceutical Technology, Defence Research Laboratory, Tezpur, Assam, India

Jamshidi ALCL
Center for Technology and Geosciences -CTG/UFPE, Av. Moraes Rego, 1235 – Cidade Universitária, CEP: 50670-901, Recife, PE, Brazil

L. Nascimento
Center for Technology and Geosciences -CTG/UFPE, Av. Moraes Rego, 1235 – Cidade Universitária, CEP: 50670-901, Recife, PE, Brazil

RJ Rodbari
Department of Sociology-DS, Islamic Azad University, Tehran-Iran

GF Barbosa
Department of Physics, Department of Exact and Natural Sciences-DENS/UFERSA, Rua da Harmonia Alto de São Manoel, CEP: 59625210-Mossoró, RN-Brazil

FLA Machado
Department of Physics -DP, Center of Exact and Natural Sciences-CENS/UFPE, Av. Prof. Moraes Rego, 1235-Cidade Universitária, CEP: 50670901, Recife- PE, Brazil

JGA Pacheco
Center for Technology and Geosciences -CTG/UFPE, Av. Moraes Rego, 1235 – Cidade Universitária, CEP: 50670-901, Recife, PE, Brazil

Barbosa CMBM
Center for Technology and Geosciences -CTG/UFPE, Av. Moraes Rego, 1235 – Cidade Universitária, CEP: 50670-901, Recife, PE, Brazil

GA Farzi
Material and Polymer Engineering Department, Hakim Sabzevari University, Sabzevar, Iran

N. Reza-Zadeh
Mechanical Engineering Department, Hakim Sabzevari University, Sabzevar, Iran

A.Parsian Nejad
Mechanical Engineering Department, Hakim Sabzevari University, Sabzevar, Iran

Thouraya Bohli
Laboratory of Research: Engineering Processes and Industrials Systems (LR11ES54), National School of Engineers of Gabes, University of Gabes, St Omar Ibn Elkhattab, 6029 Gabes, Tunisia

Nuria Fiol
Department d'Enginyeria Quimica, Agraria i Tecnologia Agroalimentaria,Universitat de Girona, Avda Lluis Santolo, 17003 Girona, Spain

Isabel Villaescusa
Department d'Enginyeria Quimica, Agraria i Tecnologia Agroalimentaria,Universitat de Girona, Avda Lluis Santolo, 17003 Girona, Spain

Abdelmottaleb Ouederni
Laboratory of Research: Engineering Processes and Industrials Systems (LR11ES54), National School of Engineers of Gabes, University of Gabes, St Omar Ibn Elkhattab, 6029 Gabes, Tunisia

Augustine K. Asiagwu
Chemistry Department, Delta State University, Abraka, P.M.B.1, Delta State, Nigeria

Hilary I. Owamah
Civil Engineering Department, Landmark University, P.M.B1001, Omu-Aran, Kwara State Nigeria

Izinyon O. Christopher
Civil Engineering Department, University of Benin, Benin, Nigeria

J. Mulopo
Council for Scientific and Industrial Research, Natural Resources and the Environment, Pretoria, South Africa

D. Ikhu-Omoregbe
Chemical Engineering Department, Cape Peninsula University of Technology, Cape Town, South Africa

Ashish Chauhan
Department of Chemistry, Dr. B. R. Ambedkar National Institute of Technology, Jalandhar 144 011 (Pb) India

Priyanka Chauhan
Department of Chemistry, Dr. B. R. Ambedkar National Institute of Technology, Jalandhar 144 011 (Pb) India

Balbir Kaith
Department of Chemistry, Dr. B. R. Ambedkar National Institute of Technology, Jalandhar 144 011 (Pb) India

Soumaya Gharbi
Research Laboratory Analytical Chemistry, Macromolecular and Heterocyclic, Ipest, Tunisia

Jameleddine Khiari
Preparatory Institute for Engineering Studies of Bizerte, Tunisia

Bassem Jamoussi
Research Laboratory Analytical Chemistry, Macromolecular and Heterocyclic, Ipest, Tunisia

Nouri Hanen
Research Laboratory: Engineering Process and Industrial System, National School of Engineers of Gabes, University of Gabes, Gabes, Tunisia

Ouederni Abdelmottaleb
Research Laboratory: Engineering Process and Industrial System, National School of Engineers of Gabes, University of Gabes, Gabes, Tunisia

Claas Wagner
ACABS Research Group, Aalborg University, campus Esbjerg (AAUE), Denmark

Kim H Esbensen
ACABS Research Group, Aalborg University, campus Esbjerg (AAUE), Denmark
Geological Survey of Denmark and Greenland, Copenhagen, Denmark

Thouraya Bohli
Laboratory of Research: Engineering Processes and Industrials Systems (LR11ES54), National School of Engineers of Gabes, University of Gabes, Gabes, Tunisia

Isabel Villaescusa
Departement d'Enginyeria Quimica, Agraria i Tecnologia Agroalimentaria, Universitat de Girona, Girona, Spain

Abdelmottaleb Ouederni
Laboratory of Research: Engineering Processes and Industrials Systems (LR11ES54), National School of Engineers of Gabes, University of Gabes, Gabes, Tunisia

Maher A El-Hashash
Chemistry Department, Faculty of Science, Ain Shams University, Abassia 11566, Cairo, Egypt

Dalal B Guirguis
Chemistry Department, Faculty of Science, Ain Shams University, Abassia 11566, Cairo, Egypt

Nayera AM Abd El-Wahed
Department of Chemistry of Natural and Microbial Products, Division of Pharmaceutical Industries, National Research Center, El Behos street, Dokki, Egypt

Mohamed A Kadhim
Chemistry Department, Faculty of Science, Anbar University, Iraq

Samira Kherici
Department of industrial organic chemistry USTO, POB 1505 El Mnaouar Oran, Algeria

Djillali Benouali
Department of physical chemistry USTO, POB 1505 El Mnaouar Oran, Algeria

Mohamed Benyetou
Laboratory of modeling and optimization of industrial systems (LAMOSI) USTO, Algeria

Selvaraj Raja
Department of Biotechnology, Manipal Institute of Technology, Manipal, Karnataka, India

Varadavenkatesan Thivaharan
Department of Biotechnology, Manipal Institute of Technology, Manipal, Karnataka, India

Vinayagam Ramesh
Department of Biotechnology, Manipal Institute of Technology, Manipal, Karnataka, India

Vytla Ramachandra Murty
Department of Biotechnology, Manipal Institute of Technology, Manipal, Karnataka, India

Asli Yuksel
Izmir Institute of Technology, Faculty of Engineering, Department of Chemical Engineering, Gulbahce Campus, Urla, Izmir, Turkey

R. Kumar
Forest Ecology and Environment Division, Forest Research Institute, Dehradun, India

N. Yadav
Forest Ecology and Environment Division, Forest Research Institute, Dehradun, India

L. Rawat
Forest Ecology and Environment Division, Forest Research Institute, Dehradun, India

MK Goyal
National Institute of Pharmaceutical Education and Research, Mohali, India

AM Al-Mayouf
Chemistry Department, College of Science, King Saud University, Riyadh, Saudi Arabia

MSA Saleh
Chemistry Department, College of Science, King Saud University, Riyadh, Saudi Arabia

A Aouissi
Chemistry Department, College of Science, King Saud University, Riyadh, Saudi Arabia

AA Al-Suhybani
Chemistry Department, College of Science, King Saud University, Riyadh, Saudi Arabia

Mojtaba Mansourpoor
Pars Oil and Gas Company, Process Engineering Unit, Asalouye, Iran

Dr. Ahmad Shariati
Gas Engineering Department, Petroleum University of Technology, Ahwaz, Iran

J. Ojediran
Agricultural Engineering Department, Landmark University, Omu-Aran, Kwara State, Nigeria

A.O. Raji
Agricultural and Environmental Engineering Department, University of Ibadan, Ibadan, Nigeria

H.I. Owamah
Civil Engineering Department, Landmark University, Kwara State, Nigeria

Mahesh Shantaram Patil
Vishwakarma Institute of Technology, Department of Chemical Engineering, Pune, India

www.ingramcontent.com/pod-product-compliance
Lightning Source LLC
Chambersburg PA
CBHW080257230326
41458CB00097B/5097